Human–Computer Interaction Series

Editor-in-Chief
Desney Tan, Microsoft Research, USA
Jean Vanderdonckt, Université catholique de Louvain, Belgium

HCI is a multidisciplinary field focused on human aspects of the development of computer technology. As computer-based technology becomes increasingly pervasive – not just in developed countries, but worldwide – the need to take a human-centered approach in the design and development of this technology becomes ever more important. For roughly 30 years now, researchers and practitioners in computational and behavioral sciences have worked to identify theory and practice that influences the direction of these technologies, and this diverse work makes up the field of human-computer interaction. Broadly speaking it includes the study of what technology might be able to do for people and how people might interact with the technology. The HCI series publishes books that advance the science and technology of developing systems which are both effective and satisfying for people in a wide variety of contexts. Titles focus on theoretical perspectives (such as formal approaches drawn from a variety of behavioral sciences), practical approaches (such as the techniques for effectively integrating user needs in system development), and social issues (such as the determinants of utility, usability and acceptability).

More information about this series at http://www.springer.com/series/6033

Aaron Marcus

Mobile Persuasion Design

Changing Behaviour by Combining
Persuasion Design with Information Design

🜨 Springer

Aaron Marcus
Aaron Marcus and Associates (AM+A)
Berkeley, CA, USA

Copyright 2014 American Diabetes Association
From http://www.diabetes.org
Reprinted with permission from The American Diabetes Association

Whilst we have made every effort to obtain permissions from copyright holders to use the material contained in this book, there have been occasions where we have been unable to locate those concerned. Should holders wish to contact the Publisher, we will be happy to come to an arrangement at the first opportunity.

Additional material to this book can be downloaded from http://extras.springer.com

ISSN 1571-5035
Human–Computer Interaction Series
ISBN 978-1-4471-4323-9   ISBN 978-1-4471-4324-6   (eBook)
DOI 10.1007/978-1-4471-4324-6

Library of Congress Control Number: 2015954954

Springer London Heidelberg New York Dordrecht
© Springer-Verlag London 2015

This work is subject to copyright. All rights are reserved by the Publisher, whether the whole or part of the material is concerned, specifically the rights of translation, reprinting, reuse of illustrations, recitation, broadcasting, reproduction on microfilms or in any other physical way, and transmission or information storage and retrieval, electronic adaptation, computer software, or by similar or dissimilar methodology now known or hereafter developed.
The use of general descriptive names, registered names, trademarks, service marks, etc. in this publication does not imply, even in the absence of a specific statement, that such names are exempt from the relevant protective laws and regulations and therefore free for general use.
The publisher, the authors and the editors are safe to assume that the advice and information in this book are believed to be true and accurate at the date of publication. Neither the publisher nor the authors or the editors give a warranty, express or implied, with respect to the material contained herein or for any errors or omissions that may have been made.

Printed on acid-free paper

Springer-Verlag London Ltd. is part of Springer Science+Business Media (www.springer.com)

Foreword

Aaron Marcus' education and experience spans the fields of science, technology, and design. He has captured his cross-disciplinary knowledge in a memorable form with this book. By addressing information design and persuasion design together in a case-based format, he has provided his knowledge in an accessible form as well, so that a variety of stakeholders interested in creating mobile applications can see compelling examples in many of the emerging persuasive application areas. Finally, by stepping carefully through the process of research and design, he has provided the reader with a useful view of design methodologies including user research, design principles, iterative prototyping, and evaluation methods.

Why Do We Need This Book?

Persuasion theory has been discussed since Aristotle's 4th Century (BCE) studies in rhetoric that described key elements in human discourse that could influence another person's attitudes, beliefs, or behaviors (Aristotle). As formal schools that studied psychology and human behavior developed, increasingly accurate and science-based models described an individual's experience of emotional, social, and rational influences. For example, the work of Albert Bandura provides significant insights into concepts such as self-efficacy and social learning (Cialdini 1984). Robert Cialdini has authored many papers and books on the role of social influence and persuasion (Bandura 1997). In 1998, BJ Fogg's paper on "Persuasive computers: perspectives and research directions "offered a more pragmatic look at how the fields that sought to understand human behavior and persuasion could converge with computer science to create a new field of persuasive technology (Fogg 1998). At about the same time, the psychology team of Carver and Scheier described motivational aspects of human behavior in their work on control systems theory (Carver and Scheier 1982). More recent work by Harri Oinas-Kukkonen describes a behavior change support system that reconciles many of the preceding theories with computer system architectures (Oinas-Kukkonen 2010). At the practical level of design and development, these theories need to be crafted together to create persuasive experiences and applications. How can this craft be taught?

As an active designer of persuasive mobile applications, Aaron Marcus has created a book that provides many examples that are useful not only because they are thoughtfully designed by using proven user-centered design methods but also because they are based on the sound body of theory and knowledge that has evolved in the discipline of persuasion.

Structure of the Book

The book's core structure is based on case studies that cover a useful range of topics that are appropriate for behavior change mobile applications. Aspects of our behaviors toward sustainability, money, health, learning, happiness, and marriage can have an important influence on outcomes, and while we cannot script ourselves for every eventuality, a case-studies approach is a well-recognized method for teaching topics in complex and contextually rich environments. Much as medical and business schools use case-based methods for learning, it is appropriate for designers and developers of mobile applications to use a similar method for learning because of the continually changing circumstances of the mobile user and the ever-changing environment of use. Each case study is designed to stand as an individual module so that designers or developers working on sustainability could use that chapter alone to inform their work without necessarily reading the other cases.

Strength of the Book

The book's introduction is particularly useful by providing a succinct and well-structured overview of key theories and methods used by the research team in creating their case studies; user-centered design, information architecture, and persuasion theory are core building blocks for creating successful persuasive applications. In each case study, a thorough assessment of relevant context is provided. For example, in Chap. 3, "The Health Machine," the reader is treated to a comprehensive discussion of the type 2 diabetes problem space, the situation of archetypical personas, usage scenarios, existing competitive offerings, relevant and persuasion design principles, a supporting information architecture, and a description of key functions for a mobile application. This case-based descriptive approach should be valuable for any design or development team that is seeking a quick and thorough briefing on the topic.

Motorola Human Interaction Research Laboratory   Tom MacTavish
Illinois Institute of Technology – Institute of Design
Chicago, IL, USA
February 2015

References

Aristotle, Treatise on Rhetoric, 400 BCE

Bandura A (1997) Self-efficacy: the exercise of control. W.H. Freeman, New York

Carver CS, Scheier MF (1982) Control theory: a useful conceptual framework for personality–social, clinical, and health psychology. Psychol Bull 92(1):111–135

Cialdini RB (1984) Influence: the psychology of persuasion, Harper Business; Revised edition (December 26, 2006)

Fogg BJ (1998) Persuasive computers: perspectives and research directions. Proceedings of CHI 1998, ACM Press, pp 225–232

Oinas-Kukkonen Behavior H (2010) Change support systems: a research model and agenda. In T Ploug, P Hasle, H Oinas-Kukkonen (eds) PERSUASIVE 2010, LNCS 6137, pp 4–14, 2010. © Springer, Berlin/Heidelberg

Foreword

The half-century history of human-computer interaction and user interface design has produced innovative results that have transformed the world. The designers, engineers, and scientists, who developed compelling user experiences that include social media, e-commerce, medical assistance, and online education, have much to be proud of.

Ubiquitous mobile devices, even in developing nations, enable a broad range of people to communicate with family, get medical help, carry out business activities, and enjoy diverse entertainment. These devices also enable collaborative communities at an unprecedented level, including political organization to counter oppression, while supporting creative projects that produce music, photography, video, art, and much more.

The list of empowerment experiences continues with professional work, research explorations, entrepreneurial initiatives, and economic transactions. Admirably these activities are largely accessible to users with disabilities, those with low literacy, and marginalized peoples giving them better opportunities in life. Of course, there are many disturbing outcomes such as racial hatred, terror-group recruiting, criminal activities, and malicious hackers. The challenges to personal relationships, loss of privacy, and threats of cyber-attacks undermine the pro-social outcomes, causing critics to raise concerns about whether ubiquitous mobile devices are an overall plus or minus.

Rather than wrestle with the current situation, inspiring visionaries are looking to the future to explore how these technologies can do more than facilitate existing human needs. They seek to develop transformed possibilities and wholly new opportunities to improve human experiences. While robotic and artificially intelligent scenarios may play a role, I prefer to emphasize the unbounded creativity of human beings and their thirst for self-actualization that Abraham Maslow described as early as 1943. He also made clear that the foundations for human success are built on physical safety, necessarily embedded in a nontoxic environment, which also enables good health. Then human beings can pursue their strong needs for esteem and love.

Maslow's philosophy and many others led me to believe that the central human quest is not for artificially intelligent masters or even partners, but for tools that assert the primacy of human agency. Therefore, my first principle is that successful future technologies are likely to be ones that promote human autonomy, initiative, independence, and control.

My second principle for envisioning future technologies is based on the belief that humans are inherently social, seeking intimate partners, compassionate friends, and supportive teams. Humans eagerly seek out helpful communities and pro-social organizations, frequently embedded in constructive civil societies that promote social justice and freedom of expression.

I am not alone in considering how human values clarify design principles that will bring a brighter future. For a half-century, Aaron Marcus has been an inspirational resource for me and many others. His visible contributions are hundreds of corporate design projects that have contributed to successful products and his extensive writings on user-experience design that have been widely influential. I think of his early and lucid understanding of cultural, gender, personality, and other individual differences that make every human unique, even as they are situated in countries, age groups, markets, religions, races, ethnicities, etc.

Aaron Marcus has also had a profound impact on others that is less visible. His relentless travels to present courses, tutorials, workshops, and other training experiences have influenced thousands of students, researchers, and professionals, including me. Marcus has shared his masterful knowledge, thoughtful principles, and practical guidelines, all delivered through seemingly endless, yet cogent examples, in ways that leave lasting impressions while productively changing behaviors.

This book collects more than 26 years of his work on mobile device design, and 8 years focusing especially on persuasion and motivation. He goes directly for global issues, such as business, energy, and environment, as well as for personal concerns over health/wellness, finances, and happiness. The examples show how perceptual and aesthetic principles, followed by carefully conducted evaluations, lead to design evolution. Marcus also provides psychological theory foundations for much of what he shows, enabling practitioners to understand general principles and learn from practical guidelines.

It is rare to find a book that is so clearly devoted to promoting human values through technology design. *Mobile Persuasion Design* is a valuable contribution, which will help attentive readers to do their share in building a better world.

University of Maryland   Ben Shneiderman
College Park, MD, USA
January 2015

Foreword

In this book Aaron Marcus combines two topics: persuasion design and information design.

"Persuasion" often is thought of in terms of advertising and propaganda. Thus, the warning in the author's Preface is well advised: "all of these techniques could be used to change people's behavior in bad ways." They would indeed deserve to be dealt with separately in a book on this specific theme.

On the other hand, "information," another familiar and well-analyzed term, sounds easy to consider in regard to design, but it, too, raises questions.

Information, we are told, together with material and energy, is one of the three ingredients that make up our world. Obviously of high importance, information is, at the same time, not a readily available commodity sold in clearly defined packages along with subject-specific use instructions.

To facilitate understanding, I may quote the definition of information, coined by the idX[1] group: "Information is the result of processing, manipulating, and organizing data in a way that adds to the knowledge of the person receiving it" who, I may add, must have an interest in acquiring the information so that he/she can apply it on purpose… with an objective or specific goal in mind. We might like to call such a person a "predisposed addressee." Already predisposed, such an addressee no longer needs to be persuaded.

Each "Machine" application, as explained in the Introduction of the book's Executive Summary, has its primary objective to change people's behavior (or, I may say, to assist people to change their behavior according to already gained insights) in regard to a particular subject matter domain and use context. Therefore, to develop "Machines" like the ones presented by Aaron Marcus, information designers need to have a wide spectrum of competencies so that they can create effective designs that predisposed people can make use of.

---

[1] idX=Information Design Exchange: Development of International Core Competencies and Student and Faculty Exchange in Information Design within the EU/US Cooperation Program in Higher Education and Vocational Education and Training, 2007-08-31.

By bridging disciplines like physics, semiotics, design, and computer graphics, Aaron Marcus succeeded in gaining insights into facts and relationships not taught in one single university degree course. Thus, he deserves to be singled out as a front-runner second to none. However, what really needs to be underlined are his very down-to-earth objectives as demonstrated by the Machines" he presents in this book. Mediating between insights of theme-specific specialists and everyday life concerns, they provide what I consider the essence of what ultimately can be achieved: information mobilizing people to accomplish critical tasks (IMPACT), with respect to ecology, health, financial matters, and many other challenges.

International Institute for Information Design   Peter Simlinger
Research and Education, Vienna, Austria
February 2015

Preface

This book is the result of 5 years of effort (2009–2014) on the part of my firm, Aaron Marcus and Associates, Inc. (AM+A), to research and analyze how mobile devices (phones, tablets, as well as Web sites accessed through desktop computers) could help change people's behavior by combining aspects of persuasion design with information design. The projects stemmed from my experience a few years earlier attending and presenting at the Persuasive Technology 2007 conference at Stanford University, Palo Alto, California. [de Kort, Y., IJsselsteijn, W., Midden, C., Eggen, B., Fogg, B.J. (Eds.) (2007). Persuasive Technology, Second International Conference on Persuasive Technology, Persuasive 2007, Palo Alto, CA, USA, April 26–27, 2007. Revised Selected Papers Series: Lecture Notes in Computer Science, Vol. 4744, Subseries: Information Systems and Applications, incl. Internet/Web, and HCI.]

Many interesting and intriguing presentations at the conference concerned software/hardware designed to help people maintain their health better by encouraging them to take their medicines properly, to help people stop smoking, and to help people with other behavior challenges. It occurred to me that I might explore this topic in the mobile realm because of our ongoing work for mobile clients such as Nokia, Motorola, etc.

I selected a first project based on an earlier work we had done in the area of sustainability, which was a popular topic at the time in 2008–2009. With the assistance of research and development designer/analyst interns, we were able to work on this and subsequent projects for about 3 months, which included all planning, research, analysis, design, evaluation, and documentation. Of course, the projects were limited in scope because of time limitations. These were "Machine" projects: the sound of the name Green Machine, the first, seemed fortuitously pleasant, and I continued the use of this group name for these concept designs of mobile applications. All the Machines were unfunded R+D with no financial help from outside clients or organizations.

I undertook the Machine projects because of my personal interest in persuasion design, information design, and mobile user-experience design, but also in the individual subject matter of the first and subsequent projects: sustainability, obesity/

diabetes, financial planning for baby boomers, storytelling, etc. In fact, all of the projects and their subject matter derived in some way from my personal history. For example, I was diagnosed as obese and suffering from the onset of type 2 diabetes in 2009. My subsequent attendance at nutrition courses, my weight loss, and my change of exercise and diet all motivated me to explore the topic further through the Health Machine. I certainly believed that we were trying to improve the world through the development of these concept designs and "giving away" the results of our efforts through publications, lectures, workshops, and tutorials about the Machines.

I have collected, edited, and mildly revised the white papers we prepared for each Machine. Please understand and be patient with certain limitations in preparing this text: The 11 Machine white papers that served as a basis for the chapters of this book were prepared over a 5-year period. Our thinking and document structure evolved over those years, e.g., becoming more sophisticated in terms of competitive product analysis and introducing marketing research studies. In addition, the world has moved on; many new applications and publications have appeared and continue to appear daily about each of the chapters and their subject matter, for example, mobile device manufacturers becoming interested in the user's health data or the availability of smart watches with many of the functions and data sets of smartphones. Many newer, revised versions of the applications we considered have appeared in the intervening years, many more sophisticated than the ones we originally studied. It is/was simply not feasible to "reevaluate" all of the products, to re-do all the competitive analysis, or to "re-research" all the discussions about each subject matter. The chapters are a chronological recording of our work. I have simply tried to coordinate and record more succinctly and clearly the work that we did accomplish so that others might benefit from our observations, evaluations, analyses, designs, and philosophy.

I regret that because of legal contexts, we have not been able to reproduce all screens of competitive product analyses. I am grateful for the assistance of Mr. James Robinson of Springer London, for his assistance in securing permissions.

Any errors remaining are, unfortunately and inadvertently, mine. I have attempted to "clean up" and repair texts prepared by others; I may have missed something. Thanks for any comments/feedback about any errors or omissions.

I am hopeful that this version of our work may yet be interesting, inspiring, usable, useful, and appealing to students, teachers, and professionals. Professions/disciplines that might benefit include, but are not limited to, anthropology, culture studies, ethnography, gamification, human factors and ergonomics experts, mobile application developers, software engineers in general, subject-matter experts, usability specialists, and user-experience designers.

In regard to how best to read this book: I suggest that readers with specific subject domain interests read the introductory Chap. 1 and then the chapters that correspond to their subject-matter interests per the title of each Machine. Others may wish to read the book from the beginning to the end, which is organized chronologically. Each subsequent Machine development team was able to analyze and benefit from the previous white papers prepared for each project.

One cautionary note: The idea of making applications that are smart and persuasive is both challenging and potentially frightening or dangerous. All of our objectives concerned matters that most people would find "good" or "beneficial," like improving one's health or financial circumstances. However, all of these techniques could be used to change people's behavior in bad ways. The world has seen examples of powerful communicators persuading people to do harm to themselves or others (e.g., smoking cigarettes or killing groups of people because they were considered sub-human). Consequently, the proper and long-term use of these techniques requires a strong ethical awareness and basis for action. Alas, the ethics of design is not included in most designers' educations. There are resources, and I suggest some in Chap. 1.

Each of the Machines is a work in progress. We have published information about them and made their ideas available through earlier publications and lectures. This book assembles all the information into one easy reference. I wish you success in taking any of these Machines further. Please let me know if we can assist in any way with your own efforts at redeveloping them for specific target markets, cultures, subject matter, and platforms.

Berkeley, CA, USA   Aaron Marcus

Contents

Glossary of Useful Terms

It seemed appropriate for all Machines to provide a glossary of terms useful for understanding the text. The glossary follows:

Co-Creation	A form of marketing strategy or business model that emphasizes the generation and ongoing realization of mutual firm-to-customer value. It views markets as forums for the firm and customer to partner together to create products.
Gamification	The use of game design techniques, game thinking, and game mechanics to enhance non-game contexts.
Goals	Desired and requisite quantified, measurable steps in the development of an application.
Information Architecture	The art and science of organizing and labeling software to support usability.
Innovation	Creation of better or more effective products or services.
Objectives	The desired actions of a user when using an application.
Open Innovation	A paradigm that assumes that firms can and should use external ideas as well as internal ideas… as the firms look to advance their technology. In such a model, firms use internal and external sources of knowledge to turn new ideas into commercial products and services. Firms employing the open innovation model attempt to draw innovations from their own employees, customers, and competitors. This contrasts with the more traditional "closed innovation," in which companies are more self-sufficient with their innovation by controlling

	the creation of ideas via their own research and development departments.
Persona	Characterizations of primary user types and are intended to capture essentials of their demographics, contexts of use, behaviors, and motivations, and their impact on design solutions. Personas are also called user profiles.
Prototype	An early sample or model built to test a concept or process or to act as a thing to be replicated or from which to learn.
Social Network	A virtual community of friends, colleagues, and other personal contacts.
SWOT Analysis	Analysis of a product's strengths, weaknesses, as well as the opportunities and threats that arise from it, as a competitive product.
Use Scenario	A list of steps, typically defining interactions between a person in a role and a system to achieve objectives and goals of development.

Chapter 1
Introduction

1.1 Introduction

This book is a recording of ongoing development of mobile-device application concept-designs that seek to combine persuasion design theory with information design/visualization theory. Each application has as its primary objective changing people's behavior in regard to a particular subject matter domain and use context. Each of these projects is discussed within a chapter that describes what has been accomplished thus far. The projects are sometimes described partly in the present tense because each of them is an evolving design, but they were developed over a period of 5 years (2009–2014). They are presented in approximately chronological order. Each project utilizes user-centered design-process techniques and persuasion design, which topics are discussed individually below and incorporated into each subsequent chapter.

1.2 User-Centered Design

User-centered design (UCD), as discussed in many books and publications (e.g., Hartson and Pyla 2012), links the process of developing software, hardware, the user-interface (UI), and the total user-experience (UX) to the people who will use a product/service. The user experience can be defined as the "totality of the […] effects felt by a user as a result of interaction with, and the usage context of, a system, device, or product, including the influence of usability, usefulness, and emotional impact during interaction, and savoring the memory after interaction" Hartson and Pyla 2012]. That definition means the UX goes well beyond usability issues, involving, also, social and cultural interaction, value-sensitive design, emotional impact, fun, and esthetics.

© Springer-Verlag London 2015
A. Marcus, *Mobile Persuasion Design*, Human–Computer
Interaction Series, DOI 10.1007/978-1-4471-4324-6_1

The UCD process focuses on users throughout all these development steps, or tasks, which sometimes occur iteratively:

- *Plan*: Determine strategy, tactics, likely markets, stakeholders, platforms, tools, and processes.
- *Research*: Gather and examine relevant documents, stakeholder statements.
- *Analyze*: Identify the target market, typical users of the product, personas (characteristic users), use scenarios, competitive products.
- *Design*: Determine general and specific design solutions, from simple concept maps, information architecture (conceptual structure or metaphors, mental models, and navigation), wireframes, look and feel (appearance and interaction details), screen sketches, and detailed screens and prototypes.
- *Implement*: Script or code specific working prototypes or partial so-called alpha prototypes of working versions.
- *Evaluate*: Evaluate users, target markets, competition, the design solutions, conduct field surveys, and test the initial and later designs with the target markets.
- *Document*: Draft white papers, user-interface guidelines, specifications, and other summary documents, including marketing presentations.

AM + A carried out most of these tasks in the development of each of the Machine concepts described in subsequent chapters, except for implementing working versions.

The above analysis describes the essential "verbs" of the profession. Over the past three decades in the UI/UX design community, designers, analysts, educators, and theorists have identified and defined a somewhat stable, agreed-upon set of *user-interface components*, or "nouns" on which the above verbs act, i.e., the essential entities and attributes of all user interfaces, no matter what the platform of hardware and software (including operating systems and networks), user groups, contents (including vertical markets for products and services), and contexts.

1.2.1   User-Interface Design Components

These UI components can enable developers, researchers, and critics to compare and contrast user interfaces that are evidenced on terminals, workstations, desktop computers, Websites, Web-based applications, information appliances, vehicles, and mobile devices. Marcus (Marcus et al. 1999; Marcus 2002), among others, provides one way to describe these user-interface components, which is strongly oriented to communication theory and to the applied theory of semiotics (Eco 1976; Peirce 1933; Innis 1985). This philosophical perspective emphasizes communication as a fundamental characteristic of computing, one that includes perceptual, formal characteristics, and dynamic, behavioral aspects of how people interact through computer-based media. Expanding upon Claude Levi-Strauss's idea of human beings as sign makers and tool makers (Levi-Strauss 2000) the theory understands a user interface as a form of dynamic, interactive visual literature as well as a suite of conceptual tools, and as such a cultural artifact. The user-interface components are the following:

Metaphors Metaphors are fundamental concepts communicated via words, images, sounds, and tactile experiences (Lakoff and Johnson 1980). Metaphors substitute for computer-related elements and help users understand, remember, and enjoy entities and relationships of computer-based communication systems. Metaphors can be overarching, or communicate specific aspects of user interfaces. An example of an overarching metaphor is the desktop metaphor to substitute for the computer's operating system, functions, and data. Examples of specific concepts are the trashcan, windows and their controls, pages, shopping carts, chat-rooms, and blogs (Weblogs). The pace of metaphor invention, including neologisms or verbal metaphor invention, is likely to increase because of rapid development and distribution through the Web and mobile devices of ever-changing products and services. Some researchers, such as David Gelernter, are predicting the end of the desktop metaphor era and the emergence of new fundamental metaphors (as cited, for example, in Tristram 2001).

Mental Models Mental models are structures or organizations of data, functions, tasks, roles, and people in groups at work or play. These are sometimes also called user models, cognitive models, and task models. Content, function, media, tool, role, goal, and task hierarchies are examples. They may be expressed as lists, tables, and diagrams of functions, data, and other entities, such as menus. They may be tree-structured, or more free-form.

Navigation Navigation involves movement through the mental models, i.e., through content and tools. Examples of user-interface elements that facilitate such movement include those that enable dialogue, such as menus, windows, dialogue boxes, control panels, icons, and tool palettes.

Interaction Interaction includes input/output techniques, status displays, and other feedback. Examples include the detailed behavior characteristics of keyboards, mice, pens, or microphones for input; the choices of visual display screens, loudspeakers, or headsets for output; and the use of drag-and-drop selection, and other action sequences.

Appearance Appearance includes all essential perceptual attributes, i.e., visual, auditory, and tactile characteristics, even olfactory in some unusual cases. Examples typically include choices of colors, fonts, animation style, verbal style (e.g., verbose/terse or informal/formal), sound cues, and vibration cues.

1.3 Information Design and Information-Visualization Design

Crucial to much effective user-experience design is gathering the data, information, knowledge, and wisdom that must be interactively explored, analyzed, displayed, understood, and acted upon. Entire professional groups are devoted to information design and information visualization (plus sonification and other rarer perceptual forms of information display). Among the organizations are the Society for

Technical Communication (STC), its conferences and publications, and the International Institute for Information Design, with its *Information Design Journal* and associated conferences such as Vision Plus.

In these professions, similar development steps, especially user studies, task analyses, and careful design of new terminology, schema, querying, forms of reply, and other systematic approaches lead to higher-level, more strategic solutions to people's needs for, desires for, and uses of information. A discussion of the field in general is contained in Marcus (2009).

A word about data and information. Computers have been called number-crunchers or data-processing machines. Nowadays, more is required, and people speak of computer-based systems for information processing, and Chief Information Officers have evolved in corporations. The following practical definitions are appropriate and useful:

- Data are significant patterns of perceptual stimuli, e.g., a collection of temperature sensations or readings.
- Information is significant patterns of data, e.g., the temperature and other weather conditions, or the traffic conditions for a particular road, for a particular day in a particular city.
- Knowledge is significant patterns of information *together with action plans*, e.g., the weather conditions for a city on a particular day, their impact on traffic patterns, and the likely alternate roads on which to drive to arrive safely and on time at a destination, with a likely best choice indicated or in mind.
- Wisdom is significant patterns of knowledge, *either in-born or acquired through experience*, e.g., the knowledge of past experience taking certain roads, the likelihood of traffic accidents or repairs along that route, and familiarity with the various route options.

Helping people make smarter decisions faster means helping them to make *wise* decisions, no matter what the subject domain, context, or personal experience and expertise of the user.

Of special interest are the means for communicating structures and processes, which may be shown in abstract or representational forms. Classically, these may be described as tables, forms, charts, maps, and diagrams. Many fine, classical, and thorough treatises have appeared in the past decades, such as the works of Bertin (1983) and others.

The preceding list of graphical communication techniques suggests an approximately increasing complexity of visual syntax. This term and approach, derived from semiotics, the science of signs (see, for example, (Eco 1976)), identifies four dimensions of "meaning:"

Lexical: how are the signs produced?
Syntactic: how are the signs arranged in space and time, and with what perceptual characteristics?
Semantic: to what do the signs refer?
Pragmatic: how are the signs consumed or used?

In the context of this book describing the Machines, metaphors may be termed the fundamental "concepts" of the Machine. The information architecture comprises the metaphors, mental model, and navigation. The discussion in the ensuing chapters about the screen designs will describe, also, the interaction and appearance, especially as the designs move from conceptual designs (so-called wire-frame versions) to perceptual designs (so-called look-and-feel versions). One unique approach of the Machines is to combine information-design and information-visualization (tables, forms, charts, maps, and diagrams) in the context of designing a persuasion process. As such, the discussion has entered the realm of visual and verbal *rhetoric*, a millennia-old system of communication that has been analyzed and practiced in the visual communication professions for centuries (see, e.g., Lanham 1991).

This book does not have scope, space, and time to explore all of these subjects and their ramifications for design. Readers are urged to explore further in the publications cited and recommended throughout the book.

1.4   Market Research and Competitive Analysis

In order to have a clearer vision of the target market for the machines, AM+A in some instances conducted qualitative research with potential customers via questionnaires, surveys, interviews, and ethnographic observations. In some cases, AM+A used the assistance of students in the graduate marketing course offered by Prof. Robert (Bob) Steiner, University of California at Berkeley, Extension Program, International Diploma Program.

In particular, AM+A carried out competitive product analyses, reviewing, briefly, the reviews and comments posted on products evaluated as well as subjective observations/comparisons of the screens and functions/data provided on Web-accessed application stores. AM+A derived these applications' major benefits and drawbacks. We assembled our analyses into "Pros" and "Cons" groupings and in some cases undertook a "strengths, weaknesses, opportunities, threats" (SWOT) analysis familiar to business schools and business managers. This in-depth analysis helped further develop initial ideas for the detailed functions, data, information architecture (metaphors, mental model, and navigation) and look and feel (appearance and interaction) of the machine prototypes.

Please note: Our analyses were conducted over several years' time, while products have continued to develop, change, and improve. I have not been able to re-conduct these analyses, so they should be interpreted as "historical" evaluations that demonstrate the kind of competitive analysis required for good user-centered design practice. I have attempted to identify and credit screen images from these products at the time; they have, in most cases, changed considerably.

1.5 Personas

User-centered design focuses attention on the users, especially the user types. These are often termed personas or user profiles. A persona is not an actual user, but a "pretend user" or a "hypothetical archetype." (Cooper 2004). Personas can assist in keeping specific users vividly in the minds of developers throughout the development process. They are "fictional" collections of key characteristics of key users but based on known facts or trends. Typically, UI development teams define one to nine primary personas.

A persona is defined by stating a set of objectives/goals distinct from other personas, such that their measures of a successful user-interface and task flow will vary according to their needs and interests. Although there are many *persons* who may use the product, the best designs tend to be focused rather than diluted. This focus can be achieved only by optimizing the design toward a few key personas. Personas are also helpful in optimizing sections of the software towards the user types that will use the product. It is important to note that one of the Machines is unique in its use of "double" personas: the Marriage Machine. AM+A is currently researching and evaluating the use of double personas in user-experience design.

Personas typically state/show the following:

- Name, age, title, slogan of the Persona.
- Image of the Persona (photograph, sketch, cartoon, etc.).
- Context (physical, social, cognitive, emotional, cultural, etc.).
- Objectives/goals (personal, professional, communal).
- Impacts on the design.

1.6 Use Scenarios

Use scenarios are a UX/UI development technique whereby, during the development of a prototype to simulate the major characteristics of a software product, UX designers write a use scenario to determine what behavior will be simulated. A scenario is essentially a sequence of task flows with actual content provided, such as the user's demographics and goals, the details of the information being worked with, etc. Note that use scenarios differ from *use cases*, a term from the world of software development. Use cases are detailed descriptions of tasks that list or describe specific functions of a software application and the data input/out related to these functions. The use case is often somewhat technical in nature and a step further away from software code itself than so-called pseudo-code. Use scenarios, on the other hand, are usually descriptions in everyday language, not in deeply technical terminology.

For example, to simulate a Print dialogue box, a scenario might state that "an account executive wants to print out a Powerpoint presentation in landscape format, duplexed, with page numbers, as quickly as possible." The prototype developer then needs to simulate, with prototyping software, how the account executive would accomplish this task using the software.

Scenarios typically should have a format that includes the following:

- A descriptive and compelling title.
- The background of the situation and the user.
- The event or information that prompts user action.
- Step-by-step listing of user actions (typically technology-independent) to reach a goal or conclusion.
- A list of user benefits demonstrated by the scenario.
- Optionally, reference to any supporting materials, such as existing screen designs, paper materials, information listings, or prototypes.
- Optionally, a description of the current methods used to accomplish the same goal. These descriptions are good for comparison, because the new scenario is meant to be an improvement upon the existing methods.

1.7 Persuasion Theory

The Machines attempt to use persuasion theory in changing people's behavior. What is persuasion theory? When it comes to the theory, or more precisely, to the science of persuasion, first of all Cialdini's (2001, 2007) work must be taken into account. He concentrates particularly on the psychological dimensions concerned in the act of persuasion: What makes an individual comply with another's request? What makes someone change or adapt new attitudes or actions? Cialdini distinguishes six basic phenomena in human behavior, which are supposed to favor positive reactions to persuasive messages of others: reciprocation, consistency, social validation, liking, authority, and scarcity. These tendencies in the social influence process are characteristic of human nature and are thus valid across national boundaries; nevertheless, cultural norms, traditions, and experiences can have an impact on the relative weight of each of the six mentioned factors, which Cialdini points out.

In addition, Fogg has made significant contributions to persuasion-related research (Fogg 2003; Fogg and Eckles 2007). In alignment with Fogg's persuasion theory, we have defined six key processes to achieve behavioral change via a Machine's functions and data:

- Attract users via traditional marketing techniques, that is, making users aware and motivating them to examine, try, and use the application.
- Increase frequency of using the application.
- Motivate users to change some habits, for example, interaction with and openness towards the subject matter and people, experience, and observation of differences in experience, and ways of documenting changes.
- Teach users how to change habits.
- Persuade users to change habits (short-term change).

- Persuade users to change general approach to objectives, people, objects, contexts, obstacles, and emotions (long-term, or life-style change).

Each step has requirements for the application.

Motivation is a need, want, interest, or desire that propels someone in a certain direction. From the sociobiological perspective, people in general tend to maximize reproductive success and ensure the future of descendants. We apply this theory in the Machine by making people understand that a determinate behavior can fundamentally enrich their daily experiences, that it can increase their knowledge and understanding, and that it can ultimately trigger a process of relationship building to improve success.

We also drew on Maslow's Theory of Human Motivation (1943), which he based on his analysis of fundamental human needs. We adapted these needs to each Machine context:

Safety and security: met by the assistance through family, friends, or advisors; and by the provision of obstacles-, fun-, human-contact-, and others-related information, tips, and advice.

Belonging and love: expressed through social sharing and support among friends and family.

Esteem: satisfied by social comparisons that display progress and destination expertise, as well as by self-challenges that are suggested by the application and that display the goal-accomplishment processes.

Self-actualization: fulfilled by being able to follow and retrace continuous progress and advancement in a personal diary.

Many publications on persuasion and technology have appeared in recent years. One such compendium is the *Proceedings* of Persuasive Technology 2015 (MacTavish 2015), the tenth anniversary conference on the subject.

1.7.1   Persuasion and Ethics

I offer a brief cautionary note. Persuasion design must have a strong underlying ethical basis; otherwise, it can be misused, e.g., in propaganda directed to inhuman treatment of others. Unfortunately, most designers are not trained in ethics. There is much discussion of "good" and "bad" in the professions, but often without substantial knowledge of ethical theory. There is one source of assistance that focuses on the ethics of design in (Becker 2011).

Another aspect to consider is that much practical persuasion theory has been used in addiction programs. There is a danger that some users will simply substitute the Machine for independent cognition, speech, and behavior, exchanging one addiction for another, namely, the benefits, answers, or rewards that the Machine itself can provide. These are perhaps extreme examples, but one should remain vigilant against misuse of the techniques by developers as well as by users.

1.8   Information Architecture

Information architecture is the structure or organization of the mental model of the application and the application's implied navigation. The mental model can be considered a designer-driven user-model, that is a designer-model that has been successfully transplanted to the user's mind. Of course, one can investigate native or inherent user models. These may serve as excellent resources for the designer-driven user-model. Just as some landscape designers let pedestrians make their own paths over the grass and define the "people's choice" for pathways, the designers of application mental models may be inspired by or influenced strongly by inherent "native" or "intuitive" organizations. However, other factors such as marketing objectives, complexity of functions, complexity of data, lack of experience in users, or lack of education may lead designers to impose conventions that they believe and/or have proven to be more useful than native or indigenous solutions.

In general, the Machines described here have a basic and consistent primary information architecture comprised of five essential components or modules:

- Dashboard: Where am I now in my journey?
- System or process overview: From where have I come? Towards what end is my journey directed?
- Focused social network connections: Not 5,000 Facebook friends, but which family and friends who care about my journey?
- Focused just-in-time tips or advice: Not 5,000,000 returns of a Google search, but what specific information can assist me just now at this point in my journey?
- Incentives: What competitions, games, awards, rewards, nostalgia stores, workshops, and other content can further motivate me and keep me on the path of my journey?

Further Reading

Becker L (2004) How do we know. Lecture summary, AIGA Education Conference, October 2004. Available at http://futurehistory.aiga.org/resources/content/2/2/6/8/documents/l_becker.pdf. Checked 11 Sept 2014

Cialdini R (2014) Influence: the psychology of Persuasion. Collins Business, an Imprint of HarperCollins Publishers, New York

Eyal N, Hoover R (2014) Hooked: how to build habit-forming products. Penguin Group, New York

Jean J, Marcus A (2009) The green machine: going green at home. User Experience, 8:4, 20–22ff., 4th Quarter 2009, available online at www.UsabilityProfessionals.org

Kool L, Timmer J, Van Est R (eds) Sincere support; The rise of the e-coach. Rathenau Instituut, Amsterdam

Marcus A (1979) New ways to view world problems. East–west perspectives. J East–west Center, Honolulu, 1(1):15–22

Marcus A (2000) User-interface design for air-travel booking: a case study of Sabre. Inf Des J 10(2):186–206. John Benjamins Publishing Co., Amsterdam. Published by the International Institute for information Design

Marcus A (2011) Health machine. Inf Des J 19(1):69–89

Marcus A (2012a) The money machine. Helping Baby Boomers Retire. User Experience 11(2), 2nd Quarter 2012, pp. 24–27

Marcus A (2012b) The story machine: a mobile app to change family story-sharing behavior. CHI 2012, Austin, TX, USA, 5–10 May 2012

Marcus A/Jean J (2009) The green machine: going green at home. User Experience 8(4), 4th Quarter 2009, pp. 233–243

Marcus A, Jean J. Going green at home: the green machine. Inf Des J 17.3:233–243

Nodder C (2013) Evil by design: interaction design to lead us into temptation. Wiley, Indianapolis

Paharia R (2013) Loyalty 3.0: how big data and Gamification are revolutionizing customer and employee engagement. McGraw-Hill, New York

Shedroff N (2009) *Design is the problem.* The future of design must be sustainable. Rosenfeld Media, Brooklyn/New York

Tscheligi M, Reitberger W (2007) Persuasion as an ingredient of societal interfaces. Interactions, pp 41–43, September–October 2007

References

Becker L (2011) Design and ethics: rationalizing consumption through the graphic image. PhD dissertation, University of California at Berkeley, 04 Sept 2011. Available through www.Amazon.com. Checked 11 Sept 2014

Bertin J (1983) Semiology of graphics: diagrams, networks, maps, Madison. Originally published in French as *Semiologie graphique*, Edition Guathier-Villars, Paris, and republished in 2011 by Esri Press, Redlands, California

Cialdini RB (2001) The science of Persuasion. Sci Am 284(2):76–81, Feb 2001. (www.influenceatwork.com)

Cialdini et al (2007) The constructive, destructive, and reconstructive power of social norms. Psychol Sci 18(5):429–434, May 2007

Cooper A (2004) The inmates are running the asylum. SAMS, Indianapolis. ISBN 0-672-31649-8

Eco U (1976) A theory of semiotics. Indiana University Press, Bloomington

Lakoff G, Johnson M (1980) Metaphors we live by. The University of Chicago Press, Chicago

Levi-Strauss C (2000) Structural anthropology (trans: Claire Jacobson, Brooke Schoepf). Basic Books, New York

Fogg BJ (2003) Persuasive technology: using computers to change what we think and do. Morgan Kaufmann Publishers, San Francisco

Fogg BJ, Eckles D (2007) Mobile Persuasion: 20 perspectives of the future of behavior change. Persuasive Technology Lab, Stanford University, Palo Alto

Hartson R, Pyla P (2012) *The UX book: process and guidelines for ensuring a quality user experience.* Morgan Kaufmann (Elsevier), Waltham

Innis RE (1985) Semiotics: an introductory anthology. Indiana University Press, Bloomington

Lanham R (1991) A handlist of rhetorical terms. University of California Press, Berkeley

MacTavish T (ed) (2015) *Proceedings*, Persuasive technology 2015, 2–5 June 2015, IIT Institute of Design, Chicago. IIT Institute of Design, Chicago

Marcus A (2002) Information visualization for advanced vehicle displays. Inf Visual 1:95–102

Marcus A (2009) Integrated information systems. Inf Des J 17(1):4–21

Marcus A et al (1999) Globalization of user-interface design for the web. In: Proceedings of human factors and the web conference, Gaithersburg, MD, 3 June 1999. http://zing.ncsl.nist.gov/hfweb/proceedings/marcus/. Checked on 16 Sept 2014

Maslow AH (1943) A theory of human motivation. Psychol Rev 50:370–396

Peirce CS (1933) Existential graphs, pp 293–470. In: Hartshome, Weiss (eds) The collected papers of Charles Sanders Peirce. See Peirce, Charles Sanders. Collected papers, vols 1–6 edited by Charles Hartshorne and Paul Weiss; vols 7–8 edited by Burks AW. Belknap Press of Harvard University Press, Cambridge, 1958–1966. URL: http://www.nlx.com/collections/95. Checked 11 Sept 2014. See also, http://plato.stanford.edu/entries/peirce-semiotics/. Checked 11 Sept 2014.

Tristram C (2001) The next computer interface. Technol Rev, December 2001, pp 53–59

Chapter 2
The Green Machine: Combining Information Design/Visualization and Persuasion Design to Change People's Behavior About Energy Consumption

2.1 Introduction

In the late nineteenth century, French mathematician Jean Baptiste Joseph Fourier introduced the first theory of global warming that later would be referred to as "the greenhouse effect" (Fourier 1824). By the 1950s, concerns about global warming and sustainable development moved into the mainstream. After more than 60 years and thanks to countless global environmental-awareness campaigns, such as Marcus et al.'s audio-visual presentation "Visualizing Global Energy Interdependence" (Marcus 1979, 2009) and Al Gore's Inconvenient Truth (Gore 2006), the issues of global warming, or environmental threats, are no longer unknown. However, this information, although alarming or frightening to some, does not necessarily lead to changes in people's behavior and their way of life. At the core are issues of sustainability.

What is sustainability? According to the US Environmental Protection Agency (EPA 2010): "Sustainability is based on a simple principle: Everything that we need for our survival and well-being depends, either directly or indirectly, on our natural environment. Sustainability creates and maintains the conditions under which humans and nature can exist in productive harmony, that permit fulfilling the social, economic and other requirements of present and future generations.... Sustainability is important to making sure that we have and will continue to have, the water, materials, and resources to protect human health and our environment."

It is one thing to make people aware; this is basic. It is another thing to lead people to action. Two vital issues need to be addressed: how to help people to reduce their ecological footprint, and how to persuade and motivate them to change their behavior. Nathan Shedroff lists five approaches of sustainable design in his book Design is the problem (Shedroff 2009): reduce, reuse, recycle, restore, and process. This chapter considers another important aspect: how to persuade people to reduce their ecological footprint through the medium of a specially designed suite of

© Springer-Verlag London 2015

A. Marcus, *Mobile Persuasion Design*, Human–Computer
Interaction Series, DOI 10.1007/978-1-4471-4324-6_2

functions and data within a mobile-phone application that we have called the "Green Machine." The Green Machine's objectives are to persuade and motivate people to change their behavior and thereby to reduce their energy consumption.

Our initial concept and objective focused primarily on household energy consumption, which represents 19 % of the US total CO_2 emissions. However, our approach can be extended to other conservation areas, such as waste and recycling, transportation, shopping, and eating. The Green Machine is intended to rely upon so-called Smart-Grid technology, an important innovation that enables users to acquire instantaneous feedback about their energy consumption (see for example, NIST 2009). Studies conducted by Sarah Darby in her article, "The effectiveness of feedback on energy consumption" (Darby 2006) show that feedback has an impact on reducing household energy consumption by about 10 % without making any important lifestyle changes. Therefore, comparing this amount to the US Energy Information Administration data, one can see that, acting together, we actually might save as much energy as the US produces by wind and solar with simple and easy changes.

2.2   The Green Machine: A Work in Progress

We believe that simply showing data visualizations, as basic Smart Grid software will enable, is not enough to make people effectively reduce their energy consumption. I have had some experience with this challenge and have been involved with showing energy information since 1978 when I was appointed to be a Research Fellow at the East–west Center, a then Federally funded research institute on the campus of the University of Hawai'i. I lead a team of five visual communicators, together with others, to research and analyze how to show global energy interdependence more effectively. We designed and produced a short non-verbal, cross-cultural, audio-visual presentation that used icons/symbols, tables, charts, maps, and diagrams, with a multi-cultural musical sound track to communicate our ideas. See Figs. 2.1 and 2.2 for exemplary imagery.

Our team of professionals included representatives of the USA, China, Japan, Iran, and India. We documented our 6-month project's deliverables in publications (see, for example, Marcus 1979) and presented our work worldwide. Although the presentation was of interest, received a design commendation, and seemed worthwhile, we did "not change the world." Years later, returning to this theme of sustainability, I felt more was needed.

We turned to persuasive techniques and the study of the behavior-change process in combination with context of use, user-interface analysis, and information visualization in order to find ways to design an interactive software application, specifically one for a mobile smartphone, that would encourage and persuade people effectively to reduce energy consumption. The result of our work, the Green Machine, is an application concept design based on five main functionalities:

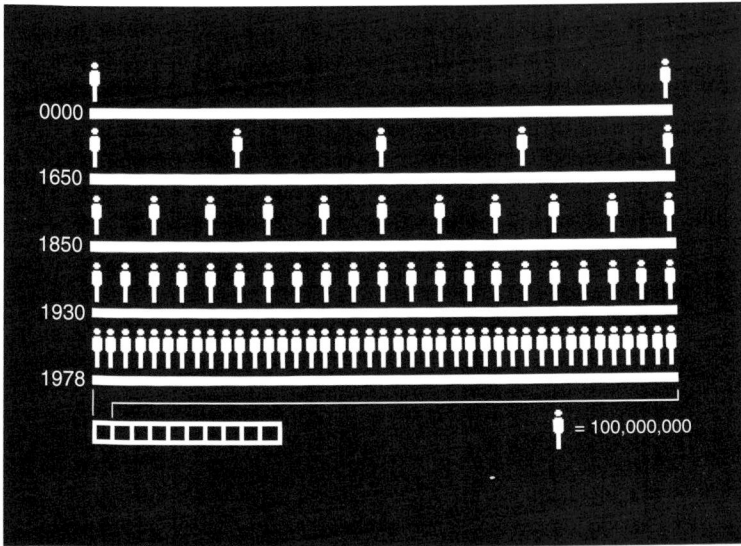

Fig. 2.1 Image from the audio-visual presentation "Visualizing Global Interdependencies," 1978, showing growth of world population over the last 20,000 years (Image: East–west Center, Honolulu, HI.)

- Provide feedback about one's energy consumption in comparison to personal goals.
- Display a vision of the future linked to that consumption.
- Enable social interactions with social networking and energy-consumption comparisons.
- Offer tips to reduce one's ecological footprint.
- Provide individual or team-based incentives to keep going, such as competitions and games.

We developed and tested a conceptual prototype to collect feedback about the application's usability, usefulness, and appeal, especially, users' impressions about the motivational aspects of the design. User test analysis led to a revised design of the Green Machine application. Figure 2.3 provides an exemplary illustration of our initial approach.

2.2.1   Background Research

We conducted secondary research in a numb~
including persuasive technology (see C~
Smart-Grid technology, and Pers~
vate people to perform ben~

~r designs,
this subject),
~jective to moti-
~y fields during the

Visualizing Global Interdependencies

The earth, a home for more than four billion people, is a place of greatly increasing diversity and complexity.

Ideas, people, and goods are moving faster and faster and intermingling.

There are rising challenges in the changing world, caused by global situations of population, food, energy, and environmental pollution.

`2 Page from the article in Perspectives, "New Ways to View World's Problems" (Marcus ~wing selected images from the audio-visual presentation and brief English narration that last ^he images and sound track (Image: East–west Center, Honolulu, HI.)

suasive a^ion is defined as "an attempt to shape, reinforce, or change ^ughts about an issue, object, or action" by Dr. B. J. Fogg, ~ive Technology Laboratories (Fogg 2003). These per- ^veloped for many different purposes, such as to

Fig. 2.3 An example screen: the Green Machine's Total Energy Use screen. The screen enables users to visualize their energy consumption in kWh, currency, and amount of CO_2 release. The screen also shows goal-setting insights and equivalent comparisons. The Calendar function and extra features would enable other kinds of comparisons to be made

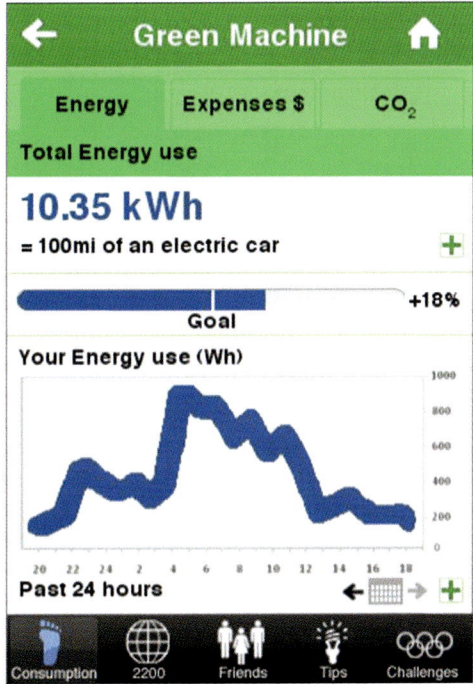

encourage people to lose weight or quit smoking, or to promote sports activities and exercise. Each application bases itself on providing feedback to users about themselves and enabling analysis to increase people's motivation and to change their lives through appropriate behavioral changes. These feedback-based persuasive applications have shown important beneficial results and have been applied to environmental sustainability. Sarah Darby (2006) conducted an important analysis of feedback in the context of household's energy consumption, which underlines that feedback about consumption is necessary for energy savings. The study shows that it reduces consumption by about 5–15 %.

Smart-Grid technology makes it possible to provide this kind of instantaneous feedback about energy consumption. The challenge of designing persuasive user interfaces oriented toward the environment, however, is significant: most people are not intrinsically motivated to care about and change their behavior, as Tscheligi and Reitberger emphasized in "Persuasion as an ingredient of societal interfaces" (Tscheligi and Reitberger 2007). On the other hand, people with high social awareness tend to be unsatisfied with minimalist feedback, as Yun showed in "Investigating the Impact of a Minimalist In-Home Energy Consumption Display" (Yun 2009). Therefore, goal setting is necessary in order to give meaning to feedback and to optimize its effectiveness. If a goal to save energy does not already exist or has not been set, feedback becomes meaningless (McCalley and Midden 2002).

Social interactions add persuasive aspects and help to increase involvement and motivation. There are two ways to support social change: (1) by mobilizing structures that help to support prominent figures and/or people and (2) by changing the belief of individuals and the culture of which they are a part (Mankoff et al. 2007). Leveraging social networks is therefore a powerful tool to integrate environmental sustainability into daily activities and social context. These results are complemented by the six "weapons of influence" in order to persuade people, as defined by Robert Cialdini (2001). These techniques are the following:

- *Reciprocation*: people tend to return a favor.
- *Commitment and Consistency*: if people commit, orally or in writing, to an idea or goal, they are more likely to honor that commitment.
- *Social Proof*: people will do things that they see other people are doing.
- *Authority*: people will tend to obey authority figures, even if they are asked to perform objectionable acts.
- *Liking*: people are easily persuaded by other people that they like.
- *Scarcity*: perceived scarcity generates demand.

An experiment by Cialdini et al. (2004) emphasized the effects of neighborhood comparison in energy savings. According to the findings, people reduced their energy consumption when they found out that their neighbors had already taken steps to curb their energy use.

Competition is another way to motivate people to increase their awareness and reduce their energy consumption. For example, the "Energy Smackdown" (see http://www.bozemanenergysmackdown.net) is an Internet-based challenge between individual households to reduce home energy consumption and CO_2 emissions.

2.2.2 Analysis

The Green Machine has two persuasion objectives: "microsuasion" and "macrosuasion," according to Fogg's terminology. The microsuasion objective is to make people reduce their household's energy consumption, and the macrosuasion objective is to change people's long-term behavior or life style. These objectives are intrinsically linked (as short-term and long-term objectives) although they exist at two different levels.

To create behavioral change through the Green Machine, we defined five key elements:

- Increase frequency of use of the application.
- Motivate reduced energy consumption.
- Educate users about how to reduce energy consumption.
- Persuade users to reduce energy consumption.
- Persuade users to change their behavior.

Each step influences requirements for the application. These steps are summarized in Fig. 2.4.

Increase frequency of using application	→	Motivate reduced energy consumption	→	Teach how to reduce energy consumption	→	Persuade users to reduce energy consumption	→	Persuade users to change behavior
• Usability • Usefulness • Appeal • Rewards		• Link between users needs and motivation • Competition and challenge • Goal setting • Persuasion issues (Fogg, Cialdini)		• Tips in context • Social interaction with advice • Consumption feedbacks related to the goal setting		• Frequent consumption feedback • Social interaction (display information, improvements)		• Long term use

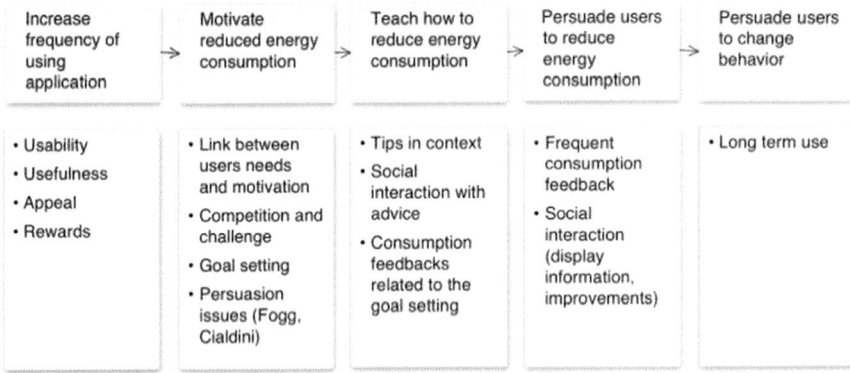

Fig. 2.4 Five-step change behavioral process that affects the Green Machine design and catalyzes specific detailed solutions

Motivation is a need, want, interest, or desire that propels someone in a certain direction. From the sociobiological perspective, people in general tend to maximize reproductive success and ensure the future of descendants. We apply this theory in the Green Machine by making people understand that every action has consequences on environmental change and on the Earth's future. We also drew on Maslow's A Theory of Human Motivation (Maslow 1943), which he based on his analysis of fundamental human needs. We adapted these needs to the Green Machine context:

- The safety and security need is met by the possibility to visualize the amount of money saved.
- The belonging and love need is expressed through membership of an eco-friendly community or belonging to a particular team in the Challenges section.
- The esteem need can be satisfied by social comparisons that display energy consumption and improvements.
- The self-actualization need is fulfilled by being able to visualize the amount of CO_2 released in the atmosphere and can also is met by enabling users to make donations to sustainability-oriented organizations.

Because setting goals helps people to learn better and improves the relevancy of feedback, the Green Machine Login page on the mobile device and/or a desktop-computer's access to an accompanying Website asks users how much money they want to save, or which friends' energy profiles they wish to look up. (Note: each mobile application assumes a likely accompanying Website, especially for large-scale data-entry tasks; these Websites are not specifically addressed in detail in this book.)

To improve learning, the application integrates contextual tips to explain how to reduce energy consumption. The application also shows tips that have been successful for other users, as well as other products or services they have tried.

Social interaction also has an important impact on behavior change. Accordingly, the Green Machine leverages social networking and integrates features like those found in forums, Facebook, Twitter, etc.

The Green Machine is intended to provide frequent feedback, including daily energy-consumption snapshots, a future-Earth metaphor, and social interactions, such as energy comparison, friendly challenges, or added comments.

We also aimed for long-term use, because, as Darby (2006) explains, it takes over three months for behavior change to become permanent.

We developed personas (Mr. and Ms. Everyone and their family) and use scenarios, as is typical for user-centered design (as described in general in Chap. 1; see Fig. 2.5).

For example, Ms. Everyone and her husband want to check their energy consumption before running a major appliance. She/he uses a device to check their current household energy use, to browse through the usage history, and to check tips on lowering usage. By lowering usage, she/he gains a higher status in her/his friends' energy-saving group, earns reward points, and "helps save the Earth." Before sleeping, she/he checks the household's total energy use for the day and compares it with her/his best friend Matthew. She/he teases him with a short message while brushing her/his teeth before retiring to bed.

To realize this use scenario, together with these personas, one must emphasize the necessity of convenient access that matches context of use, suitable to other activities, and always-on applications at people's fingertips. To achieve these characteristics, we decided to make the Green Machine a mobile-phone application. Consequently, the Green Machine would be available on a ubiquitous platform,

Fig. 2.5 Representation of the personas for the Green Machine: Mom and Dad or Mr./Ms. Everyone

somewhat globally available on the most common and well-known electronic devices in the world today. In our sketches, we have emphasized one particular mobile platform, but this is simply for ease of demonstration.

2.2.3   Design

The background research and succeeding analysis emphasized five main issues: feedback, future-Earth metaphor, social interactions, tips, and competitions/challenge. These are represented in the accompanying information architecture of the Green Machine, as shown in Fig. 2.6.

Designing the interaction and visual design was a particular challenge on the small screen of a mobile phone; therefore, special attention was given to the different components of the user interface: metaphors, mental models, navigation, interaction, and appearance (Marcus 2000), as identified in Chap. 1. In particular, the interaction mode is limited; without a regular keyboard, data must be selected to fit the small screen size, and the number of clicks and submenus must be reduced to a usable, useful minimum.

We decided to use a tabbed navigation so that every action can be achieved in less than four clicks. Users know exactly where they are in the application architecture thanks to the screen title information. The visual design of our prototype is based on typical iPhone styles (the device available in 2009), because of the

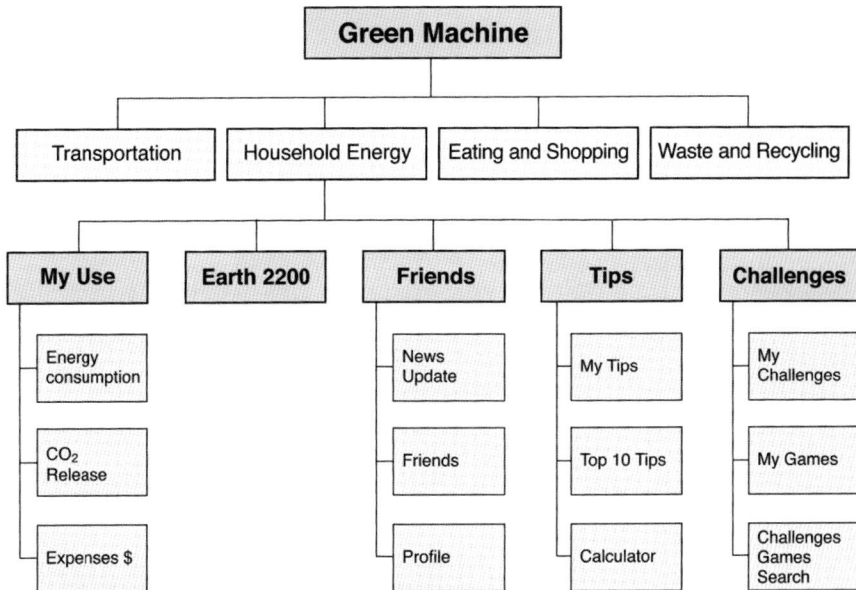

Fig. 2.6 Green Machine information architecture

product's brand image: trendy, with "early adopters" as target market, and the possibility of downloading applications.

In the user interface, the small energy-thermometer is always displayed at the top of the screen and shows the current household energy use, as shown in Fig. 2.7.

The Energy tab (Total Energy Use) displays the kWh consumed, the money spent, and the amount of CO_2 released. Users can visualize this total energy use in different time periods, such as a day, week, month, or year. This energy use is automatically compared with the user's goal settings.

Social interactions are also included. Friends' and celebrities' comparisons enable users to select one of their friends, or one of the many celebrities (e.g., President Obama or Al Gore) using the Green Machine and then to compare consumption.

The Earth in 2200 screen (see Fig. 2.8) displays simulated breaking news stories according to one's energy consumption. If there is high energy consumption, the state of the Earth is shown with dire consequences, such as increased environmental refugees, outbreak of wars, and biodiversity endangerment. For low energy consumption, we view a healthier Earth with sufficient food, water, and greater chance for peace.

A networking tab (see Fig. 2.9) is aimed at motivation through social interactions. Users can read and visualize news from their friends with regard to how much

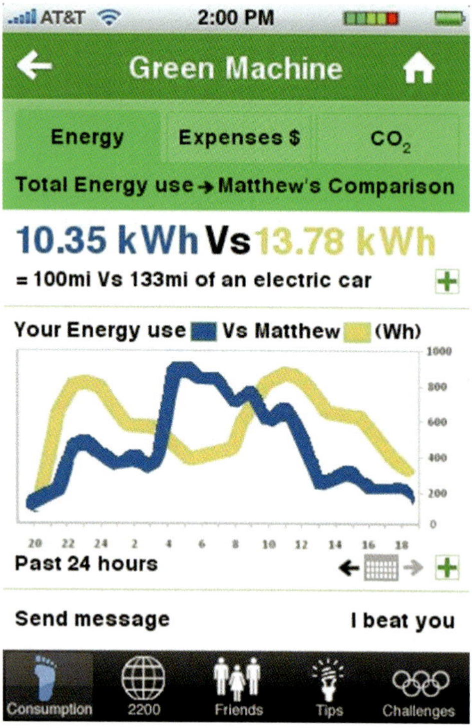

Fig. 2.7 Green machine total energy use friends' comparison

Fig. 2.8 Green machine
earth in 2200

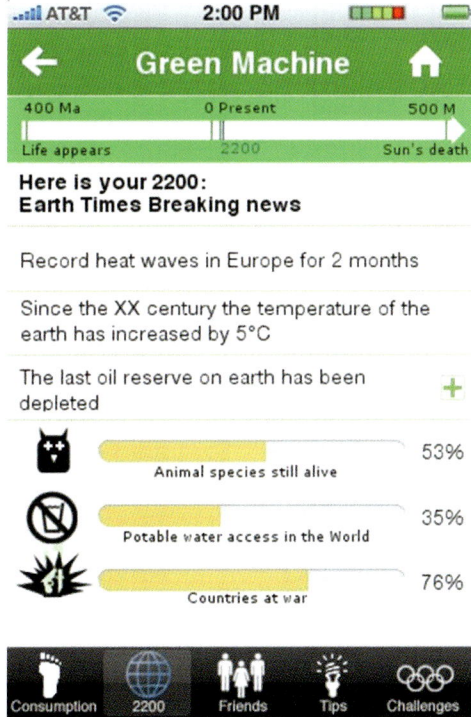

Fig. 2.9 Green machine
friend screen

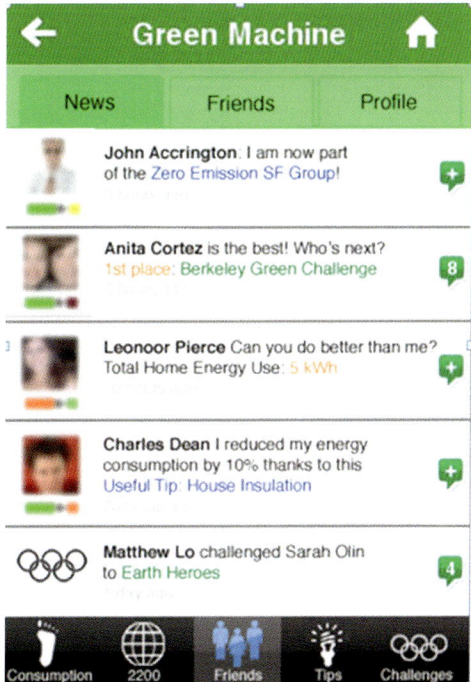

they have consumed, what were their challenges and their results, which tips were helpful, what their current energy use profile is, and to which charities they have provided donations.

The Tips tab enables users to learn how to reduce their energy consumption. The data visualization for each tip maps the cost and the amount of potential reduction. This information gives users a direct view of the impact of tips they may choose. An individual tip also shows how many friends have used it and found it helpful, their additional comments about a particular tip, its price (if it is a physical product like a light bulb), distance to the closest store to buy it, and, the overall rating given by Green Machine users.

The Challenge and Games tab has different competitions for users to reduce their energy use. Individual and team-based challenges are available to meet pro-individual and pro-social personalities. A game mode offers relevant video games to help people to reduce their energy consumption.

2.2.4 User Test

As part of our user-cantered design process, we conducted a simple user test with the limited number of initial screen designs. The primary objective of the user test was to identify usability issues with the Green Machine application's user interface. We also assessed whether users believed the application would make it easier for them to reduce their energy consumption and whether users believed the application could encourage them to make further reductions in their energy consumption.

These tests produced both quantitative and qualitative results. Sessions included free exploration to gather first impressions about the Green Machine and expectations for content and functionality. Participants then completed a list of task scenarios. Finally, they filled out questionnaires, covering both their energy consumer and green profile and their evaluation of the usefulness of the Green Machine. Assessment about people's ecological awareness was patterned on the Evaluation of Environmental Attitudes (Fernandez-Manzanal et al. 2007). We were particularly interested in whether or not users thought that this application could motivate them to reduce their energy consumption.

Following the design stage, a semi-functional simulation was developed on a PC to simulate the Green Machine's basic functionalities. We tested 20 people in Berkeley, California. They were 18- to 65-years-old and were culturally diverse. We chose men and women, students as well as lay people.

2.2.5 User Test Analysis

The respondents we tested appeared to be quite concerned about ecological problems with regard to the Evaluation of Environmental Attitudes. It is an intrinsic fact to take into account in the results analysis. However, it is interesting to test at some future point if people who are already ecologically biased could be motivated to do more and further reduce their energy consumption through this application.

In general, users' feelings toward the Green Machine application were positive and receptive to the data shown in the application. All respondents (except one who was uncertain) believed that having continuous feedback (i.e., first Green Machine screen) about their energy consumption could help them reduce their energy consumption. They thought the information provided on screens was helpful, relevant, and specific enough. An overwhelming majority say that this application would make them more self-aware and accountable for their actions. 18 respondents out of 20 (90 %) also agreed with the assumption that the Green Machine could motivate them to reduce their energy consumption. The primary motivational factors for our respondents were equally mixed across different sets of factors, such as personal challenges and competition, financial issues (i.e., saving money), fear of the future, social interactions, and information about energy use and scarcity. Moreover, 16 respondents out of 20 (80 %) thought that the Green Machine could guide them into a behavioral change and make them "greener." Figures 2.10 and 2.11 summarize the results.

The different features within the Green Machine have also been assessed. 12 respondents out of 20 (60 %) thought that the current set of features (consumption, Earth 2200, friends, tips and challenges tabs) could help them to reduce their energy consumption, with only 5 (25 %) who were uncertain and 2 (10 %) who disagreed or strongly disagreed. Although people expressed different opinions about which set of features they actually prefer, interestingly enough, about 7 respondents (35 %) mentioned the Earth 2200 screen. To the respondents, it seemed to be one of the most effective features in convincing them reduce their energy consumption, followed closely by the Tips and Challenges screens. We mention the fact that the

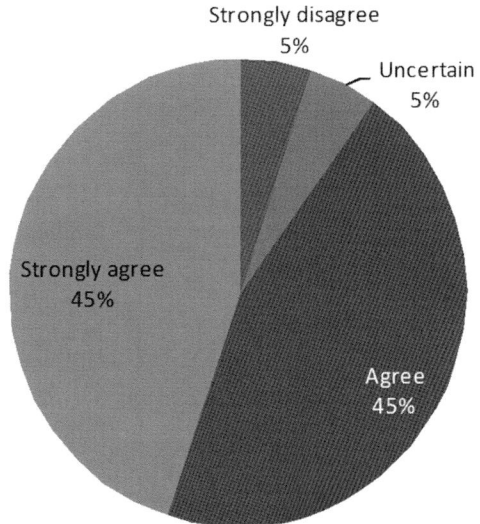

Fig. 2.10 Percentage of answers to the question, "This application could motivate me to reduce my energy consumption?"

Fig. 2.11 Percentage of answers to the question, "This application could accompany behavioral change and make me "greener" (focus on household)?"

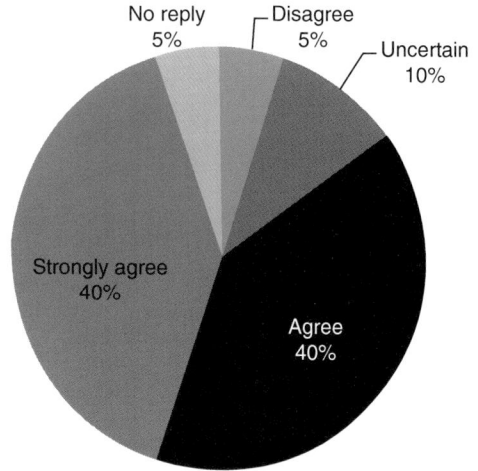

No reply 5%

Disagree 5%

Uncertain 10%

Strongly agree 40%

Agree 40%

screen presented to the respondents was a negative or dire view of the earth in 2200. Thus, we suspected that emotions and feelings of fear were stirred upon viewing it.

The user test also led us to valuable insights with regard to the application's usability. We realized that some navigation paths had to be altered to fit the users' mental model, and some icons' meanings were hard to understand, resulting from a poor references. Subsequently, we redesigned some screens and navigation paths to make sure that all the contents were clearer and more understandable, and that the application as a whole would be relatively easier to use and more compelling in the next version.

2.2.6   Redesign and Usability Improvements

In a re-design phase, we sought to improve the usability, usefulness, and appeal of the Green Machine. During the visual re-design, the user interface's metaphors and mental model were modified and significantly improved. For example, two specific Tips screens were introduced (see Fig. 2.12) and other changes were introduced (see Figs. 2.13 and 2.14).

During its first iteration, the navigational path through which users could compare their own energy consumption with a friend's was on the Consumption screen via its "Options – Friends' Comparison" tab found at the bottom right-hand side. We observed, however, that these steps were not at once obvious. Many of our testers initially tried to access the same functionality through the Friends' screen. Thus, we had to add this navigational path to fit the users' mental model while retaining the original navigational path.

Other usability improvements also included the redesign of some Green Machine icons to enhance the clarity of their meanings. For example, the "Plus" icon to

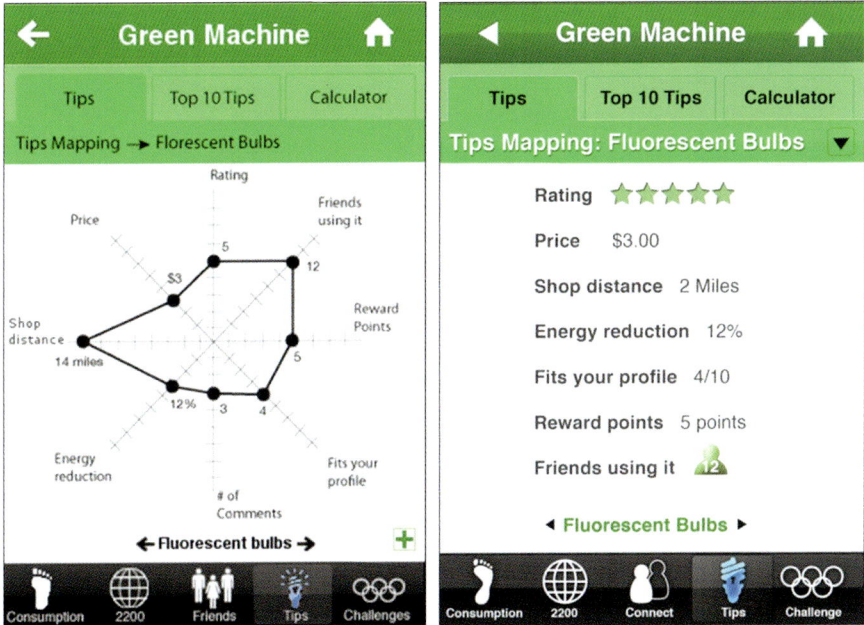

Fig. 2.12   Green machine tips screen 1st version (*left*) and 2nd version (*right*)

Fig. 2.13   Green machine
total energy use screen
version 2

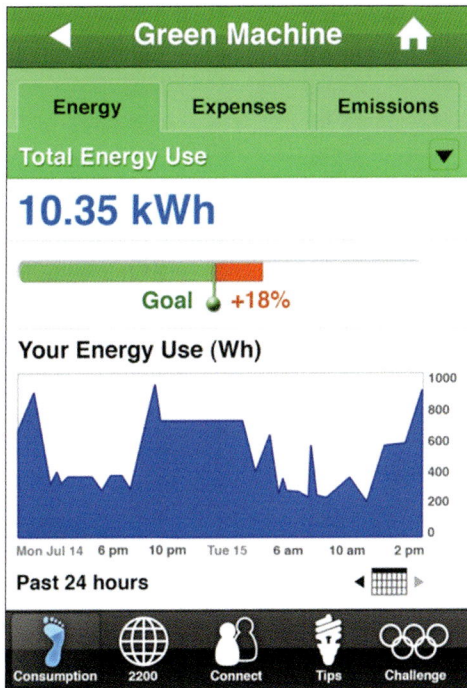

Fig. 2.14 Green machine
tips screen version 2

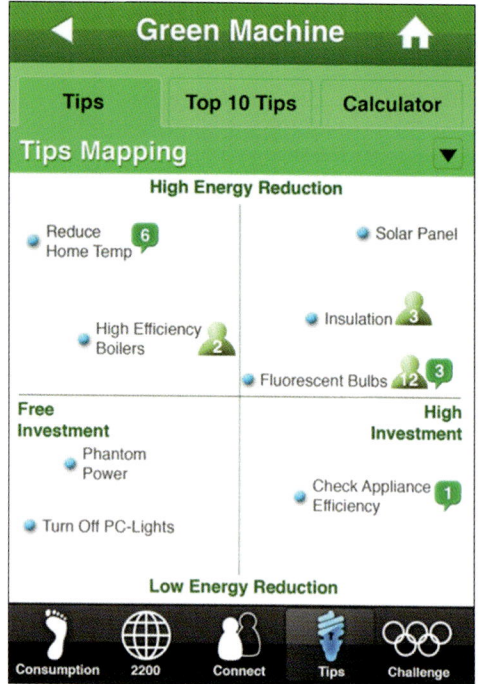

access the Green Machine's Options tab. Most of our testers could not associate the features we asked them to identify with this icon. Thus, we had to change its design and placement. The "reverse triangle" icon is now found on the top right of the screen, which renders it more visible and identifiable.

Furthermore, we also improved the overall visual consistency of the application. The five main icons found at the bottom of each screen were altered to present a more consistent look and feel.

User test analysis also enabled us to identify readily which Green Machine data visualizations were confusing or hard to understand. For example, some testers could not understand the purpose of the Energy Use equivalents found in the Consumption screens. At the same time, most were bothered by the unreadable texts found in the Tips Mapping screen. Thus, we removed the former from the main Consumption screen and moved to the Options tab while the latter was improved by placing a blue dot before the text and using a bigger font size.

After the second iteration of design, the Green Machine prototype seemed much improved. The overall visual design was cleaner, clearer, and more concise. We believed thereby that the usability, usefulness, and appeal of the new user interface were enhanced.

2.3 Green Machine 2.0 for the Enterprise

AM+A was fortunate to have an opportunity to take the Green Machine further through a project with SAP, Inc., a global enterprise-software development company. In 2010, Mr. Dan Rosenberg, SAP's Senior Vice-President for User Experience, requested a presentation about the Green Machine at its California offices. AM+A was able to explain its approach to the Chief Sustainability Officer of SAP. Following this presentation, SAP contacted with AM+A to assist SAP during the fourth quarter of 2010 to research and write guidelines for combining sustainability and persuasion, to design and evaluate four mobile-application prototypes that combined sustainability and persuasion, and to assist in the development of one enterprise software application that also combined sustainability and persuasion. In the business context, sustainability has widespread implications.

AM+A explored a number of business models, gathered and analyzed case studies, interviewed people in the US as well as a hundred people or more within SAP in several countries, and designed many detailed sketches of screens and interaction scenarios. These developments were a natural extension of the potential of the Green Machine to be applied to different user communities, use contexts, and platforms. AM+A was invited to name several of its Associates as co-inventors in three patent applications based on that work and was allowed to publish two papers explaining in some detail the planning and results of its guidelines development (Marcus et al. 2011, 2014).

2.4 Discussion and Conclusion

The work on the Green Machine is ongoing. The Green Machine seeks to incorporate persuasion and motivation for behavior change into a mobile phone application. This initial project demonstrated a possible effective use of the information from Smart Grid technology in combination with mobile technology infused with persuasive design techniques and information visualization.

Although previous research seemed to indicate that such an application would have some impact on energy consumption, it was interesting to gather actual user data and feelings. User tests demonstrated the relevancy of the motivational theory on which the Green Machine application is based. As user test results showed, every step of the behavioral change process was incorporated and realized in this application. The Green Machine application even as a concept design thus acts as a motivational, educational, and persuasive means to reduce energy consumption; we believe users with a working version of application would be encouraged to go "greener."

The application, even as a concept design, serves as a catalyst for behavioral change. Our long-term objective for the Green Machine is to create a functional working prototype, so that we can test whether it actually makes people reduce their energy consumption in the long run, under real use conditions. It will also be interesting to test it on an ecologically unbiased group to assess if the Green Machine can motivate and persuade them to change behavior. If our theories are proven to be correct, this could have significant implications on the use of Smart Grid software, which is undergoing significant expansion in the next few years in the USA and elsewhere.

We have made information about the Green Machine available since its inception to encourage others to develop further along these lines.

Further Reading

"5R" Initiative School Recycling Program (n.d.) Retrieved August 3, 2010, from saswma: http://www.saswma.org/5r.htm

(No Author) (2010) The green machine (in Korean). DesignNet 114–115 (South Korea), June 2010

Björn Stigson (2000) Sustainable development and business. Retrieved August 2, 2010, from OECD observer: http://www.oecdobserver.org/news/fullstory.php/aid/334/Sustainable_development_and_business.html

Blood AG (2008, November 5) We need sustainable capitalism. Retrieved August 2, 2010, from TheWallStreetJournal:http://online.wsj.com/article/NA_WSJ_PUB:SB122584367114799137.html

Boehret K (2010, July 28) The all-in-one remote. Retrieved July 30, 2010, from The Wall Street Journal: http://online.wsj.com/article/SB10001424052748703292704575393202060995806.html?mod=rss_Lifestyle

Brother Group Environmental Policy (n.d.) Retrieved August 3, 2010, from Brother at your side: http://www.brother.com/en/eco/idea/index.htm

Buildings & Plants (n.d.) Retrieved August 2, 2010, from Energy Star, U.S. EPA and DOE: http://www.energystar.gov/index.cfm?c=business.bus_index

CostBenefit (2010, July 30) Report finds virtual workstyles could save U.S. Business more than $400 billion a year – report quantifies benefits of workshifting for business, workers and nation. Retrieved July 30, 2010, from Environmental Valuation & Cost-Benefit News: http://www.envirovaluation.org/index.php/2010/07/29/report-finds-virtual-workstyles-could-save-u-s-business-more-than-400-billion-a-year-report-quantifies-benefits-of-workshifting-for-business-workers-and-nation

Darby S (2006) The effectiveness of feedback on energy consumption. Environmental Change Institute, Oxford University, Oxford

DOE/EIA (2008) Annual energy review 2007. www.eia.doe.gov

eCycling (2010, April 29) Retrieved August 2, 2010, from U.S. Environmental Protection Agency: http://www.epa.gov/epawaste/conserve/materials/ecycling/index.htm

ENERGY STAR Qualified Products (n.d.) Retrieved August 2, 2010, from Energy Star, US EPA and DOE: http://www.energystar.gov/index.cfm?fuseaction=find_a_product

EPA (2004, September) Retrieved August 2, 2010, from http://www.epa.gov/oppfead1/Publications/lawncare.pdf

EPA (2008, August 29) Air and radiation. Retrieved August 2, 2010, from U.S. Environmental Protection Agency: http://www.epa.gov/air/actions/

EPA (2009, October 16) Green communities. Retrieved August 2, 2010, from U.S. Environmental Protection Agency: http://www.epa.gov/greenkit/index.htm

EPA (2010, April 4) Air and radiation. Retrieved August 2, 2010, from U.S. Environmental Protection Agency: http://www.epa.gov/air/actions/drive_wise.html

EPA (2010, April 4) Information for citizens and communities. Retrieved August 2, 2010, from U.S. Environmental Protection Agency: http://www.epa.gov/oppt/p2home/pubs/citizens.htm

EPA (2010, April 9) Sustainability. Retrieved September 10, 2014, from U.S. Environmental Protection Agency: http://www.epa.gov/sustainability/basicinfo.htm#sustainability.

EPA (2010, April 9) Sustainability-basic information. Retrieved August 2, 2010, from U.S. Environmental Protection Agency: http://www.epa.gov/sustainability/basicinfo.htm

EPA (2010, August 2) Green vehicle guide. Retrieved August 2, 2010, from U.S. EPA: http://www.epa.gov/greenvehicles/Index.do;jsessionid=cf556ccfad65f73c8142a40c98a11d664300cf2376f5e0705b4e8eebec061627

EPA (2010, February 2) Adopt your watershed. Retrieved August 2, 2010, from U.S. Environmental Protection Agency: http://www.epa.gov/adopt/

EPA (2010, January 3) Polluted runoff. Retrieved August 2, 2010, from U.S. Environmental Protection Agency: http://www.epa.gov/nps/chap3.html

EPA (2010, June) Retrieved August 2, 2010, from http://www.epa.gov/region4/recycle/green-building-toolkit.pdf

EPA (n.d.) http://www.epa.gov/epahome/acting.htm. Retrieved August 2, 2010, from U.S. Environmental Protection Agency: http://www.epa.gov/epahome/acting.htm

EPA (n.d.) Protect the environment: at home and in the garden. Retrieved August 2, 2010, from U.S. Environmental Protection Agency: http://www.epa.gov/epahome/home.htm

EPA (n.d.) Protect the environment: at school. Retrieved August 2, 2010, from U.S. Environmental Protection: http://www.epa.gov/epahome/school.htm

EPA (n.d.) Protect the environment: at work. Retrieved August 2, 2010, from U.S. Environmental Protection Agency: http://www.epa.gov/epahome/workplac.htm

EPA (n.d.) Protect the environment: in your community. Retrieved August 2, 2010, from U.S. Environmental Protection Agency: http://www.epa.gov/epahome/community.htm

EPA (n.d.) Protect the environment: on the road. Retrieved August 2, 2010, from U.S. Environmental Protection Agency: http://www.epa.gov/epahome/trans.htm

EPA (n.d.) Protect the environment: think globally. Retrieved August 2, 2010, from U.S. Environmental Protection Agency: http://www.epa.gov/epahome/global.htm

EPA (n.d.) Protect the environment: while shopping. Retrieved August 2, 2010, from U.S. Environmental Protection: http://www.epa.gov/epahome/shopping.htm

Fogg BJ, Eckles D (2007) Mobile persuasion: 20 perspectives of the future of behavior change. Persuasive Technology Lab, Stanford University, Palo Alto

Grady E (2010, July 30) Inception's solar-powered set, and more. Retrieved July 30, 2010, from treehugger:http://www.treehugger.com/files/2010/07/chelsea-clinton-vegan-wedding-inception-solar-powered-set-and-more.php?campaign=th_rss&utm_source=feedburner&utm_medium=feed&utm_campaign=Feed%3A+treehuggersite+%28Treehugger%29

Gunn E (2010, July 29) Green salons that aim to match the real thing. Retrieved July 30, 2010, from The Wall Street Journal: http://online.wsj.com/article/SB10001424052748703294904575385433180709828.html?mod=rss_Lifestyle

Hall S (2010, July 29). Sustainability – bringing functions closer. Retrieved July 29, 2010, from procument: http://blog.procurementleaders.com/procurement-blog/2010/7/29/sustainability-bringing-functions-closer.html

Jean J, Marcus A (2009) The green machine: going green at home. User Experience 8(4):20–22ff., 4th Quarter 2009, available online at www.UsabilityProfessionals.org

Marcus A (2002) Information visualization for advanced vehicle displays. Inf Visual 1:95–102

Marcus A (2010) Green machine: Ein innovativer Ansatz, um Menschen zum Energiesparen zu bringen – mittels Mobile Device, Smart Grid, Information Design und Persuasion Design. Design Austria Mitteilungen (Austria), 4–7, April 2010.

Marcus A (2010) Combining information design/visualization with persuasion design, 01 June 2010, Draft Version 1.3 AM+A White Paper

Marcus A (2011a) The green machine (partially in Chinese). Submitted to user friendly/UXPA 2011, Suzhou, China, User Experience Professionals Association China, Shanghai, un-numbered pages, available from author

Marcus A (2011b) Health machine. Inf Des J 19(1):69–89

Marcus A (2012a) The money machine. Helping baby boomers retire. User Experience, vol 11:2, 2nd Quarter 2012, pp 24–27

Marcus A (2012b) The story machine: a mobile app to change family story-sharing behavior. CHI 2012, Austin, TX, USA, 5–10 May 2012

Marcus A (2013) The money machine. Inf Des J 20(3):228–246

Marcus A, Jean J (2009a) The green machine: going green at home. User Experience, vol 8:4, 4th Quarter 2009, pp 233–243

Marcus A, Jean J (2009b) Going green at home: the green machine. Inf Design J 17(3):233–243

Marcus A, Jean J. Going green at home: the green machine. Inf Des J 17.3:233–243

Maslow AH (2006) Motivazione e personalità. Ed. 11. Armando, Roma

Kahn, ME, Kotchen MJ (2010, July) Environmental concern and the business cycle: the chilling effect of recession. Retrieved August 3, 2010, from the National Bureau of Economics Research: http://papers.nber.org/papers/w16241

McCalley LT, Midden CJ (2002a) Energy conservation through product-integrated feedback: the roles of goal-setting and social orientation. J Econ Psychol 23:589–603

Myth: Paper is Better Than Plastic (2010) Retrieved July 28, 2010, from reuseit.com: http://www.reuseit.com/learn-more/myth-busting/why-paper-is-no-better-than-plastic

New Study: IT Sector Cut Annual CO_2 Emissions by 32 Million Metric Tons (2010, July 27). Retrieved August 3, 2010, from Climate Saver Computing: http://www.climatesaverscomputing.org/news/press-releases/july-27-2010

NIST (2009) NIST announces three phase plan for smart grid. National Institute for Standards and Technology. 2009-04-13. http://www.nist.gov/public_affairs/smartgrid_041309.html

Olivier Beaumais, Anne Briand, Katrin Millock, Céline Nauges (2010, May 1) What are households willing to pay for better tap water quality? A cross-country valuation study. Retrieved August 3, 2010, from http://halshs.archives-ouvertes.fr/docs/00/49/74/53/PDF/10051.pdf

SAP (n.d.) BUSINESS IN BRIEF-Market, trends, and mission. Retrieved August 2, 2010, from SAP Global: http://www.sap.com/about/investor/inbrief/markets/index.epx

SAP (n.d.) Business in brief-sustainability. Retrieved August 2, 2010, from SAP Global: http://www.sap.com/about/investor/inbrief/sustainability/index.epx

SAP (n.d.) Investor relations-business in brief. Retrieved August 2, 2010, from SAP Global: http://www.sap.com/about/investor/inbrief/index.epx

SAP (n.d.) SAP 2009 sustainability report. Retrieved August 2, 2010, from SAP Global: http://www.sapsustainabilityreport.com/overview

SAP (n.d.) SAP history. Retrieved August 2, 2010, from SAP Global: http://www.sap.com/about/company/history/index.epx

Schultz PW, Nolan J, Cialdini R, Goldstein N, Griskevicius V (2007) The constructive, destructive, and reconstructive power of social norms. Psychol Sci 18:429–434

Shedroff N (2009) Design is the problem. The future of design must be sustainable. Rosenfeld Media 2009

Smart Growth (2010, July 1) Retrieved August 2, 2010, from U.S. Environmental Protection Agency: http://www.epa.gov/smartgrowth/youth_travel.htm

Tscheligi M, Reitberger W (2007) Persuasion as an ingredient of societal interfaces. Interactions, September–October 2007, pp 41–43

Walmart (2010, July 30) The challenges and opportunities of a sustainability index. Retrieved July 30, 2010, from treehugger: http://www.treehugger.com/files/2010/07/the-challenges-and-opportunities-of-a-sustainability-index.php?campaign=th_rss&utm_source=feedburner&utm_medium=feed&utm_campaign=Feed%3A+treehuggersite+%28Treehugger%29

References

Cialdini RB (2001) The science of persuasion. Sci Am 284(2):76–81, Feb 2001. (www.influenceatwork.com)

Cialdini et al (2004) The constructive, destructive, and reconstructive power of social norms. Psychol Sci 18(5):429–434, May 2007

Darby S (2006) The effectiveness of feedback on energy consumption. Environmental Change Institute, Oxford University, UK

EPA (2010, April 9) Sustainability. Retrieved September 10, 2014, from U.S. Environmental Protection Agency: http://www.epa.gov/sustainability/basicinfo.htm#sustainability

Fernandez-Manzanal R, Rodriguez-Barreiro L, Carrasquer J (2007) Evaluation of environmental attitudes: analysis and results of a scale applied to university students. Wiley InterScience, pp 989–1009

Fogg BJ (2003) Persuasive technology: using computers to change what we think and do. Morgan Kaufmann Publishers, San Francisco, p 2003

Fourier J (1824) Remarques Générales Sur Les Températures Du Globe Terrestre Et Des EspacesPlanétaires. Annales de Chimie et de Physique 27:136–167

Gore A (2006) An inconvenient truth: the crisis of global warming. Rodale Press, New York

Mankoff J, Matthews D, Fussell S, Johnson M (2007) Leveraging social networks to motivate individuals to reduce their ecological footprints. Proc HICSS 2007

Marcus A (1979) New ways to view world problems. East–west perspectives. J East–west Center, Honolulu 1(1):15–22, Summer 1979

Marcus A (2000) User-interface design for air-travel booking: a case study of Sabre. Inf Des J 10(2):186–206. : John Benjamins Publishing Co., Amsterdam. Published by the International Institute for information Design

Marcus A (2009) Integrated information systems. Inf Des J 17(1):4–21, Figure 18, p. 19

Marcus A, Dumpert J, Wigham L (2011) User-experience for personal sustainability software: determining design philosophy and principles. In: Proceedings of design, user experience, and usability conference 2011, Orlando, FL, August 2011. Theory, methods, tools and practice. Lecture notes in computer science, vol 6769. Springer, New York, pp 172–177

Marcus A, Dumpert J, Wigham L (2014) User-experience for personal sustainability software: applying design philosophy and principles. In: Proceedings of design, user experience, and usability conference 2014, Iraklion, Crete, Greece, June 2014. User experience design for everyday life applications and services. Lecture notes in computer science, vol 8519. Springer, New York, pp 583–593

Maslow AH (1943) A theory of human motivation. Psychol Rev 50:370–396

McCalley LT, Midden CJ (2002b) Energy conservation through product-integrated feedback: the roles of goal-setting and social orientation. J Econ Psychol 23:589–603

NIST (2009) NIST announces three phase plan for smart grid. National Institute for Standards and Technology, 13 April 2009. http://www.nist.gov/public_affairs/smartgrid_041309.html.

Shedroff N (2009b) Design is the problem. The future of design must be sustainable. Rosenfeld Media, Brooklyn/New York

Tscheligi M, Reitberger W (2007b) Persuasion as an ingredient of societal interfaces. Interactions 2007:41–43

Yun T-J (2009) Investigating the impact of a minimalist in-home energy consumption display. Proc CHI 2009:4417–4422

Chapter 3
The Health Machine: Combining Information Design/Visualization with Persuasion Design to Change People's Nutrition and Exercise Behavior

3.1 Introduction

Obesity and resultant type 2 diabetes are major health concerns in the United States of America (USA) and throughout the world. A recent, significant rise in the incidence of the disease in the United States (as well as in other countries, such as China and India) has raised an alarm among public health officials and physicians. This rise is related to increased obesity among the US population. A recent article mentions that "Obese workers cost US private employers an estimated $45 billion or more annually in healthcare costs and lost labor" (Associated Press 2010). Outcomes of patients with diabetes are closely related to how well they manage their weight through diet and exercise.

For those people wishing to maintain or improve their health, currently existing health-oriented mobile phone applications provide many usable, useful functions, including medicine intake monitoring, pain management, weight loss programs, exercises tutoring, and health indicator tracking (e.g., blood pressure and heart rate). According to the American Diabetes Association, a combination of exercises, healthier food, and weight control will be more effective to alleviate or prevent diabetes. Unfortunately, seldom do the applications combine many of these above functions. They tend to be specialized. Moreover, after reviewing users' comments on current health mobile apps, the authors concluded that more adaptations and improvements could be made to better serve users' needs. Above all, these products do not provide an overall "persuasion path" to change users' short-term and long-term behavior, leading to a healthier lifestyle.

AM+A embarked on the conceptual design of a mobile phone-based application, the Health Machine, intended to address this situation. The Health Machine's objective is to combine information design and visualization with persuasion design to help users achieve their health objectives, especially regarding obesity and diabetes, by persuading them to adapt their lifestyle to include healthy eating and appropriate exercise. AM+A intended to apply user-centered design along with information

© Springer-Verlag London 2015
A. Marcus, *Mobile Persuasion Design*, Human–Computer
Interaction Series, DOI 10.1007/978-1-4471-4324-6_3

design and persuasion design to make the Health Machine highly usable and increase the likelihood of success in adopting new eating and exercise behaviors.

3.1.1 A Health Crisis

As reported by the Centers for Disease Control and Prevention, in 2007, 23.6 million people or 7.8 % of the population in United States have diabetes problems (Centers for Disease Control and Prevention 2008) (Fig. 3.1).

In addition, in 2007, at least 57 million American adults had prediabetes, i.e., impaired glucose tolerance (IGT), and/or impaired fasting glucose (IFG) (Centers for Disease Control and Prevention 2008). An even more alarming statistic is that about 1 in 7 low-income, preschool-aged children in the United States are now obese (Centers for Disease Control and Prevention 2010). This situation must be improved, because the total costs of diagnosed diabetes in the United States are so large that it already reached $174 billion in 2007 (Centers for Disease Control and Prevention 2008).

The situation can be changed. For those who have prediabetes, research has revealed that lifestyle intervention will reduce developing diabetes by 58 % during a 3-year period (of which, the reduction is 70 % among adults aged 60 years or older) (Centers for Disease Control and Prevention 2008). Studies have also shown that people with prediabetes who lose weight and increase their physical activity can prevent or delay diabetes and even return their blood glucose levels to normal (National Diabetes Information Clearinghouse 2008).

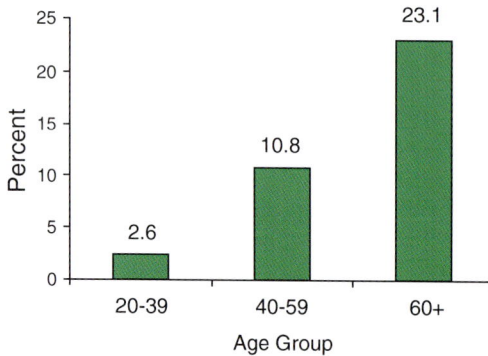

Fig. 3.1 Estimated prevalence of diagnosed and undiagnosed diabetes in people aged 20 years or older, by age group, United States, 2007 (Source: 2003–2006 National Health and Nutrition Examination Survey estimates of total prevalence (both diagnosed and undiagnosed) were projected to year 2007, from (Centers for Disease Control and Prevention 2008))

For those who have diabetes, especially type 2 diabetes (which accounts for about 90–95 % of all diagnosed cases of diabetes and is associated with older age, obesity, family history of diabetes, history of gestational diabetes, impaired glucose metabolism, physical inactivity, and race/ethnicity), self-management education or training is a key step in improving their health outcomes and quality of life (Centers for Disease Control and Prevention 2008).

As noted in Chap. 1, we applied a UCD process and a persuasion-design process to the development of the Health Machine.

3.1.2 The Health Machine's Concept

The initial idea of the Health Machine arose as the health scenarios in the United States became worse, and when AM+A noticed that most of the people with obesity and diabetes can prevent, delay, or even cure their problems through short-term and long-term behavior change. Therefore, AM+A dedicated itself to developing an innovative mobile application concept design, the Health Machine, to cope with the existing severe obesity and diabetes problems in the United States.

Moreover, the Health Machine is intended especially for those people who have low income and less education and thus may have less awareness, knowledge, and access to information about obesity and diabetes. AM+A intends, through the usable, useful, and appealing mobile device application, that the targeted users can be motivated to successful regimens of weight control and exercise and to learn about and maintain a healthier lifestyle in the long term.

The Health Machine's objectives include solving the following two critical issues:

• How can information design/visualization present persuasive information to promote sustainable, short-term, and long-term health behavior change?
• How can mobile technology assist in presenting persuasive information and promote behavior change of low-income and less-educated people?

3.2 Personas

As described in Chap. 1, personas are characterizations of primary user types intended to capture essential details of their demographics, contexts of use, behaviors, motivations, and their impact on the design solution. Based on internal analysis, for the Health Machine, we chose the following personas. They were selected in part to account for different cultures as well as different age groups and genders.

3.2.1   Persona 1: Alan Marx

Mr. Marx is 67 years old and is a small business owner. He was diagnosed with type 2 diabetes at the age of 66 after being borderline diabetic for several years. The condition is a mixture of heredity, lack of sufficient exercise, and eating portions that are too large. After surviving a mild heart attack at 57, which required triple bypass heart surgery, Mr. Marx changed his eating habits significantly and reduced fat consumption. However, his weight continued to climb until in October 2009, it reached 238 lb. for someone 6'1" in height. After being diagnosed with diabetes, assigned the use of a glucose meter, "required" to attend 5 classes on diabetes and nutrition, and urged to lose weight and increase exercise, he embarked on a sudden, massive change in health maintenance. He was motivated by fear of death and desire to achieve a workable goal. He succeeded in losing 45 lb (down to 188, three pounds from his target goal of 185) in 3 months by reducing his calorie intake to about 1500 cal, carefully noting his daily nutrients (see illustration), and increasing his daily exercise: to one hour on a treadmill at 4 mph, 100 sit-ups, and variable weight lifting (15–50 lb) three times per week. Since his initial achievement, he has become tired of constant data entry, and his weight has drifted upwards. He seeks to keep his weight below 200 lb. (His BMI range makes 185 the upward end of the range for his bone structure and height.) He seeks a mobile device that will be easier for him to monitor his nutrients, log his exercise and glucose readings, and visualize his data, which was not possible with his hand-entered data.

Alan Marx carried out a manual data-entry process by measuring the data (not all recorded at all times):

- Date.
- Glucose reading and time.
- Weight reading and time.
- Exercise, type, and duration.
- Meal type and time.
- Calories per food item.
- Fat grams per food item.
- Saturated fat grams per food item.
- Carbohydrate grams per food item.
- Fiber grams per food item.
- Protein grams per food item.
- Salt grams per food item.
- Nutrient subtotal per meal.
- Nutrient total per day.
- Comments about emotional reactions, meals, location, and other factors affecting health, behavior, and data gathering.

Such data entry is commendable and effective, but not sustainable for most people (Fig. 3.2).

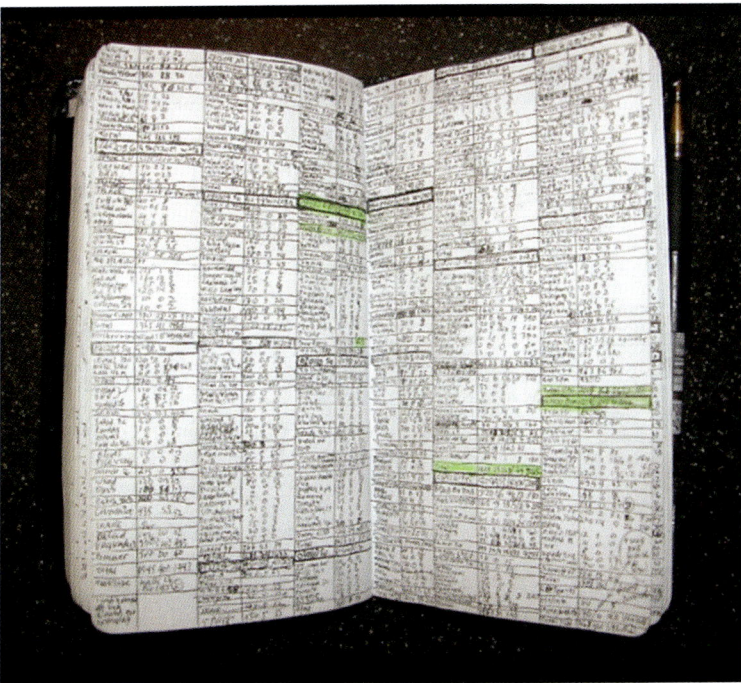

Fig. 3.2 Example of manual note-taking diary by Alan Marx

3.2.2   Persona 2: Anna White

Anna White is 57 years old and is an African-American grandmother with a grade school education who has worked as a domestic most of her life. She now weighs about 200 lb and is 5-ft, 2 in. tall. She heads a single-parent family, taking care of two children and three grandchildren. She has been diagnosed with type 2 diabetes but is not very well informed about her condition, its causes, and options available to her through a community medical care center.

3.2.3   Persona 3: Manuel Jiminez

Manuel Jiminez, 70 years old, is a legal immigrant from Mexico who started a gardening and lawn-care service now mostly managed and run by his son. He suffers from type 2 diabetes, weighs approximately 250 lb., and is 5-ft, 6 in. tall. He finds he doesn't have time to do the exercise he needs to add to his daily schedule because of continuing work pressures to contribute to the family income.

3.3   Use Scenarios

Based on the discussion of use scenarios in Chap. 1, for the Health Machine, we considered the following Health Machine use scenario for the typical users of the device. Users will want to accomplish one or more of the following tasks:

- Enter nutrient data with least effort, e.g., scanning food data labels, scanning restaurant offerings data labels, and going online to collect appropriate nutrient data.
- Review current and past data in table and chart modes, especially for new trends or for information that might change filters or goal targets for nutrition or exercise.

- Share data with other family, friends, physicians, groups, and "competitors."
- Read/react to messages or email or other communications about one's data, status, or trends.
- Compare one's own data with family, friends, and "stars," i.e., heroes/heroines.
- Read/see information about proper food preparation, how to balance low-fat and low-salt content with good taste.
- Read/see information about proper exercise, e.g., body positions for particular muscle groups.
- View future implications of current behavior.
- Read/see examples of equipment or food or restaurants that might be of interest.
- Upload or download photos relevant to one's progress as emotional or documentation markers.

The current use scenario for the Health Machine does not include accessing a Web site through a desktop PC in addition to mobile access to install or edit data. However, a complete scenario for the Health Machine would need to include desktop/tablet Web-based access.

Examples of similar total use scenarios are the Greiner et al. mobile parenting prototype developed and shown at the Spring 2010 semester reviews at the University of California at Berkeley's School (Abraham Coffman et al. 2010) and advanced health-related product designs shown by Samsung at the Usability Professionals Association annual conference in Munich, Germany, in May 2010 (Samsung Electronics 2010).

3.4   Research

3.4.1   Competitive Analysis

Before undertaking conceptual and perceptual (visual) designs of the Health Machine prototype, AM+A first studied approximately 20 most highly reviewed health iPhone applications. Through screen comparison analysis and customer review analysis, AM+A derived major "pros and cons" of these applications, which contributed to the refinement of initial ideas for the Health Machine's detailed functions, data, information architecture (metaphors, mental model, and navigation), and look and feel (appearance and interaction).

The following edited descriptions are based on texts that appear in the review of iPhone applications at the URLs cited in 3.TBD: Bibliography of URLs. AM+A added the "pros and cons." Please remember that the analysis was done several years ago and that the descriptions, pros and cons, and screen images may have changed considerably. One should check carefully for current conditions before undertaking further decisions about these products.

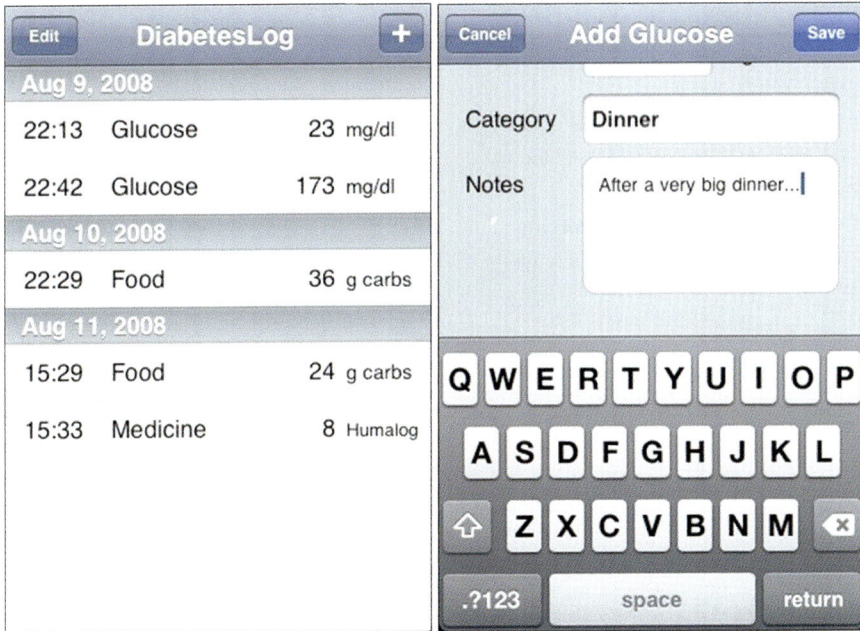

Fig. 3.3   Screen images of DiabetesLog. In 2010 (Images: Chris Ross – permission holder)

3.4.2   Diabetes-Related Applications

3.4.2.1   Diabetes Log (Fig. 3.3)

Description

The application enables diabetic users to track their glucose readings, food intake, and medicine records. The application's functions include export of records (CSV over email); user-selectable span of records to view; and recording for glucose data, medicine, and food intake. Functions are still in the planning stage at the time of review: editing of medication types and/or record classifications.

Pros

- Seems capable of replacing paper-based methods and keeps clear glucose records.

Cons

- Limited functions: only keeps logs but no charts, reports, etc.
- Needs to provide clearer classification of insulin.
- Did not provide database backup.

3.4.2.2 Track3

Description

The Tracks3 application is a personal health assistant for diabetics. It makes living with diabetes easier by tracking all of the factors that keep blood sugar balanced. Diabetics can now worry less about carbs and medication by letting the Tracks3 application assist with managing glucose levels. Tracks3 logs nutrition and exercise information as well as oral medications and insulin. It will even record different insulin types plus unique foods and exercise routines. This application does not require a connection to the Internet, so users can stay connected to their health anytime, anywhere. Tracks3 keeps at their fingertips all of the information diabetics need to stay healthy.

Pros

* Includes thorough nutritional info, e.g., carbs, calories, sodium, etc.
* Allows quick visual capture/review of last food, meds, and glucose readings at a glance.
* Good database and easy to navigate.
* Very easy setup of user's own custom foods.
* Enables users to email logs by date range in tab-delimited format.
* Customization features: ranges for high, low, goal, and time period descriptions.

Cons

* Should add ability to make notes and to group different foods into custom/ favorites meal.
* Somewhat slow to get response to email.
* Needs to allow for more visual detail.
* Little ability to copy charts to camera or to send via email.

3.4.2.3 Diabetes Pilot (Fig. 3.4)

Description

Diabetes Personal Calculator is an all-in-one insulin calculator and carbohydrate database developed for type 1 diabetics injecting premeal fast or rapid-acting insulin. Diabetes Personal Calculator has been designed for convenience and helps to remove errors and guesswork when calculating one's premeal insulin dose.

Pros

* Provides longer and more detailed review of diet for monitoring.
* Has appealing and easy-to-use charts.

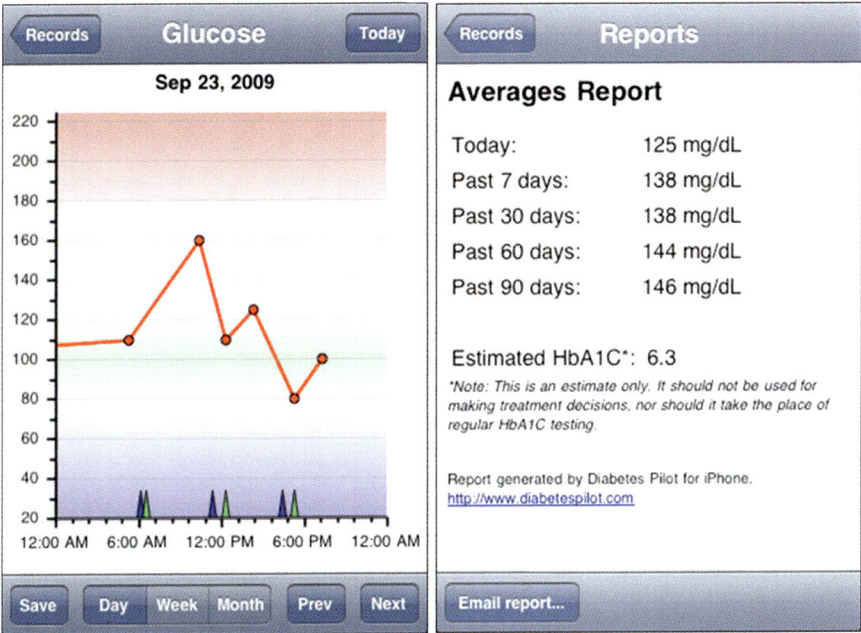

Fig. 3.4 Screen images from Diabetes Pilot in 2010 (Images: Cohesio, Inc.)

- Helpful for type 1 or type 2 diabetes patients on carb counting regimen for insulin calculation.
- Helps to generate detailed report to dietician.

Cons

- Cannot sync directly with dietician.
- Did not provide data backup function. When updating the software, all the previous data were deleted.

3.4.2.4   Glucose Buddy: Diabetes Helper 3.2 (Fig. 3.5)

Description

The product was ranked number one diabetes iPhone application in 2010. Glucose Buddy is a data storage utility for people with diabetes. Users can manually enter glucose numbers, carbohydrate consumption, insulin dosages, and activities. Then, one can view all of one's data on one's free glucosebuddy.com online account.

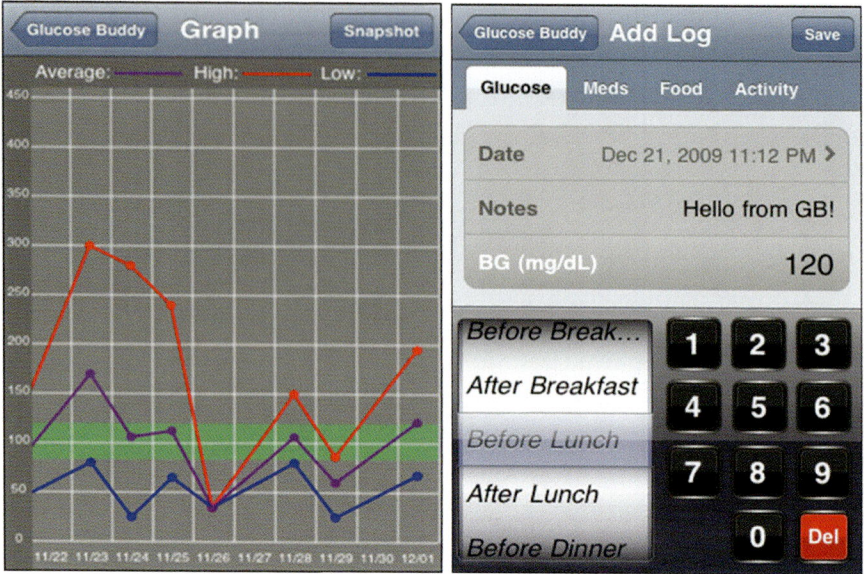

Fig. 3.5 Screen images from Glucose Buddy Diabetes Planner 3.2 in 2010 (Images: Azumio, Inc.)

Pros

- It has usable and appropriately designed log layout, including glucose measurement, time relative to meal, and space for notes.
- Developer keeps in contact with users and makes timely updates.

Cons

- It needs to add function to sort information by time of day or time related to a meal.
- Better charts need to be added.

3.4.3   Nutrition Applications

3.4.3.1   Nutrition Menu (Fig. 3.6)

Description

This product is almost an all-in-one application when it comes to calorie counting, water consumption, and keeping track of one's personal stats. One of the more appealing features is the fact that it has over 350 North American restaurants and their dishes built in, so that one can check the calories when out and about.

In addition, one can log one's calories from other foods; save favorites; keep track of one's days, weeks, and months; and display that information in different

Fig. 3.6 Screen images from Nutrition Menu in 2010 (Images: Shroomies, LLC.)

charts and layouts. There are exercises to show how many calories one burns and even a lock for the application to keep one's information private.

Pros

- Great for Weight Watchers point system and easy to track points (good incentives).
- Detailed, large database with 350+ restaurants.

Cons

- Only useful if one always eats at restaurants.
- No nutrition information of common food can be searched.

3.4.3.2 Lose it!

Description

If one is looking for an application to keep track of what one eats with the ability to add all one's favorite foods, this is a good selection, and it is free. With Lose It!, one can choose to keep in touch with friends and diet together, log one's exercise, and give oneself a daily budget of calories that one is allowed to consume.

One can even set motivators (reminders) to stay on track for good things to eat or make Lose It! remind one when one needs to log in food using the notification feature on the iPhone.

Pros

- Simple application and easy to learn.
- Large food database.
- Competition with friends serves as a good persuasion incentive for diet.
- Well-designed UI, attractive colors, and appealing custom icons.

Cons

- Only about 40 restaurant chains are included in the database.
- Only tracks calories, no other nutrients information mentioned.
- Only tracks weights in whole pounds (could be more precise).
- No water consumption and other important measurements.
- No progress or nutrient charts.
- Does not sync with Web site.

3.4.3.3   Eat This not That!

Description

If one has looked through the Eat This, Not That! line of books, this application will be surprising. One might never guess how some of the foods at a favorite restaurant ranked in the high-calorie range. What this application does is to show what one should eat instead of about a food on which one was probably thinking of splurging.

Pros

- Includes calorie information for over 23,000 menu and grocery items, and 230 exercises and activities.
- Plans customized daily calorie budget.
- Helps users to learn restaurant survival strategies and uncover best and worst foods in America.
- Visualizes user's weight loss in simple visual form.
- Builds better grocery list from thousands of graded items.
- Seems game-like, appealing, more so than similar applications.

Cons

- None yet noted among user comments.

3.4.4 Exercise Applications

3.4.4.1 Ifitness (Fig. 3.7)

Description

This is one of the more comprehensive fitness apps available and provides a wide range of information to complete the exercises and do them correctly. The application has pictures, videos, and detailed instructions for over 300 exercises, and one can customize lists of exercises to work on only the parts of the body on which one would like to focus. iFitness also lets users track their progress and store it in a profile, and it even supports multiple users, so the whole family (or friends) can use it.

Pros

- Customizes user's lists of exercises to focus on training only certain parts of the body.
- Provides large library of pictures and detailed instruction.

Cons

- Does not design an entire workout.
- Lacks the ability to keep diary of previous workouts.

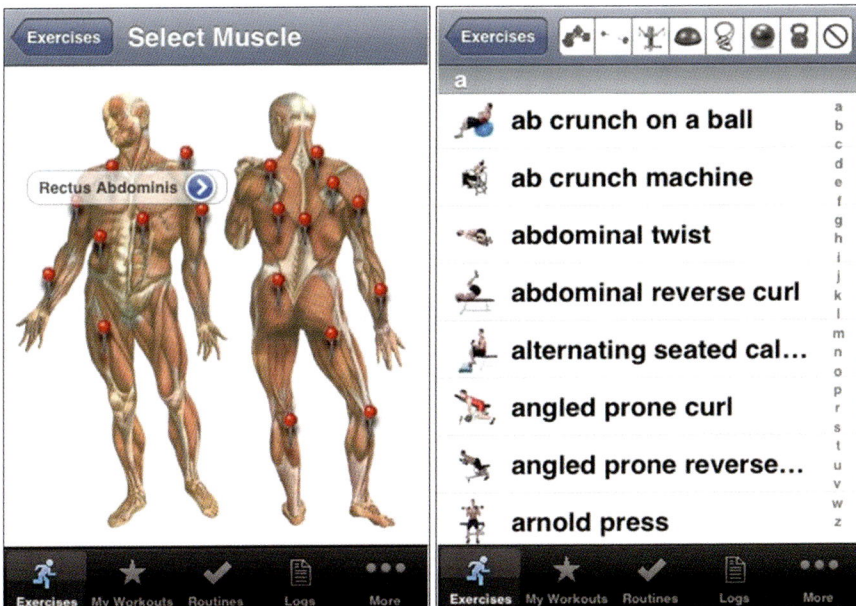

Fig. 3.7 Screen images from Ifitness in 2010 (Images: Eltima Software)

3.4.4.2   LiveStrong.com

Description

This application is another calorie counter, but with the LiveStrong.com name to back it up. Unlike many other applications with food databases that range to around 10,000 items, this application has 625,000 items in its database from which users can search and enter into their logging system. Users can set weight goals, whether they want to lose, maintain, or gain weight, and they can also set a daily intake based on their goals, current weight, and measurements. Users are also able to log fitness exercises, which calculate how many calories they burn based on their current measurements and weight.

Pros

• Large database with 625,000 items for search.
• Can set weight goals and record both calorie intake and consumption.

Cons

• Did not sync well from LiveStrong Web site to phone.
• Has a fixed database and is not easy to customize, e.g., adding new items.

3.4.4.3   iMuscle (Fig. 3.8)

Description

This interesting application is more on learning about the body and what muscles get worked rather than focusing on actual exercises. By having the entire human body in 3D, users are able to learn what muscles affect which parts of the body, which muscles are worked for different exercises, which muscles are sore, and many more possibilities.

Pros

• Can zoom in and pick muscle and provide smart exercise instructions.
• Helpful for learning exercises and showing which muscles are targeted.

Cons

• None noted in user comments.

3.4.4.4   Ease into 5K (Fig. 3.9)

Description

For users who are "couch potatoes" who watch marathons and think "I can do that… no problem….," this sets them straight, by making them follow a strictly regimented exercise routine to get in shape and be able to run for long periods of time. This

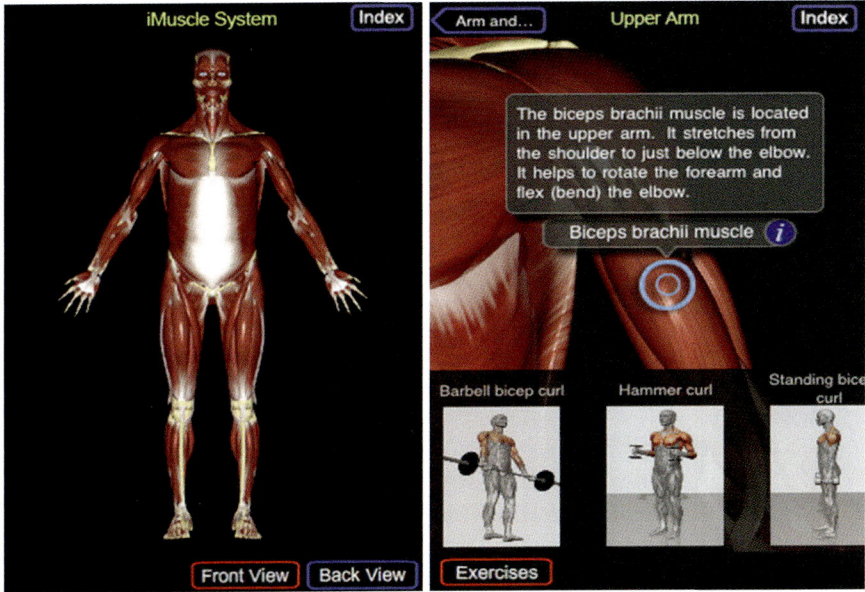

Fig. 3.8   Screen images from iMuscle in 2010 (Images: 3D4Medical, LLC.)

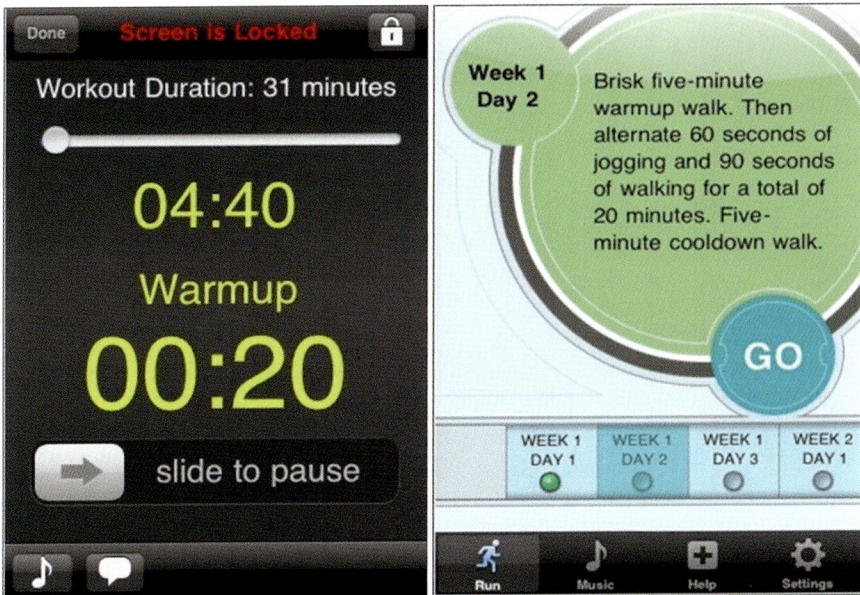

Fig. 3.9   Screen images from Ease into 5 K in 2010 (Images: Bluefin Software)

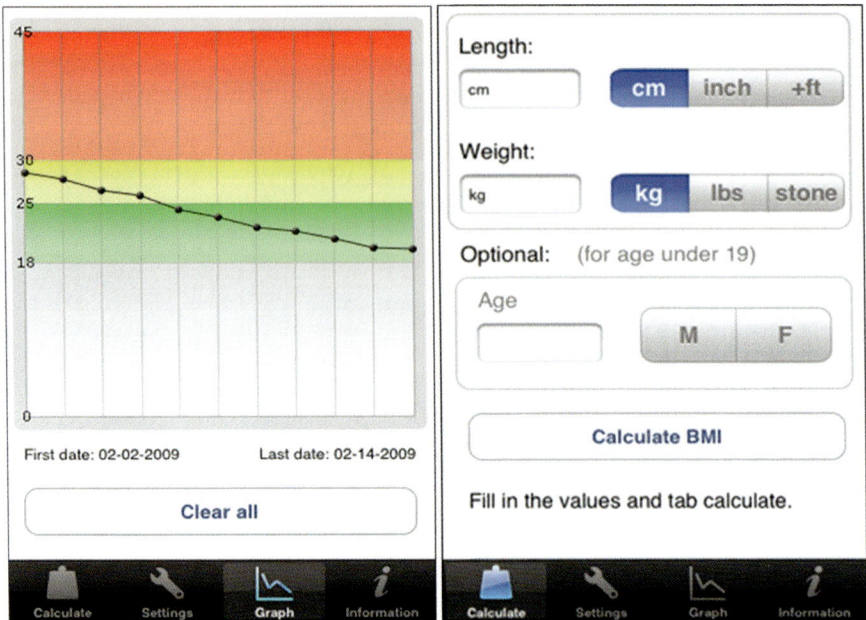

Fig. 3.10 Screen images from BMI Calculator (Images: Data Supply, Inc.)

application is designed to take a person from beginning runner status all the way up to a full-on 5 K runner in just 9 weeks while working out 3 days per week. There is also another version of this application, but for a 10 K running plan, which is designed for people after they finish the 5 K version. The calories that users burn are based on their current measurements and weight.

Pros

- Giving instructions step-by-step.
- Good to have play list of music when jogging.

Cons

- No record of data about improvements and progress.
- Cannot categorize and select/play music for separate exercise phases, e.g., warm-up, intense, moderate, cooldown, etc.

3.4.5   Other Health Applications

3.4.5.1   BMI Calculator (Fig. 3.10)

Description

This is a very simple application that measures the body mass index (BMI), which is computed from height, weight, and age and shows what range users are in:

underweight, just right, or obese. Users can see if they are getting close to obese or if they are eating right and have a correct weight. It is advisable to check every few months, because users can change their BMI and might not even know it.

Pros

- Simple application, easy to learn and use.
- Helpful for measuring BMI to know whether the user is obese, overweight, or eating correctly.

Cons

- Only provides information; not sufficient for behavior change.
- Needs to provide more incentives for lifestyle change.

3.4.5.2 Water Your Body (Fig. 3.11)

Description

The application seems simple in concept; it can keep users on track for daily required water consumption, which users routinely neglect. This application makes it very easy to choose the amount users have consumed by picking the correct icon as well as entering the amounts manually. Users can keep track of consumption over time and can be graded on their amounts taken in.

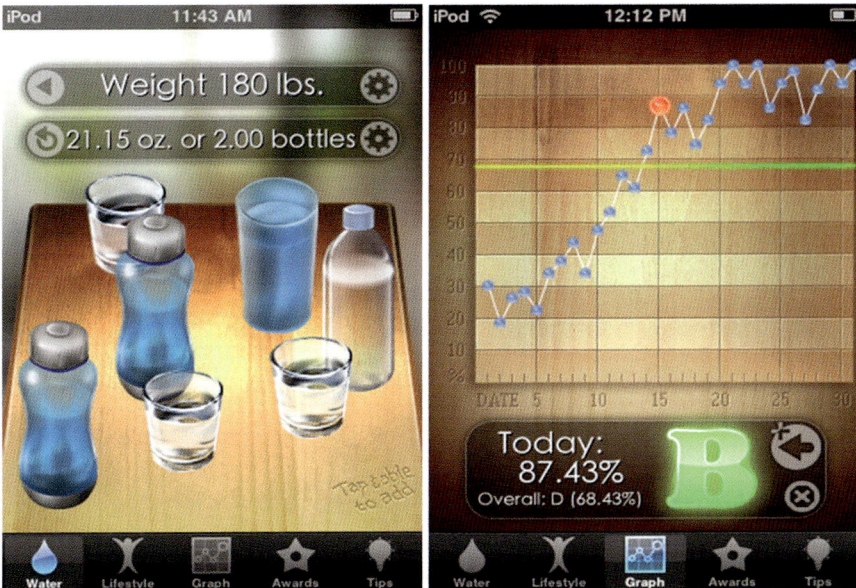

Fig. 3.11 Screen images from Water Your Body (Images: Foware.com)

Pros

- Very easy to choose the amount the user consumed by picking the correct icon as well as entering amounts manually.
- Keeps track of water consumption over time, and users are graded on the amount taken in.

Cons

- Contains some coding bugs in score-recording system.

3.4.5.3   Sleep Machine (Fig. 3.12)

Description

While many health applications are fitness oriented, few treat sleep. This application has 61 mastered stereo sounds, 54 of which are ambient sound tracks and the other 7 are music. These music tracks help users to fall asleep by providing a relaxing sound at the perfect volume to help them sleep through the night. Users can choose to let the application play throughout the night, or schedule it to shut off after a few hours when users know they are in a deep sleep.

Fig. 3.12   Screen images from Sleep Machine (Images: SleepSoft, LLC.)

Pros

- Has better "stereo" sounds.
- Easy sound-mixing user interface.

Cons

- No multiday alarm function.

3.5 Results, Analysis, and Persuasion Design

From our investigations of about 20 mobile health applications including those cited above, AM+A concluded that usable, useful, and appealing user interface (UI) design must include incentives to lead to behavior change. Good health-oriented mobile phone applications should provide group comparisons, charts, illustrations, goals, competition, and/or step-by-step instructions to motivate people to change their behavior. The proposed Health Machine needs to combine persuasion theory, provide better incentives, and motivate users to achieve short-term and long-term behavior changes, moving the user towards a healthier lifestyle.

Moreover, large, up-to-date databases are required. Users always demand large and searchable databases. Especially important is that the databases are customizable. A customizable, flexible database by which users and their network of family and friends can easily add more information is critical as a factor in increasing usage and an inevitable competitive advantage for the Health Machine. In addition, in comparison to traditional manual data-input methods, a multiple data-entry system including label scanning and database searching is required to facilitate users' data-input process.

In addition, the Health Machine must encourage and strengthen team-oriented behavior change. Based on the studies of persuasion theory, we discovered that team-oriented social comparison is a superior incentive for behavior change. Cooperation and competition within and among teams can encourage people to exercise more and carry out better diet control. Furthermore, virtual rewards (e.g., "star" designation, new skins for blogs, etc.) and real financial rewards (one corporate study (WSJ, c. June 2010) showed $500 can make a significant difference) can both be provided as motivation incentives for teams.

Last but not least, the Health Machine should be fun to use. Well-designed games will serve as an additional appealing incentive to teach, to train, and to inform users about how to select meal combinations wisely, how to exercise efficiently and effectively, and other techniques of nutrition and exercise. Also, the Health Machine should allow users to share their experience with friends, family members, and the world, primarily through Facebook, Twitter, and blogs.

Based on these concepts and available research documents, we have proposed and are developing conceptual designs of the multiple functions of the AM+A Health Machine. Subsequent evaluation will provide feedback by which we can improve the metaphors, mental model, navigation, interaction, and appearance of all functions and data in the Health Machine's user interface. The resultant improved user experience will move the Health Machine closer to a commercially viable product/service.

We believe a well-designed Health Machine will be more usable, useful, and appealing to health-conscious users, especially those having problems with obesity and/or diabetes. Our objective is to provide a mobile suite of applications that can reliably persuade people to move towards a healthier lifestyle, with consequent benefits to their own health and economy but also to those around them.

As noted in Chap. 1, we incorporated persuasion theory to accomplish the following objectives:

- Increase frequency of using application.
- Motivate changing some living habits: work out, food choices, weight control.
- Teach how to change living habits.
- Persuade users to change living habits (short-term change).
- Persuade users to change lifestyle (long-term change).

Each step has requirements for the Health Machine:

Motivation is a need, want, interest, or desire that propels someone in a certain direction. From the sociobiological perspective, people in general tend to maximize reproductive success and ensure the future of descendants. We apply this theory in the Health Machine by making people understand that every action has consequences on their health condition change and their future.

We adapted work (Maslow 1943) to the Health Machine context:

- The safety and security need is met by the possibility to visualize the amount of food expense saved.
- The belonging and love need is expressed through friends, family, and social sharings and support.
- The esteem need can be satisfied by social comparisons that display weight control and exercise improvements as well as by self-challenges that display goal accomplishment processes.
- The self-actualization need is fulfilled by being able to visualize the improvement progress of the health-matter indexes and mood and also by predicting the change of the users' future health scenarios.

3.6 Impact on Information Architecture

3.6.1 Increase Use Frequency

Games and rewards are the most common methods to increase use frequency. In Health Machine, we have developed the pet training game and the meal combo game. In terms of the rewards, the users will be awarded both virtual rewards (such as "star" nominations and new skins for blogs) and real money rewards.

In addition, we chose the social comparison as another incentive to increase use frequency. Users will form groups with families and friends and participate in competitions among different groups on diet control and exercise.

3.6.2   Increase Motivation

In the Health Machine, we set the users' future health condition predication as an important incentive for their behavior change. Through viewing their current health conditions and predicted future scenarios in the next 20–30 years, the users will have a stronger impression and awareness of their obesity and/or diabetes problems.

Because setting goals helps people to learn better and improves the relevance of feedback, the Health Machine asks users how much expense budget they want to save on healthier food, how much calories/blood glucose level they want to reduce, and how much workouts they want to accomplish. In accordance with the goal settings, the users will get suggested action plans to achieve each goal.

In addition, we created the monthly "10 Challenges." The incentive of self-accomplishment will encourage users to achieve healthy life challenges step-by-step that will generate positive impacts in the long run.

Social interaction also has an important impact on behavior change. Another remarkable component of health machine is to leverage social networkings and integrate features like those found in forums, Facebook, Twitter, or blogs. Users can send notes or messages to their social groups and share ideas with other people. The social ties will serve as an additional incentive to motivate behavior change.

3.6.3   Improve Learning

For many patients with obesity and/or diabetes, understanding their self-health management is a challenge. To improve learning, the Health Machine integrates contextual tips to explain how to eat healthier and increase exercises, the complications associated with diabetes, and ways to cope with too high/low glucose levels. Users can also update latest research articles and news about diabetes and obesity.

Moreover, we also sought to make the education process both informative and entertaining. Games were proposed to teach users how to choose the right proportion, amount, and type of food for each meal. Through playing games featuring educational information, users will learn how to eat healthier without being bored.

3.7   Information Architecture

Based on all previous steps, we constructed an information architecture for the Health Machine, as shown in Fig. 3.13.

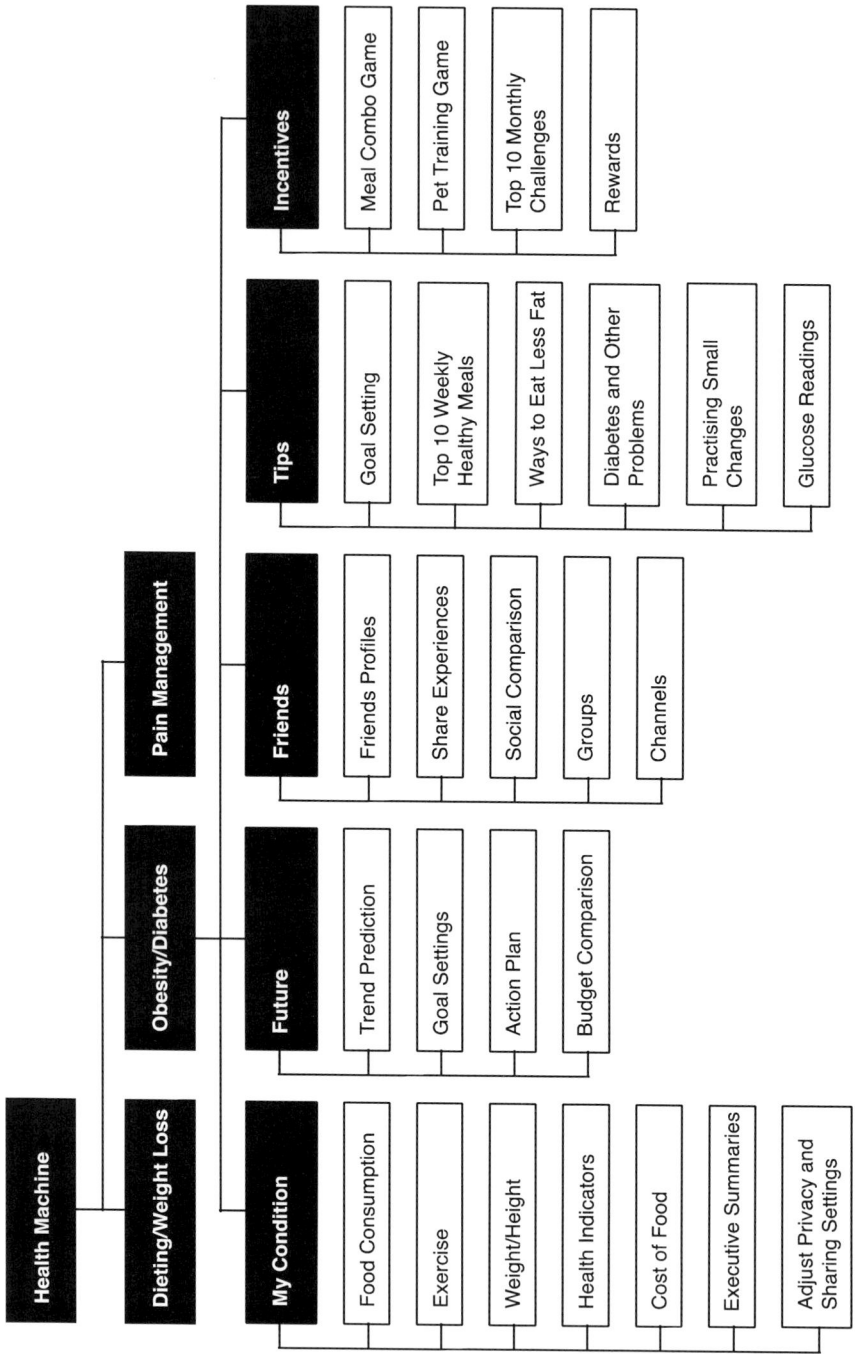

Fig. 3.13 Information Architecture of the Health Machine

The components of the information architecture are described below. Data elements are shown for many of the functions.

3.7.1   My Condition

The Health Machine will provide timely information about food in order to enable the user to make healthy choices.

The application will also assist in the recording of food nutrition consumption, such as calories, glucose, carbohydrates, etc., and making a record that can be of use to the patient as well as physicians, nurses, dieticians, and other healthcare service providers.

Exercise is another crucial factor in the equation for controlling weight. To help patients with obesity and diabetes achieve and maintain a healthy body condition, the Health Machine supports the development of healthy exercise habits, training and teaching the users with the appropriate way and amount of exercises, and enabling the integration of new activities into the user's lifestyle in a way that increases the probability of both short-term and long-term success.

3.7.2   Food Consumption

This function provides the following:

- Records nutrition components of food.
- Provides data entry of information as shown in Table 3.1.
- Sums up the nutrition intake for each meal and calculates/displays meal nutrition metrics as shown in Table 3.2.

Table 3.1 Data-entry information

Metric	Example
Time	8:00 am, 29 June 2010
Food	Milk
Amount	8 oz.

Table 3.2 Meal nutrition metrics

Metric	Example
Calories	300
Fat grams	5
Saturated fat grams	0
Carbohydrate grams	30
Fiber grams	10
Protein grams	40
Salt milligrams	250
Sugar grams	30
Customizable additional item	N/A

- Adds new food items to database and shares with other users.
- Provides ability to save selected food combinations for easy, efficient reuse.

3.7.3   Exercise

The Exercise function provides the following:

- Records exercise type and duration, with input information format shown in Table 3.3.
- Sums up the energy consumption for each exercise activity.
- Outputs energy consumption with the format shown in Table 3.4).
- Adds new exercise items to database and shares with other users (Tables 3.3 and 3.4).

3.7.4   Weight/Height

The Weight/Height function provides the following:

Records weight and height data with input information formats shown in Table 3.5
Outputs weight and height records and trend in chart format

Table 3.3 Input information formats

Metric	Example
Time	9 AM 29 June 2010
Exercise type	Jogging
Duration	1 h
Effort level	Medium
Mood	Happy
Media: music, video	Rolling Stones, "Satisfaction"
Notes	Feels easier than last week!

Table 3.4 Output information formats

Metric	Example
Calories	−1000
Deduction on glucose	−0.3
Change on blood pressure	High, −5; Low, −2

Table 3.5 Weight and height data-input information formats

Metric	Example
Date	29 June 2010
Time	10:00 AM
Weight	200 lb.
Height	185 cm

3.7.5   Health Indicators: Health-Matter Indexes Monitoring

Because daily health-matter indexes monitoring is part of diabetes management for a majority of patients, the Health Machine provides the Health Indicators function as a means of recording and reviewing glucose readings. This function enables the user to understand correlations between changes in diet and exercise and blood glucose levels. Users with diabetes also need to watch their heart rate and blood pressure. Users also should be enabled to keep track of these measures with Health Machine. Furthermore, patients will be able to use the measures for setting goals and visualizing their process that may be of significance for their self-analysis purposes. Apart from monitoring the measurable indexes, the Health Machine also allows the users to track their moods and associates them with their improvements of lifestyles, which, in general, serves as a good incentive for their behavior change. This function will also compare the record with benchmarks and personal goals by table or chart display (e.g., bar chart or line chart).

Some of these data formats are shown in Tables 3.6, 3.7, and 3.8.

3.7.6   Cost of Food

The Cost of Food function provides for the following:

• Records food expenses with data-input information formats shown in Table 3.6.
• Sums up daily food expenses per components of meals.

Table 3.6 Data-input information format

Metric	Example
Time	9 AM
Blood pressure (systolic/diastolic)	120/85
Glucose	115
Heart rate (bpm)	75

Table 3.7 Record, goal, and benchmark data formats

Indicator	Record	My goal	Benchmark/range
Calories	100	80–200	50–200
Blood pressure	Hi: 120	High: 110–130	High: 110–130
	Lo: 85	Low: 80–100	Low: 75–90
Glucose	115	90–120	50–90
Heart rate	75	70–80	60–80

Table 3.8 Food expense data-input information formats

Metric	Example
Date	29 June 2010
Time	11:30 AM
Meal type	Lunch
Expense	*$20.50*

3.7.7   *Health Thermometer*

In the user interface, the small health thermometer always displayed at the top of the screen shows the user's current health condition summary. This summary display would be computed from a mixture of health factors that would be determined in consultation with a physician. This thermometer would be consulted as often as the user might monitor the mobile phone battery level or signal strength.

3.7.8   *Executive Summaries*

The executive summary adds up daily total calories, glucose levels, and blood pressure. Diabetes literature emphasizes testing glucose several times per day and at different times. Many people test glucose once per day, e.g., in the morning before breakfast. Because exercise will reduce calories, and change glucose level and blood pressure, advanced Health Machine functions might be able to calculate the calorie/glucose/BP levels in order to spare users having to do multiple tests every day. Output of daily records data would include the formats shown in Table 3.9.

3.7.9   *Future*

3.7.9.1   Trend Prediction

Trend Prediction would consist of several subfunctions, including Goals: My future and Current health condition

Table 3.9 Data output formats

Metric	Example
Date	29 June 2010
Time	7:00 AM or 07:00
Calories	60 g
Fat grams	10 g
Saturated fat grams	4 g
Carbohydrate grams	30 g
Fiber grams	3 g
Protein grams	10 g
Salt grams	100 mg
Glucose level	100
Blood pressure	90/70
Heart rate	60 bpm
Weight	200 lb or 44 kg
Weight reduced compare to last week	+2 lb
Total food expense	$20.00

3.7.10 Goals

My future: This subfunction is special in the Health Machine. Users would first view their current health scenario in a chart format. The chart illustrates users' current health estimates based on current behavior: risk of heart attack, stroke, and diabetes in the next 20–30 years. Viewing these charts, users would have a more visual impression of their health condition and the severity of their obesity and/or diabetes challenges.

By changing the goal setting factors (weight, blood pressure, glucose, cholesterol level, etc.), users would be able to view different health scenarios of their future conditions and thereby decide the appropriate health indexes they would like to pursue and maintain.

These series of health-matter indexes or indicators would then be set as the users' long-term goals. In accordance with the goals, suggested action plans would be provided automatically by the Health Machine. Users could customize their action plans and then set them as their short-term plans.

In this way, users could expect to reach their long-term health goals though achieving the detailed action plans step-by-step. Their future health scenario would also serve as a critical incentive for them to keep going to achieve their longer-term goals.

Current health condition: The current health condition subfunction would provide the following:

- It generates users' health condition reports.
- It displays users' risk of having heart attack, stroke, hemiplegic paralysis, stupefaction, etc.
- Current data could be displayed by text, table, chart, visual image, map, or diagram, as appropriate.

An example of such health conditions is shown in an existing product in Fig. 3.14: My Health Advisor from the American Diabetes Association.

3.7.11 Goal Settings

The Goal Settings function would provide the following:

- Sets monthly health factors goals, such as weight, blood pressure, glucose, and cholesterol levels.
- Predicts future heath condition change under different goal settings.

The Health Machine would offer a means for setting goals that relate to diabetes management. For example, a user may be told by a physician that she needs to decrease her weight by 20 lbs. With the application, the user can visualize her progress towards this objective.

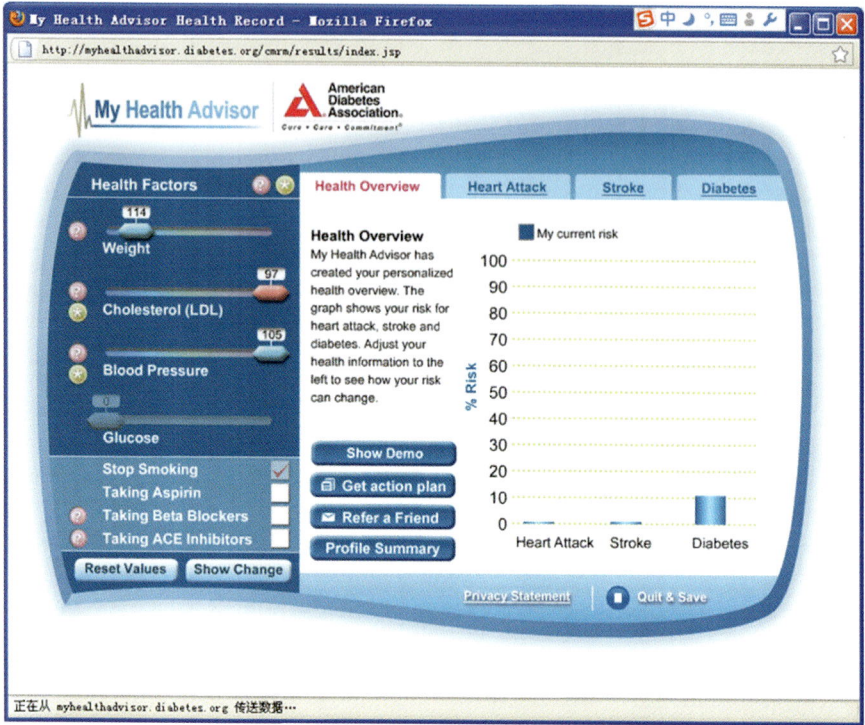

Fig. 3.14 My Health Advisor, as available in 2010 (Image courtesy of American Diabetes Association)

	Components per meal	Exercise per day
Table 3.10 Action plan format (components and specific tasks)	Calories range	Type
	Fat range	Duration
	…	Time

3.7.12   Action Plan

The Action Plan function would provide the following:

Generate recommended daily and weekly action plans based on monthly goals

Allow users to customize action plans with the action plan data formats for specific components and tasks as shown in Table 3.10

3.7.13   Budget Comparison

The Budget Comparison function would provide for the following:

• Enables weekly food expense budget setting.

- Compares actual food expense with budget.
- Demonstrates cost savings for users with diet management and control.

3.7.14   Friends

Adoption of a new, healthier lifestyle is best accomplished with the help of a support group. To encourage users to change behavior, the Health Machine would allow users to tap into social networks. Cooperation and competition within and among group members will serve as a strong social factor to motivate users' behavior change. Moreover, interactions can also consist of information sharing among selected individuals through Facebook, Twitter, blogs, and other social media.

3.7.15   Friends Profile

Establishes friends profile database and allows the participation of family members, friends, and celebrities.

3.7.16   Share Experiences

Users are likely to obtain more social support from their supporting groups (e.g., people having obesity or type 2 diabetes problems can form social support groups to share experiences and advice with each other).

3.7.17   Social Comparison

It enables individual comparisons on diet control, weight control, and exercise amount with friends, stars, people from other cities and countries, etc.

It enables group competition. Friends and family members can form different groups and hold weight loss or exercise competitions. Users are expected to receive encouragements from group members and more incentives to beat the other teams.

3.7.18   Groups/Channel

Users can watch health, exercise, cooking, and food channels through their mobile phones to learn other related information.

3.7.19   Tips

The Tips functions would provide for the following:

- Tips on personal goal settings.
- Recommendations on healthy meals everyday.
- Updated tips through the Internet.
- Knowledge of complicating diseases resulting from obesity and diabetes.
- Tips about how to eat less fat.
- Tips about how to increase exercise with little change in daily life, e.g., parking further to encourage more walking, using stairs instead of elevators, etc.

3.7.20   Goal Setting Tips

This function provides tips on setting goals for exercise and nutrition that affect the goal setting function in the future component.

3.7.21   Top 10 Weekly Healthful Meals

This function provides the following:

- Syncs with mobile phone Web site (or major diabetes/obesity research Web sites).
- Updates top 10 healthful meals weekly.

3.7.21.1   Ways to Eat Less Fat

Updates research articles on healthy and delicious diets and healthier cooking and eating styles

3.7.21.2   Diabetes and Other Problems

Provides articles about diabetes and its complications

3.7.21.3   Practicing Small Change

Provides tips about ways to increase exercise while incurring minor change in daily life, e.g., parking further to encourage more walking exercise, using stairs instead of elevators, etc.

3.7.21.4 Glucose Readings

Provides articles about symptoms of low/high glucose levels and ways to deal with such conditions

3.7.21.5 Incentives

Apart from the tips, the Health Machine would also motivate and educate users through entertainment, games, rewards, awards, and other incentives. For example, the Health Machine can offer games that can teach users to decide the right proportion and combination of food for their meals. In addition, a pet training game could be designed for some users to have a better understanding of self-health management; how they treat the virtual "pet" reflects how they treat themselves.

The 10 monthly challenges is another component that allows users to challenge themselves to achieve a series of behavior-changing goals. By accomplishing each of the challenges (e.g., stop smoking, eat an apple every day, do 10 push-ups every day, etc.), users are expected to pursue short-term behavior adaptations and later form and maintain healthier lifestyles.

Virtual and real rewards would be provided to users, associated with winning games, competitions, and challenges. Users could share their reward information with their social groups and gain more supports from family, friends, and others.

3.7.22 Meal Combo Game

Many people in the United States have lost the ability to judge correct portion sizes and to combine the right basic food groups into a meal. The meal combo game would teach users to select wisely healthful food and food combinations for breakfast, lunch, dinner, and snacks. At the beginning of the game, players can choose meal types from snack, breakfast, lunch, and dinner. Then, a plate and a food list will be shown to players. Players need to decide the portion and combination of the nutrition ingredients for the meal and then drag food items from the list to the plate to get scores. When players choose more healthful foods, they will be awarded higher marks. They may also earn extra scores from healthful food combinations. While winning a higher score, players will also learn how to eat "smart and healthy".

3.7.23 Pet Training

Some people give more care to their pets than to themselves. The virtual "pet" would track the exercise and food intake of users and demonstrate the results in terms of the effect on their pets. Users would learn to behave more healthfully as they notice the change of their pet's mood and health condition. The pet could also send tips and notes to users, to persuade them to eat healthier foods and to exercise more.

3.7.24 Top 10 Monthly Challenges

For those who really like to challenge themselves and achieve progress step-by-step, the top 10 monthly challenges would be a good option for them to realize self-accomplishment. 10 challenges would be updated from Web sites monthly and would be assigned to users. Most of them would be simple, with little change in lifestyles, such as "eat an apple every day" or "do 10 push-ups every day." After accomplishing each challenge, users could be awarded titles and/or receive virtual rewards, such as new skins for their blogs, new clothes for their pets, etc. These challenges would serve as incentives to users to make small differences in their lives that may lead to significant positive results.

3.7.25 Rewards

Both virtual and real rewards could be awarded to winners of the games and challenges. Virtual rewards may include the following:

- New skins/decorations for the users' blogs, Facebook, twitter, and mobile phone applications.
- Virtual currency that can purchase new exercise/diabetes control/healthy cooking e-book, and videos.
- Nominations of winners, such as weight loss star today, most healthy person today, muscle man, etc.

    Real rewards could include the following:

- Money rewards from the users' companies, etc. (A recent *New York Times* article reports that corporate reward campaigns typically must pay about $500 for top awards to increase better eating.) (Fig. 3.15).

3.8 Evaluation

3.8.1 Planning

AM+A planned to interview approximately six potential users (two per persona) and approximately two healthcare providers who train people who have been diagnosed recently with type 2 diabetes in order to learn what works, what doesn't,

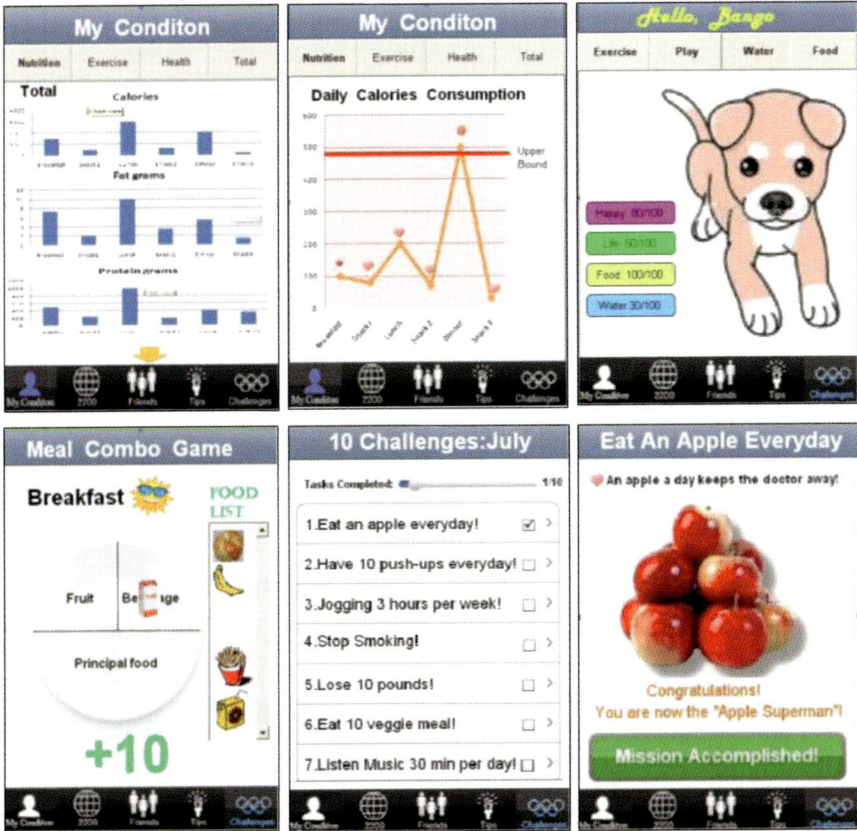

Fig. 3.15 Initial designs of some key screens of the Health Machine

and how to improve the Health Machine prototype information architecture (metaphors, mental model, and navigation) and look and feel (appearance and interaction). AM+A developed the following initial survey/questionnaire for the trainers:

- What are the age, gender, education, economic, and ethnic/racial backgrounds of your students?
- Do certain ages, gender, or ethnic groups have different kinds of health challenges? If so, how would you characterize these challenges per group?
- Do the three groups have any special difficulties or challenges to change their behavior?

  For example, not understanding diabetes or nutrition concepts, mathematical challenges of computation, difficulties to find healthier foods, difficulties to find time for exercise in stressful lives earning a basic living, family responsibilities, illiteracy, etc.

- What motivates each group best? Why?
- What might lead to short-term/long-term behavior change?

- What are the best ways to teach people?
  For example, texts, illustrations, cartoons, 3D models, instructional videos, music videos, short practice exercises, comments by famous people, comments by heroes/heroines, comments by friends, comments by family members, reminders, etc.
- What do you think would increase the frequency of using a Health Machine application for each of the groups? Why?
- Do any of the existing mobile phone applications (see screen captures) seem to have successful approaches?
- Do some have harmful or to-be-avoided qualities? Why?
- What do you think of the Health Machine sketches?
- Are important functions, data, features missing? Are some unusually likely to be effective and appealing? Are some unusually likely to be harmful, offensive, or distracting? Why?
- Any final comments about what might make the Health Machine more successful in persuading people to change their eating, nutrition, exercise, and diabetes-related behavior?

3.8.2   Practical Results of Survey

Because of legal regulations related to official healthcare patients and training providers and the requirements for obtaining necessary HIPPA documents (a process that would have delayed the project beyond tolerable limits), it was not possible to interview patients. Instead, AM + A was able to interview a skilled trainer of patients. On 21 July 2010, AM + A interviewed a registered clinical dietician (RCD) from a noted healthcare organization serving a wide demographic community in Northern California. She provided training to obese and diabetic patients. Her informal and anonymous comments regarding the survey are provided in a mixed list/text format below.

3.8.3   Demographics of Students

- Age: The majority of her students are 45+, but there are also many in their 30s.
- Gender: More women than men go to the classes.
- Education: Varies.
- Economic level: She often does not know precisely, but it seems more blue-collar than white-collar people participate in her classes.
- Ethnic/racial background: Many Asian and Hispanics, in addition to Caucasian participants.

3.8.4 Challenges per Group

A big challenge for most people is time management. They may have long commutes. They are not physically active and need to build that into their life patterns.

Another big challenge for many groups is slow eating, giving the stomach and brain enough time to work together to send and receive the "I am full" messages. Patients may not know how many times to chew food (30 times). Having a pool of challenges, which the Health Machine seems to propose, seems like a good idea, especially grouped into easy, medium, and hard challenges, or challenges for the first day, the first week, the first four weeks, because most people would not know what to select, would not know how to set up such challenges, or would not take the time to do so. She thought it best to start with small steps, like learning the division of a proper plate of nutrition: one quarter protein, one quarter starch, and one-half vegetables, with beverage and fruit "off the plate, but nearby."

Working men often seemed defensive about not getting enough exercise. They are active, but do not get enough cardiovascular exercise. For men, their job comes first, and they claim to or seem to have little or no time. In addition, they do not like to ask for help. Women care for their families and don't take care of themselves.

Many Asian and Hispanics have a typical challenge because of the role rice and corn play in their diets. Asians should be mentioned especially. Weight is not always an issue for them.

3.8.5 Special Difficulties and Challenges

Time management is a general challenge.

Regarding challenges of computation, many may have them; that is why the plate method is effective; they just have to remember how to "see" a healthy, nutritional plate. Many do have difficulty retaining information. Many people fall asleep in the classes (from workload and from the effects of diabetes).

Many participants do have difficulties understanding healthier foods. Some will think that veggie chips are better than potato chips, but the fat content is the same. Michael Polen, a nutrition teacher at UC Berkeley, has written a famous book, *The Omnivore's Dilemma* (Pollan 2006). The book is very visual and very simple in its advice, e.g., "don't eat anything with more than five ingredients" or "don't eat anything whose ingredients you don't understand or can't pronounce." These are simple maxims that do not require the patient to understand what the details are. These would make good tips, she implied.

People have difficulty calculating and relating to budgets. They don't understand that wholesome food ingredients can be cheaper.

3.8.6   What Would Increase Frequency of Use?

If it works and is easy, it will be used. If there is something catchy that shows up every time one comes to it and signs in or signs on, like simple wisdom or food facts, that would be engaging and attractive: something like "Did you know?" might help.

3.8.7   What Motivates Best?

- Something that is "personal, concrete, different" would be good.
- Men want to be active. Women want to be attractive. Weight loss might appeal to women more than men.
- For some people, feeling out of control is uncomfortable; small changes mean people feel more in control. Anything that can lead quickly to small changes would be good.
- For some, being able to live long enough to play with their children and grandchildren is a motivation.
- For some, feeling better, having more energy, being less depressed is a motivating factor.

3.8.8   What Leads to Short- and Long-Term Behavior Change?

- Helping people find out what is realistic leads to people making successful changes.
- Don't overwhelm them with choices that lead to failure.
- Suggestions or regimens must be simple, must be specific, and must be attainable.
- There must be a connected sequence of goals. For example, eating an apple a day might lead to an apple and a vegetable per day, instead of disconnected things.

3.8.9   What Techniques Teach People Best?

- 3D models, comments by heroes/heroines, comments by family members, and reminders.
- It would be good to have photos of friends.
- Don't use too many words on screens. For example, the slide show screens she saw had too much text.
- Different people learn by different means best.
- Text generally does not work alone. Illustrations are good.
- Cartoons may be only 50 % successful with people. Humor is a sensitive, complex matter.

- Short practice exercises are good.
- People like to see what famous people do.

3.8.9.1 Do Any Current Health Machine Screens Seem Good or Bad?

She felt nothing jumped out in either direction.

3.8.9.2 Any Comments on Health Machine Details?

She had no time to view them further or to discuss them.

3.8.9.3 Final Comments?

- You must get people to use it. Using it is itself a change. People like to feel in control.
- Many people do not see why they should change.
- Think about the personal positive perspectives of benefits: e.g., seeing grandchildren, marrying, seeing people being able to play golf or to be able to wipe themselves properly…they get very personal with these strong motivations.
- Wanting to change must be a tangible goal. Otherwise, this will be viewed as just another diet.

3.8.10 Impact of Evaluation

Based on the feedback from the trainer interviewed, AM+A revised the designs of the screens and updated their visual appearance.

3.9 Revised Designs

Based on the feedback, even if limited, AM+A made minor changes in the information architecture, removing one item under My Condition. With regard to screen designs, it made the controls more consistent with the platform guidelines (iPhone) and improved the simplicity of the text and imagery, in response to the interviewed nutritionist's comments. The revised screens appear in Figs. 3.16 and 3.17.

Note the My Condition screen's bar charts that look like a "health warrior's chest medals," the meal combo screen's nutrition plate, which was designed several months before the president's wife, Ms. Michelle Obama, introduced a new nutrition diagram in the form of a plate of food, the Puppy Pal: Fido screen that shows an

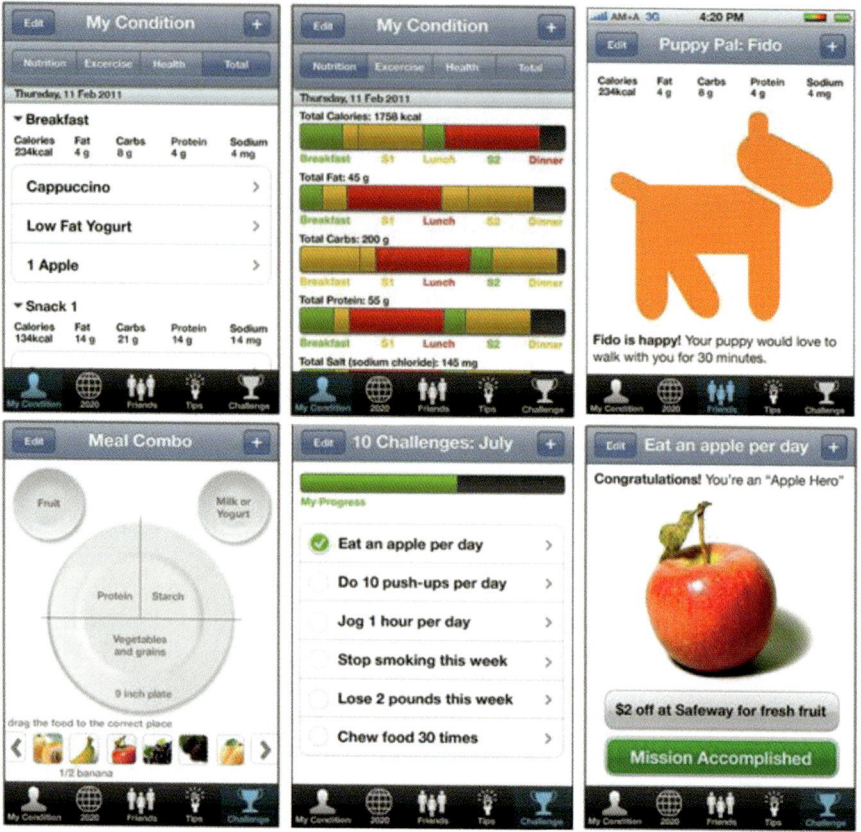

Fig. 3.16 Revised Health Machine screen designs

Fig. 3.17 Additional revised Health Machine designs

engaging pet to capture the user's emotions on behalf of better exercise and nutrition, and the Compare To screen that shows the user's breakfast in comparison to a music or film star who might license and provide a morning data feed of her breakfast statistics to her fans as an incentive to strive harder for better nutrition.

3.10 Conclusions

Based on the user-centered design process described above, AM+A plans to continue improving the complete Health Machine development process, as opportunities arise, which would require significant time and funding from an outside source.

Tasks include the following:

- Revise target personas and use scenarios for a particular market in the consumer space or in the enterprise space, or for a particular culture, e.g., North America, Europe, China, India, Africa, and elsewhere.
- Conduct user evaluations for the particular market.
- Revise information architecture plus look and feel.
- Build initial working prototypes for the iPhone, Android, or other platforms, e.g., using FlairBuilder for the iPhone or other tools.

AM+A aimed to incorporate persuasion and motivation theory for behavior change into a mobile phone application. The self-funded work on the Health Machine project is current and ongoing and was undertaken to demonstrate the direction and process for such products/services. Even though the design is incomplete, AM+A has been willing to share lessons learned and the approach in the interest of assisting the alleviation of a worldwide health challenge. At this stage, AM+A is seeking to persuade other design, education, and medical groups to consider similar development objectives and hopes they benefit from the materials provided thus far.

The author hopes this process, using self-funded concept prototypes, which inevitably limits the amount of research, design, and evaluation, will inspire others. The process has already been demonstrated successfully with a previous project, the Green Machine (Marcus and Jean 2009), versions of which were developed further by SAP, as mentioned in Chap. 2.

AM+A's long-term objective for the Health Machine is to create a functional working prototype so that it can test whether and how well the application actually persuades people with obesity and/or diabetes challenges to conduct diet control, increase exercise, and pursue a healthier lifestyle in the long run under real use conditions over a long-term period. If the theories are proven to be correct, this approach could have significant implications for obesity and diabetes control that will benefit millions of people in the United States and in other countries.

In recent years, personal healthcare products have emerged, such as Fitbit (www. fitbit.com) and Apple's iPhone 6 and 6 Plus (www.apple.com). The appearance of numerous "smartwatches" and other devices, including wearables, such as smart socks with sensors to aid monitoring Alzheimer patients getting out of bed (Sheer and Sneed 2014), means that more and more products like the Health Machine will appear. This is a "healthy" development for UX design of healthcare products.

Further Reading

Cialdini RB (2001) The science of persuasion. Sci Am 284(2):76–81, Feb 2001. (www.influ-enceatwork.com)

Fogg BJ (2003) Persuasive technology: using computers to change what we think and do. Morgan Kaufmann Publishers, San Francisco, p 2003

Fogg BJ, Eckles D (2007) Mobile persuasion: 20 perspectives on the future of behavior change. Persuasive Technology Lab, Stanford University, Palo Alto

Marcus A (2011a) Health machine. Inf Des J 19(1):69–89

Marcus A (2011b) The health machine: mobile UX design that combines information design with persuasion design. In: Proceedings of design, user experience, and usability conference, sub-conference of human-computer interface international, Orlando, FL, August 2012. Springer Publishers, London. Design, User Experience, and Usability. Theory, Methods, Tools and Practice, Lecture Notes in Computer Science, vol 6770, 2011, pp 598–607

Marcus A (2011c) The health machine (in Chinese). In: Proceedings of user-friendly 2011/ Usability Professionals Association 2014, Beijing, China, November 2011, unnumbered pages

References

Associated Press (2010) Employees earn cash for exercising more. Wall Street Journal, p D3, 2 June 2010

Centers for Disease Control and Prevention (2008) National diabetes fact sheet: general informa-tion and national estimates on diabetes in the United States, 2007. U.S. Department of Health and Human Services, Centers for Disease Control and Prevention, Atlanta

Centers for Disease Control and Prevention (2010) Obesity prevalence among low-income, preschool-aged children 1998–2008, 16 March 2010. Retrieved July 05, 2010, from www.cdc. gov: http://www.cdc.gov/obesity/childhood/lowincome.html

Coffman A et al (2010) Observations of daily living. UC Berkeley School of Information, Berkeley

Marcus A, Jean J (2009) Going green at home: the green machine. Inf Des J 17(3):233–243

Maslow A (1943) A theory of human motivation. Psychol Rev 50:370–396

National Diabetes Information Clearinghouse (2008, June) National diabetes statistics, 2007. Retrieved July 05, 2010, from http://diabetes.niddk.nih.gov: http://diabetes.niddk.nih.gov/dm/ pubs/statistics/#prevention

Pollan M (2006) The Omnivore's dilemma. The Penguin Press, New York

Samsung Electronics (2010) Persuasive healthy life. SAIT, Computer Science Lab, Future Experience Part, Toronto

Sheer R, Sneed A (2014) Teen wins big for his sock invention. Scientific American, 14 September 2014. http://www.scientificamerican.com/article/teen-wins-big-for-his-sock-invention/?WT. mc_id=SA_1014Advances. Checked 18 Sept 2014

Applications URLs[1]

http://iphone.appstorm.net/roundups/lifestyle-roundups/20-health-fitness-apps-for-iphone/
http://itunes.apple.com/us/app/abs-diet-smoothie-selector/id336855403?mt=8
http://itunes.apple.com/us/app/bmi-calculator/id292796789?mt=8
http://itunes.apple.com/us/app/bp-buddy-80-off-sale-blood/id293626282?mt=8
http://itunes.apple.com/us/app/c25k-couch-to-5k/id301233668?mt=8
http://itunes.apple.com/us/app/calorie-tracker-achieve-your/id295305241?mt=8
http://itunes.apple.com/us/app/crunchfu/id309637176?mt=8
http://itunes.apple.com/us/app/fitnessbuilder/id306287984?mt=8
http://itunes.apple.com/us/app/hundred-pushups/id301174591?mt=8
http://itunes.apple.com/us/app/id337998484?mt=8
http://itunes.apple.com/us/app/ifitness/id290451423?mt=8
http://itunes.apple.com/us/app/imuscle/id315994842?mt=8
http://itunes.apple.com/us/app/ipump-pilates/id292835472?mt=8
http://itunes.apple.com/us/app/lose-it/id297368629?mt=8
http://itunes.apple.com/us/app/mens-health-workouts/id319740615?mt=8
http://itunes.apple.com/us/app/nutrition-menu-calorie-exercise/id294692235?mt=8
http://itunes.apple.com/us/app/runkeeper-pro/id300235330?mt=8
http://itunes.apple.com/us/app/sleep-machine/id323061162?mt=8

---

[1] Retrieval date: 10 June 2010 for all URLs.

Chapter 4
The Money Machine: Combining Information Design/Visualization with Persuasion Design to Change Baby Boomers' Wealth Management Behavior

4.1 Introduction

Finding a life-enhancing and wealth-preserving way of life is a twenty-first century global challenge, especially in the USA, where baby boomers reaching 65 are considering how to manage, preserve, grow, and use their lifetime liquid and real assets. Baby boomers now constitute 40 % of the US population, controlling of 67 % of the nation's wealth. As they approach retirement, they are facing important decisions about their present and future.

Wealth management information, products, and services are available to increase financial awareness and propel changes by monitoring investments. Unfortunately, many of these tools do not focus on innovative data visualization. Furthermore, they are often targeted to the PC/Web Mobile products are beginning to appear, but many baby boomers are not as familiar as their children and grandchildren with mobile-based as well as Web-based financial management.

Lastly, these tools lack effective persuasive techniques to convert baby boomers to safe and reliable action plans for preserving wealth in the face of economic and health-related uncertainty. While communicating critical data helps build awareness, it does not automatically effect behavioral changes. The question then becomes: How can we better motivate, educate, and persuade people to manage their finances, manage asset consumption, and preserve their legacy for future generations?

The Money Machine project of 2011 researched, analyzed, designed, and evaluated powerful ways to improve behavior toward personal wealth and finances. The project intends to persuade and motivate people to reduce their asset consumption and increase the longevity of available funds by means of a well-designed mobile phone application concept prototype (with associated Web portal): the "Money Machine."

© Springer-Verlag London 2015
A. Marcus, *Mobile Persuasion Design*, Human–Computer
Interaction Series, DOI 10.1007/978-1-4471-4324-6_4

The author's firm has designed and tested analogous application prototypes in the recent past: The Green Machine aimed to change energy conservation behavior in 2009, and the Health Machine application aimed to prevent obesity and diabetes through better behavior regarding nutrition and diabetes in 2010. The Money Machine uses similar principles of combining information design/visualization and persuasion design. AM+A's presentation and this chapter explain the development of the Money Machine's user interface, information design, information visualization, and persuasion design.

Between 1 January and 31 December 2011 in the USA, an average of 7000 people each day will turn 65 years old, resulting in a recent and significant rise in their financial activities (AARP 2010). This rise has caught the notice of advisors and wealth management professionals, and baby boomers' financial, retirement, and wealth management concerns are ever more visible in such publications as the *Wall Street Journal*, the *New York Times*, and *Bloomberg Businessweek*, as well as in the monthly magazine of the socially and politically powerful Association for the Advancement of Retired Persons (AARP).

For those people wishing to maintain or improve their financial condition, current wealth-management mobile applications provide some usable, useful functions, including forecasting, electronic funds transfer (ETF) information, investment functions, portfolio review, financial "pain management," financial risk management, exercises (including "what-ifs") and tutoring, and wealth-indicators tracking (e.g., stocks, bonds, commodities, and other markets).

A combination of exercises, accessible information, and risk management will be more effective to alleviate or prevent financial disaster. Unfortunately, these applications seldom combine many of these above functions; they tend to be specialized. Moreover, after reviewing current financial mobile apps, the authors found opportunity for further adaptations and improvements to better serve users' needs. Above all, these products do not provide an overall "persuasion path" to change users' short-term and long-term behavior. Such a path is essential in order to lead users to an improved wealth-management lifestyle.

During 2010, Aaron Marcus and Associates, Inc. (AM+A) carried out the conceptual design of a mobile-phone–based product, the Money Machine, intended to address this situation. The Money Machine's objectives are the following:

- Combine information design and visualization with persuasion design.
- Help users achieve their wealth-management objectives, especially regarding longevity, stock market volatility, and estate legacy.
- Persuade users to adapt their lifestyles to include healthier consumption and appropriate risk management.

AM+ applied user-centered design along with persuasive techniques to make the Money Machine highly usable and increase the likelihood of success in adopting new wealth-management behavior.

4.2   Initial Discussion

4.2.1   *Workforce*

Baby boomers control more than 80 % of personal financial assets and more than 50 % of discretionary spending power.

In general, baby boomers are known as "worker bees," who live to work rather than work to live. They usually define their social status and personal value by their professional achievements. Thus, their retirement plans are often put on hold. Baby boomers are established career professionals, said to be at the top of world's largest, most powerful companies. According to a recent survey, 40 % of 801 baby boomers surveyed plan to work "until they drop" (AARP 2010).

Labor statistics indicate that nearly 80 million baby boomers will exit the workplace in the next decade. These employees are retiring at the rate of 8,000 per day or more than 300 per hour.

Baby boomers are responsible for more than half of all consumer spending, including 77 % of all prescription drugs, 61 % of over-the-counter (OTC) medication, and 80 % of all leisure travel. Seventy-six percent of baby boomers intend to continue working and earning past age 65, either for personal fulfillment or because they can't afford to retire (60 %).

4.2.2   *Recession*

Meanwhile, the sad truth is that the vast majority of baby boomers have not adequately saved for retirement. For many of them, their home equity was damaged by recent financial crises in the early and later years of the first decade of the twenty-first century. For others, their 401 k savings were reduced when the stock market tanked. The Federal government has already begun to pay out more in Social Security benefits than it is taking in, and, to some critics, the years ahead look downright apocalyptic for the Social Security program: 35 % of Americans already over 65 rely almost entirely on Social Security payments alone (Social Security Administration 2011).

More than any other age group, baby boomers feel their long-term prospects were damaged by the recent years of recession. More baby boomers, for instance, say they've lost money on investments and endured damage to their household finances than any other group (Newman 2010).

4.2.3   *Poor Preparation*

Many baby boomers thought rising home values would anchor their retirement plans, one reason the savings rate plummeted over the last decade. The housing bust, which has driven home values down by more than 30 % nationwide (Newman

2010) significantly harmed that plan. What's more, far fewer baby boomers have guaranteed pension plans than in prior generations, likely leaving millions of Americans on the cusp of retirement in a huge financial predicament.

4.2.4   Resistance to Change

Baby boomers are well aware of the problems facing the USA, and especially Washington. They are nonetheless resistant to change, perhaps because they have so much invested in the current system already. For instance, though a Federal sales tax could increase revenue and help pay down debt, 54 % of baby boomers oppose this plan. This proportion is higher than among those both younger and older.

4.2.5   The Money Machine's Concept

The initial idea of the Money Machine arose as the US economy worsened starting in 2008 following the recession of 2002–2003. AM+A noticed that many people's financial and wealth management problems could be prevented, delayed, or even cured through short-term and long-term behavior change. Therefore, AM+A dedicated itself to developing an innovative mobile application, the Money Machine, to cope with the continuing recession and financial complications in the USA.

The Money Machine is intended especially for those people with sufficient income and education to have amassed general financial knowledge and awareness, as well as access to information about wealth management, risk, economic factors, etc. AM+A intends, through a useful, usable, and appealing mobile device application, to motivate targeted users to adopt successful regimens of financial control. These regimens include focused saving, limiting consumption, and learning about and maintaining healthier, and more financially sound lifestyle for the long term.

The primary objective of the AM+A Money Machine is to combine information design and visualization with persuasion design and to help users achieve their wealth-management objectives, especially regarding longevity, stock market volatility, and assets legacy issues. The Money Machine aims to answer the following two critical questions:

How can information visualization and design promote sustainable change in wealth-management behavior both short- and long-term?
How can mobile technology assist in presenting persuasive information and promote behavior change of medium-high income and educated people?

4.3   User-Centered Design

As noted in Chapter 1, the user-centered design (UCD) approach links the process of developing software, hardware, and user-interface (UI) to the people who will use a product/service. UCD processes focus on users throughout the development of a product or service. The UCD process comprises these tasks, which sometimes occur iteratively:

- *Plan*: Determine strategy, tactics, likely markets, stakeholders, platforms, tools, and processes.
- *Research*: Gather and examine relevant documents, stakeholder statements.
- *Analyze*: Identify the target market, typical users of the product, personas (characteristic users), use scenarios, competitive products.
- *Design*: Determine general and specific design solutions, from simple concept maps, information architecture (conceptual structure or metaphors, mental models, and navigation), wireframes, look and feel (appearance and interaction details), screen sketches, and detailed screens and prototypes.
- *Implement*: Script or code specific working prototypes or partial "alpha" prototypes of working versions.
- *Evaluate*: Evaluate users, target markets, competition, the design solutions, conduct field surveys, and test the initial and later designs with the target markets.
- *Document*: Draft white papers, user-interface guidelines, specifications, and other summary documents, including marketing presentations.

AM+A carried out all of these tasks in the development of the Money Machine concept design except for implementing working versions.

4.4   Market Research

In order to have a clearer vision of the target market for the Money Machine, AM+A conducted qualitative research with potential customers. For its interviews, AM+A prepared and distributed a version of the following questionnaire.

4.4.1   Questionnaire

Part 1: Smartphones and Social Media

- Do you own or would you ever purchase a smartphone? If so, what applications do you own? What attracted you to these applications?
- Do you play any games on your smartphone or computer? If so, what attracted you to them?

- Are you a member of any social network (e.g., Facebook)? Have you played any social-media games? If so, what attracted you to them?

Part 2: Money Management

- What tools do you use to manage your money (e.g., financial advisement, spreadsheet software)?
- What are your short-term and long-term financial goals?
- Are you or is any member of your family planning for your retirement? If so, how? Can you describe your life after retirement? Do you know how much money would you need for this lifestyle?

Part 3: Technology and Money

- Do you use online banking or any wealth management software? What aspects of those tools do you use?
- Would you ever share financial information with your friends? With whom do you share financial information (e.g., financial goals, percentage of increase in spending, saving for college)?
- Would you consider using a smartphone application to manage your wealth? What features would you look for?

Part 4: Quick Background Information

- What is your average annual household income?
- Do you have any financial investments? If so, what sort? (You may circle more than one option.)
- Stocks, bonds, mutual funds.
- Retirement fund.
- Health insurance.
- College fund for your children.
- Real estate.
- How do you make these financial investments?
- Via a broker.
- By yourself.
- Via a bank.
- Via institutions such as Charles Schwab, Bernstein.
- If you have a broker, would you rather be able to handle your own investments without an intermediary?
- Why or why not.
- Do you conduct research to help you with your investment decisions? (You may circle more than one option.)
- Yes: online.
- Yes: print (newspaper, magazine).
- Yes: word of mouth.
- Yes: financial advisors.
- No.

- Please rank the following communication tools from 1 (being the one you use the most) to 5 (being the one you use the least):
- Phone calls.
- Email.
- Skype.
- Texting.
- Social media such as Facebook, LinkedIn.

AM+A interviewed 10 people between the ages of 48 and 65. The results from the questionnaire are shown below in Figs. 4.1, 4.2, 4.3, 4.4, 4.5, 4.6, 4.7, 4.8, 4.9, and 4.10.

4.4.2   Demographics

Age: Majority mid 50's, range from 48 to 65.
Gender: More women than men.
Education: Varies, from little to university educated.
Income level: Majority over $100,000, but ranges from less than $40,000 to this.
Ethnic/racial background: Many Caucasian and Asian.

4.4.3   Challenges Per Group

For most:

- Time management and not understanding enough about investing.
- Planning for retirement.
- Using current technology.
- Calculating budgets to see where excessive spending lies.
- Understanding investments and financial information.
- Getting financial advice from a non-paid person (i.e., friends, coworkers). People have problems with sharing financial information and/or tips.
- Figuring out Social Security figures.

What Would Increase Frequency of Use?

- People use it if it works and is easy.
- Something appealing with every sign-in or log-in, like notifications and person-alized information visualization.

What Motivates Best?

- Something personal, concrete, different.
- Some want to feel in control.

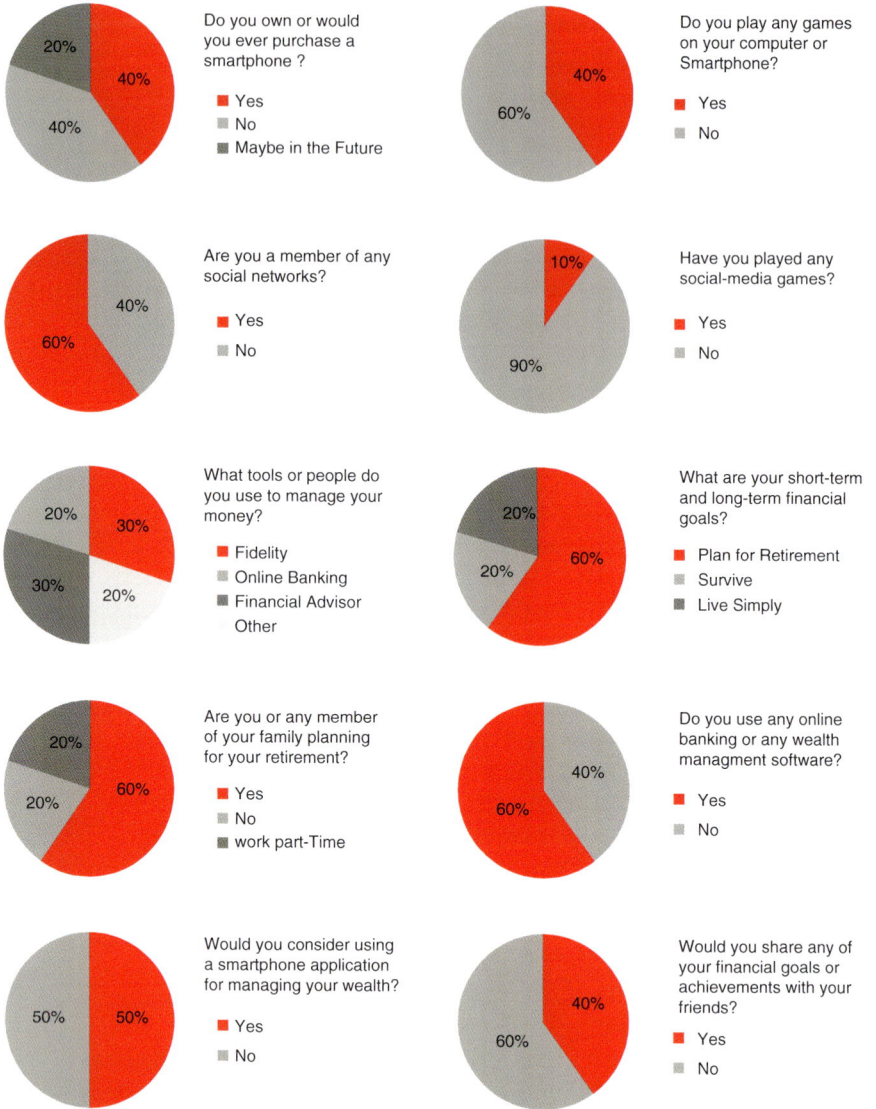

Fig. 4.1 Questionnaire results

- Anything that can lead quickly to small changes.
- For some, saving money for children/grandchildren.
- Feeling safer financially, less worried.

What Leads to Short- and Long-Term Behavior Change?

- Help people find out what is realistic in making successful changes.
- Don't overwhelm people with choices, which leads to failure.

- Offer simple, specific, attainable suggestions or regimens.
- Offer connected sequence of goals, e.g., saving for food per day leads to saving for food and a new car per day, instead of disconnected goals.

What Techniques Teach People Best?

- 3D models, comments by heroes/heroines, comments by family members, friends, photos of friends, peeking into lives of famous people.
- Don't use too many words on screens.
- Different people learn by different means; text alone generally does not work well.
- Use illustrations, but cartoons may not work with all people.
- Humor is sensitive, complex issue.
- Short practice exercises good.

Challenges to AM+A

- People must start to use Money Machine; using it: itself a change.
- People like to feel in control.
- People must see personal benefits, e.g., seeing decrease in spending.
- Wanting to change must be tangible goal; otherwise, Money Machine will be viewed as just another financial app.

Based on the data collected, AM+A constructed personas and use scenarios.

4.5   Personas

As noted in Chapter 1, personas, or user profiles are one to nine characterizations of primary user types and are intended to capture essentials of their demographics, contexts of use, behaviors, and motivations, and their impact on design solutions. For the Money Machine, the following persons were determined by analyzing available data.

4.5.1   Persona 1: Sheila Jones

- 52-year-old African-American housewife.
- College educated, married, two children.
- Proficient with iPad and Excel.
- Manages household bills with Excel.
- Annual household income: $72,000.
- Net worth: $750,000.
- Investments: 529 college fund.
- Persona image credit: AM+A.

4.5.2   Textual Summary

Sheila spends the majority of her day taking care of her kids, cleaning the house, walking her dog, and preparing food for her family. She is a proud mother and happily married to husband Darnell, a logistics manager for Matson. Sheila's two children are age 15 and 19, and starting high school and college, respectively. Sheila recently received an iPad for Mother's Day and uses it more and more every day. Her favorite applications include Martha Stewart's Cookie app and SplashShopper List Manager. The Jones' distribute their children's allowances by transferring money to gift cards. Their dog Bruno not only requires a lot of attention, but is also a significant cash drain. Sheila and Darnell are pleased to have sufficient capital to have set up a 529 for their children's college expenses. Sheila is the bookkeeper of the household and takes care of the budget. Having learned Excel in school, Sheila manages and checks her budget daily on her home computer. This routine consumes a significant portion of her day. Shelia waits to pick up her children at school, and if she could use that time to manage her books, Sheila could have an extra bit of free time.

4.5.3 Design Implications Summary

The following are implications of the above characteristics on the design of the Machine.

Objectives

- Monitor budget simply and effectively.
- Categorize money for each member of the family.
- Manage her children's 529 accounts.

Context

- User is accustomed to older technology, like Excel spreadsheets.
- Keeps track of the family finances.

Behavior

- Manages budget daily with Excel spreadsheets.
- Uses an iPad and has downloaded applications that she uses frequently.
- Physical environment: uses iPad at home and while running errands.
- Social environment: stays in touch with friends at book club and yoga classes.
- Information sharing: shares her financial information with her husband, and shares financial objectives and goals with her friends from time to time.

Design Implications

- Design the Money Machine for iPad users like Sheila.
- Create a way to partition the budget for each family member, including their dog Bruno. For example, the Money Machine could show how much money can be given to her son for his weekly allowance, and alert her if her son spends too much on sports gear.
- Create a feature where she can identify and be alerted to cash drain(s).
- Create an alert for when her kids use their gift cards.
- Have a simple information visualization to show budget monitoring.

4.5.4   Persona 2: Bethany Wilson

- 47-year-old Caucasian software engineer.
- Bachelor's in Computer Science from Carnegie Mellon, married no children.
- Proficient with iPhone.
- Is unsure how best to invest her large income Consults with financial advisor once a month.
- Annual household income: $120,000.
- Net worth: $1,500,000.
- Investments: 401 k.
- Persona image credit: AM+A.

4.5.5   Textual Summary

Bethany is 47 years old and works as a software engineer in Silicon Valley. She is married to a US Army lieutenant who is stationed abroad. She has an iPhone, and she regularly uses and downloads new applications like Yelp and Yellow Book. Bethany's income is quite substantial, particularly when supplemented by her husband's income. She plans to use her employer-provided 401 k to help with her retirement. She has hired a financial advisor, with whom she consults monthly. She would like to understand her investments better and be able to check them more frequently. She is uncertain about how much money she will need for retirement, and thus the age at which she can stop working.

4.5.6   Design Implications Summary

The following are implications of the above characteristics on the design of the Machine.

Objectives

- Learn more about investing so that she can make decisions on her own.
- Have more convenient connection with her financial advisor so that she can communicate with them more often than once a month.
- Monitor her investments while she is away from home.
- Figure out when she would be able to retire comfortably.

Context

- Is a technical person who enjoys using new technology.
- Has financial advisors, but would like to become less dependent on them.

Behaviors

- Could put some of her sizable income towards investments, but would like advice on how to invest most advantageously.
- Meets with financial advisors once a month to discuss her investments.
- Physical: uses her iPhone at work and at home.
- Social: talks to coworkers during work and at company events.
- Information sharing: shares financial information with her financial advisor and her husband exclusively.

4.5.7   Design Implications

The following are implications of the above characteristics on the design of the Machine.

- Design the Money Machine for iPhone users like Bethany.
- Provide clear investment updates and depict how these investments affect her net worth.
- Include customizable templates to help determine a retirement plan.
- Create a streamlined system for quick communication with financial advisors.
- Offer clear and simple investment tips and up-to-date articles on market trends.

4.5.8   Persona 3: Lance Richards

- 50-year-old Caucasian construction worker.
- High school graduate.
- Single, cares for his elderly father.
- Uses Android for calling and texting.
- Annual household income: $38,000.
- Net worth: $350,000.
- No investments; lives paycheck to paycheck.
- Persona image credit: AM+A.

4.5.9   Textual Summary

Lance has been working in construction, specifically street maintenance, since he graduated high school. Lance's father is currently living at home with him, recovering from complicated knee surgery. Lance owns a smartphone, but uses it only for phone calls, text messages, and voicemail. Lance works hard and budgets his money well, but had to forfeit his savings to pay for his father's hospital bills. He is now living paycheck to paycheck. Lance enjoys going to Arty's Sports Bar in Walnut Creek to grab a pitcher of beer and watch the Giants with his friends. Over a few months, his bar expenditures have begun to accumulate. He uses online banking from his computer at home to check his budget and to enroll in automatic bill payment. Lance would like to retire or work less in the next 20 years, but is unsure how to start save for his future again.

4.5.10 Design Implications Summary

The following are implications of the above characteristics on the design of the Machine.

Objectives

- Monitor his budget quickly and simply.
- Identify where he is spending too much money.
- Estimate funds to allocate for potential medical expenses, and the possible effect of these expenses on his budget.
- Determine how much money he needs in order to live comfortably in retirement.
- Calculate when he might be able to retire, or work part time.

Context

- Takes care of his father, who is recovering from surgery at home.
- Lives paycheck to paycheck.

Behaviors

- Is unaware of sources of overspending (e.g., at Arty's Sports Bar).
- Uses only the pre-installed applications on his smartphone.
- Does not have sufficient funds to start investing; is focused on saving for his immediate expenses only.
- Would like to save enough to retire or work part time.
- Physical: uses his smartphone on the job and at a bar, and checks his online bank statement from his home computer.
- Social: hangs out at local bar after work and on weekends.
- Information sharing: talks about his financial information only with his father or with representatives at his company.

4.5.11 Design Implications

The following are implications of the above characteristics on the design of the Machine:

- Provide a budgeting feature to help the user monitor spending.

- Provide a "cash drain" feature to identify specific overspending and alert the user when he/she is not sticking to the savings plan or goal.
- Make the application sufficiently user-friendly (with clear graphics and language) to convince users like Lance to use applications other than the ones already installed.
- Gamify the application with achievement-based challenges.
- Devise a way to set funds aside for emergencies (e.g., medical expenses), showing costs that commonly relate to the user's profile.

4.5.12   Persona 4: Brett Sanders

- 55-year-old Caucasian senior executive at Merrill Lynch.
- MBA from Yale; married, no children.
- Proficient with Blackberry.
- Manages his own money.
- Annual household income: $200,000.
- Net worth: $2,500,000.
- Investments: numerous.
- Persona image credit: AM+A.

4.5.13 Textual Summary

At 55, Merrill Lynch financier Brett is happily married to Angela, a retired realtor. Brett carries his company Blackberry with him at all times, whether traveling for business or golfing at the country club. Brett has been planning for retirement for some time: He has many investments, including an employer-provided 401 k that his company provided for him, and hopes to retire soon and live off these assets. Brett has great confidence in his business acumen, and manages his wealth without assistance. He manages all of his accounts and spending on his laptop. Although his laptop provides some mobility, a mobile application for Blackberry would be significantly more convenient. Brett would like to receive notifications about fluctuations in his portfolio so that he can tailor his strategy and retire soon.

4.5.14 Design Implications Summary

The following are implications of the above characteristics on the design of the Machine.

Objectives

- Check his investments several times a day to see if he needs to buy, sell, etc.
- Amass sufficient funds from his investments to retire.
- View possible changes in retirement planning strategy, based the performance of his investments.

Context

- Manages his finances from his laptop, but would also like to do so on his smartphone.
- Deals with a substantial amount of money and has a number of investments.
- Plans to retire with his investments to support him.
- Has a 401 k account from Merrill Lynch.

Behaviors

- Earns a large income.
- Devises his own investment strategy because he works in the industry.
- Physical: uses his Blackberry and laptop on business trips and at home.
- Social: spends times at the country club where he meets up with his friends.
- Information sharing: brags to friends and fellow country club members about his success financial stories.

Design Implications

- Design the Money Machine for Blackberry users like Brett, because Blackberry is popular among large businesses.

- Offer customized alerts to inform users when their investments rise or fall. Include how he might change his retirement planning (e.g., save an extra $100/ month because investment is doing poorly).
- Provide updates on general investments (e.g., AMZN is falling).
- Include charts about his current returns on investment, risk, how diverse his portfolio is, and tax efficiency.
- Include a breakdown of net worth to help him see how he is doing financially.
- Extra secure because he is dealing with actual investments accounts that he can transfer money between.

4.6   Nancy Chen

- 48-year-old Asian grade school teacher and tutor.
- Graduate degree in education from Berkeley, divorced, no children.
- Likes technology with simple functions.
- Enjoys social network interaction with friends.
- Uses Fidelity to manage her money.
- Annual household income: $42,000.
- Net worth: $465,000.
- Investments: 401 k.
- Persona image credit: AM+A.

4.6.1   *Textual Summary*

At 48 years of age, Nancy is a grade school teacher by day and tutor by night. Nancy's sense of humor has made her popular among her circle of friends and they convinced her to join several social networks. She is comfortable with new technologies such as the iPhone and the iPad, but is very selective about downloading applications. She prefers apps that are intuitive, graphically oriented, and visually customizable. A recent divorcée, Nancy is now learning to manage her own finances, without her ex-husband. She learns about financial tips through her friends and coworkers, whom she converses with both at work and via Facebook. In addition to her base knowledge and friends, she also uses Fidelity to keep track of her investments. Her objective is to work part time, while still living comfortably and being able to send money to her parents and relatives in China.

4.6.2   *Design Implications Summary*

The following are implications of the above characteristics on the design of the Machine.

Objectives

- Calculate when she might be able to work part time.
- View her wealth simply and graphically.

Context

- Juggles two jobs (teaching by day and tutoring at night), which affords her little free time.

Behavior

- Is well-versed in current technologies that do not require advanced like social networking Websites or simple smartphone applications.
- Dislikes the iPhone's small font size.
- Likes to visually customize digital apps.
- Physical: carries her iPhone at all times, including while teaching and tutoring.
- Social: loves interacting with friends and coworkers both online and offline.
- Information sharing: exchanges investment ideas with trusted coworkers.

Design Implications

- Design the Money Machine for users like Nancy, who have both an iPad and an iPhone.
- Include a feature for users to set savings goals within specific time frames (e.g., put $100/month into a savings account). At any point in time, users can track

their progress — how much they have saved, and how much they need to reach their goals.

- Offer retirement planning assistance, with tips on saving and retirement funding, as well as future projections.
- Include a social-networking feature or connection to a social-networking Website for users to share their financial triumphs. For example, Nancy could post that she saved enough for a new car on her Money Machine's share feature. Her peers could then congratulate her, or ask for tips on achieving goals.
- Include an option for a larger font size (research shows that a 14-point font is most legible for baby boomers).
- Include visually customizable graphic elements that represent her wealth (e.g., charts showing all her current assets).

4.7   Use Scenarios

4.7.1   Definition

As noted in Chapter 1, use scenarios help determine what behavior to simulate. The following use scenario topics are drawn from the five preceding personas. However, some specific examples might be relevant only to a particular age group, education level, gender, or culture. Note the general usage of the terms "objective" and "goal." An objective is a general sought-after target circumstance. A goal is more specific and is usually qualified by concrete, verifiable conditions of time, quantity, etc. For example, an objective is "I want to be rich someday." A related goal would be "I want to earn $60,000 per year by the end of 2015."

4.7.2   Financial Monitoring

- View the assets in chart format (e.g., line chart, bar chart, pie chart, or other data visualization, depending on user preference).
- Partition the user's budget into categories (e.g., set up one's own section, as well as sections for one's spouse/partner, dependent child, or dependent parent).
- Connect to investment accounts via bank, brokerage, or telecom company (e.g., AT&T, Google).
- Receive up-to-date articles, advice, and tips regarding monitoring future or current investments. This information may come from professionals, family and friends, or the general public.
- Set customizable alerts for investment fluctuations, whether up or down, positive or negative.
- Receive overspending alerts.
- Establish and maintain objectives (e.g., "I want to retire soon").

- Establish and maintain goals (e.g., specify date of retirement, allocate money for a parent, save for a car, etc.). See the ramifications of this goal on current and future budgets, including a timeline and new budget to fund the goal.
- Follow and customize pre-set financial templates (best practices) to achieve goals.
- Visualize and monitor the user's net worth.
- Plan for retirement with a variety of different tools and templates.

Security

- Share financial goals with specific friends and family.
- Alert users to any unusual activity in their accounts.

Social Media

- Post financial achievements on the users' own walls and possibly their friends' walls, similar to a merit-badge system.
- Connect with financial advisors by text messaging or chat to ask questions and react to their strategy suggestions.
- Share tips and strategies with specific friends or family.
- Import personal information from social media sites (e.g., race, sex, age). Users not connected to a social media site can add their information manually through the Money Machine.
- Resolve any urgent ethical issues.
- Set and use pre-existing achievements to help manage finances.
- Compare estimated spending with actual spending (e.g., find cash drains).
- Purchase digital items using currency earned from meeting achievements.
- Reward valuable tipsters with currency or credit for advice taken by others (e.g., recommending particular investments, creating plans to spend less or save more).
- Develop an "economy of tipsterism," likes and dislikes, bribes and no bribes, objective vs. biased opinion, etc.

4.7.3   Specific Persona Use Scenarios

4.7.3.1   Sheila's Use Scenario

Homemaker Sheila manages the household expenses for her family of four (plus dog Bruno). She likes using the Money Machine because it consolidates her financial information into one simple application. Sheila really enjoys the different view settings within the Money Machine; she prefers the spreadsheet setting, since she is accustomed to Excel. She also monitors her children's 529 plan on the Money Machine. The Money Machine helps Sheila keep track of her children's spending. Last week she was alerted to her son's $300 purchase at the Apple store. When she followed up with him, her son explained that he needed to buy Photoshop for his computer, along with a new mouse. The Money Machine also helped Sheila with the family's aging dog Bruno, who has a number of veterinary expenses. With the Money Machine's partitioning function, Sheila can allocate funds for Bruno's veterinary bills. Because her day is packed with running errands and keeping the house in order, Sheila likes using the Money Machine on the go, when she has a spare minute. Sheila received an iPad as a gift recently, and was able to manage her books while waiting for her dentist appointment the other day.

4.7.3.2   Sheila's Behavior Changes

The Money Machine gives Sheila more free time during the day. Now she keeps track of her finances while she is out running errands or waiting at the dentist's office.

The Money Machine helps her audit her children's spending. The Money Machine alerts Sheila via email of any large purchases, and allows her to monitor her children's accounts. This function helps her save money and teach her children proper spending habits.

The budgeting feature helped Sheila realize that even though she loves her dog, he is a sizable expense. Sheila uses the Money Machine to set aside and monitor a budget for emergency veterinary bills.

Sheila now views her 529 whenever and wherever she likes. Consequently, she understands its performance better, and can make more informed investment decisions.

Asking for and receiving tips has made Sheila more decisive in her account management.

4.7.3.3   Bethany's Use Scenario

Bethany has a few investments and is still looking for new investment opportunities. She has hired a financial advisor to help improve her investment strategy. Bethany now uses the Money Machine to learn more about investing. She reads market updates through the Money Machine when she has a free moment. She also uses the Money Machine to see how and when she can retire. Bethany entered her desired retirement age, her expected income in retirement, and other necessary information. Based on the information provided, the Money Machine determines how much Bethany should save in order to reach her retirement goals. The Money Machine alerts Bethany of significant fluctuations in her investments and calculates possible adjustments to her budgets accordingly. Checking her current net worth on the Money Machine gives Bethany a clear and simple picture of her investment strategy and her progress toward retirement. When Bethany has questions about her investments, she either asks one of her friends for clarification or messages her financial advisor — all through the Money Machine. The Money Machine provides a quick and easy way for her friends and advisors to respond.

4.7.3.4   Bethany's Behavior Changes

Bethany gets investment help from friends and financial advisors whenever and wherever she needs it.

She has learned more about investing and is more confident in her financial decisions.

The Money Machine's retirement planning features let Bethany know when she should be able to retire comfortably.

Bethany makes more informed financial decisions because she knows her current net worth.

4.7.3.5   Lance's Use Scenario

Lance was living paycheck to paycheck after paying for his father's post-surgery medical bills. Since acquiring the Money Machine, Lance has been able to save some capital for his goal of working less when he gets older. The budget-partitioning feature in Money Machine has helped Lance allocate sufficient funds to care for his father, who is living at home. He particularly appreciates the Money Machine's overspending alerts on his smartphone. Lance used to spend too much money at the bar; now the Money Machine gives Lance a comfortable spending limit when he goes out. The Money Machine achievement system alerts Lance's friends that he has reached a certain milestone. His friends' congratulations after he saved up $2,500 made Lance feel proud, strengthening his motivation. Also, he received enough achievement points to download an SF Giants digital theme for the Money Machine. Lance continues to set goals for himself through the Money Machine.

4.7.3.6   Lance's Behavior Changes

Lance was living paycheck to paycheck, but has now begun a savings strategy with the Money Machine.

The Money Machine allows Lance to allocate funds for specific needs. Now Lance sets aside a certain percentage of his monthly income for his father's care and treatment.

Lance's overspending (i.e., on beer) has ebbed, thanks to smartphone alerts from the Money Machine. Now he has lost 15 lb, along with his beer belly. He is also getting attention from the ladies and has set up a date this Saturday at the Giants game.

Lance has set up a template for his goal of working less as he grows older. The template notifies Lance if he is going off track, helping him to stay on the right path.

Lance receives support from his friends when they see his Money Machine achievement badges on his Facebook wall. This encouragement helps him stay committed to saving money and making smart financial decisions.

4.7.3.7   Brett's Use Scenario

Brett is constantly on the move. He often travels for business and always has his Blackberry with him. Brett uses the Money Machine to check on his numerous investments when he has free time. Brett likes to monitor his investments to determine whether he is still on track to retire. The Money Machine also shows Brett his current net worth, which Brett likes to check every day to see how he is doing financially. The application notifies him of any substantial changes in his portfolio, so that he can make adjustments if needed. He also uses the Money Machine to look at his current return on investments, as well as his portfolio's risk level, diversification, and tax efficiency. Brett keeps an eye out for future investments, and uses the Money Machine to monitor these potential opportunities. Brett takes pride in his financial successes, and often makes a substantial amount of money on his investments. The Money Machine lets him brag about his latest victory by posting his accomplishments on his wall and friends' walls.

4.7.3.8   Brett's Behavior Changes

Brett stays on top of his investments with Money Machine alerts, which help him keep up with the market, make faster decisions, and maximize his investments.

Brett checks on his banking and investment accounts while away from home. The greater mobility gives him more freedom to make transactions.

The Money Machine lets Brett compete with his friends and compare financial goals. Competitive by nature, Brett finds himself striving to perform even better with his investments and financial accomplishments.

The retirement planning eases Brett's mind about how much he should be saving. When his investments fluctuate, the Money Machine updates his savings schedule to help him stay on track with his retirement goal.

4.7.3.9   Nancy's Use Scenario

Since Nancy's divorce, she has been learning financial self-reliance. Her first decision was to download the Money Machine on her iPad. With the Money Machine, she can easily monitor her assets anywhere she goes, either with her iPad at home or iPhone at work. This access and ease of use gives her increased confidence in her personal life path. The multifeature money management application has shown Nancy that handling her finances does not need to be difficult. The Money Machine helps her translate complex financial data from different bank accounts into simple pie graphs, which is her graphic representation preference. Nancy cannot imagine herself retiring completely, but she would like work one job (instead of two) in the future. The Money Machine is working with her step-by-step to realize this goal. Using the financial information she has already entered, the Money Machine calculates that she could quit one of her jobs within 5 years, as long as she spends 15 %

less on monthly groceries or earns 3 % more from her investments. Nancy is delighted to have found these modifications very practical, and not too challenging. When Nancy has free time, she also enjoys the Money Machine's social network feature. She posts her financial achievements (e.g., saving 20 % on monthly groceries) on her wall, where her friends can congratulate her or ask for advice on reaching their own goals.

4.7.3.10   Nancy's Behavior Changes

Because of the Money Machine, Nancy was able to figure out when she could start working part time and still live on a reasonable income.

The Money Machine alerts her to changes in her portfolio, and how these changes may affect her retirement plan and timeline. With this information, Nancy makes well-informed decisions about altering her investment strategy.

Nancy can stick with her retirement plan, now that she has set up goals with precise deadlines and dollar amounts. This specificity helps her evaluate whether her expectations are realistic and practical.

Overspending notifications help Nancy stay on track with her plans.

Posting her achievements to a social networking site and seeing her friends' supportive and encouraging comments further motivate Nancy to save money and stick with her financial strategies. Nancy's achievements also earn her credits, with which she can purchase customized skins for the Money Machine.

The Money Machine's user-friendliness, including the larger font size option for baby boomers, makes Nancy more comfortable with using iPhone application. She now considers using other applications (i.e., Green Machine, Health Machine, Story Machine) from the Money Machine creators, because of their focus on ease of use and attention to detail.

4.8   Competitive Analysis

Before undertaking conceptual and perceptual (visual) designs of the Money Machine prototype, AM+A studied approximately 20 financial and wealth management Websites and iPhone applications in 2011. Through screen comparison analysis and customer review analysis, AM+A derived these applications' major benefits and drawbacks. This in-depth analysis helped further develop initial ideas for the Money Machine's detailed functions, data, information architecture (metaphors, mental model, and navigation) and look and feel (appearance and interaction). The applications are grouped approximately by overall functional characteristics.

4.8.1 Money Management Websites and/or Associated Mobile Applications

4.8.1.1 Mint.com

Who Were They in 2011?
- Created by Aaron Patzer (Masters in Computer Science and engineer from Princeton and Duke) in September 2007.
- A patent-pending categorization technology that automatically identifies and organizes transactions made in virtually any bank, credit, investment, brokerage, or retirement account.
- A proprietary search algorithm that finds savings opportunities unique to each user.
- Acquired by Intuit in 2009.
- Number of users? 2008: 500,000. 2011: more than 5 million.
- Pulls financial data from users' bank accounts and categorizes the data: savings/checking/investments/auto/mortgage.
- Handles all types of investments: 401 k, mutual funds, brokerage accounts, and IRAs (users can shape up their IRA).
- In terms of security uses the same 128-bit encryption and physical security as banks. Monitored and verified by TRUSTe, VeriSign and Hackersafe, and supported by RSA Security.
- Works as a read-only device: can organize and analyze but not move money around.

What Did It Let Users Do in 2011?
- Make a budget based on income and spending. Automatic budgeting/customization.
- Auto-categorization: allows users to label their transactions however they want. Every transaction is tag-able and filterable.
- Set up goals.
- Customize tips on savings/investments based on user spending behavior and user income.
- Detect and notify the unusual activities that happened to a user's account 24/7 via emails and text messages. This seemed to be an important feature, because it is being repeated twice (with the same picture and description) under both "timely alerts" and "safe and secure" sections.
- Help users choose the best kind of credit card, investment, and insurance that fit their lifestyle (i.e., judging from their spending habit).
- See how other people's spending habits in the user's area are like through "MintData".
- "MintLife" is an award-winning blog that contain articles with personal finance advice, money management tips and help with financial planning.

- "MintAnswers" will let the user ask the question about their personal financial situation and problem(s). The people who will answer the user's question will be an expert and other users (i.e. works similarly to Yahoo answers).

Web Strategy

- Extremely detailed Website: any question a user might have is answered.
- A very useful tool is the "trends": users can create graphs and charts based on their income/spending etc.
- Mint Canada has just become available but without the French language feature yet. But otherwise, American banks only.
- QR code technology, which has become more popular recently, can be used to install Mint application to an Android phone using its barcode scanner application.

User-Friendliness

- User-friendly: free, easy to create. YouTube videos and tutorials.
- "Adding an account": Clear pictures and well-defined step-by-step control flow.
- "Creating a budget": Adding a new budget for uncategorized items is not very clear (hitting a plus sign). Automatically creates a budget based on what the user spends, which is convenient, but it is not clear how to delete a budget, only how to modify one.
- "Timely alerts": described clearly and set up in an easy to use manner.
- Seeing trends: very user-friendly visualization to help the user picture how their money differs each month. Picking time period is not immediately clear on how to do. "Graphs to try" is a nice feature.
- Goals: linking goal to specific accounts is a nice feature to easily allow the user to adjust their different budgets.
- Recommendations: doesn't clarify how easy it is to actually switch banks, credit cards, etc.

Website Visual Analysis

- Green as the main color palette suggests money and sustainability.
- Having a small icon (included symbols like a piggy bank or a golden egg) representing each category in the Website contributed to its user-friendliness. Without a lengthy description, it allows the user to be able to visualize what each category is about right away just by browsing through.
- Bar charts, line/fever charts, and pie charts are the key elements that show information to the user. They became more effective when combined with a time-based video tutorial under "helpful graphs" menu.
- Simple language (i.e., words usage) makes the information easy to understand (Figs. 4.2 and 4.3).

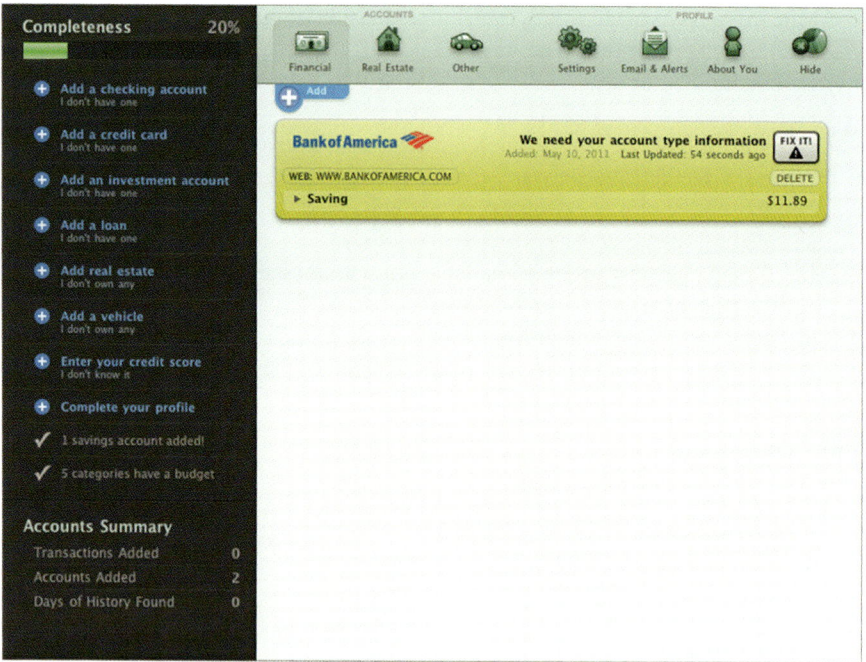

Fig. 4.2 Web screen of Mint.com. (Images: Intuit, Inc.)

Fig. 4.3 Exemplary screens of mobile version of Mint.com (Images: Intuit, Inc.)

Mobile Strategy

- Mobile applications are available for both iPhone (free and password protected, Best Finance App by the 1st Annual App Awards, Best Online Money Application by Bestcovery and one of the TopWeb Apps by Webware) and Android platforms.
- However needs to be used as a complement to mint.com and not by itself as most of the set-ups are done via the Website.
- American banks only (not available for Canadian banks).

Mobile Visual Analysis

- Bold numbers, horizontal bar charts, and color-coding (i.e., red, yellow, green for different degrees of financial significance) are the main infographical elements of the mobile device.
- Words displayed on the screen are short and not descriptive, assuming that the user already knew how Mint works.

Primary Competitors

- Quicken.
- Microsoft Money.
- SmartMoney.
- Yodlee MoneyCentre.
- Mvelopes.

4.8.2 Tips

4.8.2.1 Find Savings

Mint helps users save on home loans, credit cards, bank accounts, and more. Analysis of the best credit card (for example) based on: monthly credit card spending, the type of spending (groceries/fuel/restaurants, etc.), and the commitment to pay their credit card every month, and finally their credit score. After this analysis, it determines which credit card from which bank would be more useful to the user. This analysis is available for any banking operation, investment operation, and insurance operation.

4.8.2.2 Achieve Your Goals

The user enters a "goal" (buy a home, save for retirement, get out of debt, pay off your credit card) of money they need, a date, and the accounts they will want to use to reach that goal. It allows the user to track their progress but also gives them customized advice on how to reach their goal based on their behavior.

4.8.2.3   Ways to Invest

Depending on the user's investment style (active/low maintenance/hands off) it gives them a link to the best investment advisor such as Betterment, Vanguard, LendingClub, Folio Investing. Mint makes recommendations for how the user can save more money as they use the application.

4.8.2.4   Social Networking

Presence on Facebook (fewer than 100,000 followers in 2011).
  Twitter
  SWOT Analysis

Strengths

- Up-to-date status of user spending.
- Useful tips and tools (charts, budgeting).
- Safe and secure.
- Free.
- User-friendly.

Weaknesses

- Read-only device. It doesn't manage money.
- It doesn't take into account brokerage and categorizes it as "Business purchase."
- There is no way users will detect bank errors if they don't double check their account.
- No proactive planning and organizing.
- No persuasion behavior change.

Opportunities

- Develop a heavier investment oriented app.
- Develop social networking strategy.
- Develop the mobile strategy (should exist by itself and not only in association with the Website).
- More comprehensive investments to be done.
- From a read-only device to an "active" device.
- Wider market.
- More persuasion theory–based content.

Threats

- Security might be an issue in terms of persuading people to use the app.
- Brokers might be an issue if they view the app as a strong competition.

4.8.2.5 mvelopes.com

Who Were They in 2011?

- Online "envelope budgeting system" in existence for over 10 years.
- Product of Finicity corporation (founded in 1999 and headquarters in Utah): a leading Internet and mobile software services company specialized in the development and delivery of financial productivity applications for personal and business use.
- Paying online money management.
- Proactively takes control of the user's money: divides their income into "envelopes" and knows the impact of user's spending.
- They promise users that they "will take control" of their money, and manage their spending within their income.

What Did It Let Users Do in 2011?

- Divide users income into different "envelopes" to control and manage their spending and allowing them to set money aside for holidays/getting out of debt.
- Create a spending plan.
- Allows users to import the Quicken transaction data.
- Online bill pay: up to 15 payments/month.
- Envelope transfer.
- Reporting: cash flow report.

Web Strategy

- User-friendly Website with a "quick tour": What makes us different?/what is envelop budgeting?/envelop budgeting cashless tour/time to get started.
- No precise explanation on how they have access to users' finance: do they pull the data from their bank accounts?
- US market only.
- There are two different Websites that are active. One is accessed through mvelopes.com; but the other one was accessed via user reviews URLs.
- This "second Website" is less clear, more confusing, the tutorial video is 6:12 long.

User-Friendliness

- Does not look easy to use: too many options available and not clear what the order of completing each step should be.
- The software automatically making adjustments with overspending could be helpful to some people but might confuse others if they use other ways to look at their budgeting. Why is there no indication of when the program does this, especially if the user would not want it?

Website Visual Analysis

The Website's main element of information visualization is storytelling. The home page contained two avatars instructing how this Website works through different stories/scenarios that involved the envelopes concept. However, it is hard to verify the validity of these stories; they could be fictitious.

The Website also allows users to share "success stories" in both written and visual (video) forms. These stories serve as an inviting and a motivating tool for the new user (Fig. 4.4).

Mobile Strategy

- Mobile version of the online product, but no thorough description of it.
- iPhone app available but no Android currently (in the blog section, on 5 May 2011, the writer said the Android app is in final beta testing and will launch "very soon").

Mobile Visual Analysis

- Focused on functionality, but not esthetics. The user will be able to recognize the function of each feature through words and numbers. However, nothing distinguished this application to any other applications; it looks like a typical table that one can draw or program easily.
- The colors seem very basic and bland.
- The visual characteristic of the application also has no connection to the mvelopes.com, which could create confusion whether the two are supposed to function similarly.

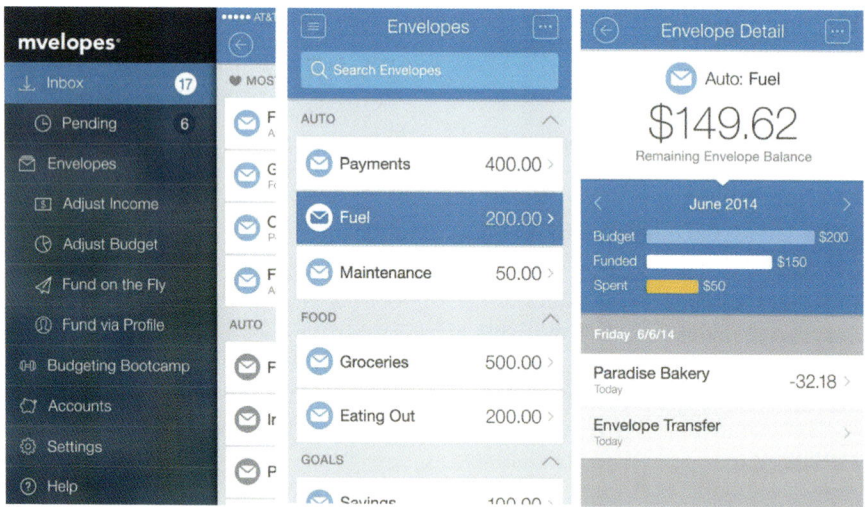

Fig. 4.4 Example screenshots of the mvelopes app as of 2015 (Image courtesy of Finicity, Inc., images used as requested by company)

Primary Competitors

- Quicken.
- Microsoft Money.
- SmartMoney.
- Yodlee MoneyCentre.
- Mint.com.

Tips

- Free debt analysis.
- Credit card management feature.
- Personal coaching with live chat window.

Social Networking

- Practically no strategy: 89 followers on Facebook. Presence on Twitter.
- Keeps the users updated with new information through blogs.

SWOT analysis

Strengths

- 30-day free trial.
- Allows users to pay bills online.
- Users can cancel their free membership anytime they want.

Weaknesses

- No mention of investments management.
- Paying system: $189.60 for a 2-year plan, $129.60 for a 1-year plan, $39.60 for 3-month plan.
- No precise explanation as how it actually works: how do they set up this management system: from users' bank accounts? Do users enter their own amount and then divide it however they want? Is there an analysis of their spending?
- Mobile strategy is unclear.
- Practically nonexistent social media strategy.
- After 15 online payments, they charge $0.50 per payment.
- No alert: "you have $xxx left in this envelope."
- Can users cancel their subscription? If yes do they get a refund?

Opportunities

- Develop social networking.
- Develop financial management.

Threats

- Price might be an issue.
- Users don't get a refund if they are not happy with the software.

4.8.2.6   You Need a Budget (YNAB)

Who Were They in 2011?

- Created by Jesse and Julie Mecham. Regular "couple next door."
- Paying budgeting software.
- It allows users to create their budget. They enter their own amounts and assign funds to different budgets.

What Did It Let Users Do in 2011?

- Budgeting.
- Unclear about whether users have access to their bank account from this platform.

Web Strategy

- No quick tutorials, just long ones (8 min and more). Regarding the user interface: the product documentation just presents snapshots of each way of using the Website. No real explanations of some features. For example: "YNAB will make small adjustments when you overspend": what does this advice mean?
- The users' reviews were more useful than the actual Website.

User-Friendliness

- The Website is not really user-friendly; it is hard to tell where is what. Users have to dig around to find FAQs and information on the company.
- The paycheck feature is really nice.
- The software has clear defined tabs and nice visualization that shows the user what they can do. Flow with obvious steps to take once a user picks a task. Available tasks also very clear and right amount of information shown so that the user is not overwhelmed.
- Webinars available.

Website Visual Analysis

- The 2.5-min video introduction to how YNAB works is useful (although the country-music background can be distracting). In the video, the Web summarizes how the software works into four simple rules: give every dollar a job, save for a raining day, roll with the punches, and stop living paycheck to paycheck.
- It is hard to navigate through the Website because it is one linear page with links to all the images. The information seemed scattered; the Website does not separate different aspects of the software into categories, which would be much easier to navigate.
- Beside tables, the software also contains bar charts, pie charts, and line graphs, which make the information easier to visualize. However, it is difficult to know that the software through the Website because of how the Website is laid out (Figs. 4.5 and 4.6).

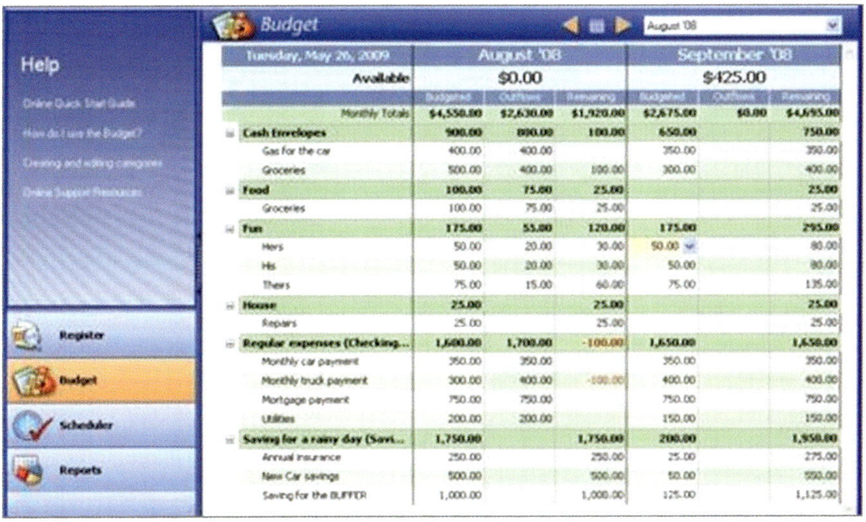

Fig. 4.5 Example Webpage design of YNAB in 2011 (Images: YouNeedABudget.com)

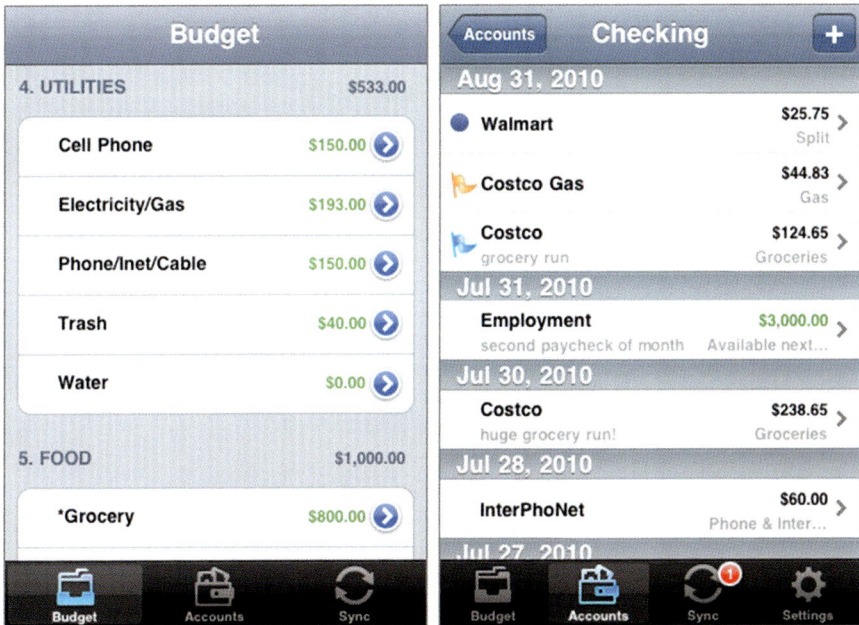

Fig. 4.6 Mobile screen designs of YNAB in 2011 (Images: YouNeedABudget.com)

Mobile Strategy

- iPhone app available: the video tutorial is probably the best. 3 buttons: budget/accounts/sync (synchronize with users' desktops).
- No Android version of the application.
- Users must enter all of the amounts themselves (at the grocery store, users enter their spending based on their receipts).

Mobile Visual Analysis

- The main design element of the application is a table with bold letters for budget categories and color-coding numbers for amount of money spent (i.e., green is safe, red is danger).
- No distinguishable feature. It looks similar to a typical table, with even fewer details than mvelopes' application.

Primary Competitors

- Quicken.
- Microsoft Money.
- SmartMoney.
- Yodlee MoneyCentre.
- Mvelopes.
- Mint.com.
- wesabe.

Tips

- Free online classes.
- Free manual.

  YNAB "methodology" based around four rules of cash flow:

- Users live off of the previous month expense.
- They assign every dollar to a "task."
- They save for "rainy days."
- If they overspend in one category, the budget available for the next month will auto-adjust based on that overspending.

Social Networking

- No strategy. Presence on Twitter?

SWOT analysis

Strengths

- It's a software not a platform.
- 7-day free trial.
- Tutorials available online.
- 30 day-refund policy available.
- Contains internal search engine.

Weaknesses

- Budgeting tool only.
- No tips/no managing/no proactive/no investment strategies.
- Website is not user-friendly.
- Paying software: $60.
- Month-to-month tool: no "long-term" tool/investment/budgeting.
- Good for a light budgeting use.

Opportunities

- Android application.
- Social networking strategy.
- Develop investment strategy.
- Develop a more user-friendly Website.
- Develop more friendly tutorials.
- Long-term tool.
- Wider markets.

Threats

- Additional fees for additional features?
- Security.
- Too much manual entry, not intuitive enough: customers might get bored.

4.8.2.7   Quicken

Who Were They in 2011?

- A brand of Intuit Corporation, created in 1983.
- Intuit bought Mint.com.
- The software is both PC- and Mac-compatible.
- Personal finance software.
- Basic money management.
- Invest and track growth (with financial help, advice and tips).

What Did It Let Users Do in 2011?

- Basic money management with Quicken Deluxe or Mint.com.
- Create a long-term saving plans with Quicken Deluxe at $59.99.
- Invest and track growth with Quicken premier at $89.99.
- Pay bills online.
- Comprehensive investing and planning tools: track, analyze and optimize investments (401 k, IRA, brokerage).

Web Strategy

- User-friendly.
- Pleasant interface, easy to find the right tab.

- Videos, tutorials, tips, FAQs.
- All presentation videos have the script below for deaf people.

User-Friendliness

- Paycheck feature is really nice.
- Clear defined tabs and nice visualization that shows the user what they can do. Flow with obvious steps to take once a user picks a task. Available tasks also very clear and right amount of information shown so that the user is not overwhelmed.

Website Visual Analysis

Clean design and well organized. The Website separates each feature into clear, different categories. Very easy to browse through and find out the specific piece of information.

Since the software has many versions, and each version contains much details, it would be hard to explain all on the Website without taking too much space. To save the user some time and energy to read through a lengthy description of each feature, including the overall tutorial, the Website contained many interactive links that explain each aspect of the software with simple graphics and language (Fig. 4.7).

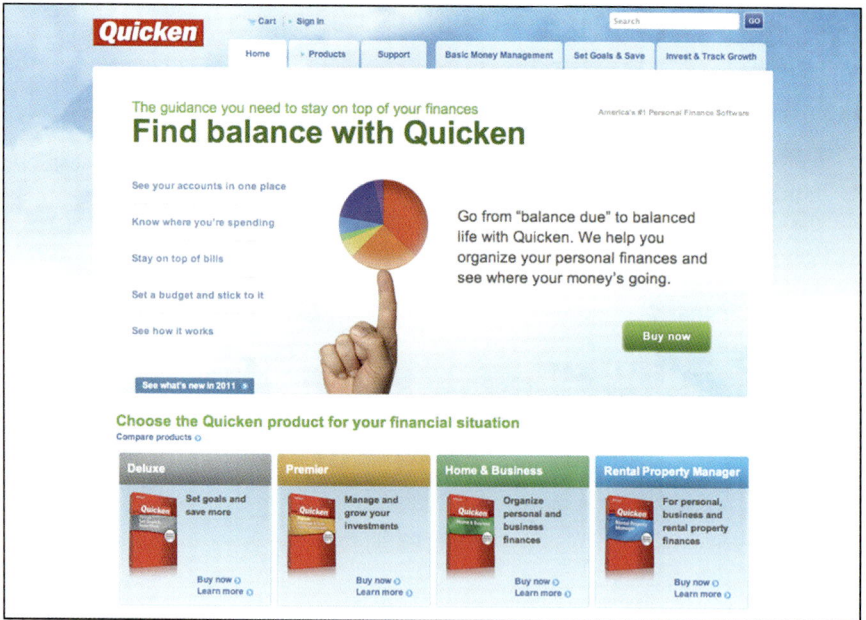

Fig. 4.7 Website design of Quicken in 2011 (Image: Intuit, Inc.)

Mobile Strategy

- Their iPhone app is the Mint.com app.

Primary Competitors

- Microsoft Money.
- SmartMoney.
- Yodlee MoneyCentre (tracking device).
- (Mint.com).

Tips

- Tips for how/when to pay bills.
- Tips to teach users children to save/manage their finances as they grow.
- Tips to maintain a healthy credit scores.
- Shows users how to minimize taxes on their investments.

Social Networking

- Facebook (16,000 followers in 2011), YouTube, Twitter, blog, and Live community.

SWOT analysis

Strengths

- Parent company's software.
- Different versions of the software available according to users' use: Deluxe (set goals and helps them manage their money), Premier (Deluxe + Manage and grow their investments), Home and Business (Premier + helps organize business/personal finances), Rental property Manager (Home and Business+) organize personal and property rental finances.
- A lot of tips and advice.

Weaknesses

- Can be quite expensive if users want to focus on investments.
- Hidden fees.
- Downloading problems.
- Users have identified user interface problems.
- Lots of ads.
- Upgrades are expensive.
- Investments allowed seem limited.

Opportunities

- Develop a mobile app that is more in accord with their more in-depth investment strategy (their app is Mint.com).
- Persuasion theory.
- New markets (outside of North America).

Threats

- The different versions and updates to download are annoying to customers.
- Nostalgia existing because of Microsoft Money. Someone could "recreate" software similar to Microsoft Money.
- Difficulty synchronizing with various banks.

4.8.2.8 Microsoft Money

Microsoft Money was no longer available in 2011. According to the Website, "more users shift their attention to full-service offerings provided by banks and brokerages, and demand for a comprehensive personal finance toolset has declined."

4.8.2.9 SmartMoney.com

Who Were They in 2011?

- Part of the *Wall Street Journal.*
- Free Website: users need to create a username and password.
- They provide a wide array of choices, advice, articles to help with investment strategy.

What Did It Let Users Do in 2011?

- Article topics: Invest, Spend, Borrow, Plan, Retirement, Taxes.
- Under the "Tools" section users have a variety of calculators for Autos, Bonds, College planning, Debt management, Economy, Estate Planning, ETFs, Health care, Insurance, Investing, Marriages and relationships, Mutual funds, Real Estate, Retirement, Stocks, Taxes.
- Advice about elder care, 401 k, as well as how to borrow, how to plan.
- Portfolio of stocks.
- Watchlist.
- Stock screens, stock quotes.
- Advise about brokers.
- Retirement planner.
- How-to guides.

Web Strategy

- An extremely comprehensive Website that would be the *main* competitor for The Money Machine in terms of what they do on their Website.

User-Friendliness

- Very friendly user interface, easily accessible, clear and easy to use.

- How-to Guides are in a weird location on the site, not under a specific category but placed randomly on each different investment main page, could be placed better.
- No obvious place to go for first time investors seems site is more geared towards people already with investments and knowledge and looking to increase their knowledge with recent findings and articles.
- Has some helpful tools dealing with how well users are coming along in retirement and investments, but these aren't in obvious places either; one discovers them after clicking through several pages.

Website Visual Analysis

- A newspaper-like layout, titles in series, and the gold-gray color scheme, which corresponded to that of *Wall Street Journal*, suggest the sophisticated aspect of the Website (has to do with target audience).
- Similar to *Wall Street Journal*, this Website and its design seemed to aim toward an older, more educated, and wealthier audience (e.g., a 50-year-old college graduate with $210,000 annual personal income) (http://sales.marketwatch.com/newspaper/about/audience).
- Like a newspaper, the user of the Website is not expected to read or know everything that is on each page or category; the Website provided the user with a number of choices (e.g., article titles) in which the user will pick according to their interest.

Mobile Strategy

- None.

Primary Competitors

- Financial Times.
- CNN Money.

Tips

- How-to guides.
- Alerts and newsletter.
- Calculating tools.

Social Networking

- Link on their Website to Twitter (several Tweets a day).
- Facebook less than 10,000 followers in 2011 and RSS feed.
- Newsletters.
- Alerts.
- Video center.

SWOT Analysis

Strengths

- Credentials: part of *WSJ*.
- Very comprehensive Website.
- Free.
- Articles, news, and updates available in real time.
- Print version of a monthly magazine available.
- Video guides.
- How-to guides.
- Possibility to create your portfolio and to track your investments.

Weaknesses

- Their social networking strategy seems a little weak; no forums/blogs, for example, where users could exchange tips.
- No mobile application.
- It is not a software application but rather a Website with content.

Opportunities

- Develop a mobile app.
- Develop the social networking strategy.

Threats

- As it is part of the *WSJ*, it might become a paying app.

4.8.2.10 Wealth Management Software

ASI Wealth Manager

Who Were They in 2011?

- Investment planning platform.
- Manage household investments through balance sheet approach.
- Pricing and point of sale are unclear.
- Their latest product is called Goalgami.

What Did It Let Users Do in 2011?

- Analyze the household balance sheet and cash flows to delivering personalized advice to investors.

Web Strategy

- The demo video is the most useful tool to learn about the software. Similar to the Website and the software, the text-oriented demo design makes it difficult to follow at times. The demo is also not updated, because their latest product, Goalgami, has a totally different design.

User-Friendliness

- Clear categories and sub-categories to help user take whatever action they would like.
- Information shown clearly in a way that tells the user what actions they can take.

Website Visual Analysis

- The first image(s) that we see on the first page of this Website is the view of office buildings, skyscrapers, and people in corporate dress. This set of images implies the audience of this Website (i.e., sophisticated, investors, corporate-level) (Fig. 4.8).
- This Website relies on descriptive text, but very few visuals.
- Paragraphs and bullet points are used to describe each section and sub-section of Website in detail. However, it is very difficult to digest all the information because of the overwhelming amount of text and link on each page. When combined with the financial terms, this layout can make the Website more complicated than it should.

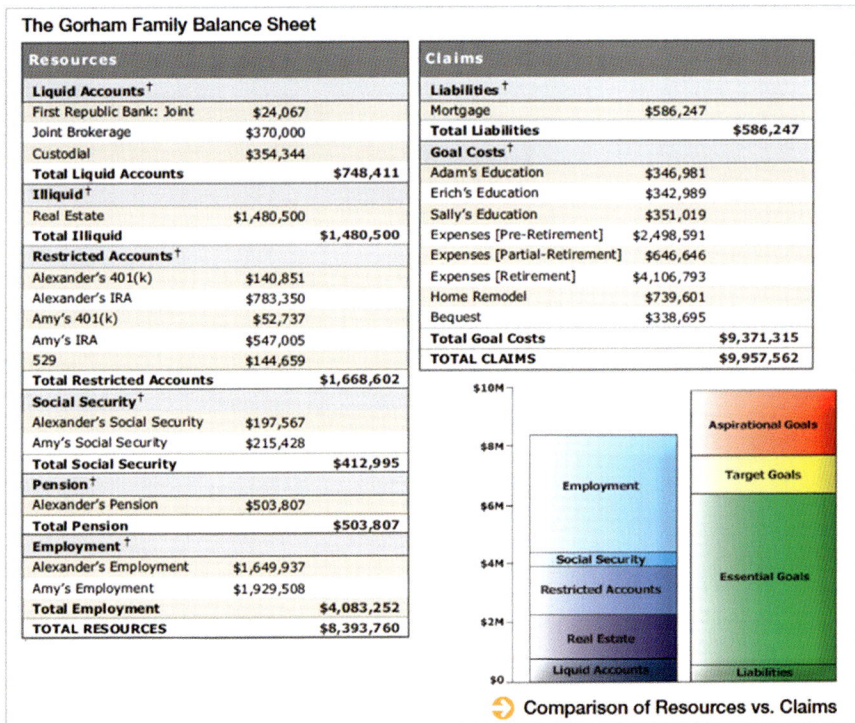

Fig. 4.8 Example Website page from ASI Wealth Manager in 2011 (Image: Advisor Software, Inc.)

- The color scheme (de-saturated, monotone colors) and the format of the Website (text-oriented, minimal visual) seemed old-fashioned and outdated; it is as if this Website (and the software) was designed in the 1990s, not recently.
- The software demo video seemed to aim towards PC users.

Mobile Strategy

- None.

Primary Competitors

- SmartMoney.
- Yodlee MoneyCentre (tracking device).
- Mint.com.

Tips

- None.

Social Networking

- No presence on Facebook.
- Presence on Twitter with 568 followers (as of 20 May 2011).

SWOT Analysis

Strengths

- Descriptive Website.
- Clear target audience.
- Is a software.
- 14-day free trial.

Weaknesses

- Not user-friendly.
- Not compatible with Apple products.
- No mobile strategy.
- Outdated design.

Opportunities

- Develop mobile app.
- Develop social media.
- Using pre-existing designs that AM+A did for them.
- Develop an Apple-friendly Website.
- Develop a user-friendly Website.

Threats

- Stepping on their original software.

4.8.2.11 Goalgami

Who Were They in 2011?

Allow user to engage in the advice process by self-exploring goals and creating their own goal plan.

A user-friendly version of ASI Wealth Manager, created by ASI Advisor software, Inc.

What Did It Let Users Do in 2011?

- Identify and organize financial resources and goals.
- Generate a personal household balance sheet.
- Measure financial resources vs. their goals over a life span.
- Calculate to what extent their resources can fund their goals.
- Explore how different life scenarios will affect their balance sheet.
- Create their own financial plan and send it directly to their financial advisor.
- Get instant feedback that informs their financial decisions and priorities.

GoalgamiPro: Workflow that allows an advisor to create a household balance sheet report for a client to augment an existing client assessment process. Designed for collaboration with the client using Goalgami (in development).

Goal-Based Proposal Solution: Workflow that allows the advisor to create a household balance sheet and risk analysis (capacity and tolerance) for a client, and map client accounts to an advisor-input model portfolio based on the risk analysis, asset allocation, and asset location (in development).

Goal-Based TAMP: Same experience as the Goal-Based Proposal Solution, but for advisors wishing to outsource the investment model creation and servicing. Sub-advisory services provided by Advisor Partners, a Registered Investment Advisor wholly owned by ASI (in development).

Web Strategy

- One of the main competitors of the Money Machine because of shared target audiences and functions.
- Today vs. retired is a very nice feature.
- Neat interactive features to make the application fun.
- What-if feature is nice for considering making important life-changing decisions.

User-Friendliness

- Clear categories for actions.

Website Visual Analysis

- The Website has a unique logo design and concept (an origami) (Fig. 4.9).
- The main communication tool came in the form of questions and answers.
- Images on the Website are symbolic (e.g., money on one side of a seesaw, with the title "Life's a balancing act. Let's get your act together."

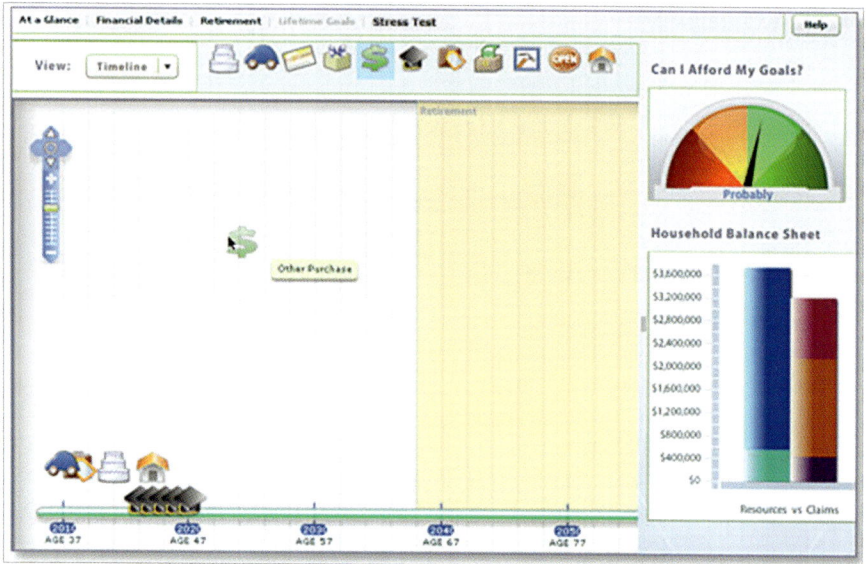

Fig. 4.9 Example Website page from Goalgami in 2011 (Image: Advisor Software, Inc.)

- User-friendly design with clear categories: What is it? Why use it? How do I use it? Community. About.
- A lot of images, including the Website title, are low resolution, which caused the Website to look amateur.
- Distinguish different plans for different age groups, genders, marital status, levels of income, and retirement status.
- Contains demo, which shows how the software works, but the demo contains no audio; text alone is difficult to follow (but much better than ASI Wealth Manager because of the improved infographics).

Mobile Strategy

- None.

Primary Competitors

- SmartMoney.
- Mint.com.

Tips

- Advice is given for specific questions in the form of FAQs.

User Reviews

- No user reviews.

Social Networking

- No presence on Facebook, but presence in Twitter, with over 1,000 followers.

SWOT Analysis

Strengths

- Different plans for various groups of audience.
- Much more user-friendly than ASI Wealth Manager.
- Unique logo and software strategies.

Weaknesses

- Still Beta-testing.
- Unfinished, unpolished.
- The link to the demo, which is the most informative tool for software instruction, is hidden in the corner.

Opportunities

- Develop a mobile app.
- Develop a community/social networking.
- Develop a more investment oriented Website.

Threats

- Very close competition to the Money Machine.

4.8.2.12 Ameritrade

Who Were They in 2011?

- American online broker.

What Did It Let Users Do in 2011?

- Common and preferred stocks, futures, ETFs, option trades, mutual funds, fixed income, margin lending, and cash management services.

Web Strategy

- Not user-friendly, but relied on the already-existing frequent user (more than 6 million US customers, and many more internationally).
- Contains platforms comparison that compares different services to match the user's needs.
- Also has TD Ameritrade UK.

User-Friendliness

- Tools and information are in well-labeled places on the Website.
- Easy to navigate and lots of resources.

Website Visual Analysis

- Green as the main color palette represents money and sustainability.
- Contained line chart and table to visualize the data; assume that the user is familiar with the stock-market graphics.
- Emphasis on its popularity and positive feedback. The main image on the homepage shows the 5-star ratings from SmartMoney with the link to another page that contains awards and accolades.
- Many categories and links that are text-oriented. Infographics are evident, but users have to go through links to get to them.

Mobile Strategy

- Free download.
- Has mobile trading apps and mobile web trading.
- Compatible to iPad, iPhone, Android, Blackberry, and Windows phone.
- In contrast to TD Ameritrade Website which relies heavily on text description, the mobile version has infographics like color coded line chart and bar chart as their main elements.

Primary Competitors

- Stockspy.
- Stock signal.
- Mastock.

Tips

Questions and concerned about anything on the Website or that the Website has addressed can be answered through phone call, email, fax, and mail.

Social Networking

- Presence on Facebook pages, with the most popular page having 334 "likes".
- Presence on Twitter with over 900 followers.

SWOT Analysis

Strengths

- Already popular: a lot of based customers.
- Very detailed.
- Excellent research, analysis and trading tools.
- Provide a lot of options for various Website usage.
- Presence on social-network.
- Good customer service with hotlines and various language options.
- Free to download mobile application.
- The mobile version is user-friendly, up-to-date, and compatible with many platforms.

Weaknesses

- Can be expensive.
- Not a lot of banking service.
- Not user-friendly Web design.

Opportunities

- Develop banking services.
- Develop tips and challenges.

Threats

- Very popular Website.
- Strong competition for the Money Machine.

4.8.2.13 Charles Schwab International

Who Were They in 2011?

- American online investment firm.

What Did It Let Users Do?

- Open or transfer an account.
- Manage investments (Common and preferred stocks, ETFs, option trades, mutual funds, bonds and fixed income, market insight).
- Perform active trading.
- Manage banking (checking, savings, mortgages, home equity).
- Receive guidance about investing.
- Estimate savings needs for a college fund or retirement.
- Have an expert manage their portfolio.

Web Strategy

- User-friendly, use professional colors and pictures of happy people.
- Phone numbers and branch locator on most of the pages for extra help.

User-Friendliness

- Tools and information are in well-labeled places on the Website.
- Easy to navigate and lots of resources.

Website Visual Analysis

- Brown and gray as the main color palettes represent professionalism and expertism.
- Contained line graph to visualize the data; assume that the user is familiar with the stock-market graphics.
- Emphasis on its new features. There are pictures of these features followed by a collection of the latest news stories of the company.

- Many categories and links that are text-oriented. Infographics are evident, but users must go through links to get to them.

Mobile Strategy

- Free download.
- Available in two languages.
- Very positive reviews.
- Compatible to iPad, iPhone, iPod touch, Mobile Web, and Android.
- Deposit checks by taking a picture.
- View transaction histories and keep an eye on accounts.
- Track investments with real-time quotes, charts, news, and more.
- Trade stocks, ETFs, mutual funds, and options.
- In contrast to the Website which relies more on text description, the mobile version still has a lot of text, but also includes more infographics like color coded line graphs and bar charts.
- Upcoming features: PIN login, pay bills, multitasking, streaming quotes, branch locator.

Primary Competitors

- E Trade.
- Merrill Lynch.
- JP Morgan.
- Goldman Sachs.

Tips

Questions and concerns about anything on the Website or that the Website addresses can be answered through phone or email.

Social Networking

- Presence on Facebook pages, with 14,205 fans.
- Presence on Twitter with 4,501 followers.
- YouTube channel with 404 subscribers.

SWOT Analysis

Strengths

- Already popular: a lot of based customers.
- Many links for advice or help that a user may be looking for.
- Excellent research, analysis, and trading tools.
- Lots of options to navigate through and use the Website.
- Presence on popular social networks.
- Good customer service with hotlines and branch locator.
- Free to download mobile application.
- The mobile version is user-friendly, up-to-date, and compatible to many platforms.

Weaknesses

- Having the company manage a portfolio or meet and give advice can be expensive.
- No current banking service on mobile (planned for the future though).
- Posting bad user reviews could have a negative impact on the company.

Opportunities

- Develop banking services on mobile.
- Develop an application for more mobile platforms.
- Work on customer service so there are less negative reviews.
- Offer free initial services to get more customers.

Threats

- Very popular Website and mobile application.
- Strong competition for the Money Machine.

4.8.3 Money Management iPhone Applications

4.8.3.1 iFinance for Mac

What Was It in 2011?

- Apple app available for iPhone, iPod Touch, iPad.
- Tool for "managing personal finances."
- Available in 12 languages.
- Size: 32.2 MB.
- Price: $29.99.
- Password protected and encrypted financial personal data.

What Did It Let Users Do in 2011?

- Overview of finances (record spending and income, categorize, charts and reports to optimize).
- Import existing data or digital bank statement. Files supported: .CSV; .QIF; .OFX; MT-940.
- Convenient data entry: recurring transaction frequency can be set up, transfers will automatically appear, foreign currency operations can be displayed in users' own currency at the current exchange rate.
- Budgeting: set up limits for monthly spending.

User-Friendliness

- Easy to read sidebars with categories.
- Tasks that the user can accomplish are clear and so are the places to find them.

Visual Analysis

- The financial information is expressed through pie charts, line charts, and a color-coding system.
- The grayish brown color scheme with texts and numbers causes the application to feel clinical and not up-to-date (especially the line-chart page) (Fig. 4.10).

Tips

- Charts and reports: interactive 3D charts/numbers for the reports.

Social Networking

- None.

Primary Competitors

- iBank.
- mPortfolio.
- Mint.

SWOT Analysis

Strengths

- Users can add their own transactions on the go.
- Inexpensive compared to other programs (such as Quicken).
- Maintenance of finances easily.
- Budgeting feature with charts and reports.

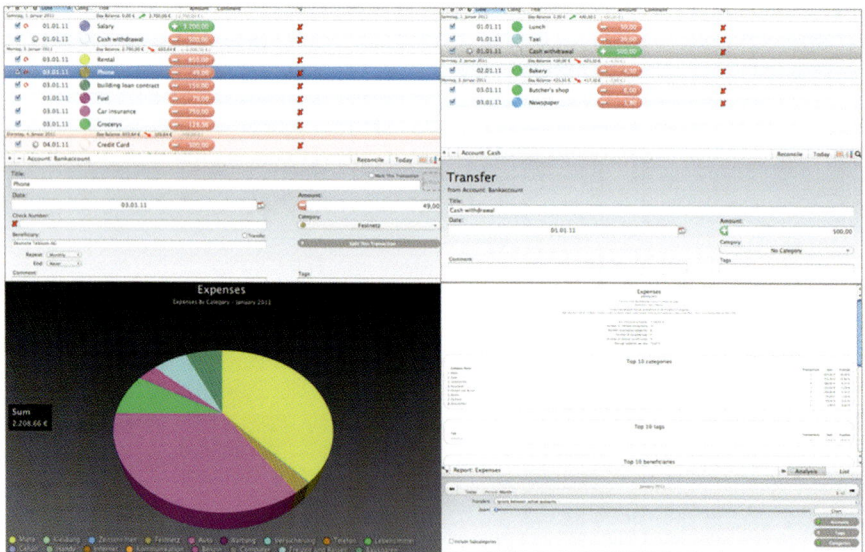

Fig. 4.10 Example Website pages from iFinance in 2011 (Images: Synium Software)

Weaknesses

- No synchronization with financial institution.
- User interface and usability are hard to understand. There is room for improvement.
- Access to investments?
- Only a budgeting tool.
- Does not register Money Market funds.

Opportunities

- Social networking.
- Develop financial investments.
- Improve user-interface.

Threats

- Price.
- Competition (Quicken).
- Security and protection of data. Moving money around.

4.8.3.2   iBank

What Was It in 2011?

- Apple app available for iPhone, iPod Touch, iPad.
- Tool for "managing personal finances."
- Available in only two languages.
- Size: 15.9 MB.
- Price: $59.99.
- Password protected.

What Did It Let Users Do in 2011?

- Full set of money management features.
- Enter and edit transactions.
- Download online account data.
- Track investments.
- Split/schedule/categorize transactions.
- Accounts for checking's/savings/credit cards/loans and investments.
- Devise specific budgets.
- Export to tax software.
- Generate, print, and save detailed reports.
- Powerful investment tracking for bonds, stocks, mutual funds, retirement accounts (IRAs or 401 k): can track buys, sells, splits, dividends, and ROI. Create an investment summary report for portfolio performance.
- Tax management feature: Categorize tax-related or tax-deductible expenses.

User-Friendliness

- Computer Software: Clear steps for user to get started.
- Computer Software: Minimal tasks in each category to help user see what they can do and what is important.
- Home screen on mobile app has icons that are not clear what they represent.
- Less visualization on mobile app, looks a lot more cluttered.

4.8.3.3   Visual Analysis

A small icon representing each category helps the user visualize the kind of information that each category contains. Other types of data are expressed through table, pie chart, line graph, bar chart, and color-coding system which are also useful in information visualization (Fig. 4.11).

Tips

- With iBank mobile (sold separately), users can enter transactions on the go.
- Online tutorials.
- User forums.
- Attach digital receipts as PDFs to any transaction.

Social Networking

- None.

Primary Competitors

- Quicken/mint app.
- iFinance.

SWOT Analysis

Strengths

- Easy to switch for Quicken to this app.
- Ability to organize all the account details from various banks into one location.
- Does everything Quicken does for less expensive.
- On-the-go.
- Fast to install.

Weaknesses

- The investment accounts section is not user-friendly.
- Doesn't work well with too many accounts.
- Chase bank is not supported.
- Cannot change sizes/ font/ colors.
- Credit card feature is not optimum.
- Bugs and crashing have been reported.

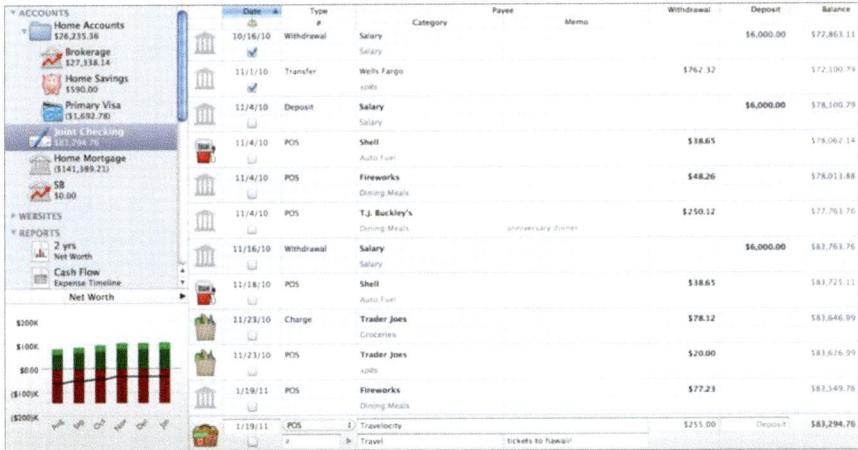

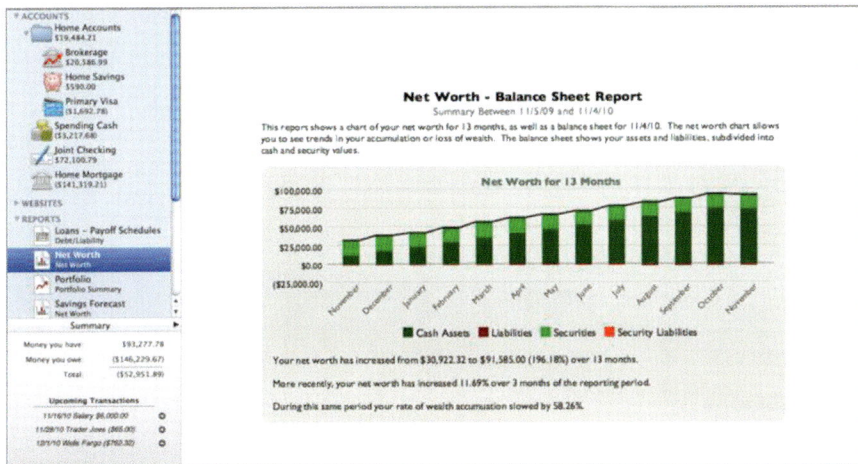

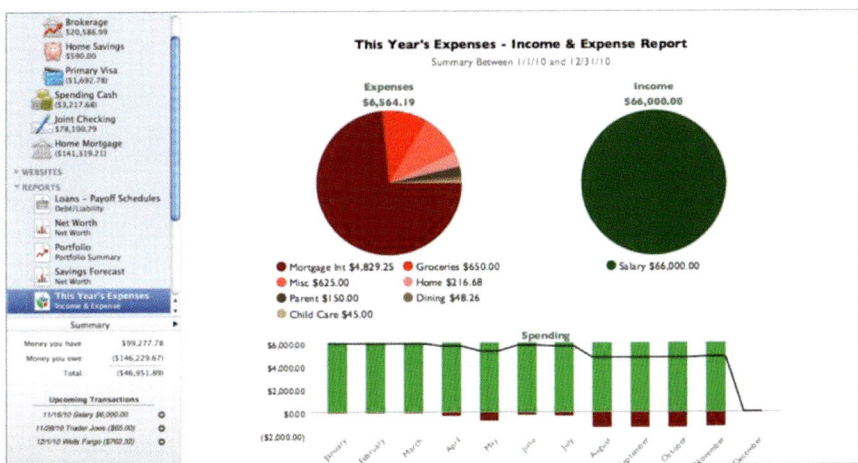

Fig. 4.11 Example Website pages from iBank in 2011 (Images: IGG Software, LLC.)

Opportunities

- Develop social networking.
- Financial investments.
- Expand the market to more banks.

Threats

- Security and date protection.
- Bad customer service.

4.8.3.4   Property Evaluator: Real Estate Investment Calculator

What Was It in 2011?

- Apple app available for iPhone, iPod Touch, iPad.
- Real estate investment software.
- Available only in English.
- Size: 1.4 MB.
- Price: $49.99.

What Did It Let Users Do in 2011?

- Information about the property.
- View performance projections.
- Comparison between properties.
- PDF of the projections.
- Add your company's logo on the cover page and header of PDFs reports.

User-Friendliness

- Lots of nice features that appear easy to find and read.
- Too many new pages for each category, there should be more in-screen popups or other ways to show the data.

Visual Analysis

- Simple table design with bar chart that represents, for example, the cost of rent over time.
- Analysis, prediction, and comparison of numbers (money) are the main element that this application has to incorporate; perhaps, there can be more emphasis on typographic treatment of the numbers other than them being red, regular, or bold (Fig. 4.12).

Tips

- None available.

Social Networking

- None.

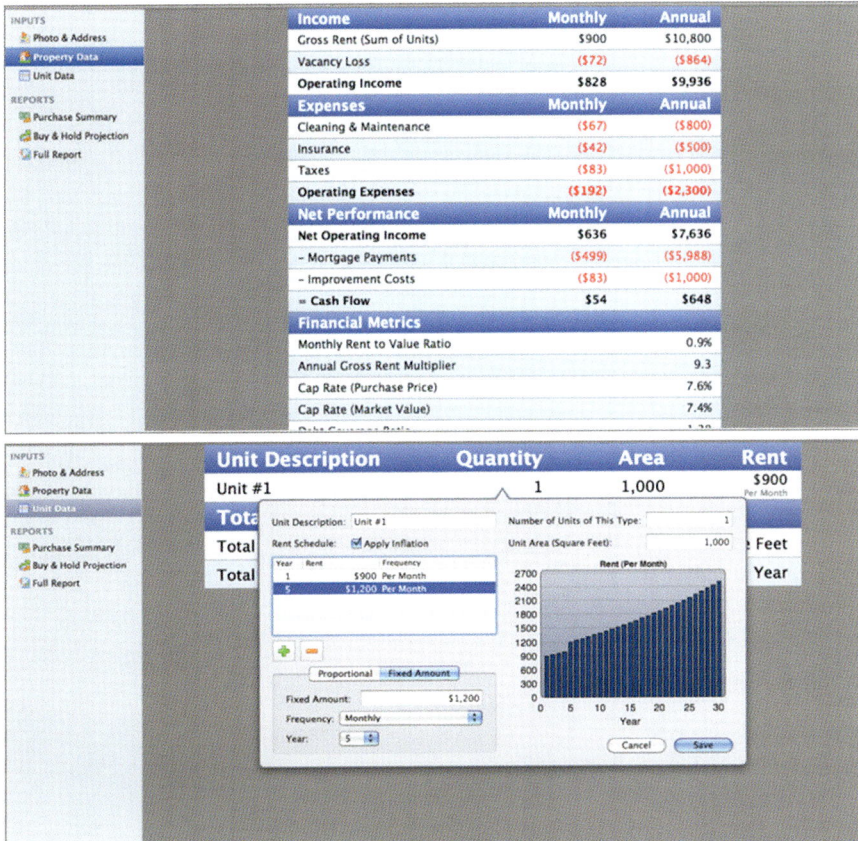

Fig. 4.12 Example Website pages for Property Evaluator in 2011 (Images: Real Estate Tools)

Primary Competitors

- iPerito evaluator: just to estimate the real estate value.

SWOT Analysis

Strengths

- Friendly interface.
- Easy to use.

Weaknesses

- Friendly to non-real estate agents?
- Very few users.
- Very specific group of user. Difficult to see the growth potential over time.

Opportunities

- Integrate this knowledge as a feature of a broader application on investments.
- Develop tips and challenges.
- Develop social networking strategy.
- Add more languages.

Threats

- Language might be too "real-estate" oriented.
- How to integrate it for non-real-estate specialists.

4.8.3.5   mPortfolio

What Was It in 2011?

- Apple app available for iPhone, iPod Touch, iPad.
- Creating/managing and tracking stock portfolios.
- Available only in English.
- Size: 9.5 MB.
- Price: $24.99.

What Did It Let Users Do?

- Unlimited investment portfolio tracking.
- Watch stock list so users can keep updated on stock performance.
- Automatic price updates from free Internet Website.
- Real time RSS feed.
- Charting and stock performance visualization.

User-Friendliness

- Updates shown oddly; just plaintext, could be visualized nicer.
- Looks clear on how to use but there do not seem to be many features.

Visual Analysis

- The application seemed one-dimensional; there's only one main page that oper-ates the whole time.
- There is more focus on functionality, but not the esthetic or the user-friendly aspects of this application.
- The design seemed to be useful to those who are familiar with numbers and typography (format) used in stock market. However, it does not encourage the new users who are not familiar to stock market format to use this application.
- The bottom left corner of the page provides superfluous content and advertise-ments that have no connection to the application (Fig. 4.13).

Tips

- None available.

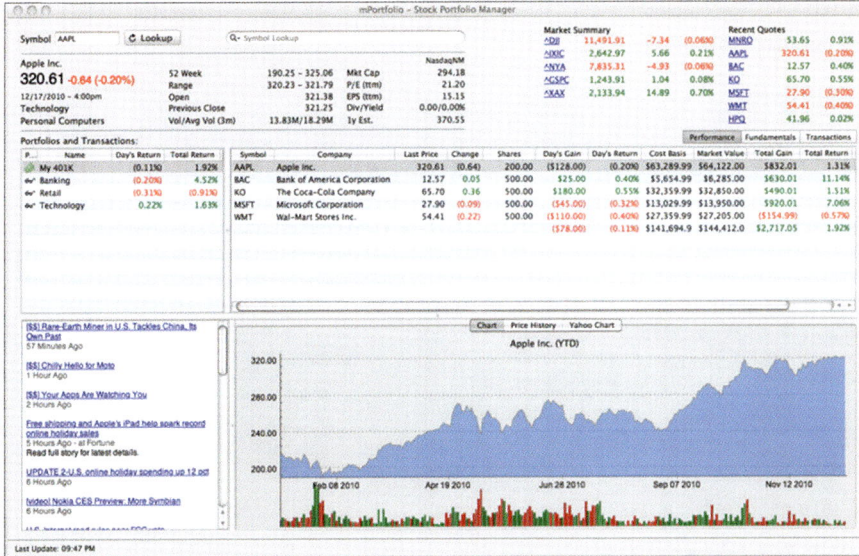

Fig. 4.13 Example Website page for mPortfolio in 2011 (Images: Meg Software)

Social Networking

- None.

Primary Competitors in 2011

- Investoscope.
- Stock-spy.
- Stock signal for Mac.
- iFinance.

SWOT Analysis

Strengths

- Good tracking device.
- Good for small investors.
- Easy to enter information.

Weaknesses

- Doesn't handle cash.
- Doesn't handle 50+ stocks.
- Does not delete a stock.
- Limited functionality.

Opportunities

- Develop heavy investments.
- Develop a more user-friendly interface.

- More languages available.
- Handle multiple portfolio.

Threats

- Security and data protection.
- Too much competition and similar apps available for less expense.

4.8.3.6   Investoscope

What Was It in 2011?

- Apple app available for iPhone, iPod Touch, iPad.
- Portfolio tracker for individual investors.
- Available only in English.
- Size: 4.7 MB.
- Price: $59.99.

What Did It Let Users Do in 2011?

- Track unlimited number of portfolio in any currency.
- Automatically downloads latest quotes.
- Users can attach RSS feeds to their stocks, funds, and other investments.
- Several accounting methods.
- Portfolio charts and reports including capital gain income and performance reports.
- Import transactions from OFX and CSV files.
- Support most world market including North America, Europe, Australia, and South America.

User-Friendliness

- Computer software has sidebar with categories, good visual aids, and direction on how to look at different items.
- Hard to determine mobile application because no demos available.

Visual Analysis

- Contained pie charts, tables, and line graphs as the main information visualization tools.
- While the monotone gray can be interpreted as seriousness and sophistication, it lacks visual punch that makes this application memorable; the gray becomes boring and bland (Fig. 4.14).

Tips

- None.

Social Networking

- None.

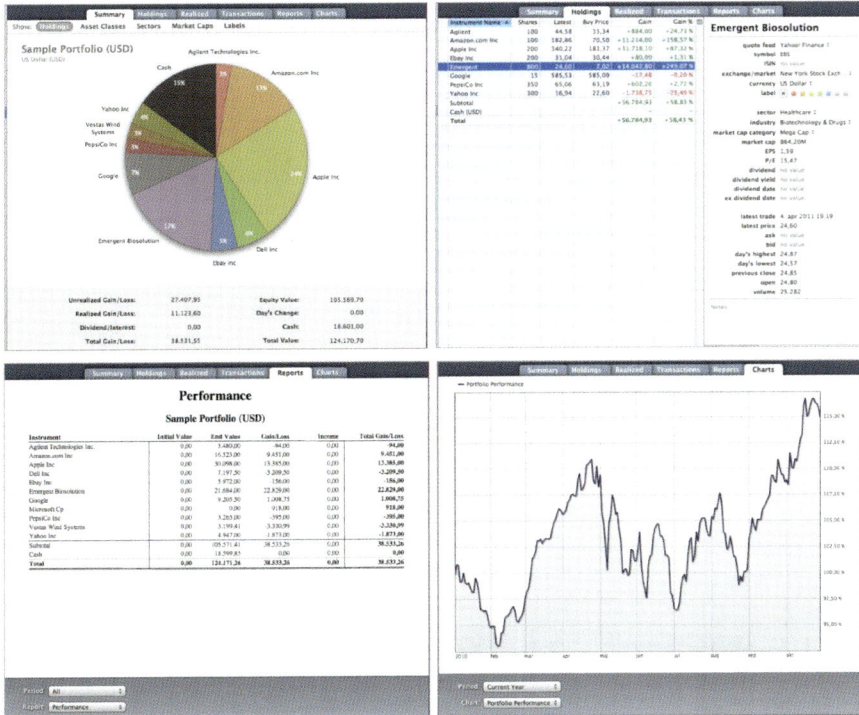

Fig. 4.14 Example Website pages from Investoscope in 2011 (Images: Investoscope Software)

Primary Competitors

- mPortfolio.
- Stock spy.
- MaStock.

SWOT Analysis

Strengths

- Highly customizable.
- Allows to track portfolio easily.
- News feed.
- Completely manual.
- Quick updates.

Weaknesses

- No heavy investments.
- No tips on what and where to finance.
- No portfolio analysis available.
- Customer support is inefficient.

- No bank access.
- Limited features.

Opportunities

- Develop heavy investment.
- Networking strategy.
- Life cycle strategies.
- Wider market.
- Portfolio analysis.
- Tips and challenges.

Threats

- Too manual: people might get bored.
- Customer service.
- Protection of data.
- Price.

4.8.3.7   Mastock

What Was It in 2011?

- Apple app available for iPhone, iPod Touch, iPad.
- Stock market analysis and portfolio management software.
- Available only in English and French.
- Size: 22.5 MB.
- Price: $49.99.

What Did It Let Users Do in 2011?

- Upload latest quote from internet.
- Locally stored historical data.
- Technical analysis.
- Manage quotes uploaded from files.
- Standard indicators (Ichimoku and Points & Figures) on a daily, weekly, monthly, quarterly, or yearly basis.
- Ability to define trend lines, Fibonacci retracements, Fibo lines, Elliott waves.
- Semi-log or linear cycle.
- Perform Japanese candlesticks analysis.
- Navigate through historical data.
- Automatically generate stock split.
- Research and modify data in a spreadsheet.
- Embedded RSS news reader.

User-Friendliness

- Too hard to tell from screenshots how the userflow would work, no available demos.

Visual Analysis

- The main visual element of the application is the overlap sets of line graphs. These graphs are sometimes illegible when there are a number of overlaps in one spot.
- There is a room for design improvement; the graphics of these lines seemed raw and not well-thought out (i.e., to make it easier to read and more esthetically pleasing) (Fig. 4.15).

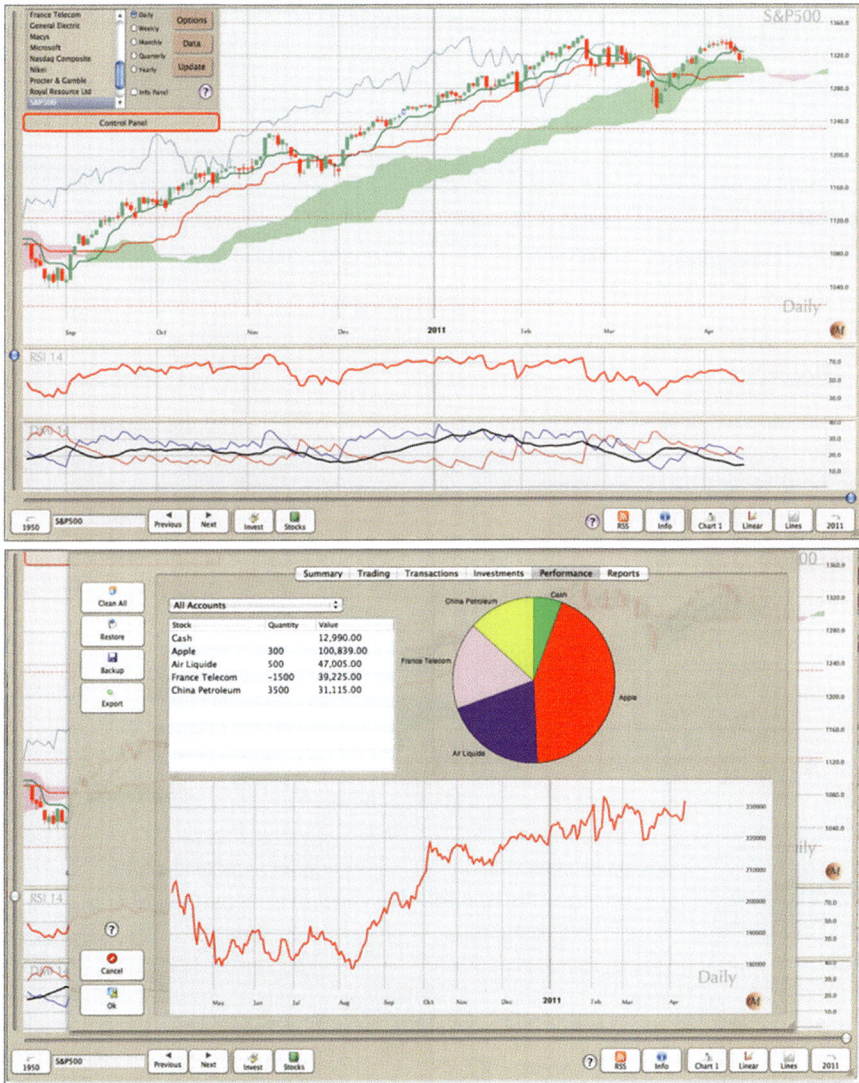

Fig. 4.15 Example Website screenshots of Mastock in 2011 (Images: Michael Montagne)

Tips

- None.

Social Networking

- None.

Primary Competitors

- Stockspy.
- Stock signal.

SWOT Analysis

Strengths

- Looks like a strong and useful tool.
- Friendly interface from the screenshots available.

Weaknesses

- Not enough credit.
- No users/no reviews.
- Expensive.

Opportunities

- Develop a "yahoo finance" stock charts analysis: 20, 50, and 200 days interval.
- Develop a forecasting tool for stocks.
- Develop networking strategy.
- Develop tips and challenges.

Threats

- Security and data protection.
- Price.
- Competition.

4.9   Results

From our investigations of about 20 wealth management applications for mobile, Web, and desktop, including those cited above, AM+A concluded that usable, useful, and appealing UI design must include incentives in order to prompt behavior change.

The proposed Money Machine needed, also, to incorporate persuasion theory. As noted in Chapter 1, persuasion theory is a discipline that has evolved over the years in relation to marketing and technology developments. AM+A incorporated the work of B. J. Fogg and Robert Cialdini into the Memory Machine as discussed below.

This powerful combination motivates users to change their behavior in the short and long term, and to achieve healthier, more financially sound lifestyles. A good wealth management application for mobile devices such as smartphones and tablets should help users set goals, provide dynamic charts and illustrations, host competitions, and provide step-by-step instructions to motivate behavior change.

Extensive, up-to-date, searchable databases are another priority. These databases must be sufficiently user-friendly and flexible for users and their network of family and friends to add and update information easily. This adaptability is critical to maintaining and increasing usage, and will inevitably provide a competitive advantage for the Money Machine. Similarly, the Money Machine must liberate users from traditional (and cumbersome) manual data entry, furnishing a varied data-entry system (e.g., document scanning, database searching) to facilitate the data entry process.

The Money Machine must also encourage and strengthen team-oriented behavior change. Studies of persuasion theory show that social comparison is a superior incentive for behavior change in general. For finance in particular, co-operation and competition within and among teams can encourage greater restraint and financial control. Virtual rewards (e.g., "star" designation, a new skin for the application) provide strong motivation, and financial or other awards can prompt a significant change (Wang 2014) in behavior.

Last but not least: the Money Machine should be fun to use. Gamification will serve as further incentive for users to learn about selecting wise investment/expenditure combinations, controlling risk efficiently and effectively, among other fiscal techniques. (See Marcus 2011). The Money Machine should allow users to share these experiences with friends, family members, and the world, primarily through Facebook, Twitter, and blogs.

Based on these concepts and available research documents, we proposed and continue ongoing development of conceptual designs for the multiple functions of the AM+A Money Machine. Subsequent evaluation will help us improve the metaphors, mental model, navigation, interaction, and appearance of all functions and data in the Money Machine's user interface. The resulting improved user experience will move the Money Machine closer to being a commercially viable product and service.

A well-designed Money Machine will be more usable, useful, and appealing to money-conscious users in comparison with current products/services, which tend to be too narrowly focused, especially those users having challenges with expenses and/or savings. Our objective is to provide a set of functions in an application that can reliably stimulate people toward healthier financial lifestyles, with consequent benefits not only to their own economy, but also to those around them.

4.9.1 Persuasion Theory

In alignment with Fogg's persuasion theory (Fogg and Eckles 2007), we defined five key processes to create behavioral change via the Money Machine's functions and data:

- Increase frequency of using application.
- Motivate changing some living habits: save, plan, invest.
- Teach how to change living habits.
- Persuade users to plan short-term change.
- Persuade users to plan long-term change.

Each step has requirements for the application.

Motivation is a need, want, interest, or desire that propels someone in a certain direction. Humans' sociobiological instinct is to maximize reproductive success and ensure the future of descendants. We apply this theory in the Money Machine by prompting users to understand that every action has consequences on their current and future economic condition.

We also drew from Maslow's *A Theory of Human Motivation* (Maslow 1943), which he based on his analysis of fundamental human needs. We adapted these to the Money Machine context:

- The *safety and security need* is met by the possibility to visualize the amount of expenses saved.
- The *belonging and love need* is expressed through friends, family, and social sharing and support.
- The *esteem need* is satisfied by social comparisons that display improvement in financial control and skill, as well as by self-challenges that display goal accomplishment.
- The *self-actualization need* is fulfilled by the ability to visualize improvement of financial indexes and mood, and to predict change in users' future economic scenarios.

4.10 Impact on Information Architecture

4.10.1 Increase Use Frequency

Games and rewards are among the most common methods to increase use frequency. We have developed several financial game concepts for the Money Machine. In terms of rewards, the users might be awarded both virtual rewards (such as "star" designations and new skins for blogs) as well as real money rewards.

In addition, we chose social comparison as another incentive to increase use frequency. Users can form groups with family and friends, and participate in various group competitions. Although these competitions are based around financial controls and exercises, users do not have to reveal personal financial data, which most prefer to keep securely private.

4.10.2   Increase Motivation

In the Money Machine, we set the users' potential financial conditions as an important incentive for their behavior change. Viewing their current versus predicted economic status over the next 20–30 years gives users greater understanding of the strengths and weaknesses to their financial strategy.

Because setting goals improves learning outcomes and provides quantitative performance data, the Money Machine asks users to set time-based goals for budget reduction (e.g., saving on consumer items or maintenance services, both required and optional), savings, and retirement. Users receive suggested step-by-step plans of action in order to achieve each goal.

In addition, we created the concept of 10 challenges for each month. In meeting these challenges, users are making short-term financial behavior change that will generate long-term positive impacts.

Social interaction is another strong motivator of behavior change, both through community support and informal competition or comparison. The Money Machine leverages social networking by integrating features found in forums and on blogs, Facebook, and Twitter. Users can send messages to and share ideas with their social groups. These social ties serve as additional incentive to motivate behavior change.

4.10.3   Improve Learning

For many people with significant financial challenges, understanding long-term wealth management is crucial. To improve learning, the Money Machine integrates contextual tips on the following topics:

- Consuming more wisely.
- Increasing financial control.
- Tackling complications associated with debt and poor investments.
- Coping with principle burn rates that are either too high or too low.

Users can also choose to receive updates on latest research articles and news about wealth management.

We seek to make the education process entertaining as well as informative. Proposed games teach users to choose the right proportion of investments, amount and type of each instrument, etc. Through playing games featuring educational information, users learn to manage their wealth like a skilled financial advisor, without getting bored.

4.11   Information Architecture

Figure 4.16 shows the information architecture we designed. The following sections discuss individual components of the information architecture.

4.11.1   Components of Information Architecture

4.11.1.1   My Use (Accounts)

Main Function

- Enables multiple data-entry.
- Displays user's financial records and budgets in tables and trend charts.

Budgets

- Records amount of money allocated for specific categories.
- Calculates and displays money used so far in each category compared to goal amounts of money.

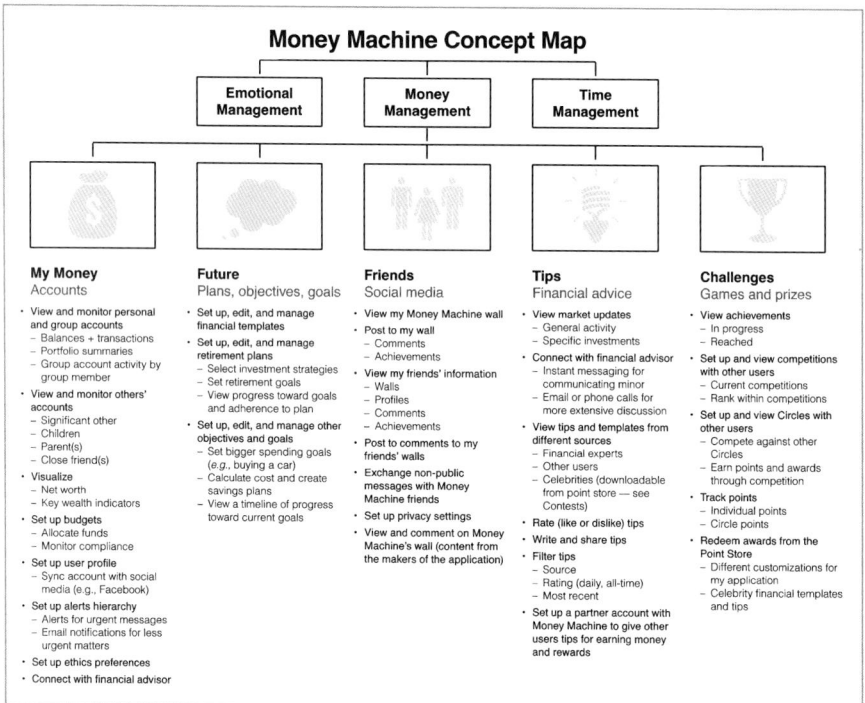

Fig. 4.16 Money Machine Concept Map or Information Architecture showing five key components for Money Management

- Provides ability to see graphs comparing past budgets and more detailed transactions.

Personal Accounts

- Displays user's transactions and changes in personal accounts.
- Provides ability to see graphs comparing past personal account transactions and changes.

Group Accounts

- Displays transactions and changes in group accounts by each group member.
- Provides ability to see graphs comparing past group account transactions and changes.

Net Worth

- Displays overall net worth.
- Provides ability to see more detailed expansion of contributing factors.
- Provides ability to see graphs comparing past new worth.

Financial Advisor

- Enables quick communication between user and financial advisor.

Settings

- Enables user to set notification settings. Notifications are shown in email and on the landing page. They are less urgent than alerts and the user checks them on their own time.
- Enables user to set alert settings. Alerts are important to know about right away and are received via text or through red exclamation points on top of the Money Machine application icon. Examples of alerts are: overspending, investment increase/decrease, account monitoring.
- Enables user to set any other related preferences for subcategories in "my use."

4.11.1.2   My Future (Plans and Goals)

Financial Templates

- Displays collection of financial templates.
- Alerts when plans have changed based on assets changing.
- Enables user to fill out a new financial template or view an existing, already filled out financial template.

Goal Setting

- Records amount of money and time allocated for specific goals.
- Calculates and displays money put so far towards each goal compared to total amount needed.

- Provides ability to see graphs of money/time for goals or a more detailed textual view.

Retirement Planning

- Displays collection of retirement planning calculators.
- Alerts when retirement plans have changed based on assets changing.
- Enables user to fill out a new retirement calculator or view/edit an existing, already filled out retirement calculator.

My Friends (Social Media)

- View and make social media achievement postings

Announcements

- Displays announcements of posts from friends on the Money Machine.
- Enables user to post own updates.
- Enables user to comment on posts or give them a "thumbs up."

Profile

- Displays user's profile.
- Includes user's picture, comments from friends, and updates from user.

Friends

- Displays list of friends in the Money Machine.
- Displays privacy settings next to each friend.
- Allows search for a friend, or clicking on a friend to go to their profile.

Money Machine Wall

- Displays the main Money Machine profile/wall.
- Includes updates and tips from the Money Machine company.
- Allows comments from users of the app.

4.11.2   My Tips (Financial Advice)

Market Updates

- Displays updates on the financial market.
- Displays investments based on user interest.

Tips

- Displays feed of tips from experts, celebrities, and other users.
- Displays rating on tips and allows users to pick "thumbs up" or "thumbs down."
- Allows filtering based on popularity, type of person posting the tip, time of posting.

- Allows search for tips on specific subjects.

Money Machine Partnership

- Displays information about forming a partnership with the Money Machine company.
- If user is already in partnership, displays updates on the partnership and opportunities to each more.

4.11.2.1   Contests (Incentives)

Achievements Collection

- Displays pictures of achievements.
- Displays progress on each achievement through a progress bar.
- Enables clicking on picture to receive more information about a specific achievement.

Competitions

- Displays current competitions and progress of each user involved.
- Alerts about changes of user ahead in competition.
- Allows starting of new competitions.

Point Store

- Displays featured objects purchasable with points.
- Displays list of all objects and associated cost in points.
- Allows purchasing of objects with accumulated Money Machine points.

Circles

- Displays list of circles.
- Displays user's joined circles.
- Allows looking at specific pages for already joined circles.

4.12   Designs

4.12.1   Initial Sketches and Revised Designs

AM+A sketched metaphor concepts, information architecture concepts, and specific screen design concepts before rendering detailed diagrams and screens. The following are representative sketches (Fig. 4.17).

Website AM+A analyzed and designed a prototype Website for the Money Machine. The Website helps facilitate data entry during set-up (Fig. 4.18).

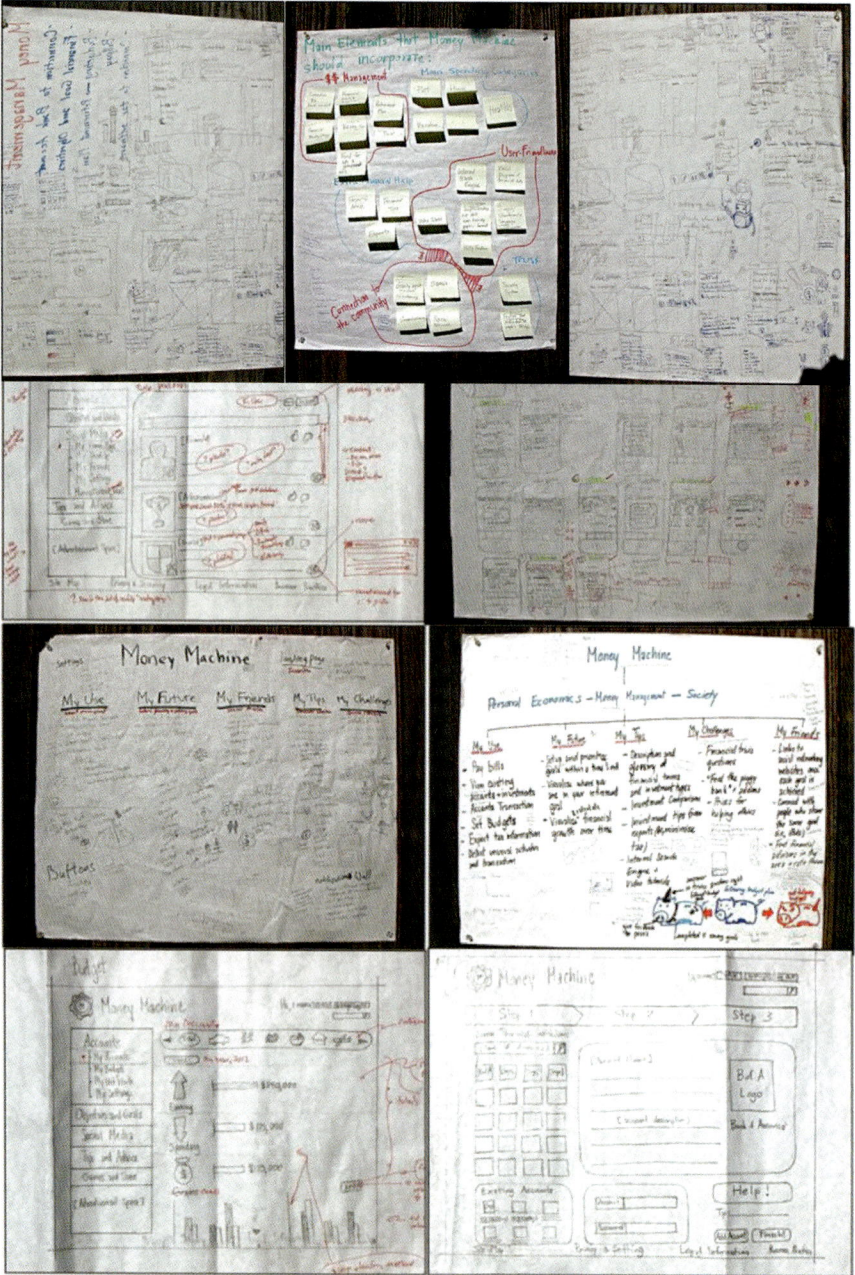

Fig. 4.17 Initial Sketches of Money Machine Screens

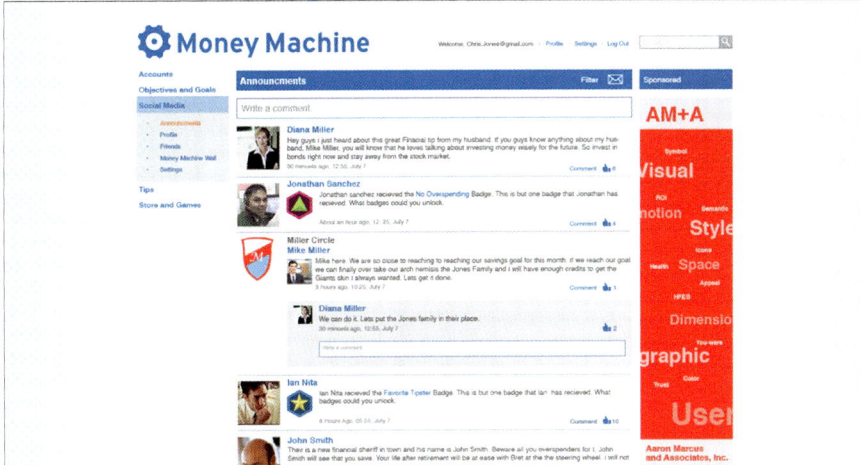

Fig. 4.18 Initial design for the Money Machine Website to accompany the mobile app

4.12.2 Examples of Initial Sketches and Revised Screen Designs

AM+A prepared initial concept sketches based on the information architecture. We used these key Money Machine screens to refine our initial sketches into thorough visual UI design. Below, the initial concept sketches are shown at left, with the refined iterations at right.

4.12.3 Landing Screen

User-selected priority information is displayed at the top of the screen, with arrows for scrolling at either side. The bottom half of the screen displays market updates as well as social networking and competition notifications.

4.12.4 My Money/Submenu

This My Money screen typifies submenus throughout the Money Machine (Fig. 4.19).

Fig. 4.19 Money Machine submenu design screens

4.12.5   My Money/Budgets

User-filtered budget categories are displayed at the top. "Total" displays total earnings, expenses, and savings for a given timeframe (weekly, monthly, yearly, lifetime, etc.). Users can view specific budgets by clicking on the associated icon above (Fig. 4.20).

4.12.6   My Money/Retirement Planner

This sample chart illustrates users' current assets, allowing them to view a their retirement plans graphically and create a more intuitive understanding of factors like risk, savings rates, and tax management (Fig. 4.21).

4.12.7   My Money/Group Accounts

Group accounts lets users manage shared funds easily. Selecting a group account member from the top bar displays the total money in the account, versus that member's contributions and expenses for a given timeframe (weekly, monthly, yearly, lifetime, etc.) (Fig. 4.22).

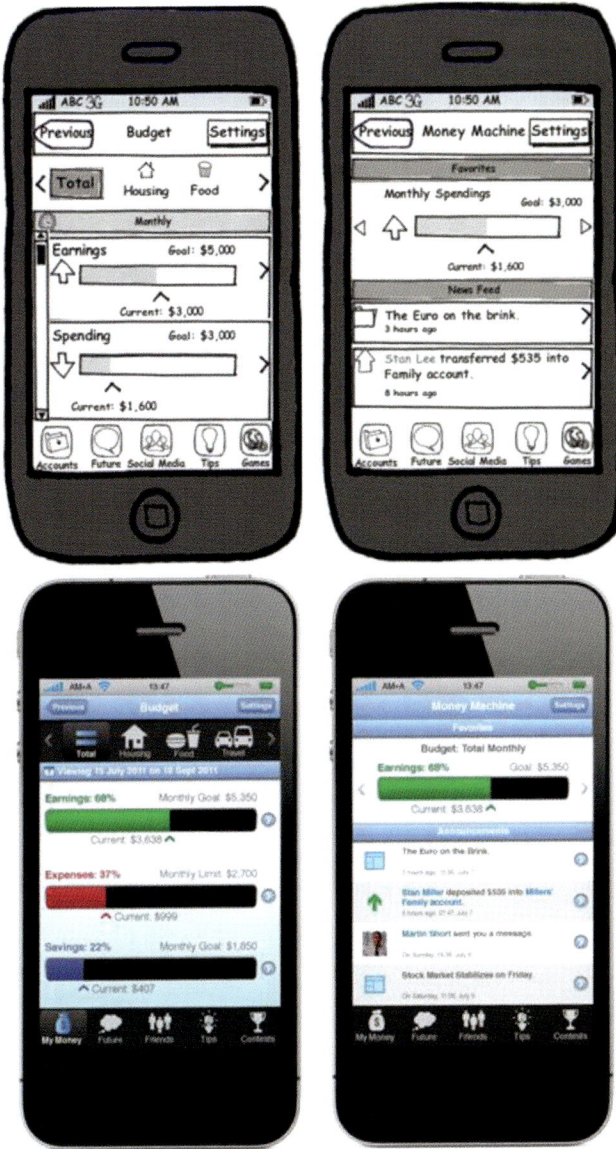

Fig. 4.20 Money Machine Budget screens

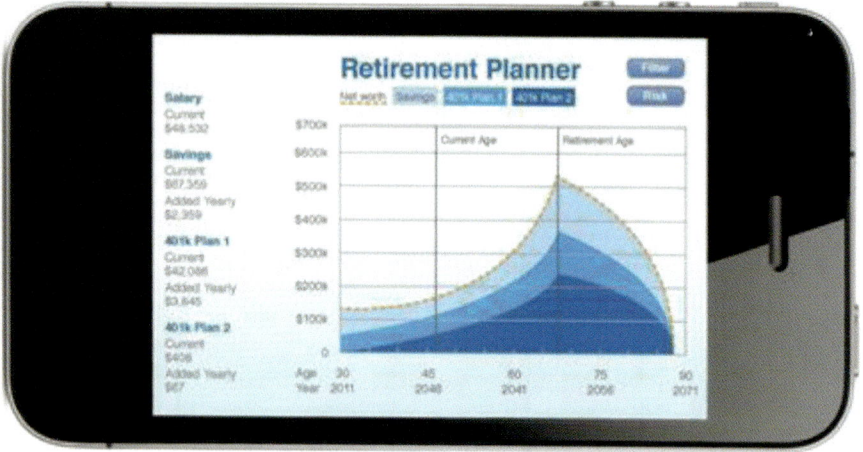

Fig. 4.21   Retirement Planner screen design

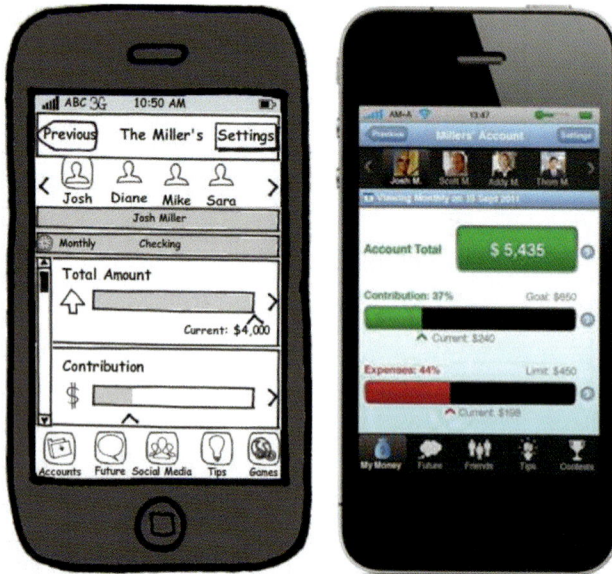

Fig. 4.22   Group Accounts screen designs

Fig. 4.23 Financial Goal screen designs

Under the heading of Future, users set financial goals (acquisition, savings, etc.), each of which has an associated status bar showing the users' progress toward their goal. Users can access further details, such as expected date of accomplishment, by clicking on the arrow at right (Fig. 4.23).

4.12.8 Friends | Main Page

This screen displays a search-enabled list of users' friends on the Money Machine. Users click on a friend's name or icon to view a complete profile. A privacy indicator sits next to each friend, representing that friend's access to sensitive financial information (red signifies zero access, yellow signifies limited access, and red signifies complete access) (Fig. 4.24).

4.12.9 Friends | Announcements

Announcements falls under the heading of Friends, the Money Machine's social media network. This is a list of posts from friends and circles on the application. The user can give posts a thumbs up. The posts can be filtered by person or subject through clicking on the filter button. The inbox button will take the user to his/her messages (Fig. 4.25).

Fig. 4.24 Friends main page screen designs

Fig. 4.25 Friend News Feed/Announcements screen designs

4.12.10 Tips | Main page

This is a news feed of financial tips from users on the application. Tips can be filtered by person or subject through clicking the filter button. Users can write their own tip by clicking on the "write" button. Users can also search through the tips. Each tip can be rated with a thumbs up or down, and more detail about a tip can be seen by clicking on the button to the right of it (Fig. 4.26).

4.12.11 Contests | Achievements (Incentives)

This is the wall of the user's achievements. Each achievement is represented by a badge, and the progress toward it is shown in a bar underneath the badge. Achievements are earned through various methods that show the user is improving financially (Fig. 4.27).

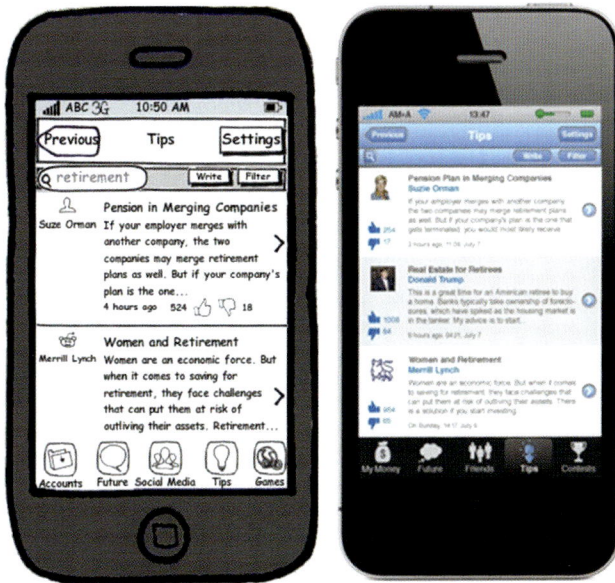

Fig. 4.26 Friend Tips screen designs

Fig. 4.27 Achievements screen designs

4.13 Evaluation

AM+A planned to interview approximately ten people (two per persona) and approximately two financial advisors with expertise in wealth management behavior change. Through these interviews, we hoped to learn what works, what doesn't, and how to refine the Money Machine prototype further, from information architecture (metaphors, mental model, and navigation) through look and feel (appearance and interaction). At the present time, AM+A not able to complete such evaluations.

4.14 Next Steps and Conclusions

Using the user-centered design approach described above, AM+A plans to continue to improve the complete Money Machine development process, which would require significant time and funding from an outside source. Tasks include the following:

- Revising personas and use scenarios.
- Conducting user evaluations.

- Revising information architecture and look and feel.
- Building initial working prototype (e.g., for iPhone, or other tools, platforms).
- Evaluating the Money Machine across different cultures.
- Developing the Money Machine for business use as well as personal use.
- Researching and developing improved information visualizations.

AM+A aimed to incorporate information design and persuasion theory for behavior change into a mobile phone application. This self-funded preliminary effort on the Money Machine project is current and ongoing, and was undertaken to demonstrate the direction and process for such products and services. Though the design is incomplete, AM+A is willing to share the approach and lessons learned, in the interest of helping alleviate worldwide wealth-management challenges. AM+A continues to seek to persuade other design, education, and financial groups to consider similar development objectives. We hope that our process and concept prototypes (the self-funding of which inevitably limited the amount of research, design, and evaluation) will inspire others, and that they benefit from the materials provided thus far.

The process has already been demonstrated successfully with a previous project, the Green Machine (Marcus and Jean 2009), versions of which have been considered and used by SAP for enterprise software development (Marcus et al. 2011, 2014).

AM+A's long-term objective for the Money Machine is to create a functional working prototype to test whether the application can actually persuade people with wealth-management challenges to exercise greater fiscal control, increase healthier wealth-management habits, and pursue a more financially sound lifestyle under real use conditions over the long term. If the theories are proven to be correct, this approach could have significant wealth-management implications that will benefit millions of people in the USA and abroad.

Acknowledgments The author thanks his AM+A Associates, Designer/Analysts: Hélène Savvidis, Catherine Isaacs, Chris Chambers, Carlene O'Keefe, and Tim Thianthai, for their significant assistance in planning, research, design, analysis, and documentation of the Money Machine. The author also thanks Lucas Lima, AM+A Designer/Analyst for his persona drawings.

The author acknowledges subsequent publications about the Money Machine:

Marcus A (2012) The money machine: helping Baby Boomers Retire. User Experience Magazine, 11:2, pp 24–27

Marcus A (2013) The money machine. Inf Des J 20(3):228–246

Further Reading

AARP Bulletin, July–Aug. 2011, 52nd ed., sec. 6. Print.
Ameritrade. Trade Architect. Web. http://www.tdameritrade.com/tradingtools/tradearchitect.html
AVA Book (2005) Designing for small screens. AVA Publishing SA

Budvietas R (2011) Baby Boomers use social media differently than generation X and Y, 12 Jan 2011. Web. http://www.sexysocialmedia.com/baby-boomers-use-social-media-differently-than-generation-x-and-y/

Carracher J (2011) How Baby Boomers are embracing digital media, 6 Apr 2011. Web. http://mashable.com/2011/04/06/baby-boomers-digital-media/

Cialdini RB (2001) The science of Persuasion. Sci Am 284:76–81

Coughlin J (2011) Why US Baby Boomers are slow to buy smart phones: Technophobes or value buyers?, 13 Apr 2011. Web. http://bigthink.com/ideas/38209

Credit Loan. Too big to fail: inside America's economic downfall. Web. http://www.creditloan.com/blog/2011/06/28/too-big-to-fail-inside-americas-economic-downfall/

Fidelity (2010) Fidelity enhances iPhone and iPod touch mobile app with access to workplace retirement plans, 22 June 2010. Web. http://www.fidelity.com/inside-fidelity/employer-services/fidelity-enhance-iphone-and-ipod-touch-mobile-app

Hartman B (2011) Retirement plan – alarming statistics of middle-income Baby Boomers postponing retirement reported, 2 June 2011. Web. http://starglobaltribune.com/2011/retirement-plan-alarming-statistics-of-middle-income-baby-boomers-postponing-retirement-reported-9202

Holiday Touch. Video games: no longer just for kids. Web. http://www.holidaytouch.com/Retirement-101/senior-living-articles/activities-and-lifestyle/video-games-not-just-for-kids.aspx

Junior Achievement. Junior Achievement Programs. Web. http://www.ja.org/Programs/programs.shtml

Lampa J (2009) 10 ways financial advisors can attract Baby Boomers in 2010, 25 Nov 2009. Web. http://www.producersweb.com/r/pwebmc/d/contentFocus/?pcID=99aebaf67fa527fec8cffb4286c7aa70

Jones M, Gary M (2006) Mobile interaction design. Wiley, Hoboken

Juha L, Antti A, Pertti H, Ilkka S (2007) Personal content experience, managing digital life in the mobile age. Wiley, Hoboken

Lighthouse International. Big type is best for aging Baby Boomers. Web. http://www.lighthouse.org/accessibility/design/accessible-print-design/big-type/

Merrill Lynch. What matters most to you? Web. http://www.totalmerrill.com/TotalMerrill/pages/WhatMattersMostToYou.aspx

Norman J (2010) Boomers joining social media at record rate, 16 Nov 2010. Web. http://www.cbsnews.com/stories/2010/11/15/national/main7055992.shtml

Orlov L (2010) AARP: Baby Boomers are not comfortable with the internet – really? 10 June 2010. Web. http://www.ageinplacetech.com/blog/aarp-baby-boomers-are-not-comfortable-internet-really

Perez S (2010) Boomers slowly joining the mobile web, 4 Mar 2010. Web. http://www.readwriteweb.com/archives/boomers_slowly_joining_the_mobile_web.php

Rao L (2011) Flurry: time spent on mobile apps has surpassed web browsing, 20 June 2011. Web. http://techcrunch.com/2011/06/20/flurry-time-spent-on-mobile-apps-has-surpassed-web-browsing/

Samsung Electronics (2010) Persuasive money life. SAIT, Computer Science Lab, Future Experience Part, Toronto

Silverman K (2011) Medical Monday: do Baby Boomers use social media? 7 Feb 2011. Web. http://blog.ogilvypr.com/2011/02/medical-monday-do-baby-boomers-use-social-media-this-past-weekend-i-went-with-my-family-to-my-great-aunt%E2%80%99s-96th-birthday-the-whole-family-met-in-long-island-coming-in-from-other-parts-of/

Simlinger P (2011) Editor. Inf Des J. Special Issue on Financial Information 19(3):203–279

Six J (2010) Designing for senior citizens/Organizing your work schedule, 17 May 2010. Web. http://www.uxmatters.com/mt/archives/2010/05/designing-for-senior-citizens-organizing-your-work-schedule.php

University of Maryland. Visualization. Web. http://www.cs.umd.edu/hcil/research/visualization.shtml

Usability Professionals' Association. Web. http://www.upassoc.org/

Weiss S (2002) Handheld usability. Wiley, Hoboken

References

AARP (2010) Approaching 65: a survey of Baby Boomers turning 65 years old. AARP, Washington

Fogg BJ, Eckles D (2007) Mobile persuasion: 20 perspectives on the future of behavior change. Persuasive Technology Lab, Stanford University, Palo Alto

Marcus A (2011) Gaming the user experience. User Exp Mag 10(4):32

Marcus A (2012) The money machine: helping Baby Boomers retire. User Exp Mag 11(2):24–27

Marcus A (2013) The money machine. Inf Des J 20(3):228–246

Marcus A, Jean J (2009) Going green at home: the green machine. Inf Des J 17(3):233–243

Marcus A, Dumpert J, Wigham L (2011) User-experience for personal sustainability software: determining design philosophy and principles. In Proceedings of design, user experience, and usability conference 2011, Orlando, FL, August 2011. Theory, Methods, Tools and Practice. Lecture Notes in Computer Science, vol 6769, 2011, pp 172–177. Springer Publishers, New York

Marcus A, Dumpert J, Wigham L (2014) User-experience for personal sustainability software: applying design philosophy and principles. In: Proceedings of design, user experience, and usability conference 2014, Iraklion, Crete, Greece, June 2014. User Experience Design for Everyday Life Applications and Services. Lecture Notes in Computer Science, vol 8519, 2014, pp 583–593. Springer Publishers, New York

Maslow AH (1943) A theory of human motivation. Psychol Rev 50:370–396

Newman R (2010) Why Baby Boomers are bummed out, 29 Dec 2010. US News and World Report

Social Security Administration (2011) Fast facts & figures about social security, 2011. SSA Publication No. 13–11785. SSA Office of Retirement and Disability Policy, Office of Research, Evaluation, and Statistics, Washington, DC

Wang SS (2014) In tests, scientists try to change behaviors. Wall Street Journal, 28 July 2014. Retrieved from Wall Streen Journal Online, http://online.wsj.com. Checked 04 October 2014

iPhone Mobile/Website Money Applications URLs[1]

http://www.advisorsoftware.com/products/ASIwealthManager.asp
http://www.goalgami.com/content/index.php
http://www.iggsoftware.com/ibank/
http://investoscope.com/
http://www.ipadappsdude.com/update-wealth-manager-finance/
http://itunes.apple.com/us/app/property-evaluator-real-estate/id335518202?mt=8
http://mastock.michelmontagne.com/
http://www.microsoft.com/money/default.mspx
https://www.mint.com/
http://mportfolio.umich.edu/
http://www.mvelopes.com/
http://quicken.intuit.com/
https://www.schwab.com/public/schwab/home/welcomep.html
http://www.smartmoney.com/
http://www.syniumsoftware.com/ifinance/
http://www.tdameritrade.com/welcome1.html
http://www.youneedabudget.com/

---

[1] Retrieval date: 07 July 2011 for all URLs.

Chapter 5
The Story Machine: Combining Information Design/Visualization with Persuasion Design to Change Family-Story Sharing Behavior

5.1 Introduction

Finding a life-enhancing way to share family wisdom (history, lore, facts, traditions, and beliefs) among generations is a twenty-first century global challenge, especially in the USA, where geographically distributed, asynchronous, merged families, with diverse cultural heritages constitute a significant portion of the population. Many in the older generation are not as familiar with mobile- and Web-based tools to manage their media and storytelling as their children and grandchildren. Nevertheless, they want to and need to share important stories about the past, perceptions about the present, and predictions of the future.

Story/memory management products, and services, are available to increase people's awareness of family history, personalities, and issues, and to encourage awareness and change, but they do not focus on innovative data visualization, and they lack persuasive effectiveness to convert family members to preserve family wisdom. Communicating family facts, concepts, and emotions helps build awareness and identity, but does not result automatically in effective behavioral changes. The question then becomes: How can we better motivate, persuade, educate, and lead people to manage their stories, media, and time, thereby preserving their legacy for future generations?

The Story Machine project of 2011 researched, analyzed, designed, and evaluated powerful ways to improve intergenerational story-sharing behavior by persuading and motivating people to increase story generation, and to increase story sharing with other family members by means of a well-designed mobile application (for smart phone, tablet, and associated Web portal) concept prototype: the "Story Machine."

The author's firm previously designed and tested similar concept prototypes: the Green Machine application in 2009, oriented to persuading home consumers to make energy conservation behavior changes; the Health Machine application in 2010, oriented to avoiding obesity and diabetes through better behavior regarding

© Springer-Verlag London 2015
A. Marcus, *Mobile Persuasion Design*, Human–Computer
Interaction Series, DOI 10.1007/978-1-4471-4324-6_5

nutrition and exercise; and the Money Machine in 2011, targeted to baby boomers and oriented to their managing their wealth better. The Story Machine uses similar principles of combining information design/visualization with persuasion design. AM+A's presentation and this chapter explain the development of the Story Machine's user interface, information design, information visualization, and persuasion design.

Grandparents, parents, children, and grandchildren live in geographically dispersed locations, with different, asynchronous schedules, sometimes time zones apart, with increasingly multicultural mixtures of ethnic backgrounds, with busy schedules, and with many computer-generated media to reduce the time spent communicating with each other.

In some cases, by the time the grandchildren are old enough to appreciate the personalities and wisdom of the grandparents, one or both grandparents have passed away. People lead increasingly isolated and lonely lives, despite the surface connectivity of social media.

Unfortunately, seldom do the social media applications combine many of the functions of specialized applications for storytelling, story gathering, and story sharing. After reviewing users' comments about current family storytelling, family history, and media apps for Websites and mobile devices, the authors concluded that more adaptations and improvements could be made to better serve people's needs. Above all, most of the products reviewed do not provide an overall "persuasion path" to change users' short-term and long-term behavior, leading to improved story sharing, and in some ways, a richer lifestyle.

Aaron Marcus and Associates, Inc. (AM+A) has embarked on the conceptual design of a mobile-phone/tablet-based product, the Story Machine, intended to address this situation. The Story Machine's objective is to combine information design and visualization with persuasion design to help users achieve their story making, story management, story sharing, and story-absorbing objectives, especially regarding acquiring wisdom from every generation, building stronger links, preserving family history, increasing depth and breadth of identity, and preserving the family legacy of stories by persuading users to adapt a story-oriented lifestyle to include healthier consumption of stories and appropriate media management.

AM+A intends to apply user-centered design along with persuasive techniques to make the Story Machine highly usable and to increase the likelihood of success in adopting new story-sharing behavior. The direction of storytelling is not necessarily only from the older generations to the younger ones. The younger generations have stories to share, as well, which can educate, entertain, and inspire older generations.

In the Jewish tradition, one says of a deceased family member, "may her/his memory be a blessing." The living are blessed with the memories and wisdom of those who have preceded us. The Story Machine aims to increase the number of blessings that all family members can derive from each other, whether living or dead.

5.2   Initial Discussion

In these times of globalization and intense migration (for studies, work, family relocation), families are more and more dispersed, living in different cities, states, countries, even continents. Communication with family members becomes a challenge because of time differences or incompatible working schedules, or because most of the younger generations, especially parents with small children, lead very busy lives. Communicating is done across borders of culture, time, space, and discipline, and is often facilitated by modern technology and by the increasing use among older generation of social media. The Internet allowed Twitter, Facebook, Skype, and other software products/services to become the typical tools used by many to be able to communicate, chat, and send/receive email, documents, photos, and videos.

According to one source [Editors, Grandparents.com], grandparents represent about one-third of the USA population with 1.7 million new grandparents every year, and they control 75 % of the wealth of the country [Grandparents, 2011]. Grandparents want to be more and more involved in the life of their children and grandchildren. They are living longer. Many of them have a Facebook account (the baby boomers are the fastest growing segment of this social media platform) and use it to "connect" with their children and grandchildren. 75 % of them are online and about 45 % of them use social networking platforms. Grandparents are younger than ever (43 % of them became grandparents in their fifties) and they enjoy spending time with their grandchildren. About 66 % of them travel with their grandchildren for holiday and family vacations, about 81 % have their grandchildren for summer holidays and 72 % of them consider that being a grandparent is the single and most important satisfying thing in their lives.

However, despite the increasing desire of communicating, often barriers arise (e.g., time, money, motivation, and distance). Those living in different time zones and with up to 8 or 10 h of time difference will know how difficult it can be to coordinate a conversation, regardless of media.

Another issue that must be addressed is the role and usability of social media. If these social media are useful when it comes to share pictures (present or past), they are a little bit more challenging when it comes to share past family stories. The social media platforms are more about the "now" moments and less about the past moments. If Facebook is a useful platform for grandparents to see what their grandchildren are up to, it doesn't enable the younger generations to have a more deep and meaningful conversation with the older generations about stories from the past. Pictures are an important medium or trigger to storytelling, but the tools available on the Internet do not necessarily facilitate a conversation between a grandparent and her/his grandchild triggered by a photo.

Storytelling is an important aspect of social and family interaction but is often used only in "real" time, if at all.

5.3 User-Centered Design

As noted in Chap. 1, user-centered design (UCD) approach links the process of developing software, hardware, and user-interface (UI) to the people who will use a product/service. UCD processes focus on users throughout the development of a product or service. The UCD process comprises these tasks, which sometimes occur iteratively:

- Plan: Determine strategy, tactics, likely markets, stakeholders, platforms, tools, and processes.
- Research: Gather and examine relevant documents, stakeholder statements.
- Analyze: Identify the target market, typical users of the product, personas (characteristic users), use scenarios, competitive products.
- Design: Determine general and specific design solutions, from simple concept maps, information architecture (conceptual structure or metaphors, mental models, and navigation), wireframes, look and feel (appearance and interaction details), screen sketches, and detailed screens and prototypes.
- Implement: Script or code specific working prototypes or partial "alpha" prototypes of working versions.
- Evaluate: Evaluate users, target markets, competition, the design solutions, conduct field surveys, and test the initial and later designs with the target markets.
- Document: Draft white papers, user-interface guidelines, specifications, and other summary documents, including marketing presentations.

AM+A carried out all of these tasks in the development of the Story Machine concept design except for implementing working versions.

5.4 Market Research

In order to have a clearer vision of the target market for the Story Machine, AM+A conducted qualitative research with potential customers.

AM+A carried out market research in order to find out more about the motivations and behaviors of people when it comes to use of technology, family bonds, family physical interaction, and family legacies. AM+A submitted two questionnaires, one for the generation of baby boomers aged 47 and older, the second one for the younger generations. Each of the questionnaires had a total of 50 respondents. The objective of these questionnaires is to understand as much as possible about the needs and wants of people regarding their memory/story sharing. The questionnaire and results shown in Figs. 5.1, 5.2, 5.3, 5.4, and 5.5. TBD follow, with analysis interspersed.

5.4.1 Questionnaire 1: For People 47 and Older

1. *How old are you?*

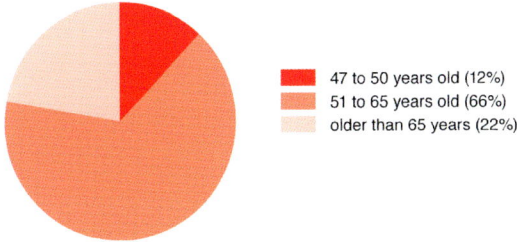

47 to 50 years old (12%)
51 to 65 years old (66%)
older than 65 years (22%)

Data were gathered from 50 people aged 47 and older, with the majority aged 51 to 65.

2. *Do you use any storing or audio-video recording applications on your smart-phone (e.g., photo album)? Which ones? What attracted you to them?*

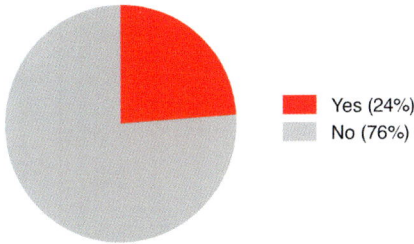

Yes (24%)
No (76%)

The majority of the respondents are not using any storing or audio-video recording applications on their smartphones. The ones that do said that compactness (i.e., they can multitask by using one small phone) and convenience are what attracted them to use these applications on their Smartphones. The possible objects that they would store in these applications are selected songs, movies, videos, and photographs (i.e., of friends, family, and memorable locations).

3. *On a scale from 1 to 5, how close are you to your children (1 being not at all and 5 being extremely close)?*

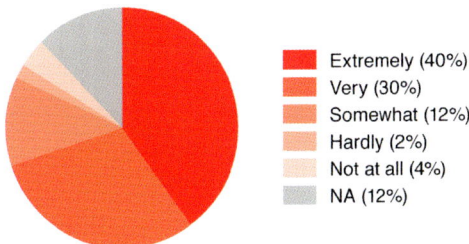

Extremely (40%)
Very (30%)
Somewhat (12%)
Hardly (2%)
Not at all (4%)
NA (12%)

Of the people who have children, 20 out of 44 said that their relationships are "extremely close."

4. *On a scale from 1 to 5 how close are you to your grandchildren (1 being not at all and 5 being extremely close)?*

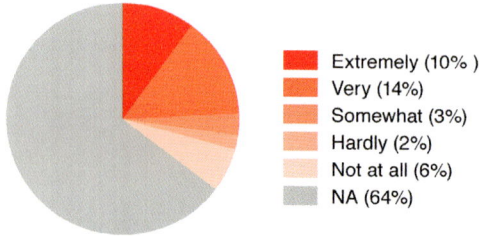

Most of the respondents do not have grandchildren. Those who do said they are quite close to their grandchildren. This fact is interesting because on the younger-generation survey, most said that they are not close to their grandparents at all. Perhaps, this has a lot to do with generational differences in perception of closeness.

5. *How often do you interact with your children (in person or via phone, Skype, or email)?*

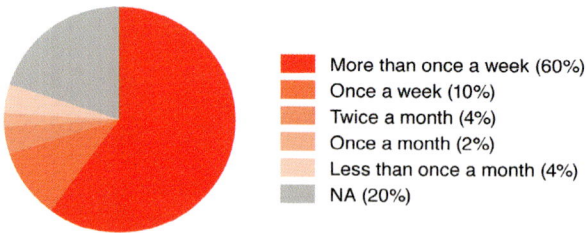

30 out of 40 respondents who have children interact with their children very often even when their children live far away.

6. *How often do you interact with your grandchildren (in person or via phone, Skype, or email)?*

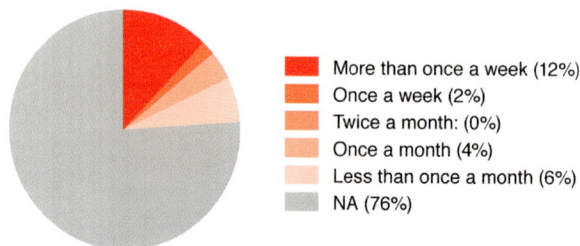

Of 12 respondents who are grandparents, 6 interact with their grandchildren more than once a week.

7. *Do some of your family members live in another city, state, or country?*

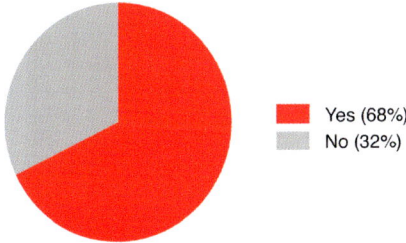

Yes (68%)
No (32%)

8. *Do your family members speak another language (in addition to your native language)?*

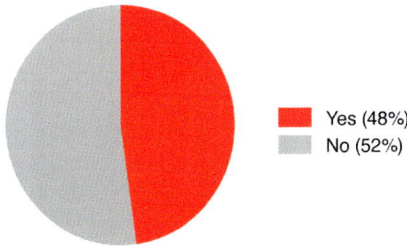

Yes (48%)
No (52%)

9. *How or what tools do you use to preserve memories of your loved ones (e.g., printed photos, Websites, family films, social media, etc.)? How do you store family videos/photos (e.g., albums, DVDs, hard drives, etc.)?*

   Note: the total will not add up to 100 % because respondents may use numerous tools to preserve memories.

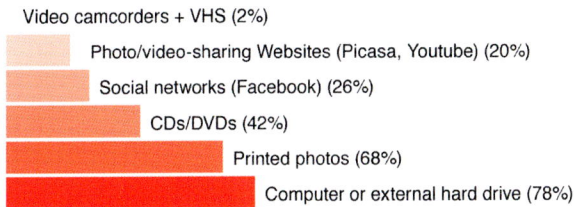

Video camcorders + VHS (2%)
Photo/video-sharing Websites (Picasa, Youtube) (20%)
Social networks (Facebook) (26%)
CDs/DVDs (42%)
Printed photos (68%)
Computer or external hard drive (78%)

10. *Do you think it is important to preserve and share memories? Why or why not? What kind of stories would you like to hear from older family members who are still alive or who have already passed away?*

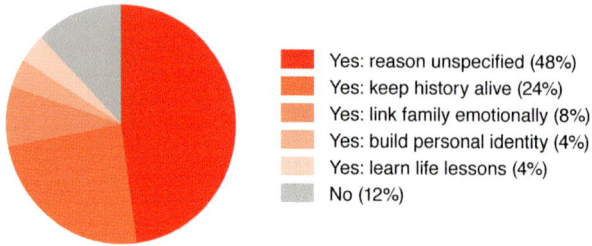

Only 2 people responded that memories help build their unique identity and contribute to who they are today. This small response rate is interesting as this was most popular reason for the younger-generation survey.

Similar to the importance of memories, there are also patterns in the kind of stories that the respondents want to hear from generations above them:

- 27 out of 50 respondents said they would like to hear stories about how and where the generations above them were raised, and the important events that they have been through (e.g., wars). 12 of these 21 respondents said they were interested to hear stories in order to compare and contrast their lives with others.
- 8 respondents said they would like to hear about the important decisions their past generation made, and 4 respondents would like to hear about the older generation's relationship with their loved ones, especially their love stories.

11. *What is the legacy or wisdom that you would like to share with younger generations once you have passed away?*

Note: the total will not add up to 100 % because some respondents may want to share more than one legacy or wisdom, while others do not want to share anything.

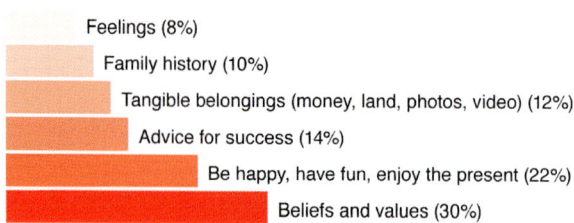

- 15 respondents said they would like sharing their beliefs and values, and making sure the generations after them are good citizens.
- 11 respondents said they would like to tell the next generation to be happy, have fun, and enjoy the present.
- Of the 7 respondents who would give advice on how to be successful, 2 wanted to tell the younger generation to learn from the respondent's mistakes, and 2 wanted the younger generation to know that education is the key to success.

12. *Would you ever consider leaving behind a message for your loved ones once you have passed away? Would you be interested in a family member leaving a message for you?*

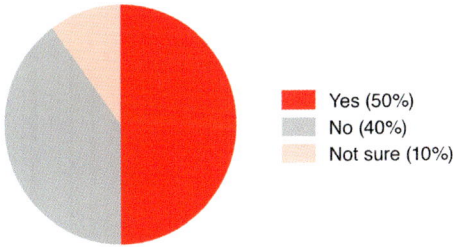

Yes (50%)
No (40%)
Not sure (10%)

Some respondents gave additional comments, suggesting possible types of message such as it being a regret they could not say during their lifetime, a warning to prevent them from making mistakes, or the message of appreciation that will act as a reminder to the younger generation.

13. *In a mobile application that would allow preserving and sharing memories between generations, what features would you want to see?*

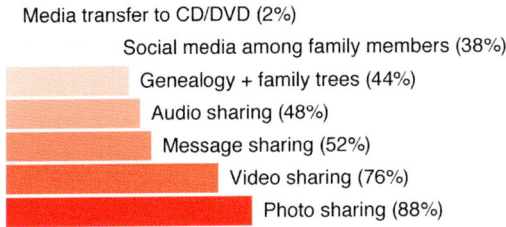

Media transfer to CD/DVD (2%)
Social media among family members (38%)
Genealogy + family trees (44%)
Audio sharing (48%)
Message sharing (52%)
Video sharing (76%)
Photo sharing (88%)

5.4.2 Questionnaire 2: For People Younger than 47

1. *How old are you?*

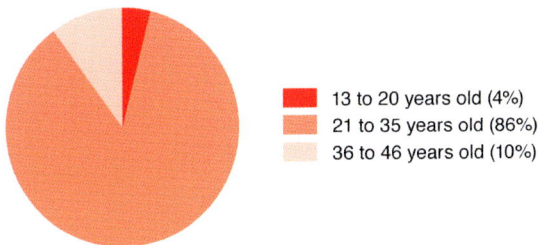

13 to 20 years old (4%)
21 to 35 years old (86%)
36 to 46 years old (10%)

Data were gathered from 50 people from the ages of 13 to 47, with the majority (43) being between the ages of 20 and 35.

2. *Do you use any storing or audio/video recording applications on your smart-phone (e.g., photo album)? Which ones? What attracted you to them?*

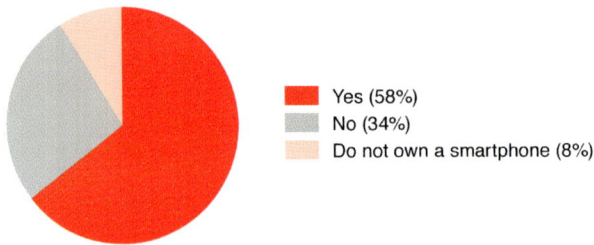

Yes (58%)
No (34%)
Do not own a smartphone (8%)

Most respondents (27 out of 29) use the storing and recording applications that come with the phone such as Photo Album and Voice Memos. An application that stands out in this survey is Instagram. This application allows the users to take, customize, and enhance the graphics of their photographs. One user mentioned an application called Hipstamatic, which also works in similar way. A couple of users mentioned (secondary features of) social network applications, like Facebook to store photos.

3. *On a scale from 1 to 5, how close are you to your parents? (1 being not at all and 5 being extremely close).*

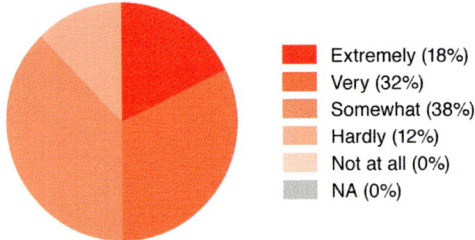

Extremely (18%)
Very (32%)
Somewhat (38%)
Hardly (12%)
Not at all (0%)
NA (0%)

4. *On a scale from 1 to 5 how close are you to your grandparents (1 being not at all and 5 being extremely close)?*

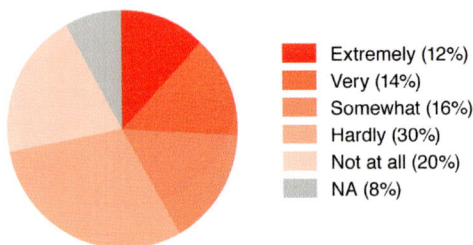

Extremely (12%)
Very (14%)
Somewhat (16%)
Hardly (30%)
Not at all (20%)
NA (8%)

In contrast to the relationship between the respondents and their parents, the relationship between the respondents and their grandparents are leaning towards nonexistent. Some gave the reason that they did not even remember what their grandparents looked like because they had already passed away when they were young or even before they were born.

5. *How often do you interact with your parents (in person or via phone, Skype, or email)?*

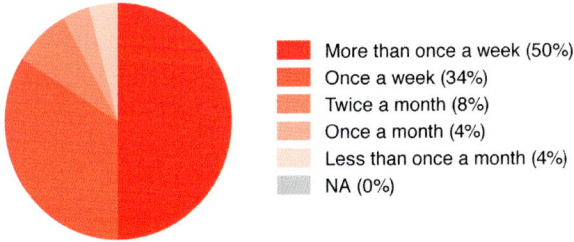

- More than once a week (50%)
- Once a week (34%)
- Twice a month (8%)
- Once a month (4%)
- Less than once a month (4%)
- NA (0%)

6. *How often do you interact with your grandparents (in person or via phone, Skype, or email)?*

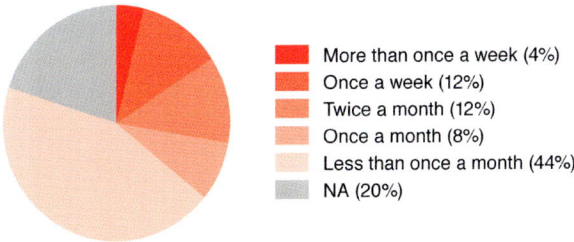

- More than once a week (4%)
- Once a week (12%)
- Twice a month (12%)
- Once a month (8%)
- Less than once a month (44%)
- NA (20%)

In contrast to the interaction between the respondents and their parents, the interaction between the respondents and their grandparents are quite rare. One respondent said that one of the reasons is because their grandparent does not feel comfortable with using technologies like email or Skype, which made the communication harder. One respondent also stated that culture plays a major role in the children and grandparents' interaction; e.g., in a number of Chinese families, grandparents and grandchildren interact with one another more because they live together in the same household.

7. *Do your parents/grandparents live in another city, state, or country?*

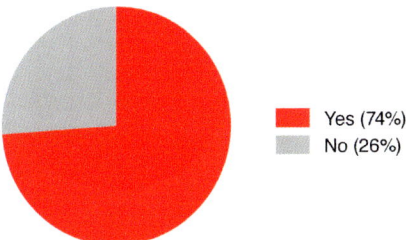

- Yes (74%)
- No (26%)

4 out of 50 respondents said their parents do live in the same city as them, but their grandparents do not.

8. *Were your parents or grandparents raised with a different culture from yours?*

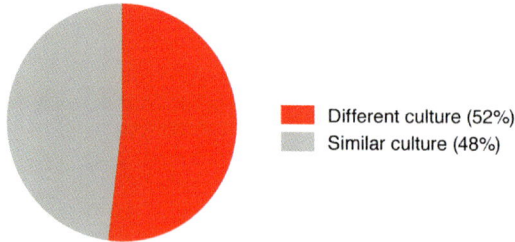

Different culture (52%)
Similar culture (48%)

3 out of 50 respondents said they were raised with the same culture as their parents, but not the same culture as their grandparents.

9. *Do they speak another language (in addition to your native language)?*

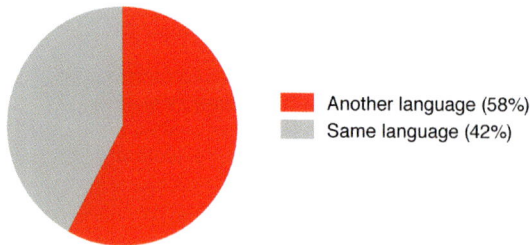

Another language (58%)
Same language (42%)

Twenty-one said they spoke the same language as their parents (even though they may know another language), but not their grandparents.

10. *How or what tools do you use to preserve memories of your loved ones (e.g., printed photos, Websites, family films, social media, etc.)? How do you store family videos/photos (e.g., albums, DVDs, hard drives, etc.)?*

Note: the total will not add up to 100 % because a respondent may use more than one tool to preserve memories.

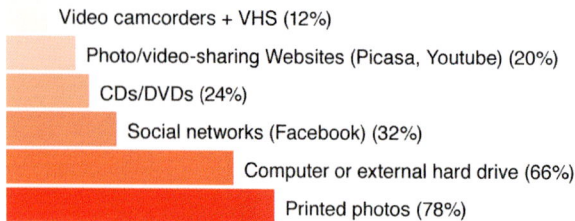

Video camcorders + VHS (12%)
Photo/video-sharing Websites (Picasa, Youtube) (20%)
CDs/DVDs (24%)
Social networks (Facebook) (32%)
Computer or external hard drive (66%)
Printed photos (78%)

11. *Do you think it is important to preserve and share memories? Why or why not?*
    What kind of stories would you like to hear from older family members who are
    still alive or who have already passed away?

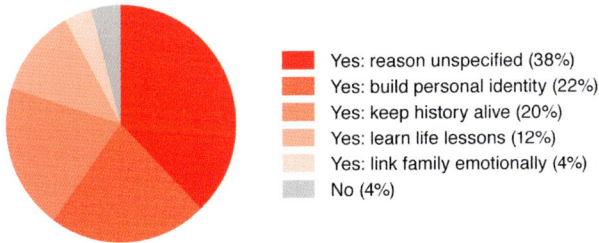

Yes: reason unspecified (38%)
Yes: build personal identity (22%)
Yes: keep history alive (20%)
Yes: learn life lessons (12%)
Yes: link family emotionally (4%)
No (4%)

Similar to the importance of memories, there are also patterns in the kind of
stories that the respondents want to hear from generations above them:

- 21 out of 50 respondents said they would like to hear stories about how and
  where the generations above them were raised, and the important events that
  they have been through (i.e. wars).
- 12 of these 21 respondents said they were interested to hear this in order to compare and contrast their lives with them.
- 4 respondents said they would like to hear about the important decisions their past
  generation made, and 3 respondents would like to hear about their love stories.

12. *What legacy or wisdom would you like to inherit from older generations?*

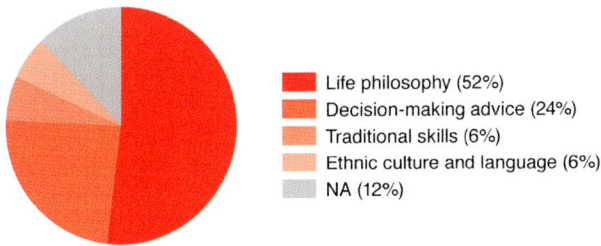

Life philosophy (52%)
Decision-making advice (24%)
Traditional skills (6%)
Ethnic culture and language (6%)
NA (12%)

Respondents cited sewing and cooking techniques as traditional skills.

13. *Would you be interested in a family member leaving a message for you after*
    they have passed away? What if you received it 10 years after they were gone?

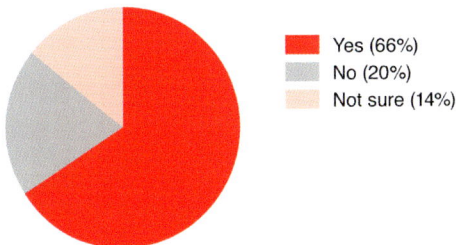

Yes (66%)
No (20%)
Not sure (14%)

Of respondents interested in receiving time-delayed messages, 11 indicated that the time period should be shorter than 10 years, while 1 respondent said it should be 20 years after. Some gave additional comments suggesting the possible types of message, such as its being a reminder, a final goodbye, or the kind of message that requires the recipient's maturity in order to understand.

14. *In a mobile application that would allow preserving and sharing memories between generations, what features would you want to see?*

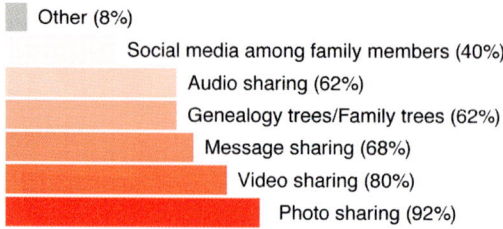

Other (8%)
Social media among family members (40%)
Audio sharing (62%)
Genealogy trees/Family trees (62%)
Message sharing (68%)
Video sharing (80%)
Photo sharing (92%)

Other suggested features included synching to a Website where one can view everything on a big screen, like Evernote.

5.4.3 Follow-Up Survey

A follow-up survey was conducted with younger respondents to elicit further information, as summarized below.

1. *How do you define "family"?*

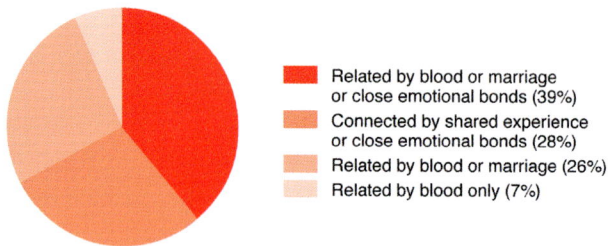

Related by blood or marriage or close emotional bonds (39%)
Connected by shared experience or close emotional bonds (28%)
Related by blood or marriage (26%)
Related by blood only (7%)

Two thirds (67 %) of respondents extended the definition of family beyond traditional lines of blood and marriage to include people with whom they have shared experiences or to whom they feel close emotionally (e.g., best friends).

2. *On a scale of 1 to 5, how well do you feel you know your family's history (1 meaning hardly at all, 5 meaning very well)?*

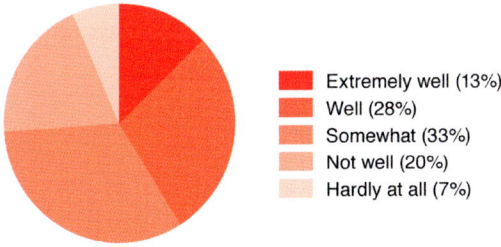

According to the respondents, some of the following reasons affected their answers:

The stories that they heard about their family history came in the form anecdotes and story fragments.
The family history got lost during world events like World Wars and revolutions.
Some family members intentionally kept their family history a secret.
People learned about their family history, but forgot it as time went by.
People who are racially mixed have a hard time of keeping track of both (or more) sides of the family with such different cultures.
They are immigrants, or are adopted.

3. *In your opinion, what makes for a good story? Can you give some examples of the elements of a good story, or the best way(s) to tell a story?*

Note: the total may not add up to 100 % because a respondent may list more than one element or way to tell a good story.

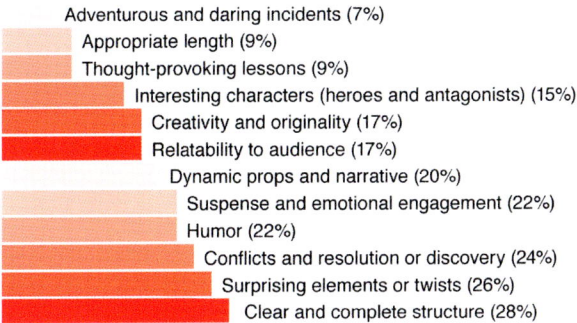

4. *How often do you share stories or memories with your family?*

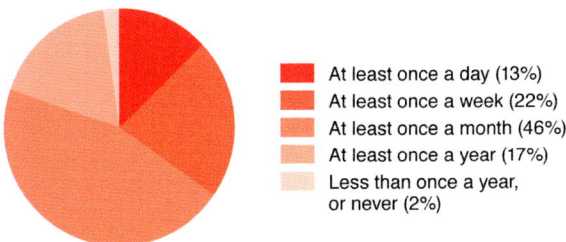

5. *Do you have favorite storytellers in your family?*

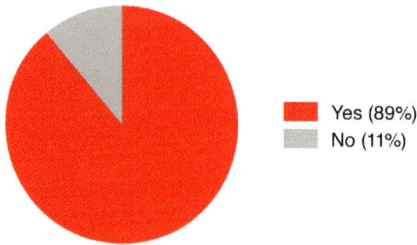

Yes (89%)
No (11%)

6. *If so, what makes them such good storytellers?*

Note: the total number will not add up to 100 %, because a respondent may list more than one reason, while some did not give any, when it comes to the quality of their favorite storytellers.

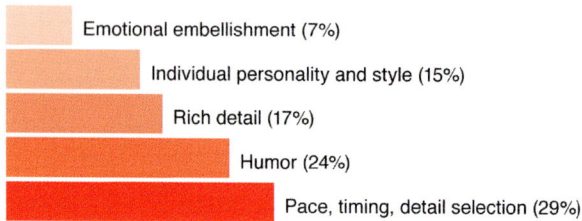

Emotional embellishment (7%)
Individual personality and style (15%)
Rich detail (17%)
Humor (24%)
Pace, timing, detail selection (29%)

7. *If you were to share a memory or story about yourself with someone from another generation, would you go about it differently?*

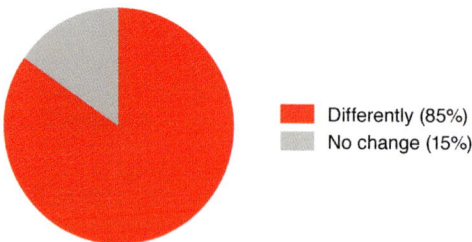

Differently (85%)
No change (15%)

8. *If so, how does this differ from your communication with people your own age?*

Note: the total will not add up to 100 % because one respondent may list more than one method, while some did not give any, when it comes to ways to communicate with people of different generations.

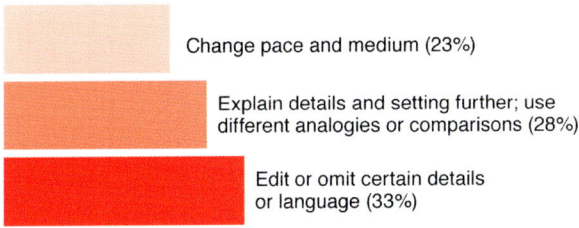

An example of a change in medium would be communicating through a social networking site rather than over the phone or through letters.

9. *What sort of stories or facts would you definitely not want to hear or tell?*

Note: the total number will not add up to 100 % because one respondent may list more than one subject, while some did not give any.

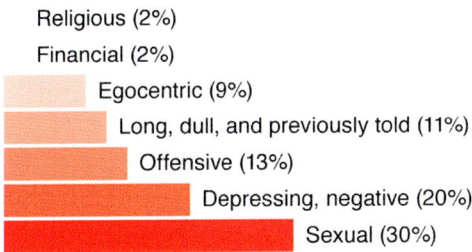

Respondents were particularly sensitive to not offending their listeners, and not wanting to listen to overly prideful, egocentric stories.

10. *What sort of stories or facts would you definitely want to hear or tell?*

Note: the total will not add up to 100 % because a respondent may list more than one subject, while some did not give any.

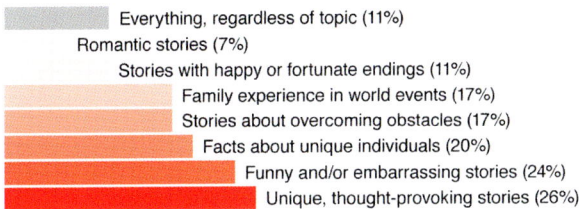

11. *In your opinion, what are some barriers to storytelling?*

Note: the total will not add up to 100 % because a respondent may list more than one barrier, while some did not give any.

No barriers (4%)
Technology (2%)
Language differences (9%)
Generational and maturity differences (11%)
Fading memory (13%)
Lack of story structure and organization (15%)
Lack of time for storytelling (17%)
Inaccuracy (17%)
Unwillingness to share personal details (20%)
Listeners' lack of attention/interest (33%)

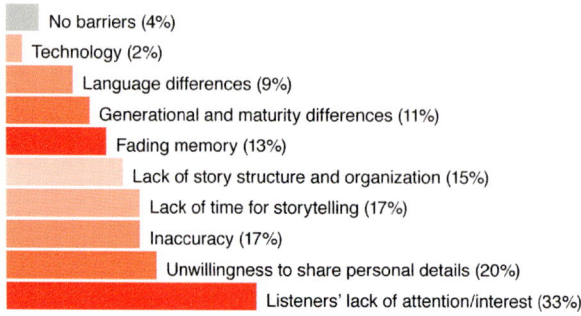

Of the 8 respondents who listed inaccuracy as a barrier to storytelling, 5 indicated that inaccuracy can prompt one to overly embellish details, while 3 stated that inaccuracy can result in a loss of important facts.

12. *Are there things you'd like to know about your friends or family, but haven't asked? If so, what has kept you from asking these questions?*

Note: the total will not add up to 100 %, because one respondent may list more than one reason.

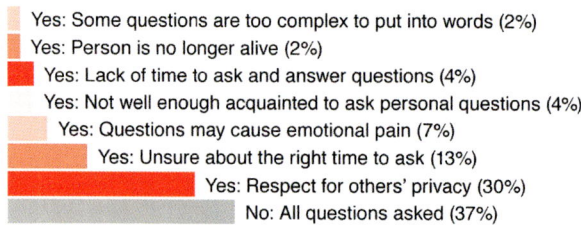

Yes: Some questions are too complex to put into words (2%)
Yes: Person is no longer alive (2%)
Yes: Lack of time to ask and answer questions (4%)
Yes: Not well enough acquainted to ask personal questions (4%)
Yes: Questions may cause emotional pain (7%)
Yes: Unsure about the right time to ask (13%)
Yes: Respect for others' privacy (30%)
No: All questions asked (37%)

Propriety is a substantial barrier to story sharing. Of those respondents with lingering questions, many did not want to violate others' privacy or cause emotional pain. Others did not know how to phrase their questions well, or were unsure of the appropriate time to ask them, often forgetting them when the time comes. Still others have missed the chance entirely, with their family members having already passed.

13. *Would you want to know the background story or the world events at the time of the story?*

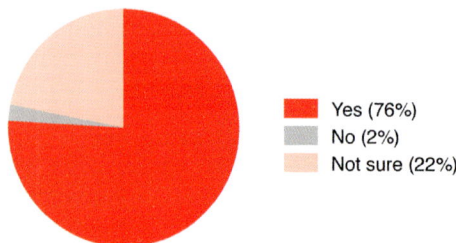

Yes (76%)
No (2%)
Not sure (22%)

14. *Can you give some examples of these types of stories or events?*

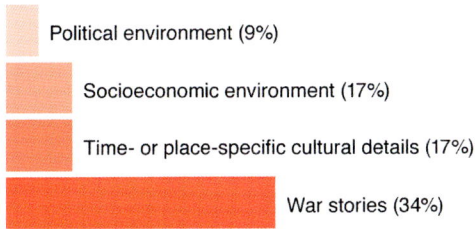

Political environment (9%)

Socioeconomic environment (17%)

Time- or place-specific cultural details (17%)

War stories (34%)

Respondents cited the Great Depression as a particular point of socioeconomic interest. Respondents interested in place-specific stories wanted to learn more about the culture in their family's home country. Respondents interested in time-specific stories wanted to learn more about the culture during the time of the story (e.g., the 1960s). Of the 12 respondents who indicated they would like to hear war stories:

- 8 were interested in stories from World War II.
- 2 were interested in stories from the Korean War.
- 1 was interested in stories from the American Civil War.
- 1 was interested in any war story.

15. *What tool(s) would you use to record a story or memory?*

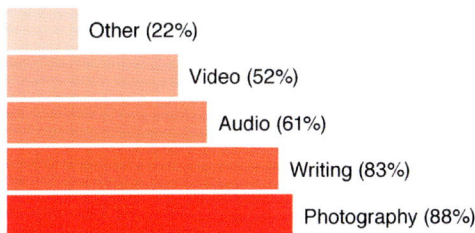

Other (22%)

Video (52%)

Audio (61%)

Writing (83%)

Photography (88%)

5.5 Personas

As discussed in Chap. 1, personas are characterizations of primary user types and are intended to capture essentials of their demographics, contexts of use, behaviors, and motivations, and their impact on design solutions. Personas are also called user profiles. Typically, UI development teams define one to nine primary personas. For the Story Machine, after internal discussion, we developed the following personas to provide a reasonable range of ages, genders, professions, etc.

5.5.1   Persona 1: Angela Adams

- Age: 20.
- Race: Caucasian.
- Occupation: College Student.
- Education: Liberal Arts at University of California, Santa Barbara.
- Proficiency: Uses her iPhone and computer often, and is not afraid to try new technologies.
- Interaction with Family and Friends: Through social media, emails, and phone
- Grandparents: All of her grandparents are living.
- Concern: Wants to be able to share memories in a much easier way than using existing social media websites like Facebook.
- Persona image credit: AM+A.

5.5.1.1   Textual Summary

Angela is 20 years old and has always been very social. She currently goes to UCSB (University of California, Santa Barbara) where she is a member of five academic clubs, including the monthly UCSB newspaper where she is the chief editor and photographer. Angela has 1,024 friends on Facebook and is constantly uploading pictures using the application Instagram. One of Angela's favorite pastimes is "Photoshopping" her friends' pictures and sharing them on Facebook. Needless to say, Angela is attached to her smartphone and feels naked without it. Not only is Angela popular at school, she is also very close with all of the members of her family. In fact, last week it was her grandfather's birthday, and Angela was responsible for the video recording. She uploaded the video on YouTube for the rest of her family to see, although Angela was a bit perturbed by the fact that she could not limit the video to be viewed by just her family and she had to cut her videos to 10 min apiece.

5.5.1.2   Design Summary

Objectives

- Preserve memories in the form of photographs and videos via Smartphone, while being able to customize and share them with friends in social network.
- Limit the viewers to personal family circles.

Context

- Attends college.
- Constantly surrounded by people from being popular and in many academic clubs.
- Records videos and posts them on YouTube.

Behavior

- Uses Smartphone for simple photo editing and posting photos on Facebook.
- Uses Smartphone daily to keep in touch with friends and family.
- Uses Photoshop to edit photos, records videos, and puts both of these kinds of media online.
- Uses Facebook and YouTube.

Design Implications

- Include family network but also a way to share media with friends (could be just link to Facebook).
- Have features similar to Photoshop.
- Include video sharing that is private to personal family network and can be uploaded with over 10 min of footage.

5.5.2   *Persona 2: David Lee*

- Age: 35 years old.
- Race: Asian.
- Marital Status: Just married.
- Occupation: Lives in New York as an investment banker.
- Proficiency: Uses his iPhone and computer often.
- Interaction with Family and Friends: Close to family; parents live in New Jersey, grandparents live in Hong Kong.
- Grandparents: Only 2 grandparents alive, on his mother's side.
- Concern(s): Wants to be able to record his family stories in an organized manner. Wants to learn from his grandparents' experiences.
- Persona image credit: AM+A.

5.5.2.1   Textual Summary

A 35-year-old overachiever, David Lee, went to Princeton and after his graduation was immediately accepted into a high-end internship at a hedge fund. He soon became partner a few years later. David just married his wife, Monica Lin, and has no children. He currently lives in New York and his grandparents live in Hong Kong. David is very close to his family and likes to connect with them often, but the distance between him and his grandparents makes talking over the phone quite expensive. Skype has helped them keep in touch without breaking the bank and allows David to actually see his grandparents. Sometimes it is hard for David's grandparents to familiarize themselves with new technology because most new technologies are in English, which they barely speak. In these conversations David tries to compare himself to his grandparents because he wants to learn from their experiences. He actively maintains his grandparents' legacy by keeping their memories recorded and organized. Each picture and video has a great story behind it, and David tries to write notes on the back of each photograph. With all of the photographs and documents in paper form, however, he is cluttering up his tiny hallway closet.

5.5.2.2   Design Summary

Objectives

- Video chat with grandparents frequently, easily, and without costing a lot of money.
- Learn and record stories from grandparents, keep them stored so he can hear/see them again.
- Compare personal timeline with the timeline of family members.

Context

- Spends most of the time at work or at home.
- Lives in New York.

Behavior

- Uses Skype and video chat frequently.
- Keeps print photos and writes on them to record the stories behind them.
- Keeps videos on tape in the closet.

Design Implications

- A feature to simplify cross-cultural communication.
- A video chat with recording and file sharing capabilities.
- File sharing with tagging by audio or text.
- A personal timeline that is comprised of all documents, photos, and videos.

5.5.3   Persona 3: Clarence Owens

- Age: 50 years old.
- Race: African American.
- Occupation: Civil servant.
- Education: High school graduate.
- Marital Status: Divorced his wife ten years ago.
- Proficiency: Has a Smartphone and uses it for calling and a few other applications.
- Interaction with Family and Friends: No children. Has siblings who live nearby and have children themselves. Likes being in constant touch with family because he lives alone.
- Grandparents: Deceased.
- Concern: Not being able to leave his legacy behind because he has no children.
- Persona image credit: AM+A.

5.5.3.1   Textual Summary

Clarence is 50 years old and works as a civil servant. He has no children and divorced his wife ten years ago. Clarence does have many siblings who live close by and have their own children. He likes to be in constant touch with his siblings and their children because he lives alone. He often uses his Smartphone to call a niece or nephew to catch up with them. Clarence also uses a few other applications on his phone, including Photos, Camera, Voice Memos, Calendar, and YouTube, and he spends a lot of time on the Internet. He often receives emails with pictures or videos of family and saves them on his hard drive. Sometimes Clarence forgets specific things about the photos he took of his family and wishes he could remember the stories behind each one without having to write it all down on the back of the photo or in a Word document. Clarence loves to share his own photos and knowledge with his family too. He has always been a great cook and uses a cookbook that has been handed down from generation to generation in his family. His younger sister often calls him about recipes, which he has to scan and then send over email, making the task quite daunting. Clarence loves his family and does not want his legacy to die with him. Because he has no children of his own, Clarence would like to leave behind messages and money to his nieces and nephews.

5.5.3.2 Design Summary

Objectives

- Be in constant touch with siblings and nieces/nephews.
- Easily share photos, videos, and other information with family members.
- Create a message to leave behind for his family when he passes.

Context

- Lives home alone with no children or wife.
- Works part-time and has a lot of free time.

Behavior

- Uses Smartphone for calling and a few memory saving applications (photos, videos, voice notes).
- Uses email to share photos and videos, saves onto hard drive.
- Prints out some photos to hang around the house.

Design Implications

- Allow sharing of things such as cooking recipes (besides photos and videos).
- Have an easy way to stay in constant touch with family (e.g., have a continuous feed with text messages, videos, photos, updates, etc.).
- Be able to tag photos and videos by text, voice, or other media.
- Include a way to create a slideshow of photos and videos and then add audio so that a whole story can be told.

5.5.4   Persona 4: Linda Suarez

- Age: 65 years old.
- Race: Hispanic.
- Marital Status: Married to Carlos.
- Occupation: Retired.
- Education: College graduate.
- Proficiency: Has Smartphone, but has a hard time viewing and reading the screens. Capable of using basic features on the computer like checking email. Prefers old-fashioned printed photos over digital ones.
- Interaction with Family and Friends: Has four children living in different states, and all 5 grandchildren living abroad
- Grandparents: deceased
- Concern(s): Wants to be able to keep in touch with all of her children and grandchildren, but it seems like an impossible task because of the distance. Organizes family events and trips once in a while, but always forgets to keep track of when they happened and if there were any memorable instances. Fears that her children and grandchildren will not know of their roots, yet is not sure if she has already shared her family stories with all of her children.
- Persona image credit: AM+A.

5.5.4.1   Textual Summary

At 65 years of age, Linda Suarez lives with her husband Carlos in a retirement home in Palm Springs. She has four children: Luis (45), Marta (42), Brad (38), and Jerry (35), as well as many grandchildren. As much as she likes to keep in touch with her children and grandchildren via email, she finds it difficult to be up-to-date with what is going on in their lives because they all live in different locations around the world. Linda owns an iPhone, but she only uses it to call people most of the time because her eyesight does not allow her to read what is on the iPhone's screen well. Once every couple of years she organizes a family gathering or trip, where all of her family comes together and strengthens their bonds as a family. However, because she is always busy organizing and running the event, she also forgets to record what happened during the events. As she looks back, sometimes she even forgets when memorable events, like their family trip to Tahoe, actually happened. When it comes to preserving memories, she prefers tangible, printed photographs over any form of digital storage; for her, what is not tangible, is not real. As she is growing older, she is wondering and concerned whether her children and grandchildren will know anything about their roots, and whether their bonds will be as strong after she passes away.

5.6   Use Scenarios

5.6.1   Definition

As described in Chap. 1, use scenarios are a UI "usage stories" that are used during UI development. In the development of an initial concept prototype, a use scenario helps determine what behavior to simulate. These descriptions are good for comparison, as new scenarios are meant to be an improvement upon existing methods.

5.6.2   General Use Scenario

The following use scenario topics are drawn from the four preceding personas. However, some specific examples might be relevant only to a particular age group, education level, gender, or culture.

Video Memories

- Record memories visually and verbally.
- Record an audio message to attach to a photo, or provide narration to a silent video clip.

Personal Timeline

- Organize information graphically through the Personal Timeline feature. The Timeline logs memories and mementos in an intuitive, chronological structure.
- Compare Personal Timelines with friends and family, weaving a complex and multileveled story (e.g., when Dad was graduating from high school, Mom's father got a new job, and Mom's family moved to Ann Arbor).

- Enhance Timelines with anecdotal and general history: See Nana's chronicles of 1963 ("Jackie looked so lovely that day") alongside facts from Wikipedia timelines ("JFK assassinated").
- Use the Reminders function so that important events never go unnoted.

Video Chat

- Connect with family and friends over video chat (including group chat).
- Record videos or conversations for later use.
- Create a transcript of the audio.
- Use recordings to tag other media in collection.
- Use the cross-cultural communication feature to help translate application into different languages.

Social Media

- Create different social networks for friends and families.
- Share photos, videos, special dates, and documents with specific networks.
- View feeds and comment on posts from family and friends within networks.

Photography Software

- Use features similar to Instagram (e.g., change the decade: Use different effects on a photo by selecting a time period – the picture will then be colorized like it was taken in the corresponding era).
- Upload or print photos with the touch of a button.

Family Trades

- Preserve and share family recipes in the Cookbook.
- Learn and practice family skills (e.g., Dad's carpentry, Grandma's embroidery).
- Record and distribute family knowledge (e.g., the purple flowers that bloomed in Papa's garden in springtime were called bluebells).

Media Specific to Time/Place

- Site-specific media require the recipient's presence at a certain location (tracked by GPS) in order to view a message, photo, or video.
- Time-specific media require the recipient to wait until a predefined moment to view a message, photo, or video. Time-specific media can also be used to compose and send messages in advance, in order to celebrate birthdays, user-provided data (e.g., "In what year were Mimi and Pop-Pop married?"), or anniversaries.

Gamification

- Generate trivia from creating a family-oriented "Newlywed Game".
- Accrue experience points for meeting achievements (e.g., scanning in old photos, identifying people and places in photos, adding a branch to the family tree). Use these experience points to earn digital prizes and titles.

General Behavior Change

- Provide people with more time because the Story Machine allows the user to organize and keep record of memories in one place.
- Make people feel comfortable switching between digital and physical forms of media, as the data from their Smartphone can be transferred to a hard drive or a printer.
- Strengthen the bond among family members because the user will be able to know what other families are doing without being in their proximity.
- Make people aware and become more knowledgeable about their family roots and wisdom that are being passed on from generation to generation.
- Increase happiness, reduce depression, and reduce stress through the behavior changes that lead to wiser memory management.
- Help people rely more on themselves to preserve memories instead of hiring people to do it for them (i.e., event photographer).
- Make people aware of the value of their irreplaceable memories and how the memories can get lost without a system to preserve them.
- Provide people with an engaging, playful, interactive way to preserve memories, and provide a reward or incentive system for knowing their family history.
- Show people that preserving memories is not as difficult as they might think and influence them to preserve more.
- Persuade people that disciplined memory preservation can lead to successful happiness management by showing how memories and history can be fun, and by making the whole process of using the application very clear and simple.
- Reduce anxiety about using technology to preserve memories through a simple, fun, easy-to-read, graphical user interface. The Story Machine can act as a gateway for people to open up more to new technologies and other applications (i.e., the Health Machine, the Money Machine) in the future.
- Encourage people to continue to preserve memories wisely and maintain the practice by using social media awards and games.

5.6.3   Personas Use Scenarios

The following use scenarios are oriented to the individual personas described above. In addition to a general use scenario, these specific ones can provide additional insights to developers concerning functions, data, features, benefits, and specific behavior changes.

5.6.3.1   Angela's Use Scenario

With the Story Machine, Angela's smartphone becomes not just her friend, but her best friend. Instead of uploading her photographs to her computer, and customizing them in Photoshop, Angela only needs to use the Story Machine's photo customization feature to manipulate any photographs she takes. She can make her photographs and videos look as vintage, as stylized, or as dramatic as she wants. After she uploads these customized pictures and videos to Facebook, her friends always give her big "wow" comments, as the quality of these pictures and videos are impeccable. When Angela has free time, she also plays games on the Story Machine to unlock more photo-customizing options to make the graphics of her photos even more unique. The Story Machine also draws Angela closer to her family, as it gives her the opportunity to become the family's trusted photographer and video recorder when it comes to family events, like her grandfather's birthday. Unlike YouTube, the Story Machine lets her select specific family members and friends to view her videos, which can be of any length and size. With the Story Machine, her nephew's wedding video that lasts an hour can be uploaded and shared as a single item.

5.6.3.2   Angela's Behavior Changes

Angela will have more time on her hands now that the photo customization process is simpler; instead of uploading the photos to a computer and doing Photoshop work before re-uploading them to Social networking Websites, she can customize the photos using the Story Machine and upload them to social networking Websites right away. Angela will get to know her family better as she gets to record and edit the photos and videos of her family events.

5.6.3.3   David's Use Scenario

David wants to learn more about his grandparents and keep records of their legacy. Saving all of his video conversations over Skype with them, as well as photos and videos sent through email and the mail, was becoming very tedious. David started using the Story Machine to help with these tasks. Now he receives videos and photos straight to his phone, already tagged with text, audio, or other pictures. He no longer has a closet filled with unorganized photos and videos. David also uses the Story Machine to video chat with his grandparents. He can automatically store the videos on his phone, or download them to his computer. He also often uses the feature to transcript his videos to text or just audio, so that he can tag other media in his collection with these. The Story Machine has allowed David to begin compiling a timeline of important events in his grandparents' lives, including pictures, videos, and even documents. David has shared the timeline with his grandparents through the application so that if he misunderstands something, they can edit it, and they can add material themselves as well. David has started to create a timeline of his own life, and enjoys comparing it to the one of his grandparents' lives to see how their lives have been different.

5.6.3.4   David's Behavior Changes

David no longer has to keep a closet full of pictures and videos because everything is on his phone. The Story Machine also provides an easy way to download everything to his computer or hard drive, so he will never need a huge storage space in which to keep everything. The Story Machine allows David to easily tag photos and videos together, or with text or audio; he no longer has to allocate huge amounts of time to go through his photos and write descriptions on them. He, also, does not have to worry about forgetting information about a photo, because everything is in the application if it exists. David's grandparents are able to share more with David using the Story Machine, and he is learning a lot more about them, especially because they can all share and edit the timeline depicting their own lives.

5.6.3.5   Clarence's Use Scenario

Before getting the Story Machine, Clarence had to constantly check his email and call his family to catch up with them. Occasionally, he would call them at inconvenient times, which made him feel bothersome, but he enjoyed being updated about their lives often. With the Story Machine, Clarence sees a continuous feed about the family involving pictures, videos, text, and other documents. He feels he is less intrusive, because the family posts things when it is convenient for them. He also has a better idea of when is a good time to call or video chat with the shared calendars feature. Clarence uses the Story Machine to tag different media sent to him by their relevant text/pictures/audio/video pairs, and when his phone gets close to being full he downloads them to his hard drive. Clarence also uses the Story Machine to take photos of his recipes and then easily shares them with his family through the application. When he has free time he enjoys creating stories for his nieces and nephews. The Story Machine allows him to compile slideshows with pictures and videos, and then record an audio file to describe the story behind them all. Clarence also stores messages on the Story Machine, which he would like to leave behind for his family when he passes away. He hopes that all of the information he shares with them through the application will keep his legacy alive.

5.6.3.6   Clarence's Behavior Changes

Clarence does not have to spend as much time when sending recipes to his family. He just takes a picture with his phone and sends this through the application, instead of tediously scanning the recipe and emailing the resulting picture. Clarence can stay updated about his family through his phone and not feel so bothersome by calling them too often. He also has a fun way to share his own stories through slideshows with the family, and again need not worry about inconveniencing them, because they can look at the stories on their own time. Clarence no longer feels worried about forgetting information behind a certain photo or video, because each is tagged with its specific information. Because Clarence is able to share all of his stories and messages through the Story Machine, he feels much better about his legacy being left behind. He also knows that his messages to his family for after he passes away will be securely delivered and at the right time.

5.6.3.7 Linda's Use Scenario

As the sun is setting, it is typical to see Linda sitting on her porch, thinking about what the relationships among her children and grandchildren will be like without her. With the Story Machine, she no longer has to worry about that. The Story Machine allows all of her family members to share what is going on with their lives via instant videos and photo sharing. For Linda, pictures definitely speak louder than words, as her aural and visual senses are starting to deteriorate with age. With video and photo sharing, she will no longer have to pick up her eyeglasses to read the small text in the emails or texts from her children and grandchildren. As she is viewing the video of Alberto, one of her grandchildren, playing with his toy that Linda gave him for a birthday, she can also tag her comments using just the tip of her finger; once she touches the screen while the video is playing, the video will pause and allow her to write a message that will serve as a reminder of happy moments she sees. Even after she passes away, her children and grandchildren will be able to see her comments from time to time as they watch the old videos. Memories will never be lost. Knowing that her husband Carlos is not as technically savvy as she is, she always loves to share these happy videos with him by transferring and viewing them on their computer or television. Carlos and Linda can also pause the video at any moment to capture a screenshot and print it out on their printer right away in order to put them in a photo album. Linda no longer has to spend her evening worrying about her children and grandchildren. With the Story Machine, Linda's (and Carlos') evenings will be filled with laughter and joy as they never get bored watching their memorable family videos.

5.6.3.8 Linda's Behavior Changes

Linda will no longer have to worry about her children and grandchildren not having recorded memories of her and their roots in one place. Linda will be able to keep in touch with all of her children and grandchildren using just one tool, the Story Machine. The Story Machine is versatile; the data on it can be transferred and viewed on different types of devices such as smartphones, computers, tablets, and TV screens. Linda becomes familiar with technology by using these simple features and is less threatened by all of the modern media. She begins to use her phone and the Story Machine application more often.

5.7   Competitive Analysis

Before undertaking conceptual and perceptual (visual) designs of the Story Machine prototype, AM+A first studied approximately 10 memory/family iPhone applications as well as Android-based applications on other platforms.

5.7.1   Memory Management Websites

5.7.1.1   Memory-Of.com

Who Were They in 2011?

- Created by Tal, initially for his brother James, in 2004.
- Committed to providing resources and support to those in mourning of lost loved ones.
- Number of users: over 35,000 visits each day, 68,000 personal memorial Websites and more than 2.9 million "candles" lit.
- In 2007 expanded to include grief counselors, advice, community forums, and real-world projects and initiatives.

What Was It in 2011?

- Allows people to create a Website devoted to a loved one that has passed away.
- Creating and editing the memorial is free. Having the Website is free for 2 weeks, and then the user has to pay for monthly, annual or "everlasting" maintenance. Part of what a user pays is donated to the International Federation of Red Cross and Red Crescent Societies.

What Did It Let Users Do?

- Make a Website for a loved one.
- Light an online candle in remembrance of the loved one.
- Place up to 300 photos on their Website.
- Create a slideshow with specific pictures.
- Share the date and other information about the loved one's funeral.
- Share music that the loved one liked.
- Publish audio and video clips of the loved one.
- Share writing that the loved one created.
- Create a timeline of important events that happened in the loved one's life.
- Build a family tree.
- Have a mailing list of people to be emailed when the Website is updated.
- Allow visitors to contribute to the Website, and edit their content if wanted.
- Customize the Website and layout design.

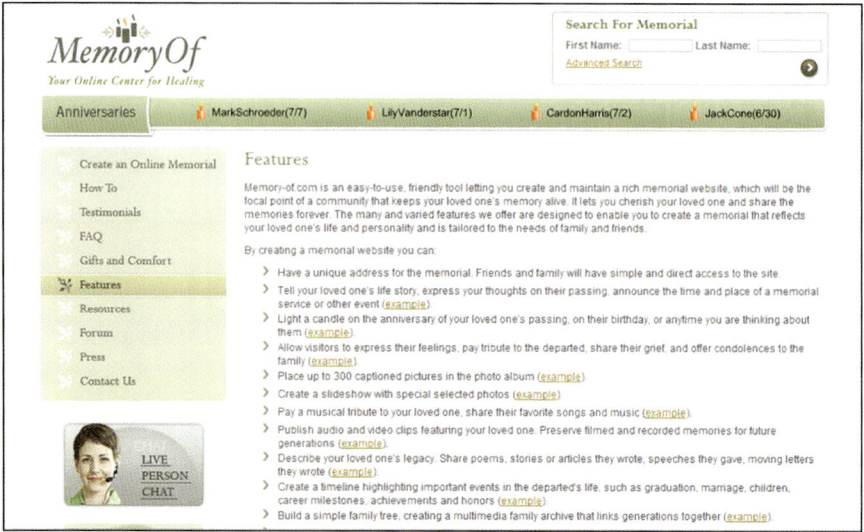

Fig. 5.1 Web screen of MemoryOf in 2011 (Image: Life Stage Media)

Web Strategy

- Extremely detailed Website: any question a user might have is answered.
- A huge number of features for users to include on the Website they make.
- Help for users that are having difficulty coping with the loss of their loved one.
- Free to set up the Website, and small fees for keeping it.
- Users get their own Web address.
- Online store where people can buy gifts for people related to the person who passed away.
- Forum for users to talk to each other.

User-Friendliness

- User-friendly.
- Easy-to-use features.
- Sign up is simple.
- Site is organized well with clear categories, and information is easy to find.
- Many resources available to users.

Visual Analysis

- Gray as the main color palette suggests sophistication, also gives the site softness.
- Sidebar with well-defined categories helps the user navigate the Website.
- Examples of each feature let the user see what they would be able to do before deciding to get a Website. The examples also serve as a template for the user to follow if they want.

- Simple language (i.e., words usage) makes the information easy to understand.
- Scrolling bar at the top displays dates of death in memory of the people who have passed, building a sense of community among users.

Mobile Strategy

- None.

5.7.2 Tips

Feature Examples

- Examples of what users can do with their Website help them not only picture how they will have it look, but also show them what they can do.

Resources

- Users can receive help if they are having trouble coping with the loss of their loved one.

User Reviews

- Testimonials category devoted to user reviews.
- All of the testimonials are positive, and many say that having a site is very inexpensive, helps people connect around the person that has passed away, and helps greatly with the grief.

Social Networking

- Users share the Website with people on an email list.

5.7.2.1 SWOT Analysis

Strengths

- Resources for users grieving.
- Many features for users to add to their Website.
- Low price.
- User-friendly.

Weaknesses

- Not free.
- Only for keeping the memory of people who have already passed away.
- Users have to set up and maintain their site by themselves.
- No persuasion behavior change.
- No mobile.

Opportunities

- Develop social networking strategy besides email.
- Develop the mobile strategy (should exist by itself and not only in association with the Website).
- More features could be added.
- Make maintaining the site cheaper or free.
- Wider market.
- More persuasion-theory-based content.

Threats

- Security might be an issue since people are making their own Websites, leaves room open for hacking.

5.7.2.2 MemoryHub

Who Were They in 2011?

- Company for online storage of videos and photos.
- Founded by Michael London, who wanted to digitalize all of his photos and found no easy or cheap way to do so. Combined with StashSpace, a company for video conversion, and Denevi Digital Imaging, a digital conversion business.

What Was It in 2011?

- Website where people can store their videos and photos.
- Keeps users' videos and photos safe where they can access them anytime.
- What their users can do.
- Upload videos and photos for storage.
- Create gifts and other mementos from photos.
- Convert videos and photos to be digitalized.
- Scan photos and even negatives.
- Edit photos.
- Share photos.

Web Strategy

- User-friendly Website with a list of categorized features that the user can use.
- Friendly design with simple blue colors and pictures of animals.
- Slideshow highlighting specific features.
- Live chat service for people with questions.
- Lots of information about the company, FAQs, testimonials, and social networking connections.

User-Friendliness

- Looks easy to use: labels for each feature, under which there are the steps needed to use that feature.
- A lot of information is displayed on some of the options that takes a while to read and makes it somewhat confusing.

Visual Analysis

- Website's main element of information visualization is a large slideshow with features that they have. There are cute elephants and clouds at the top of the site as part of the logo, which make the site look very friendly. There are easy-to-follow categories that the users can click on to find what they're looking for.
- Pricing information is easy to find. It does not look like the company is trying to hide any fees.
- Website also allows users to share "success stories" in written form. These stories serve as an inviting and a motivating tool for the new user.
- There are pictures of people enjoying the action of whatever category is clicked on. This imagery makes the site seem friendly and look like it is easy to do what the person clicked on.

Mobile Strategy

- None.

Tips

- Explanation of how they perform each feature, and what would be best for the user (e.g., what type of video conversion).
- Personal coaching with live chat window.

User Reviews

- Testimonials category on the Website.
- Very positive reviews, said to be life-changing, especially because people can view things they no longer thought they could.

Social Networking

- 67 followers on Facebook.
- Twitter link is broken.
- Blog available.

5.7.2.3 SWOT Analysis

Strengths

- Allows users to pay for services online.
- Have basically any feature a user would want in terms of preserving photos and videos.
- Simple explanations of what they do and which kind of feature might be right for the user.
- Live chat for people with questions.
- Many great testimonials.

Weaknesses

- None of the features are free.
- Social networks are old and not involved.
- No mobile strategy.

Opportunities

- Develop social networking.
- Develop mobile strategy.
- Have free or trial features.

Threats

- Price might be an issue.

5.7.2.4   MyFamily

Who Were They in 2011?

- Owned by ancestry.com.
  (Ancestry.com retired the website on 30th September 2014)

What Was It in 2011?

- Family sharing Website platform.
- Create its own social media but also connect to other social media platforms.

What Their Users Could Do

- Create a family group.
- Have groups within a family group.
- Family tree.
- Add a family blog/create discussions.
- Connect with social media platforms (Facebook, Twitter, Tumblr).
- Family calendar for organizing events.
- Upload pictures, videos, and recordings of stories.

Web Strategy

- Paying platform but reasonably priced (30 dollars per year per family but does not say how many members a family can count).
- Doesn't allow to access if you don't pay.

User-Friendliness

- Easy to use and to sign up.
- Allows instant signing up via Facebook account.
- Provides a "tour"/guide on how to use the Website and the service.
- Very easy to navigate through the different pages of the Website.

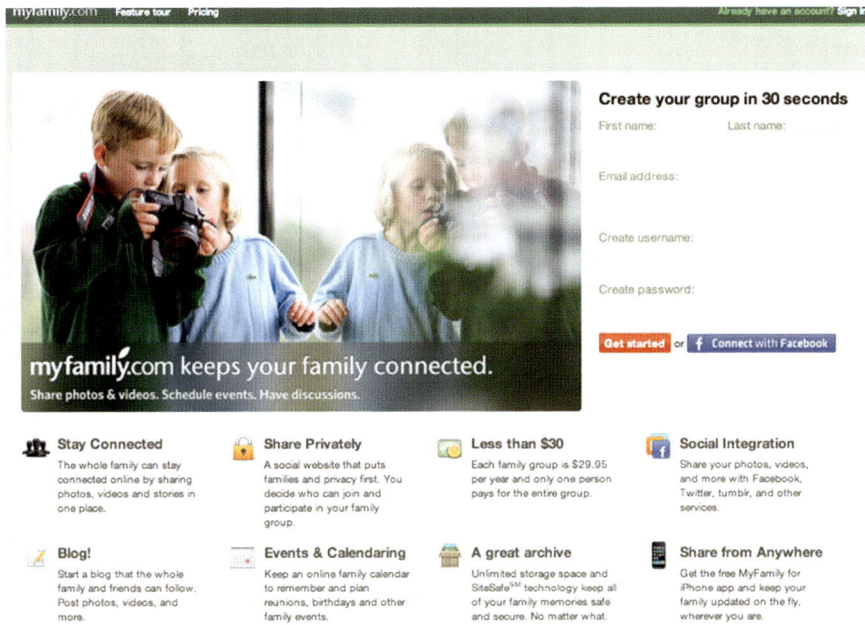

Fig. 5.2 Web screen of MyFamily in 2011 (Image: Ancestry.com, LLC.)

Fig. 5.3 Mobile screens of MyFamily in 2011 (Image: Ancestry.com, LLC.)

Visual Analysis

- Appears simple to navigate.
- All the information available at one glance.
- Picture on the home page with the tag line is a good mental model for what the Website sells.
- Color green is not really user-friendly and does not associate with family concept.

Mobile Strategy

- Application available for iPhone.

Tips

- None.

User Reviews

- None.

Social Networking

- Allows you to connect to the existing social media platform.
- Twitter account with 700 followers.
- No Facebook page.

5.7.2.5   SWOT Analysis

Strengths

- Allows user to share memories through a variety of media (pictures, videos, recordings).
- Very complete Website.
- Provides security of information.
- Provides an important storage to archive your stories and memories.
- Reasonably priced.
- Create a personalized URL.

Weaknesses

- Not free.
- No persuasion theory included.

Opportunities

- Expand their mobile strategy to Android.
- Develop their social media strategy although they are relatively active on Twitter.

Threats

- Price might be an issue.

5.7.2.6 Living Memory

Who Were They in 2011?

- Created by James Murray.
- Record life experiences online.

What Was It in 2011?

- Site allows users to record life experiences online and share them with friends and family.

What Their Users Could Do in 2011

- Create an online profile.
- Share moments, events, stories, and photos with the world or only certain people.
- Record a timeline of a child to preserve their childhood.
- Leave a record of themselves for future generations.

Web Strategy

- Pictures of example photo books, timelines, family trees, and other things the user can do.
- Tour of all the features that are available to the user if they sign up.
- Looks like it is free to sign up, but doesn't say.

User-Friendliness

- Bright colors and lots of pictures make the site seem very friendly.
- Lots of space with well-designed buttons makes the site look modern.

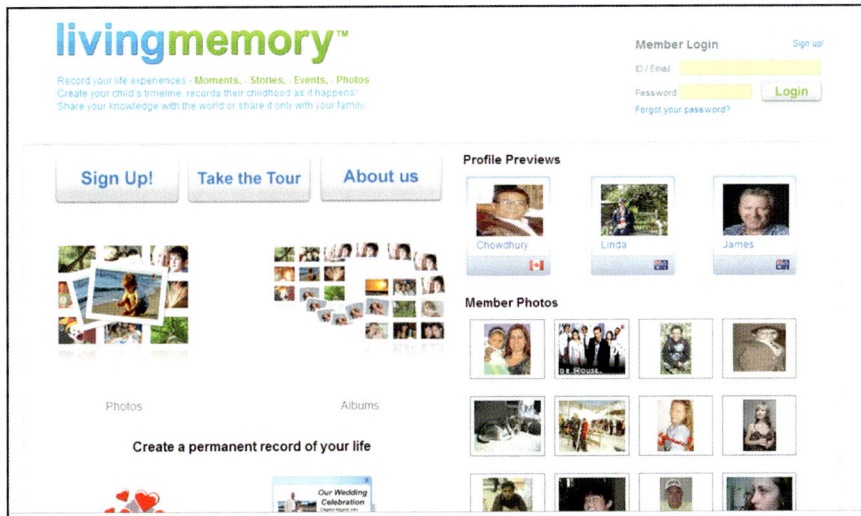

Fig. 5.4 Web screen of LivingMemory in 2011 (Image: Living Memory International Pty Ltd.)

- The tour of the site has clear defined tabs and nice visualization that shows the user what they can do. Flow with obvious steps to take once a user picks a task. Available tasks also very clear and the right amount of information is shown so that the user is not overwhelmed.

Visual Analysis

- The tour introduction of different things the user can do is useful.
- Nice spacing and colors.
- Pictures and brief descriptions of specific features help highlight the strengths of the site.

Mobile Strategy

- None.

Tips

- None.

User Reviews

- None on the actual Website.
- Overall positive reviews.
- Many people think it's a unique and fun Website, with good ideas driving it.

Social Networking

- Has its own social network with user profiles. Profiles can be shared with everyone on the site or only specific people, such as only family members.

5.7.2.7   SWOT Analysis

Strengths

- Lots of good features for recording things happening in a person's life.
- Preserves everything for younger generations to look at eventually.
- Free.
- Has preset questions for users that are signed up, to encourage them to write more about their lives and save more information for younger generations.

Weaknesses

- Can't try out any of the features without signing up first.
- No outside social networking.
- No mobile aspect.
- Many of the picture previews on the main page are blurry and/or stretched to a weird shape.

Opportunities

- Mobile application.
- Outside social networking strategy.

- Develop video tutorials that show people who are not signed up more about what the site can do.

Threats

- Additional fees for additional features?
- Security (there is a big picture on the homepage that says "security" but has no more information and is not clickable).

5.7.2.8   Memory Miner

Who Were They in 2011?

- Apple software created by John Fox in December 2004, part of the company GroupSmarts.
- Tool for linking digital media.

What Was It in 2011?

- A software that allows users to take photos and link them to each other based on people, time, and place.

What Their Users Could Do in 2011

- Look at photos over time using a dragging tab on a timeline.
- Search for photos based on time, people, or place.
- Tag photos by time, people, and place.
- Search for specific photos of a period in a specific person's life.
- Add captions to markers on a photo.
- Zoom in on markers in a photo to look at different captions.
- Record audio and/or video about a photo.
- View where a photo was taken on Google maps.
- See a map of photos of certain people, to see how they have traveled over time.
- Create slideshows.
- Import and edit photos.
- Publish photo collections, and add optional password protection.

Web Strategy

- Demo video that goes over all the features (very helpful and detailed).
- Big buttons to try a lite version or buy the full version.
- Information about the company and blog is very detailed.

User-Friendliness

- Software has a lot of features that would take some getting used to.
- Some of the features do not look very intuitive.

Visual Analysis

- Standard Apple software colors and format.
- Shows all of the features, users have a lot of control.

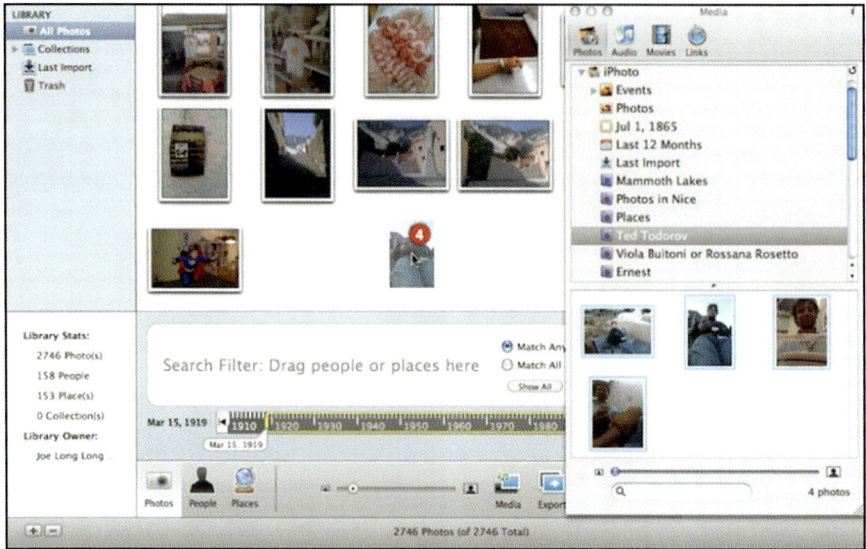

Fig. 5.5 Web screen Memory Miner (Image courtesy of GroupSmarts, LLC.)

Mobile Strategy

• None.

Tips

• Video demo on the website.

User Reviews

• None on the actual Website.
• On Mac update, 4 ratings give it 3.5 stars.
• Reviewers like it but think a lot of features could be added/improved.

Social Networking

• Users can upload to Flickr.

5.7.2.9   SWOT Analysis

Strengths

• Lots of good feature ideas, like tagging by place and time.
• Preserves everything for younger generations to look at eventually.
• Free trial.
• Demo to highlight features and explain how to do things.

Weaknesses

- Full version is not free.
- No social networking besides Flickr.
- No mobile aspect.
- Software looks confusing to use until the user has experimented for a while with it.

Opportunities

- Mobile application.
- Outside Social networking strategy.
- Improve interface so it's more intuitive.

Threats

- Additional fees for additional features?
- Competition that is easier to use.

5.7.2.10   eFamily

Who Were They in 2011?

- Website similar to Facebook but for a family-only network.
- Acquired Famiva and built a service on top of this (Famiva was another family social networking site).

What Was It in 2011?

- Allows people to create a family network similar to a network on Facebook, but private within the family.

What Their Users Could Do in 2011

- Connect and share with extended family in a private and secure space.
- Build a family tree.
- Upload HD photos and videos.
- Create family events.
- Share reminders and status updates.
- Create a custom web address.
- Share blogs and stories.
- Create a profile.

Web Strategy

- Main page has lots of positive reviews from official sites and users.
- Tour and pictures of features.
- Blog with FAQs.
- Nothing else to look at unless you sign up (free).

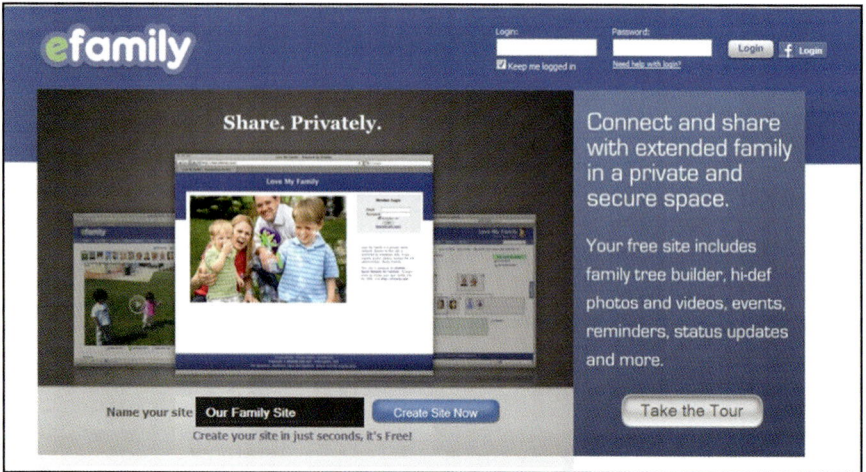

Fig. 5.6 Web screen of eFamily in 2011 (Image courtesy of efamily.com, LCC.)

User-Friendliness

- Main site is hard to navigate.
- Not many options for someone not signed up.
- For someone who is signed up, the site looks easy to use (similar to Facebook).

Visual Analysis

- Blue and white colors similar to Facebook, implying social networking and friendliness (Fig. 5.6).
- Main points are large, easy to read, catch the user's attention.
- Pictures help highlight the features for new users.
- Simple language (i.e., words usage) makes the information easy to understand.

Mobile Strategy

- None.

Tips

- Tour of features.
- Blog with FAQs.
- Feedback forum.
- Email support.

User Reviews

- Positive reviews shown on the main Website from users and other Websites.

Social Networking

- Twitter: 31 followers.
- Blog that is updated frequently.
- Email.

5.7.2.11   SWOT Analysis

Strengths

• Privacy in just having family networks.
• Many features for users.
• Free initially.
• User-friendly (similar to Facebook).

Weaknesses

• Once the user reaches the content limit they have to pay to add more.
• Only for sharing within the family.
• No persuasion behavior change.
• No mobile.

Opportunities

• Develop the mobile strategy.
• More features could be added.
• Allow users to add more content for free.
• More persuasion-theory-based content.

Threats

• Security might be an issue since people are making their own networks, leaves room open for hacking.

5.7.2.12   NPR StoryCorps

Who Were They in 2011?

• Company that allows people to record a story and then puts the story online for others to hear. Also stores all stories in an archive.
• Nonprofit organization that takes donations.
• Partners with NPR, which showcases stories on its radio programs.

What Was It in 2011?

• System for people to record and share their stories with close friends or the whole world.

What Their Users Could Do in 2011

• Record a story in a booth location.
• Receive CD of recorded story.
• Add story to national archive.
• Add story to online Website for visitors to listen to.
• Use question generator to get questions to answer in an interview.
• Rent a story kit: record two to four interviews in any location.
• Follow a "Do-It-Yourself" guide to record own interviews with own equipment.
• Listen to other peoples' stories online.
• Donate to the organization.

Web Strategy

- Main page showcases the causes that the company supports.
- Sidebar with pictures of features.
- Featured stories.
- Current events the company is involved in.

User-Friendliness

- Sidebar makes features easy to navigate.
- Lots of helpful guides related to the company's work.
- Quick buttons to popular features.

Visual Analysis

- Orange and white colors make the site very bright and friendly.
- Main points are large, easy to read, catch the user's attention.
- Pictures help highlight the features for new users.
- Simple language (i.e., words usage) makes the information easy to understand.

Mobile Strategy

- iPhone application: free, English only.
- Listen to weekly story and other favorites.
- Share stories via Facebook, Email, Twitter.
- Get tips for recording own story.
- Create a list of questions for next interview.
- Average rating 3/5 stars.

Tips

- Highlight of features.
- Instructions for different ways to record a story.
- Email support.

User Reviews

- Positive reviews shown on the mobile application.

Social Networking

- Facebook: 26,563 followers.
- Twitter: 10,748 followers.
- Blog that is updated frequently.
- Email.

5.7.2.13   SWOT Analysis

Strengths

- Allows users to record and share personally or with everyone.
- Many features on mobile for users.

- Lots of help with telling stories and questions to generate better stories.
- Stories showcased on Website and popular ones played on NPR.
- Free.
- User-friendly Website.

Weaknesses

- Only mobile application for the iPhone.
- No marketing strategy – established company and idea but not well-known.

Opportunities

- Develop the mobile strategy.
- Develop marketing strategy.
- More features could be added.
- More persuasion-theory-based content.

Threats

- Security might be an issue if users make personal stories on their phone through the mobile application.

5.7.3 Memory Management Phone Applications

5.7.3.1 Public Speech

What Was It in 2011?

- Apple application available for iPhone, iPod Touch, iPad.
- Tool for "timing and recording speeches".
- Available in English.
- Size: 8 MB.
- Price: Free.

What Their Users Could Do in 2011

- Time speeches.
- Record and save speeches.
- Play previous speeches.

User-Friendliness

- Large print, big buttons.
- Features are well-labeled and simple.

Visual Analysis

- Simple colors: black, green for go, red for stop (Fig. 5.7).
- No amazing design, looks like it was one of the first iPhone applications made.

Fig. 5.7 Mobile screens of Public Speech in 2011 (Image: USA Interactive)

User Reviews

- 1 review, says the application is quick and simple to use.
- 41 ratings give it 3 stars.

Tips

- None.

Social Networking

- None.

5.7.3.2   SWOT Analysis

Strengths

- Users can add their thoughts on the go.
- Free.
- Look at old recordings easily.
- Everything is very simple.

Weaknesses

- Not many features.
- Does not look like there is an easy way to share or export recordings.

Opportunities

- Social networking.
- Add more features.
- Improve user-interface.

Threats

- Competition with more features, quick way to export recordings, or a prettier design.
- Security and protection of data.

5.7.3.3   Ancestry.com – Mobile App

What Was It in 2011?

- Apple app available for iPhone, iPad.
- Tool for managing family trees.
- Available in 5 languages.
- Size: 12.4 MB.
- Price: Free.
- Links to account on the Website.

What Their Users Could Do in 2011?

- Link to ancestry.com account.
- Access family trees.
- Upload photos.
- See shared trees and records.
- Add vital information, immediate family members, life events, notes or new ancestors.
- Take photos of important people or objects and add them to a tree.

User-Friendliness

- Well-labeled buttons.
- Intuitive.

Visual Analysis

- Layout of the application is simple and easy to follow (Fig. 5.8).
- Nice colors (mellow green) and neat designs (family trees look like they are pinned to a board).

User Reviews

- 1851 ratings give the current version 3 ½ stars.
- 20603 ratings give all versions 2 ½ stars.
- Most people are positive and think the mobile application is a nice complement to the Website, but the Website is still necessary.

Tips

- Only on the Website: learning center, message boards, hire an expert, get help.

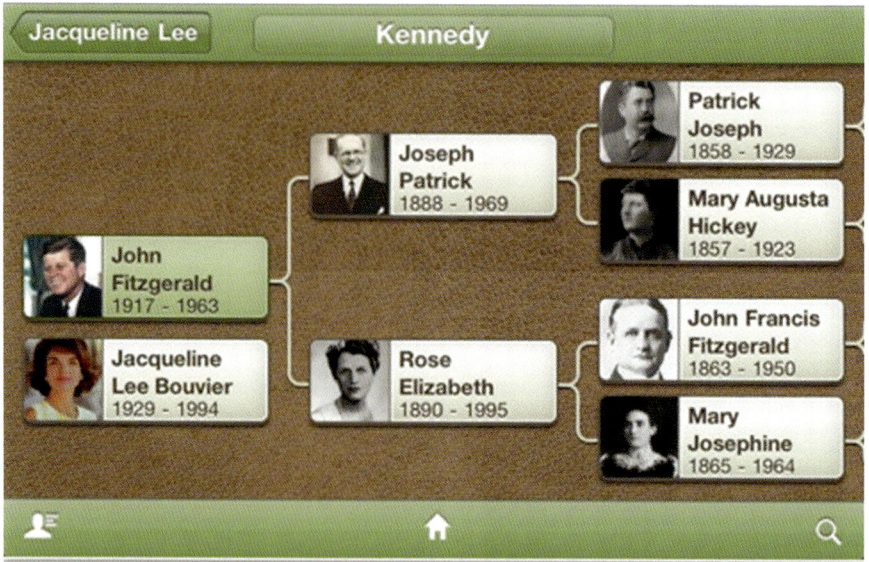

Fig. 5.8 Screenshot of the Ancestry.com app in 2011 which was reviewed (Image: Ancestry.com, LCC.)

Social Networking

- 167,786 fans on Facebook.
- 9,975 followers on Twitter.
- Blog updated frequently.
- YouTube channel.

5.7.3.4   SWOT Analysis

Strengths

- Guides users through the process of finding their family tree using a huge database of records.
- Everything is displayed clearly.
- Free.
- Take and link photos to people or family trees.
- Share trees with other users.

Weaknesses

- Mobile is limited (Website account needed).
- Crashing reported in reviews.
- No video recording.

Opportunities

- Add more features to mobile.

Threats

- Security of information.
- Continued crashing or loss of information/records.

5.7.3.5   Blurb Mobile

What Was It in 2011?

- Apple app available for iPhone, iPod Touch, iPad.
- Create and share short media stories.
- Available in 3 languages.
- Size: 6.4 MB.
- Price: Lite version is free, full version is $1.99.

What Their Users Could Do in 2011

- Create visual stories.
- Import photos, videos, and audio assets.
- Have direct access to the iPhone's camera and camera roll.
- Edit media: rotate, crop, scale, drag and drop sequencing.
- Attach audio clips to images
- Add a text caption to an image.
- Share a story on Facebook, Twitter, or through email.
- Create and share from a social setting, travel location, a personal moment, or an important event.
- Organize photos and video by stories.
- Tag a story by location.

User-Friendliness

- Lots of nice features that appear easy to find and read.
- Clear steps to take for each feature.

Visual Analysis

- Good layout of graphics: title of each story also previews the pictures in it (Fig. 5.9).
- Different designs available for the user to use in their stories.
- White and blue colors make the application look modern and friendly.

User Reviews

- 104 ratings give it 3.5 stars.
- Easy and fun to use.
- Mostly positive reviews.

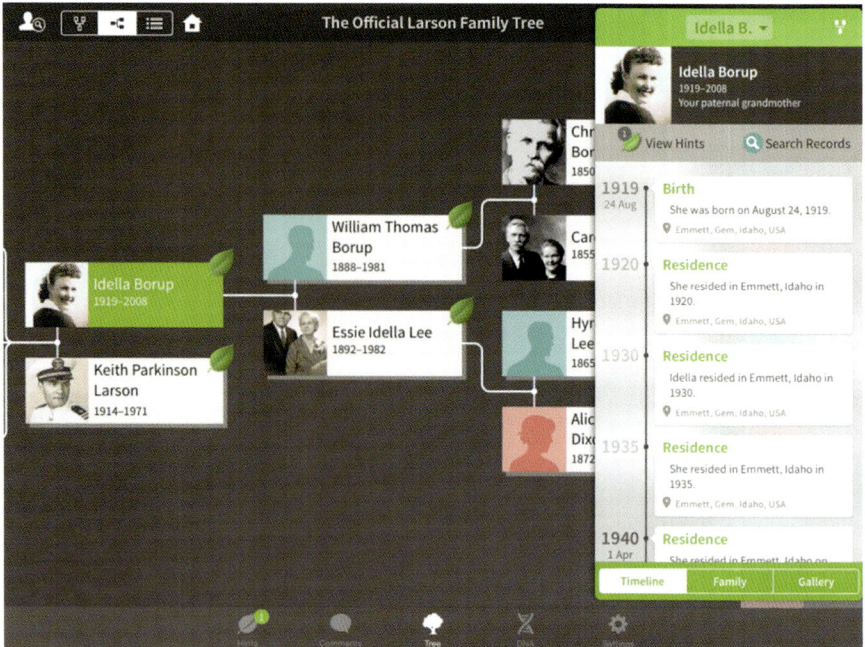

Fig. 5.9 Current mobile screen of Ancestry.com app in 2015 (Image: Ancestry.com, LCC, included at the request of the company)

Tips

- Videos on the Blurb Website to see what they can look like.
- Help on the Website.
- Settings on mobile.

Social Networking

- Users can share their stories on Facebook, Twitter, or through email.
- The Blurb company throws events at photography festivals.

5.7.3.6   SWOT Analysis

Strengths

- Friendly and easy to use interface.
- Unique mobile features.
- Good design.

Weaknesses

- Full version is not free.
- Limited to sharing on Facebook, Twitter, or email.

Opportunities

- Make completely free.
- Develop more features.
- Add more ways to share (e.g., mobile to mobile).
- Add more languages.

Threats

- Competition with more features.
- Security of stories that users make/share.

5.7.3.7 Familair

What Was It in 2011?

- An airline prototype created by a student at California College of the Arts, San Francisco, CA (Familiar 2011).
- Aimed towards families who travel together and family's trips planner.

What Their Users Could Do in 2011

- Recommend flights, hotels, and services.
- Set up polls for family members to vote on.
- Provide "Double Check" feature at the airline's kiosks, which is the digital checklist that let the users check on other family members (i.e., whether they have arrived the airport) as well as the objects that the users needed to bring with them.
- Provide a "Familair Service Screen" on the airplane where the users can use to communicate and play interactive games with other family members even though they are on different plane.
- Allowed the users to check how much time they have to wait until their luggage arrived the baggage claim area.
- Grouped together the family luggage using the family's "Pod Color".
- Allowed the user to upload the pictures from the trip and send postcards to other family members using the Familair website.

User-Friendliness

- Nice and bold features that appeared easy to find and read.

Fig. 5.10 Example screenshots of Blurb Mobile app in 2011 (Images: Blurb, Inc.)

Visual Analysis

- Graphics are simple, bold, and straight forward, which could be because the designer use it to explain the concept of how Familair works (Fig. 5.10).
- White and blue colors make the application look modern, friendly, and easy on the eyes.

5.7.3.8   Family Wallet

What Was It in 2011?

- A project proposed by a group of UC Berkeley New Product Development Course students called Living Memories (no relation to the product name) (Aziz et al. 2011).
- Family Wallet is a smartphone application to capture and share memories in the form of videos and photos while learning from past generations.

What Their Users Could Do in 2011

- Allow family members to connect to each other through a family network.
- Notifies family members when notes, pictures, or videos of them have been uploaded.
- Allows users to share family memories in the form of notes, pictures, and videos.
- Allows users to view media shared by other family members.
- Allows for private data sharing among family members.
- Allows users to share media only meant to be seen by other family members.
- Allows users to edit media sent to each other.

Visual Analysis

- There were no final version of visuals available to analyze (Figs. 5.11 and 5.12).

Fig. 5.11 Web screens of Familair (Image: California College of the Arts, San Francisco)

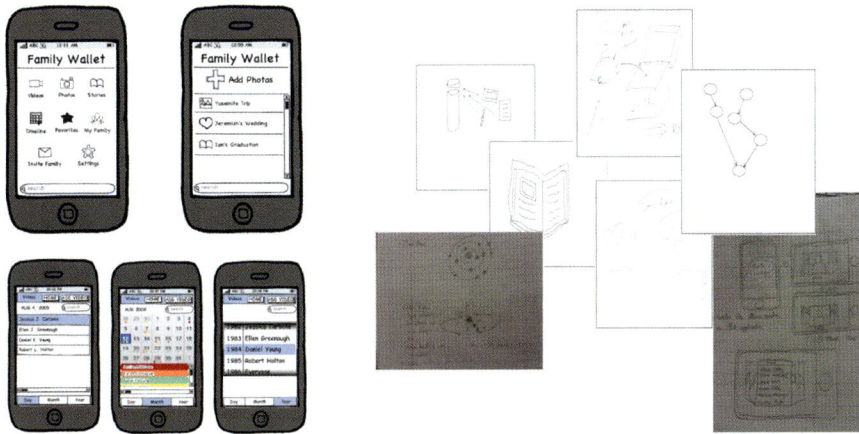

Fig. 5.12 Sketches of Family Wallet (Images: Living Memory student group, University of California, Berkeley)

5.8   Results

From our investigations of the mobile and Website memory management/ family bonding applications including those cited above, AM+A concluded that usable, useful, and appealing user-interface (UI) design must include incentives to lead to behavior change. Good storytelling and story sharing oriented mobile-phone applications should provide games, easy use of media and/or step-by-step instructions to motivate people to change their behavior. The proposed Story Machine needs to combine persuasion theory, provide better incentives, and motivate users' to achieve short-term and long-term behavior change towards a Story Machine everyday user.

Last but not least: the Story Machine should be fun to use. Well-designed games will serve as an additional appealing incentive to teach, to train. Also, the Story

Machine should allow users to share their experience with friends, family members, and the world, primarily through Facebook, Twitter, and blogs.

Based on these concepts and available research documents, we have proposed and are developing conceptual designs of the multiple functions of the AM+A Story Machine. Subsequent evaluation will provide feedback by which we can improve the metaphors, mental model, navigation, interaction, and appearance of all functions and data in the Story Machine's user interface. The resultant improved user experience will move the Story Machine closer to a commercially viable product/service.

We believe a well-designed Story Machine will be more usable, useful, and appealing to Memory-conscious users, especially those having problems communicating. Our objective is to provide a mobile suite of applications that can reliably persuade people to move toward an easier storytelling habit in order to allow the communication between multigenerational asynchronous family members.

5.9   Persuasion Theory

As discussed in Chap. 1, in alignment with Fogg's persuasion theory (Fogg and Eckles 2007), we defined five key processes to create behavioral change via the Story Machine's functions and data:

- Increase frequency of using application.
- Motivate changing some story sharing habits: gather, manage, share.
- Teach how to change story sharing habits.
- Persuade users to plan short-term change.
- Persuade users to plan long-term change.

Each step has requirements for the application.

Motivation is a need, want, interest, or desire that propels someone in a certain direction. Humans' sociobiological instinct is to maximize reproductive success and ensure the future of descendants. We apply this theory in the Story Machine by prompting users to understand that every action has consequences on their current and future family legacy.

We also drew from Maslow's *A Theory of Human Motivation* (Maslow 1943), which he based on his analysis of fundamental human needs. We adapted these to the Story Machine context:

- The *safety and security need* is met by the possibility to visualize the amount of stories created and who can view them.
- The *belonging and love need* is expressed through friends, family, and social sharing and support.
- The *esteem need* is satisfied by social comparisons that display improvement in story sharing accomplishments and skills, as well as by self-challenges that display goal accomplishment.

- The *self-actualization need* is fulfilled by the ability to visualize improvement of story sharing accomplishments and social connection indexes and moods, and to predict change in users' future story sharing scenarios.

5.10 Impact on Information Architecture

5.10.1 Increase Use Frequency

Games and rewards are among the most common methods to increase use frequency. We have developed several financial game concepts for the Story Machine. In terms of rewards, the users might be awarded virtual rewards (such as "star" designations and new skins for blogs) as well as concrete rewards or benefits regarding family mementos.

In addition, we chose social comparison as another incentive to increase use frequency. Users can form groups with family and friends, and participate in various group competitions. Although these competitions are based around story sharing controls and exercises, users do not have to reveal large amounts of personal data, which some may prefer to keep securely private.

5.10.2 Increase Motivation

In the Story Machine, we set the users' potential storytelling conditions as an important incentive for their behavior change. Viewing their current versus predicted storytelling status over the next 20–30 years gives users greater understanding of the strengths and weaknesses to their story sharing strategy.

Because setting goals improves learning outcomes and provides quantitative performance data, the Story Machine asks users to set time-based goals for media and/or story reduction (e.g., saving and maintenance, both required and optional) and legacy planning. Users receive suggested step-by-step plans of action in order to achieve each goal.

In addition, we created 10 monthly challenges. In meeting these challenges, users are making short-term story sharing behavior changes that will generate long-term positive impacts, change their family status, and entitle them to benefits and/or rewards.

Social interaction is another strong motivator of behavior change, both through family support and informal competition or comparison. The Story Machine leverages social networking by integrating features found in forums and on blogs, Facebook, and Twitter. Users can send stories, messages, requests for history, and tips, and can share ideas with their family groups. These family ties serve as additional incentive to motivate behavior change.

5.10.3 Improve Learning

For many people with significant social/family challenges; understanding long-term story sharing management is crucial. To improve learning, the Story Machine integrates contextual tips on the following topics:

- Consuming stories more wisely.
- Increasing storytelling/sharing control.
- Tackling complications associated with poor family relationships.
- Coping with losses or unexpected family changes.

Users can also choose to receive updates on latest research articles and news about storytelling/sharing management.

We seek to make the education process entertaining as well as informative. Proposed games teach users to choose the right proportion of text/imagery, amount and type of each story, etc. Through playing games featuring educational information, users learn to manage their story sharing like a skilled storyteller without getting bored.

5.11 Designs

5.11.1 Initial Sketches

AM+A sketched metaphor concepts, information architecture concepts, and specific screen design concepts before rendering detailed diagrams and screens. The following are representative sketches (Figs. 5.13 and 5.14).

5.12 Information Architecture

The following is an initial diagram of the information architecture for the Story Machine. In this diagram the information architecture is called a concept map and uses list structures rather than more typical boxes and links (Fig. 5.15).

5.12.1 Components of Information Architecture

5.12.1.1 Record Now: Record Stories or Memories Instantly, at Any Time

Profile: Personal Settings

- Update personal profile.
- Link to other social media (Facebook, Goodreads, Last.fm, etc.) to import profile data, likes, and dislikes.

Fig. 5.13 Mobile prototype screen concept designs of Family Wallet (Images: Living Memory student group, University of California, Berkeley)

- Use Story Machine tutorial.
- View personal timeline of stories and life events.
- Compare timelines with family members, celebrities, historical events.
- Adjust privacy and sharing settings.

Vault: Story and Memorabilia Repository

- View Stories (favorites, most viewed, recently viewed, recommended).
- Browse, search, or upload stories, images, written media, audio, video.
- Tag uncategorized media and memorabilia.
- Organize memorabilia for a story.

Family Lore

- View and contribute to family trees, including templates for nontraditional families.
- View and contribute to pet memorials.
- View and contribute to family atlases, a geographical visualization of a family's migration over time.

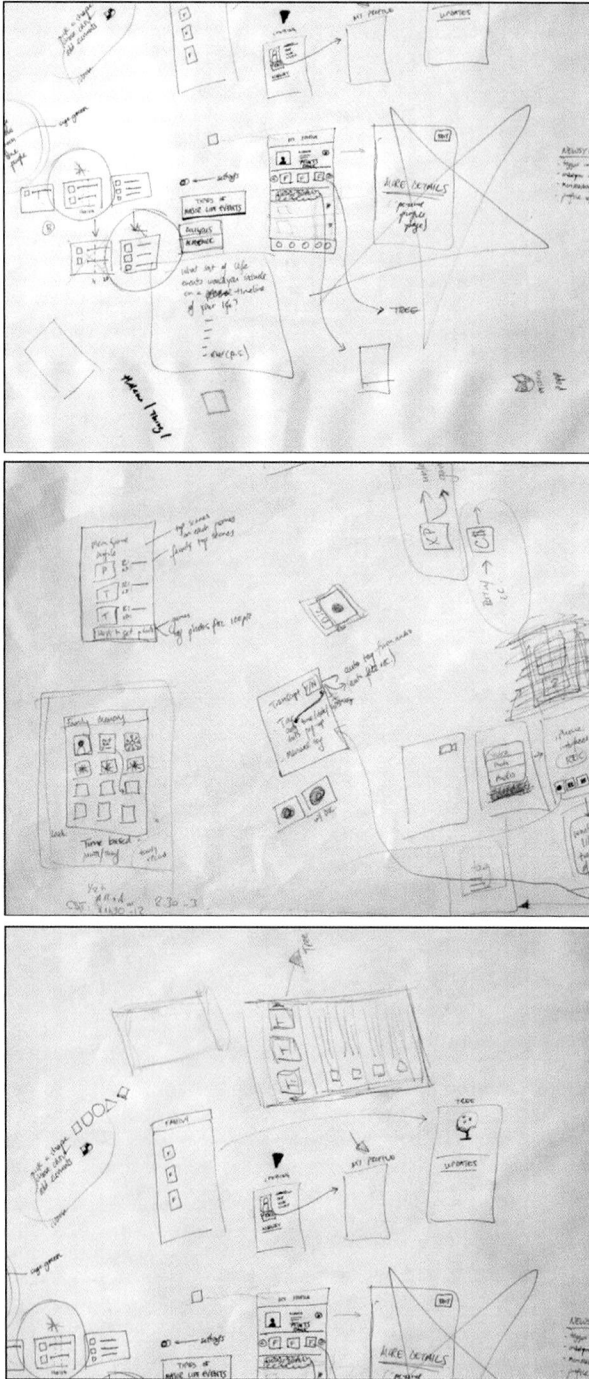

Fig. 5.14 Initial concept-design sketches of Story Machine Mobile application

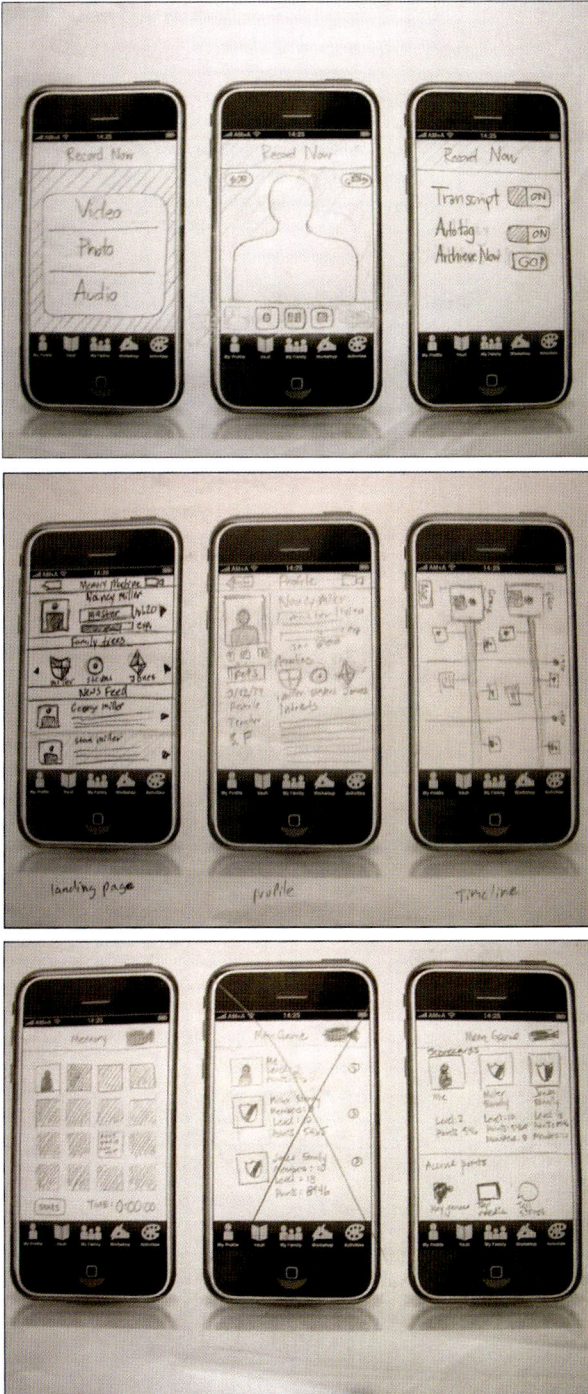

Fig. 5.15 Initial screen designs of Story Machine Mobile application

- View and contribute to family glossary, a dictionary of family-specific terms and expressions.
- View and contribute to family ritual book, a record of unique family whistles, handshakes, knocks, etc.
- View and contribute to family cookbook, a collection of family recipes.
- View and contribute to family skill book, a collection of family trades such as knitting patterns or woodworking projects.

Family; Social Media

- View family trees, family members' profiles and updates (including new stories or media).
- Ask questions and request media from family members.
- Answer questions and grant media requests.
- Contribute to others' stories using written, audio, or visual comments.
- Befriend and share media with other families (for instance, a family that grew up on the same block).
- Adjust privacy/sharing settings.

Storymaker: Guides and Templates

- View story templates, prompts, questions.
- Learn storytelling techniques such as effective structure and active listening.
- Catalog and view story triggers.
- The application supplies general triggers (e.g., "the scent of salt air," "the taste of rosemary," popular song titles), and users are encouraged to contribute and share their own story triggers, because they may be equally effective for family members.
- Create stories to be made available only at a specific time or location.
- Record audio and video conversations.

Activities: Games and Creativity

- Play Memory-themed games.
- Single-player games (concentration, puzzles).
- Multiplayer games (trivia games, hangman).
- StoryGame (app-based multiplayer game).
- Transform photos using photo filters and adding accessories or embellishments.
- Use templates to generate family media such as crests, scrapbooks, mottoes.
- Browse store for merchandise customized with newly created family crests, etc.

5.12.2   Initial Screen Designs

Based on the information architecture, AM+A has prepared initial concept sketches of some key screens of the Story Machine. These were used for internal evaluation, together with a questionnaire, to elicit feedback from a few potential users. All comments were considered in the redesigned screens that follow (Fig. 5.16).

Story Machine Concept Map

Money Management

Memory Management
- Record instantly
- Tag new stories

Health Management

Profile
Personal settings
- Update personal profile
  – Link to other social media (Facebook, Goodreads, Last.fm)
  – Import profile data
  – Import likes + dislikes
- Use Story Machine tutorial
- View Timelines
  – Personal timeline
  – Comparison timelines:
    - Family members' timelines
    - Celebrity timelines
    - Historical timelines
- Adjust privacy/sharing settings

Vault
Memorabilia repository
- Access stories
  – Favorites
  – Most viewed
  – Recently viewed
  – Recommended
- Explore, organize + house story material
  – Stories
  – Images
  – Written media
  – Audio and video recordings
  – Uncategorized + untagged materials
- View + contribute to family lore
  – Family tree
  – Pet memorial
  – Atlas
  – Glossary
  – Ritual book
  – Cookbook
  – Skill book
- Upload
  – Audio + video
  – Chats + transcriptions

Family
Social media
- View
  – Family trees
  – Family members' profiles
  – Profile updates (including new stories)
- Request
  – Information (ask Questions)
  – Media (ask for photos, etc.)
- Answer
  – Questions
  – Media requests
- Comment on others' stories
  – Written comments
  – Audio or visual contributions
- Befriend + share media with other families

Storymaker
Guides and templates
- View
  – Story templates
  – Story prompts/questions
  – Storytelling triggers: personal + general
- Learn
  – Storytelling techniques
  – Active listening
- Catalog
  – Personal memory triggers
  – Starred memory triggers
- Create stories to be made available only
  – At a specific time
  – At a specific location
- Record
  – Audio conversations
  – Video conversations

Activities
Games and creativity
- Play story- and memory-themed games
  – Single-player games (concentration, puzzles)
  – Multiplayer games (trivia games, hangman)
  – StoryGame (app-based multiplayer game)
- Transform photos
  – Use photo filters
  – Add accessories or embellishments
- Use templates to generate family media
  – Scrapbooks
  – Crests
  – Mottoes
- Browse store for customized family merchandise

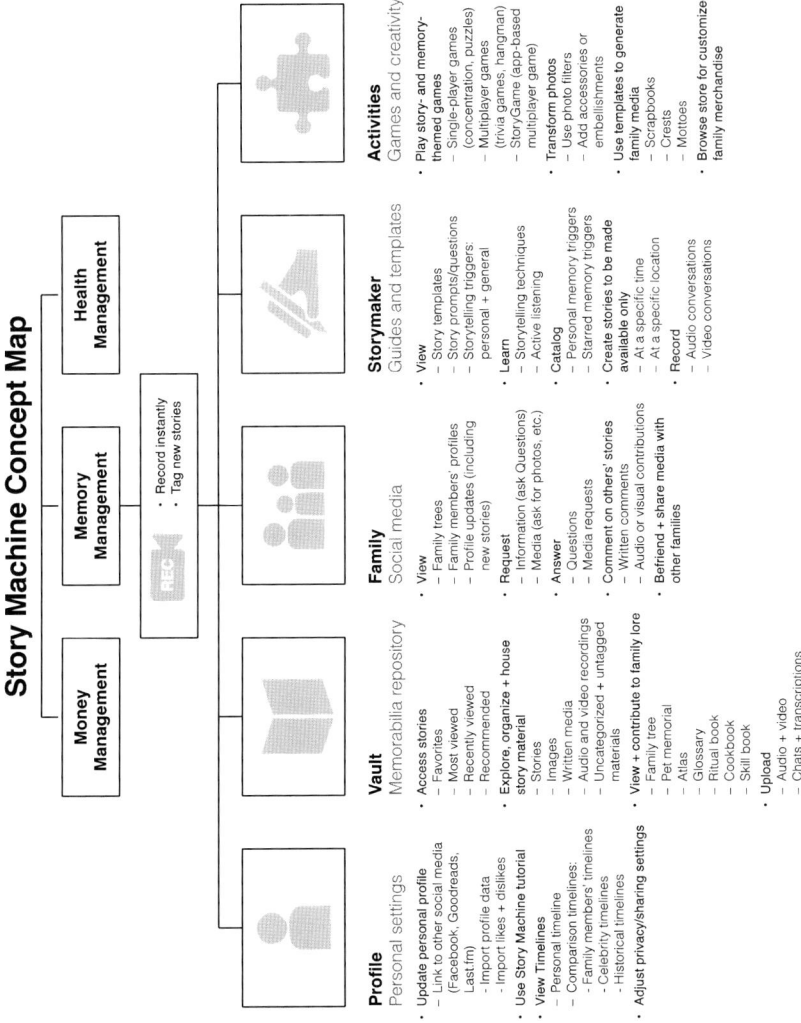

Fig. 5.16 Depiction of the Information Architecture of the Story Machine, which focuses on memory management, not on other possible functions such as money management and health management, which are not discussed in this chapter. There may be slight differences in nomenclature/titling between this diagram and subsequent screens due to ongoing development design-changes

5.12.3 Revised Screen Sketches and Designs

Based on internal review, AM+A prepared revised concept sketches of some key screens of the Story Machine. These were used together with a questionnaire, to elicit feedback from a few potential users. The screen designs show examples for both smartphones and tablets, such as the iPhone and iPad, and are not intended as an exhaustive description of the full user experience with the Story Machine.

Esthetically, the Story Machine extrapolates from the concept of the Family Tree, using trees, branches, and leaves as a larger visual metaphor for family, memory, and recording. In addition to its biological identity, a *leaf* also signifies a sheet of paper, or a page from a book. We intend the Story Machine to be a collection of leaves, which together create a rich, varied, and self-sustaining library of family memories and lore. The ginkgo leaf, with its ancient history and memory-enhancing ability, serves as the Story Machine's symbol.

5.12.3.1 Landing Page: Introduction to the Story Machine

Record Now Button: Represented by a video symbol at the top right of the application, this button is always present. Users tap this button to record audio, video, or take a photo at any time.

Logging in, the user sees his or her profile photo, display name, and status in the application-wide StoryGame (including rank, level number, points total, and number of stories told). Below is the user's Favorites, an easily accessible, scrollable list of favorite stories or media. At the bottom is the News section, showing activity related to the user (new story or media postings, updates from the user's family members, etc.) (Fig. 5.17).

5.12.3.2 Record Now: Never Lose a Train of Thought

Tapping the Record Now button — which is always present on the upper right of the application — takes the user to a menu where he or she can choose to record audio or video, or take a photo. The user may then elect to have the Story Machine transcribe speech (using Dictation), tag and categorize media automatically (using Autotag), and use the iPhone's GPS coordinates to identify the recording with his or her current location (Fig. 5.18).

After these preliminary decisions, the user is taken to a recording screen. When ready, the user taps Record (in the example below, the user is recording video). For time-based media such as audio or video recordings, a timer at the bottom right shows recording time. The user may pause, resume, or end the recording session at any point (Fig. 5.19).

Because the user has enabled Dictation, Autotag, and GPS, the Story Machine automatically generates a list of tags and buzzwords (seen at left) while recording.

Fig. 5.17 Initial screen designs of the Story Machine used for internal evaluations

At the end of the recording session, the user may choose from and add to this list of tags, provide the story with a title, and choose a screen capture or photograph to serve as the story's icon. For those events or stories with no accompanying visual media, the Story Machine provides a set of icons. The default story icon is the Story Machine ginkgo leaf.

5.12.3.3   Profile/Submenu: View and Edit Information

Using this submenu, users can view and edit the various pieces of information that go into a user profile. They may also monitor and edit privacy settings in order to determine which information other Story Machine users can view.

- *Information* consists of basic statistics such as name, date of birth, current and past residences, etc.
- With the *Family Tree* users can build and explore their genealogy.
- In the *Timeline* users can build and explore their personal history visually and temporally.

Fig. 5.18 Revised screen
design of Story Machine
Landing Page

- With *Recipes* users can contribute their own family recipes and peruse recipes from other family members.
- *Rituals* is an index and record of behavior specific to a family, such as a specific whistle, handshake, or other practice.

Users may place additional categories (e.g., Atlas, Glossary) in this quick-reference section (Figs. 5.20 and 5.21).

5.12.4   Profile/MyTimeline; Access Stories in an Organized, Visual Manner

Below is an individual timeline showing events throughout the user's lifetime, by both year and age. At the most zoomed-out level, timelines show only those stories and events that the user has tagged as major life events. As the user zooms in, the small tick marks along the x-axis come into focus, revealing further layers of stories and media.

Fig. 5.19 Revised screen designs of Story Machine Record Now function

When the user taps a story, a label pops up to provide simple background information, including a story's title and date, the date it was told and by whom, as well as a few identifying tags. Tapping the label takes the user to the story profile (see "Vault" for further information).

In some cases, several versions of the same story may exist. The multiple labels shown below indicate the presence of multiple versions. Tapping the labels lets the user compare vers (Fig. 5.22).

5.12.5 Profile/Compare Timelines: View Similarities and Differences in Life Paths

Users may choose to compare timelines with other users (including family members or celebrities with public timelines) or with a timeline of historical events. In the screens shown below, the user is viewing her timeline alongside her cousin's.

In the first example, the user is viewing timelines by year, which displays concurrent events in her and her family's lives (e.g., "When I was entering fifth grade, my cousin was just learning to walk"). In the second example, the is viewing timeline by age, to easily contrast her timeline with her grandmother's, revealing the similarities and differences in their life paths (e.g., "Nana got married at 20; when I was that age, I was studying abroad in Italy and very much single") (Fig. 5.23).

Fig. 5.20 An additional
revised screen design of
Story Machine Record
Now function

5.12.5.1   Profile/My Family Tree: Visualize Genealogy

An important aspect of the Story Machine is the visualization of one's lineage. The default view for one's Family Tree shows the user, his or her parents, grandparents, and great grandparents, along with photographs and dates of birth and death, from left to right. To explore his or her lineage further, the user simply taps an ancestor's name or photograph. The Family Tree will re-orient itself with the ancestor's name at the far left, and his or her parents, grandparents, and great-grandparents to the right (Fig. 5.24).

5.12.5.2   Vault/Search: Find the Right Story

In order to search stories in the vault, users can search by tag or keyword. The user below is searching for a story about her relative Emily. Additionally, the user is in the mood for a happy story and has entered this criterion as well. Users may choose to filter their search to reveal all media, or just a specific type (story, photo, recording, video, etc.). The user below is searching among stories.

Fig. 5.21 Revised screen
design of Story Machine
Profile function

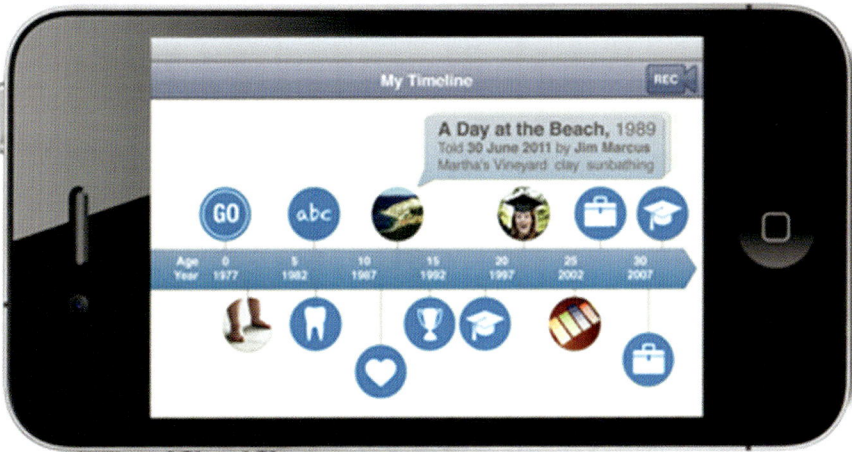

Fig. 5.22 Revised screen design of Story Machine Profile/My Timeline function

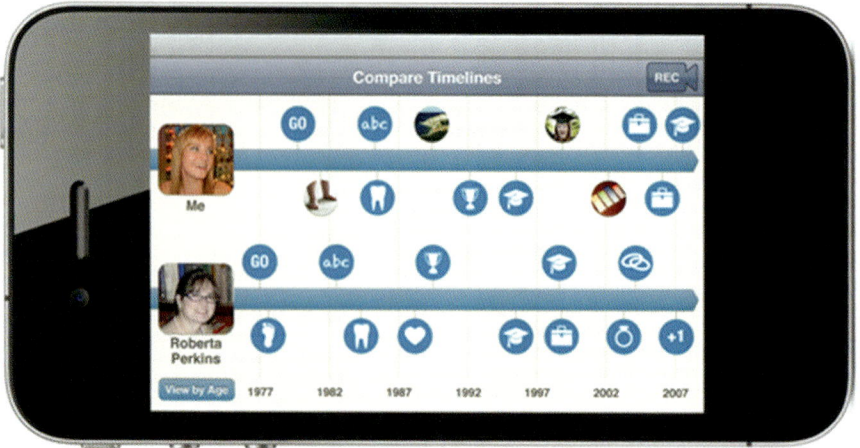

Fig. 5.23 Revised screen design of Story Machine Profile/Compare Timelines function

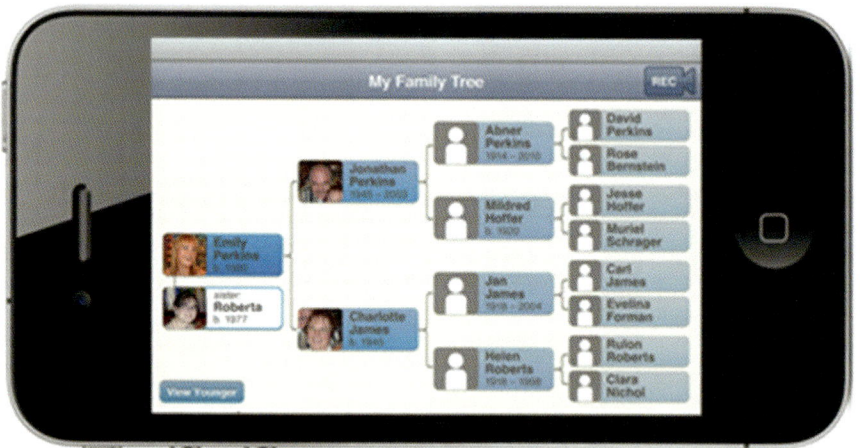

Fig. 5.24 Revised screen design of Story Machine Profile/My Family Tree function

For the story "Fourth Grade," which took place in 1988, the storyteller did not upload any accompanying visual media. As such, the default icon of a ginkgo leaf, symbolizing memory, appears below (Fig. 5.25).

5.12.5.3   Vault/Story Profile: Explore Stories More Closely

Selecting the story he/she would like to view brings the user to this detailed informational screen. The Story Profile below shows the title and date of the story, and whether or not the user has marked it as a favorite. Users may view the story itself by tapping the chevron at right. Users seeking more detailed information may continue to peruse this screen.

Fig. 5.25 Revised screen
design of Story Machine
Vault/Search function

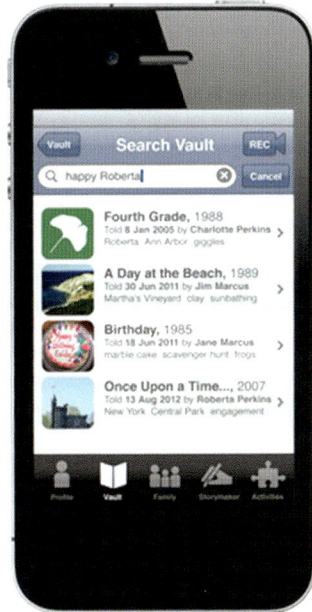

The Story Profile indicates that this story is a Major Life Event for 2 people. Major Life Events are given greater priority in Personal Timelines. Users have the option to tag and categorize stories or media after recording or uploading; designating Major Life Events is just one of the categorical options available. By tapping the boldface "2 people," the user can view these people's profiles, and choose to view this event within their timelines.

Below lies further information, identifying the storyteller, the date the story was told, and a full list of identifying tags to help users sort and categorize media. Tapping any one of these tags generates a new search.

Lastly, the Story Profile also provides story visualization. The example below contains a Story Map, which charts the development of the story as a collaboration among users. This map shows that Roberta Isaacs initiated the story, but that three more users commented on or added to this story, and that yet another user commented on or contributed further. Stories — particularly family stories —rarely consist of a single, linear perspective. Rather, they are shifting composites of multiple recollections. With the Map visualization, the Story Machine honors this network of contributions and contradictions and encourages further collaboration. Users may tap any of these nodes to see individual contributions.

Another option for story visualization is a word cloud, showing incidence of certain words or tags in the story itself (Fig. 5.26).

Fig. 5.26 Revised screen
design of Story Machine
Vault/Story Profile
function

5.12.6   Vault/View Story: Enjoy a Trip Down Memory Lane

In the following screens, the user gets to experience a story from the Story Machine. The story below consists of images and recordings, with a transcription below. The user simply swipes through the story to progress at his or her own pace.

This format is only one possibility with the Story Machine. With the templates and tools in the Storymaker section, users can combine a variety of media (photography, audio recordings, video, scanned images) to create a piece that suits both the storyteller and the story being told (Fig. 5.27).

5.12.7   Vault/Family Lore: Cookbook: Recipe

Just one example of the Family Lore housed in the Vault, the Cookbook is a record of cherished family recipes. (Other types of Family Lore include, but are not limited to, Family Rituals, Family Skills, Family Glossary, and Family Atlas.) In the

Fig. 5.27 Revised screen designs of Story Machine Vault/View Story function

example below, we see the name of the recipe, the creator of the recipe, as well as the name of the person who contributed or uploaded the recipe to the Story Machine. In this case, one person created the recipe, but a different family member uploaded the recipe into the Family Cookbook. Tapping the boldface names takes the user to that person's profile.

The top area also lists the number of stories in which the recipe appears. Tapping the boldface lets the user peruse a list of these stories. Additionally, the recipe provides an overall star rating and number of reviews; tapping the rating or reviews provides more detail. Lastly, there is a quote or summary about the recipe from the contributor, as well as the option for the user to add his or her own comments or contributions.

Below is the recipe as entered by the contributor. A speech bubble indicates a comment from another user (e.g., "I like to use shallots instead of scallions" or "A little extra cheese in the middle layer doesn't hurt"). Next, users may add items in the ingredient list to a shopping list, or email the list to themselves or others as a reminder.

Fig. 5.28 Revised screen
design of Story Machine
Vault/ Family Lore
function

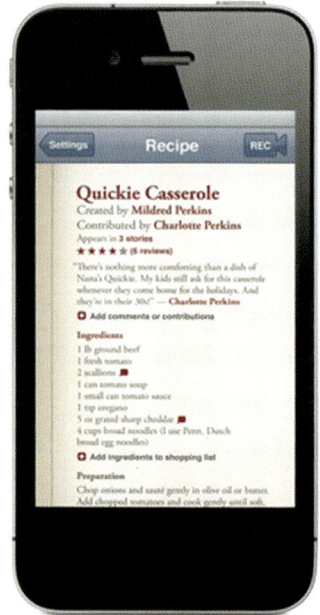

The background shown below is just one of several options available in addition to the default Swiss design. As users advance within the application-wide game, they can unlock even more options for visual customization (Fig. 5.28).

5.12.8 Family/Search

Finding out about one's family is easy with the Story Machine, by using its social network, or Family. Simply type in part of your family member's name, and all network members whose names match will appear in this filtered list that provides a photo, name, age, current residence, and profession. To view a full profile, users tap the chevron at right (Fig. 5.29).

Fig. 5.29 Revised screen
design of Story Machine
Family/Search function

5.12.9   Family/Profile: A User's Public Profile

In the top section of a user's public profile, the default settings display his or her name, relation to user (tap statement for an explanation and diagram), current residence, age and date of birth, rank, level, and total points accrued in the application-wide StoryGame, as well as number of stories recorded. Icons underneath the user's photo also indicate social networks to which he or she subscribes. Tapping the chevron at right provides this and additional information (occupation, previous residences, place of birth) in further detail.

Below is a scrollable list of more detailed information, including access to the user's family tree, personal timeline, stories (favorites, recently viewed, and most viewed), cookbook, and news (e.g., recent contributions or comments). Users can customize which of these informational pieces are available publicly, and to whom.

The next section is a scrollable list of the user's families. Families may include traditional bonds of blood and marriage, but the Story Machine is flexible, and allows for unconventional family models that include bonds of friendship and

Fig. 5.30 Revised screen
design of Story Machine
Family/Profile function

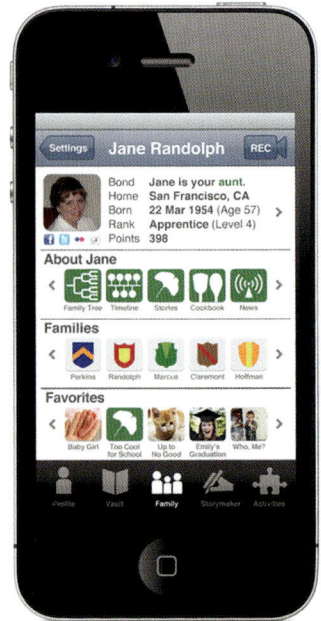

shared experience. In the example below, none of Jane's relatives bear the name "Claremont." Instead, it is the name of the user's college dormitory, and she counts 7 former roommates among her Claremont family members.

Last is a scrollable list of Jane's Favorites, so that the user might learn a little more about Jane by getting a clearer picture of the memories she holds dear. This list is the same as the list Jane views when she opens the application (except for those favorites she has chosen not to publicize) (Fig. 5.30).

5.12.10   Storymaker/Templates: 20 Questions

The Storymaker section of the Story Machine helps users organize and tell the best possible stories, with templates (story prompts and structural rubrics), techniques (active listening and ways to inject humor), and triggers (names of places or historical events, as well as sensory triggers like scent and taste).

Fig. 5.31 Revised screen
design of Story Machine
Storymaker function

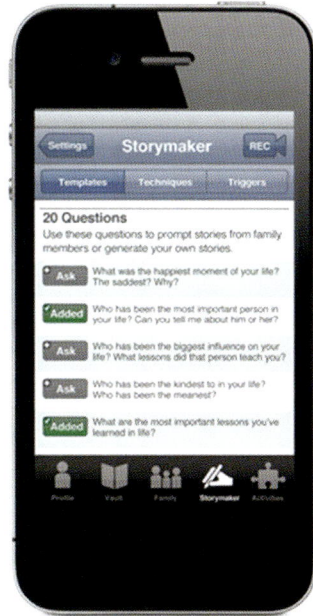

In the example below, the user has chosen to peruse a list of 20 questions, or prompts for a story. One can use these questions to elicit narratives from family members or to prompt one's own stories. The user can select among these questions to build a customized storytelling guide or interview template (Fig. 5.31).

5.12.11   Storymaker/Record: Create a Story

Here the user is recording a video chat with a family member, using the questions selected from above as a guide and structure. When the user would like to move on to a new topic, he or she simply taps the chevron at right, and the next question appears along the top. For easy review and editing, the recording is automatically indexed by time, as well as by topic. Should the conversation return to a previous topic, the user simply taps the chevron at left until he or she sees the appropriate question along the top (Fig. 5.32).

Fig. 5.32 Revised screen design of Story Machine StoryMaker/Record function

5.12.12   Activities: Play, Compete, and Create

The Activities section of the Story Machine provides users with incentives, competition, and fun. It comprises five subsections:

- View: Monitor individual and team ranks within the application-wide StoryGame.
- Earn: Learn ways to accrue more points (e.g., tagging and categorizing media, filling out user profiles, recording stories, playing single-player or multiplayer memory- and story-themed games).
- Play: Find and play single-player or multiplayer memory- and story-themed games.
- Create: Design your family memorabilia (e.g., family crest) or edit photos and with decade-specific customizable templates.
- Shop: Purchase items emblazoned with your personal family crest or special photo.

5.12.13   Activities/View

Here the user can monitor his or her level and point total within the application-wide StoryGame. As users reach certain goals (filling out their profiles, recording their first stories), they accumulate shields. Similar to merit badges, shields give the user the opportunity to boast of their achievements and incite competition with other players.

Users' families are also in competition with one another. Families pool their resources, and compete against each other in a team-oriented manner.

Fig. 5.33 Revised screen
design of Story Machine
Activities/View function

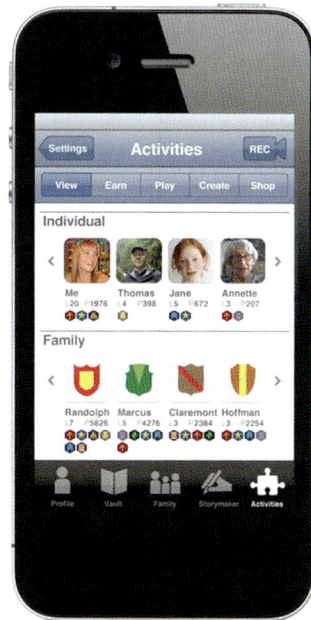

Individual and family point totals are displayed in two discrete scrollable lists. Tapping an individual or family takes the user to the individual or family profile for more detailed information (Fig. 5.33).

5.12.14   Activities/Play/Memory

In the ongoing application-wide StoryGame, players can accrue points by partici- pating in discrete single- or multiplayer games. Below is an example of one such game available in Activities.

This game of "Memory" is similar to the classic visual matching game, but with a slight variation. Instead of matching identical pairs, the user matches a family member's profile photo with his or her name and birth date. Other options include date of birth, current residence, or trivia such as favorite color.

The inclusion of these data will serve to strengthen the user's knowledge and understanding of his or her family and history. Because the Story Machine's social network is already filled with such information, the application can generate this game — and multiple iterations of it — without any extra data entry on the part of the user. Another such app-generated game is a family-themed "Newlywed Game," in which players try to guess information (e.g., "What is Grandma's favorite food?") about their relatives. Higher scores translate to higher ranks — and greater privi- leges — within the StoryGame.

Fig. 5.34 Revised screen
design of Story Machine
Activities/Play/Memory
function

At any point during the game, the user may tap Pause (though the screen may be covered during pauses to prevent cheating), Stats (to view his or her high scores, and family-wide high scores), or Help (to view options or brush up on rules). The timer on the bottom right helps players keep track of their speed (Fig. 5.34).

5.12.15 *Activities/Create/Family Crest*

Among the more exciting creative activities in the Story Machine is the Family Crest, a simple five-step process in which users can develop their own heraldry:

- Shape and Pattern: Choose from a scrollable selection of shapes and patterns with semantic meaning.
- Color: Select from an entire spectrum of colors. Swatches of previously used and/or suggested swatches are visible to the left for easy access.
- Imagery: Add mascots, symbols, objects to your crest.
- Name: Style your family name with colors and typefaces.
- Motto: Craft and style your family axiom.

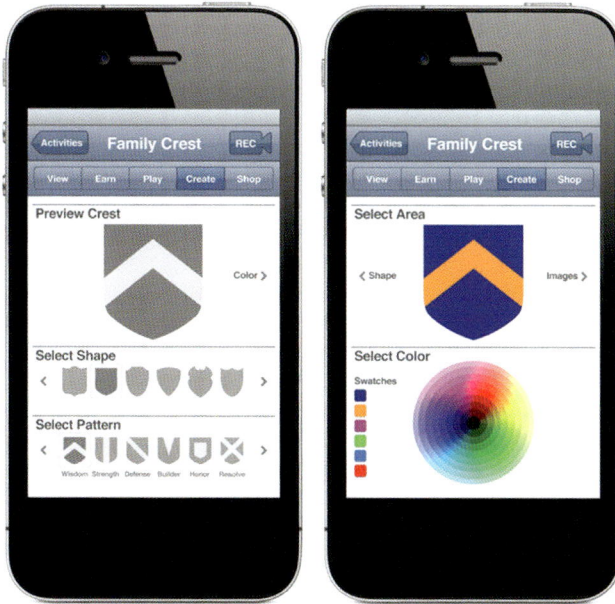

Fig. 5.35 Revised screen designs of Story Machine Activities/Create/Family Crest function

Once users have created this crest, they can use it as their family symbol. In the case of multiple crests, the family member with the highest StoryGame rank chooses the family crest (Fig. 5.35).

5.13   Evaluation

5.13.1   Plan

AM+A planned to interview approximately ten people (two per persona) and approximately two advisors with expertise in story-management behavior change. Through these interviews, we hoped to learn what works, what doesn't, and how to refine the Story Machine prototype further, from information architecture (metaphors, mental model, and navigation) through look and feel (appearance and interaction). Unfortunately, time ran out on these evaluation steps, but in the future, we can accommodate them as circumstances permit.

5.14   Next Steps and Conclusions

Using the user-centered design approach described above, AM+A plans to continue to improve the complete Story Machine development process, which would require significant time and funding from an outside source. Tasks include the following:

- Revising personas and use scenarios.
- Conducting user evaluations.
- Revising information architecture and look and feel.
- Building initial working prototype (e.g., for iPhone, or other tools, platforms).
- Evaluating the Story Machine across different cultures.
- Developing the Story Machine for business use as well as personal use.
- Researching and developing improved information visualizations.

AM+A aimed to incorporate information design and persuasion theory for behavior change into a mobile phone application. This self-funded work on the Story Machine project is current and ongoing, and was undertaken to demonstrate the direction and process for such products and services. Though the design is incomplete, AM+A is willing to share the approach and lessons learned, in the interest of helping improve family communications worldwide and overcome challenges to the loss of family legacies.

At this stage, AM+A is seeking to persuade other design, education, family-oriented companies and nonprofit organizations, museums, cultural heritage and historical heritage organization, and governmental groups to consider similar development objectives. We hope that our process and concept prototypes (the self-funding of which inevitably limited the amount of research, design and evaluation) will inspire others, and that they benefit from the materials provided thus far.

The process has already been demonstrated successfully with a previous project, the Green Machine (Marcus and Jean 2009), versions of which have been considered and used by SAP for enterprise software development (Marcus et al. 2011).

AM+A's long-term objective for the Story Machine is to develop a functional working prototype, to test whether the application can actually persuade people with story collection, management, and sharing challenges, to exercise greater preservation of family legacies, to increase healthier story-management habits, and to pursue a more culturally sound lifestyle under real use conditions over the long term. If the theories are proven to be correct, this approach could have significant implications that will benefit millions of people in the USA and abroad.

5.15   Story Machine 2.0

In 2011, AM+A gave a 1-week workshop at the Institute of Design, Illinois Institute of Technology, called Mobile User-Experience Design. The subject matter of the workshop revolved around taking the Story Machine further. Graduate students researched, analyzed, designed, and evaluated further aspects of a Story Machine, called Story Machine 2.0. The following is one of the reports, prepared by Raphael D'Amico, Russell Flench, Kwame Green, and Ted Pollari, and is presented with their permission and acknowledgement.

The Story Machine 2.0's goal is to change storytelling behavior, to persuade people to capture the stories we naturally tell one another in order to preserve them for future enjoyment or for sharing with others. Students were given leeway in their approach to the problem, free to use the information architecture and screens already developed by AM+A or to re-envision the application completely. In this Appendix, the team describes their concept. The workshop featured a final-deliverables review via Skype with one of the Story Machine 1.0 team members, Catherine Isaacs, who critiqued the work of the student team.

5.15.1   Problem Space

What defines a family? While there are many answers to this question, there is one partial answer that resonates strongly for many: shared experience and the sharing of experience which, over time, becomes shared history. As Story Machine 1.0 reports, in a recent survey, 81 % of US respondents had family members living out of state or outside of the USA. Now, more than ever, families are experiencing the dramatic effects of dispersion as members move apart for a wide variety of reasons. From school to new jobs and even upon retirement, sometimes for marriage, divorce, or even natural disasters, families commonly experience the effects of distance.

Facing these growing distances and growing frequency of intrafamily dispersion, the modern family experiences fewer opportunities to share experiences and personal histories. Qualitatively, this reduces the shared experiences of intimacy and weaken the very bonds that make family, family.

Our team saw a clear opportunity for design explorations surrounding this question: How might we enhance a dispersed family's ability to share history and enhance their experiences of emotional intimacy?

5.15.2   Existing Solutions

Communication across distances is not a new need and there are many existing solutions: from postcards and letters to long-distance telephone calls and now emails, instant messaging, video chats, and Facebook friendships. However, none of the existing solutions, with the exception of Facebook, allow for collecting or sharing a rich record of the interactions or the ability to add narrative to externally collected artifacts such as pictures or videos. While Facebook allows users to share and comment on photos and videos, it does not offer affordances for saving chats or video chats nor does it make it easy to share those interactions with other family members.

5.15.3   Personas

For this project we made the following personas:

Scott

Scott is 41 years old and lives in Washington DC. He's married and has two children – a daughter, Marlie and his newborn son, Jacob. Jacob's namesake, Scott's paternal grandfather, was a memorable man, always up to something, always with an opinion and usually a big, warm smile. Unfortunately, he passed away three years ago, widowing his wife of 62 years, Ruthie, and leaving a lot of memories for all that knew and loved him.

Scott loves his wife, Ani, and his two young children but his job demands a lot of travel and this means Scott is away from his family quite often and he rarely gets to see his own parents or his grandmother, Ruthie. When he's not away on business, Scott's focus is all on his family.

Ruthie

Ruthie is in her golden years. She's a sprightly 87 years old, still trying to keep herself busy while she continues to enjoy retirement. When she and her late husband, Jacob, retired, they moved to Boca Raton, Florida. It was a tough choice, but they both wanted to get away from the D.C. area where they'd lived and worked for years.

Ruthie still lives by herself, but she's got a lot of friends in the senior community where she lives. She tries her best to stay in touch with her loved ones and wishes they'd call or visit a bit more often. She's considered moving back to the D.C. area, but she'd be giving up all the friends she's made over the last 20+ years, living in Florida.

5.15.4   Use Scenario

Scott wants for his new son to meet his (paternal) grandmother, Ruthie, but like so many families today, they are separated by a significant distance. Living in Florida, Ruthie is over a thousand miles away from Scott. Nevertheless, Scott wants his son, Jacob, to know about just what an amazing man his namesake, Jacob, Ruthie's late husband, was. Even though Jacob is just a newborn, Scott knows that they don't have too many more years with Ruthie. While he doesn't like to think about losing his grandmother, Scott wants to be sure his son hears all about Ruthie's husband, and he wants it to come directly from Ruthie, herself.

Ruthie recently gave Scott several family photo albums which he had professionally digitized and then uploaded to the Story Machine app on an iPad. The last time they visited, Scott set up an iPad for Ruthie. She doesn't use it much, but she's curious and wants to see what all the excitement is about.

Scott hopes that with the new Story Machine app, now Ruthie will be able to easily look at the pictures snapped throughout her life and that the app will help them both bridge the distance between Washington D.C. and Boca Raton, Florida.

Fig. 5.36 Story Machine
2.0. Opening screen design

5.15.4.1   Solution

After a series of focused explorations, our team concluded that the essential components for allowing family members to share history and develop richer emotional connections was to help them actively and passively share the experience of reminiscing, both live and retrospectively.

Our concept focuses on how to allow family members curate collections of memory artifacts and connect them with stories told by individuals in their family network. Those stories are captured live through a video-chat interface that makes it easy for users to capture and highlight key segments while connecting the recording segments to specific pictures/memory artifacts. When not in use for the active video-chat/story telling activities, this concept app runs in a passive mode which presents an evolving display of photos/memory artifacts, displayed simultaneously to all members of the family network through their Story Machine 2.0 app.

In the following figures, we present a basic use scenario and screen flow on a tablet showing the core functions of our concept as used by Scott and Ruthie (Fig. 5.36).

5.15.4.2   Passive Screen

Story Machine 2.0 begins as a passive screensaver-like display which cycles through images in the Memory Vault. These may be of people or artifacts, and the app uses metadata to create the most emotionally meaningful arrangement of images, ideally

Fig. 5.37 Story Machine
2.0. Selected image design

synchronized across each family member's devices. If you can see it at any given moment, so can the rest of your family (Fig. 5.37).

5.15.4.3   Grandma Ruthie Calling

Tapping a picture of a person immediately calls that person. It is also the way one can respond to an incoming call, in this case from Grandma Ruthie (Fig. 5.38).

5.15.4.4   Conversation

Accepting the call immediately connects you to Ruthie. Along the top is the Family Strip, showing one's most frequently contacted relatives. On the bottom is the Memory Strip, here showing the last image Scott and Ruthie discussed. When one taps the current image in the Memory Strip, Story Machine 2.0 begins recording and storing the conversation within that image as metadata (using speech-to-text to make it searchable), so the story behind it is stored for posterity (Fig. 5.39).

5.15.4.5   Family Tree View

Dragging the previous view's family strip downwards expands the list into a display of the detailed family tree, including family members who have passed away.

Fig. 5.38 Story Machine
2.0. Grandma Ruthie in
conversation (design)

Fig. 5.39 Story Machine
2.0. Family strip (*top*) and
memory strip (*bottom*)

Fig. 5.40 Story Machine 2.0. Family tree view designs

Fig. 5.41 (*left*) Story Machine 2.0. Selecting media items to add to Memory strip. (*right*) Selecting stories to add

Ruthie mentioned her own father, Scott's great-grandfather. There he is, in black and white (Fig. 5.40).

5.15.4.6 Memory Vault

Dragging the memory strip upwards from the main conversation screen pulls up the Memory Vault, which aggregates the images, videos, and other media uploaded to Story Machine 2.0 (Fig. 5.41).

Fig. 5.42 Story Machine
2.0. Selecting other images

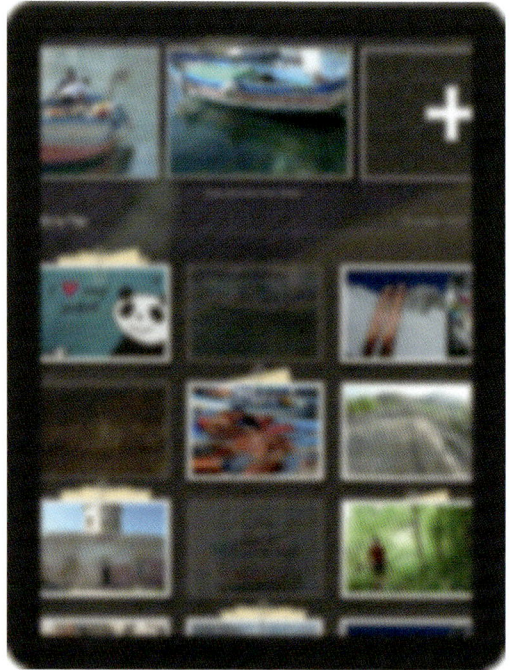

This is where you can select media items and pull them up to the Memory Strip to add to the current conversation, and view the stories already attached to them (e.g., as notes clipped to the image, highlighted in the illustration to the left).

Scott selects an old picture of a Sicilian church to review, and sees other members of his family have already discussed it. He could tap those text stories to hear the recording of that conversation, highlighted in the illustration (Fig. 5.42).

5.15.4.7   Digging Deeper

Scott decides that he wants to learn more about the boat he'd been discussing earlier, and retrieves a different picture of it, one which captures a different side, into the Memory Strip, to prompt a story and have it permanently stored with that image (Fig. 5.43).

5.15.4.8   Ending a Conversation

After a fruitful conversation, Scott and Ruthie say goodbye to each other and see a summary of the media they've discussed, in this case the two pictures of the boat. Scott can see the parts of the conversation that were recorded (Fig. 5.44).

Fig. 5.43 Story Machine
2.0. Ending a call

Fig. 5.44 Story Machine
2.0. Passive screen
showing synchronized
display of photos

5.15.4.9   Passive Screen

After the conversation is finished, Scott and Ruthie set their tablets back into their charging docks. Shortly, the passive display mode resumes, showing both a shared synchronized display of photos, updating them as time passes. As they go about their day, they occasionally catch glances of the display and think about the stories they've shared.

5.15.4.10   Conclusion and Future Directions

We believe that the concept, as presented, represents a strong opportunity to help reconnect members of the modern family who may live across town or across the globe. In all cases, an increased sense of shared experience can help strengthen the family bonds that we share with our loved ones.

The user interface presented is intended to be a simple one, usable by almost anyone. There are numerous specific details that require attention, but the purpose of the concept was specifically to focus on the key functionality of the app and we believe that it presents a powerful vision. There is also a clear opportunity to develop a platform to drive a number of interconnected systems and devices.

One could easily envision the utility of an expert/"family historian" user interface, possibly hosted on a dedicated Website. Another opportunity would be an even simpler device that focused solely on the passive-display modes, allowing families to easily make the shared memory experience a larger, ever-present part of their home or desk environment.

5.15.4.11   Process Documentation

The following figures show sketches made to document the development process (Figs. 5.45, 5.46, and 5.47).

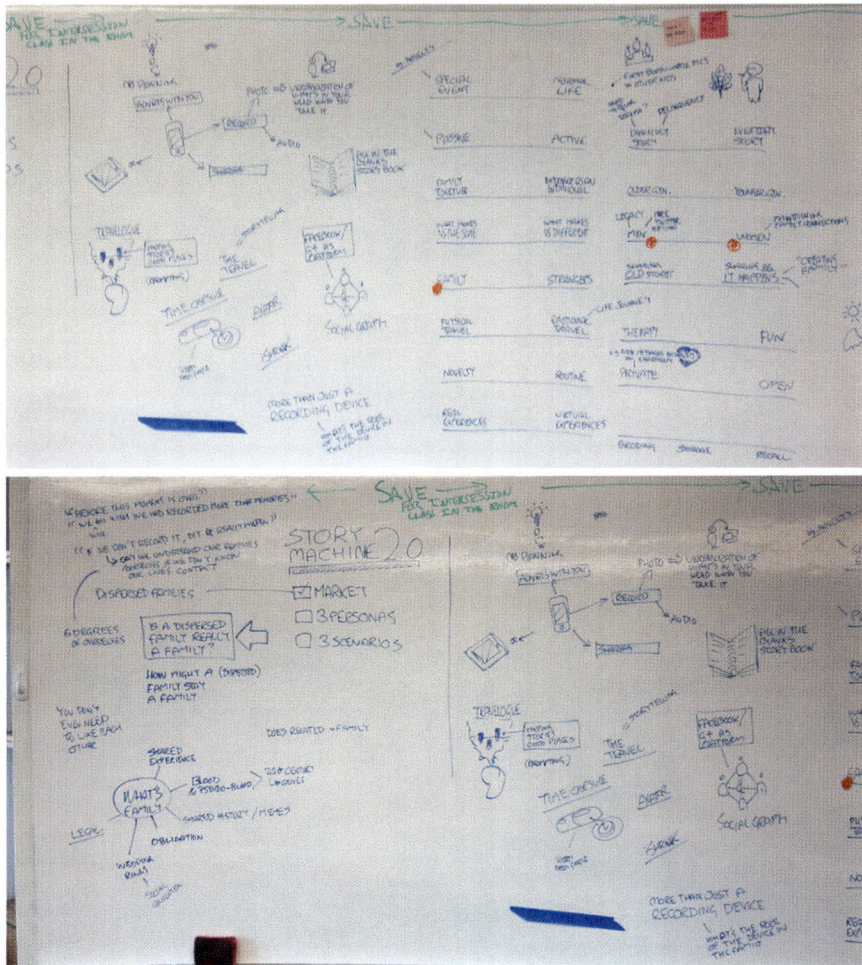

Fig. 5.45 Early diagrams of discussions

Fig. 5.46   Concept generation using persuasive design framework

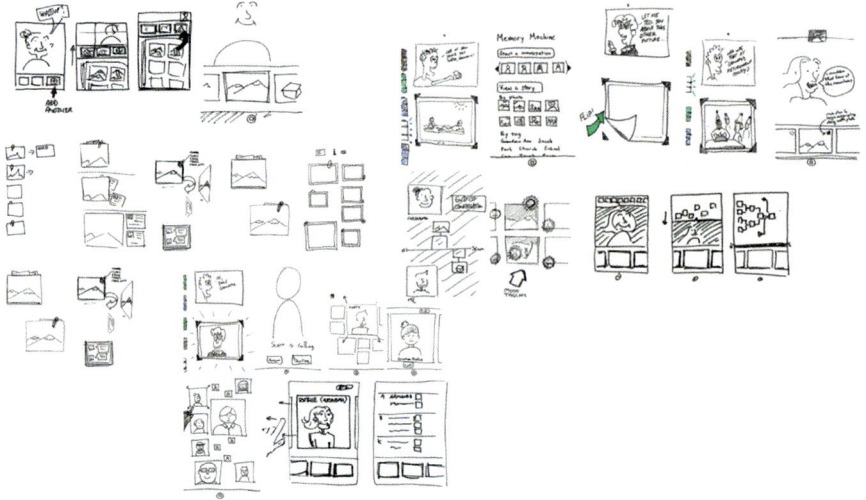

Fig. 5.47   Concept sketches for Story Machine 2.0

5.15.4.12   Process Documentation

The following figures show sketches of process documentation (Fig. 5.48).

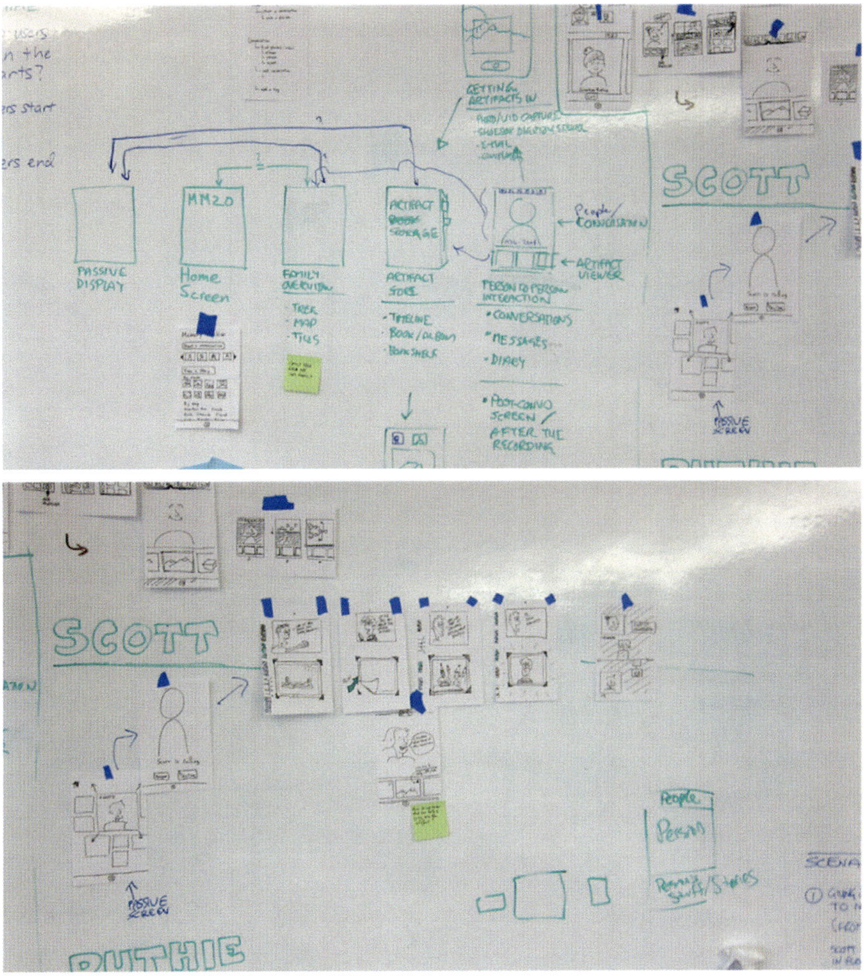

Fig. 5.48 Story Machine 2.0. Sketches of navigation/flow concepts

Acknowledgments The author thanks his AM+A Associates, Hélène Savvidis, Catherine Isaacs, Chris Chambers, Carlene O'Keefe, and Tim Thianthai, for their significant assistance in planning, research, design, analysis, and documentation related to this project.

Photography Imagery Acknowledgment Creative Commons photography in this chapter by Rachael E. C. Acklin, James Capaldi, Wenliang Chen, Dave Cooper, Mark Evans, Will Folsom, Dale Gillard, Mark Heard, Andrew Karmy, Powel Loj, Aaron Marcus, Eugene Oden, Dermot O'Halloran, Windell Oskay, Ruthanne Reid, Joe Shlabotnik, Pavel Tcholakov, Kevin Tostado, Mike Youngquist.

Further Reading

Ancestry.com [mobile app]. (Version 2.0.3). Retrieved 6 September 2011 from http://itunes.apple.com/us/app/id349554263

Ancestry.com [mobile app]. (Version 2.0.3). Retrieved 6 September 2011 from http://itunes.apple.com/us/app/id349554263

Baker M (2011) OMG! My grandparents R my BFF! Wall Street J (79) :D1, 9 May.

Blurb Mobile [mobile app]. (Version 1.3). Retrieved 6 September 2011 from http://itunes.apple.com/us/app/id430933688

Editors (2014) Surprising facts about grandparents. http://www.grandparents.com/food-and-leisure/did-you-know/surprising-facts-about-grandparents. Accessed 14 October 2014. Origoinally retrieved 15 July 2011 from http://archive.grandparents.com/surpising-facts-GPs.pdf

eFamily [web app]. Retrieved 6 September 2011 from http://www.eFamily.com

Fröhlich S, Kleis C (2007) Alles Über Meine Mutter. Krüger, Frankfurt

Living Memory [web app]. Retrieved 6 September 2011 from http://www.livingmemory.com.

Memory Miner [web app]. Retrieved 6 September 2011 from http://www.MemoryMiner.com

MemoryHub [web app]. Retrieved 6 September 2011 from http://www.MemoryHub.com

Memory-Of.com [web app]. Retrieved 6 September 2011 from http://www.memory-of.com

Museum of London Street Museum. Museum of London Website (2010) Retrieved 22 July 2011 from http://www.museumoflondon.org.uk/Resources/app/you-are-here-app/index.html

MyFamily [web app]. Retrieved 6 September 2011 from http://www.myfamily.com

NPR StoryCorps [web app]. Retrieved 6 September 2011 from http://www.storycorps.org

Ozgurol O, Rizvanoglu K (2011) How to improve user experience in mobile social networking: a user-centered study with Turkish mobile social network site users. Presented at design, user-experience, and usability track, human-computer interaction international, Orlando

Public Speech [mobile app]. (Version 1.2). Available from http://itunes.apple.com/us/app/id358055883

Ramachandran S (2011) Playing on a tablet as therapy. Wall Street J :D1, 26 July 2011

Rao L (2011) Flurry: time spent on mobile apps has surpassed web browsing. TechCrunch Website. Retrieved 6 September 2011 from http://techcrunch.com/2011/06/20/flurry-time-spent-on-mobile-apps-has-surpassed-web-browsing

Schiesel S (2011) Brave new world that's as familiar as the machine it fights with. New York Times, 16 July, p C1

Steel E, Angwin J (2011) Device raises fear of facial profiling. Wall Street J :A1–A2, 13 July 2011

References

Aziz SA, Dickey J, Fernando A, Twomey I (2011) Family Wallet (Unpublished scholarship). University of California, Berkeley

Familiar (2011) (Unpublished scholarship). California College of the Arts, San Francisco

Fogg BJ, Eckles D (2007) Mobile persuasion: 20 perspectives on the future of behavior change. Persuasive Technology Lab, Stanford University, Palo Alto

Marcus A, Jean J (2009) Going green at home: the green machine. Inf Des J 17:233–243

Marcus A, Dumpert J, Wigham L (2011) User-experience for personal sustainability software: determining design philosophy and principles. Presented at design, user-experience, and usability track, human-computer interaction international, Orlando, 9–14 July 2011

Maslow A (1943) A theory of human motivation. Psychol Rev 50:370–396

Chapter 6
The Travel Machine: Combining Information Design/Visualization with Persuasion Design to Change Travel Behavior

6.1 Introduction

The travel and tourism sector is one of the fastest growing economic sectors in the world. Numerous analysts estimate that, in the future, travel and tourism will be the most important industry in the service economy and the main job engine for many countries. Already by 2021, travel and tourism is predicted to account for 69 million more jobs, almost 80 % of which will be in Asia, Latin America, the Middle East, and Africa (WTTC 2012).

Particularly in the developed world, making trips to foreign countries has become both a luxury and a necessary good. That means that even in times of economic crisis and recession, people might travel less frequently, for shorter periods or to closer destinations, but, in general, they are very reluctant to completely renounce their holidays abroad. For this reason, despite the economic downturn in 2009–2011, the growth of the travel and tourism sector continued in 2012, so that international tourist arrivals could reach the milestone of one billion in 2012 (UNWTO 2012). Most notably in the developed world, an increased standard of living as well as numerous technological advancements significantly facilitated the process of traveling and have opened the world of travel and tourism also to people with lower incomes. Especially the great progress in the field of information and communication technologies (ICTs) fundamentally changed, restructured, and enriched the sector, giving rise to e-Tourism (Buhalis 2003).

However, there is a tendency among some analysts of the history of traveling to describe a rather devolutionary trend of tourist life: from the old, "golden" days of the Grand Tour in the seventeenth and eighteenth century to the prepackaged and standardized holiday experiences of the present day (Löfgren 2002: 279). In fact, if one thinks about an Ernest Hemingway traveling on his own in the early twentieth century through Paris, Rome, and Florence in order to absorb the essence of European civilization, writing down his impressions, thoughts, and cultural experiences in his legendary Moleskin notebooks, then one realizes that many things have changed

© Springer-Verlag London 2015
A. Marcus, *Mobile Persuasion Design*, Human–Computer
Interaction Series, DOI 10.1007/978-1-4471-4324-6_6

even since then. The classical tourist of the late twentieth and current twenty-first century is more likely to move around within a group, to follow well-beaten tracks, and to record her/his itinerary in a series of often randomly taken pictures, so that the idea of traveling as a process of intercultural learning and exchange and as a process of personal development and self-discovery no longer seems to be the main motivation when going abroad. Visiting the sights and monuments of another place has sometimes become a kind of treasure hunt, a list of places and objects that have to be seen and ticked off in the "to-do list" or "bucket list" upon arriving at a certain destination.

At the same time, the technological inventions and innovations of recent years have contributed, at least to some extent and mainly in the developed world, to a growing reemergence of independent travelers, who benefit from the assistance and guidance provided by new ICTs. Nevertheless, this development does not necessarily mean a return to the culturally enriching and mind-broadening travel experiences of the Grand Tour. When looking for an appropriate place to have dinner, for example, it is enough for the traveler of today to "consult" the mobile application of the respective destination, which will immediately furnish the address of the closest restaurant together with some special offers and/or discounts, from which she/he might benefit. It becomes less important, perhaps, whether this restaurant is a really local and traditional one or whether it offers basically the same dishes the traveler could enjoy as well in her/his home town.

The primary objective of the Travel Machine is therefore to change the above-described travel behavior, to ease, enhance, and enrich the experiences of a person going abroad, to foster the process of intercultural understanding and learning, and thus to contribute, at least partially, to a shift from leisure to cultural tourism.

Leveraging the increasing trend towards use of mobile devices in the field of tourism (Wang/Xiang 2012: 308), we planned to research and design a mobile application, the Travel Machine, which should assist and accompany a traveler in the discovery of a destination or, better yet, in the living and learning of a foreign culture. By combining information design with motivation and persuasion theory, with a particular focus on the works of Maslow (2006) and on Fogg's captology (Fogg 2003), the use of the Travel Machine should prompt the traveler to change her/his travel behavior and to make out of her/his trip a deeper and more personally enriching experience. It is above all this dimension of personal, intercultural learning that should distinguish this application from the majority of mobile apps that already exist in the field of travel and tourism and that predominantly are limited to booking and reservation functions or to the mere provision of information and location-based services (LBS) (Wang/Xiang 2012).

The success and effectiveness of this approach, i.e., the combination of information design with persuasion design in order to promote behavioral change of mobile application users, has already been studied and realized in previous projects of the author's company: the Green Machine, the Health Machine, the Money Machine, the Story Machine, the Innovation Machine, and the Driving Machine (Marcus/Jean 2009; Marcus 2011, 2012a, b). They have been/will be described in previous/succeeding chapters. All of these past Machine projects and the current Travel

Machine rely on the design of the application through a user-centered user-experience development process.

6.2 User-Centered User-Experience Development

As described in Chap. 1, the user experience (UX) can be defined as the "totality of the […] effects felt by a user as a result of interaction with, and the usage context of, a system, device, or product, including the influence of usability, usefulness, and emotional impact during interaction, and savoring the memory after interaction" (Hartson/Pyla 2012). That definition means the UX goes well beyond usability issues, entailing also social and cultural interaction, value-sensitive design, emotional impact, fun, and aesthetics (Hartson/Pyla 2012).

The user-centered design (UCD) approach links the process of developing software, hardware, and user interface (UI) to the people who will use a product/service. UCD processes focus on users throughout the development of a product or service. AM+A carried out most of the UX UCD tasks (see Chap. 1) of planning, research, analysis, design, evaluation, and documentation, except for implementation.

Over the past two and one-half decades in the user interface (UI) design community, designers, analysts, educators, and theorists have identified and defined a somewhat stable, agreed-upon set of *user interface components*, *i.e.*, the essential entities and attributes of all user interfaces, no matter what the platform of hardware and software (including operating systems and networks), user groups, and contents (including vertical markets for products and services).

6.3 Persuasion Theory

As described in Chap. 1, persuasion theory can identify how to attract and motivate users and to help change their behavior in the short term and long term. As described earlier, AM+A used Cialdini's theory (2001a, b), Fogg's theory (Fogg 2003; Fogg/ Eckles 2007), and Maslow's *Theory of Human Motivation* (Maslow 1943) in developing the Travel Machine. Each step has requirements for the application.

Motivation is a need, want, interest, or desire that propels someone in a certain direction. From the sociobiological perspective, people in general tend to maximize reproductive success and ensure the future of descendants. We applied this theory in the Travel Machine by making people understand that a determinate travel behavior can fundamentally enrich their trip experiences abroad, that it can increase their intercultural knowledge and understanding, and that it can ultimately trigger a process of self-development and improve self-understanding.

We also drew on Maslow's *Theory of Human Motivation* (Maslow 1943), which he based on his analysis of fundamental human needs. We adapted these to the Travel Machine context:

- The safety and security need is met by the assistance of a Travel Advisor and by the provision of destination-related information and tips.
- The belonging and love need is expressed through friends, family, and social sharings and support.
- The esteem need can be satisfied by social comparisons that display travel progress and destination expertise, as well as by self-challenges that are suggested by the Travel Advisor and that display goal accomplishment processes.
- The self-actualization need is fulfilled by being able to follow and retrace the continuous travel progress and advancement in the personal travel diary.

6.4   Information Architecture Impact

The above considerations have the following impact on the information architecture of the Travel Machine:

6.4.1   Increase Use Frequency

Games and rewards are the most common methods to increase use frequency. In the Travel Machine, we have developed a vocabulary game and a game testing destination-related knowledge. Moreover, we developed a third game that is related to the accomplishment of challenges suggested by the Travel Advisor. In terms of the rewards, the users will be provided virtual rewards (such as "expert" statuses and new skins for blogs) and titles (such as "X is now a four-star Japan expert"), which they can communicate, for example, via Facebook or Twitter. In addition, we chose social comparison as another incentive to increase use frequency. Users will be able to consult other travelers' diary entries, and they will compete with them on the degree of expert status and on the number of accomplished challenges.

6.4.2   Increase Motivation

Because setting goals helps people to learn better and improves the relevance of feedback, one of the main features of the Travel Machine is the mobile Travel Advisor, which assists the traveler by providing information, by giving suggestions and ideas, by posing challenges, and by offering regular feedback. The Travel Advisor motivates; it pushes the user towards the experience, the experimentation, and the observation of cultural idiosyncrasies at a destination, by suggesting a range of different activities in the form of challenges or goals. The incentive of self-accomplishment will encourage users to achieve these challenges, which, in turn, will change and improve their understanding of intercultural issues and which will enrich their travel experiences. This process of intercultural learning and self-discovery is also likely to change the users' approach towards foreign countries, cultures, and people in the long term.

In addition, the creation of a personal travel diary will increase users' awareness of their steady travel progress and will motivate them to add further material in the form of visual or textual records of objects, experiences, and encounters.

Social interaction also has an important impact on behavior change. Therefore, another important component of the Travel Machine is to leverage social networks and integrate features like those found in forums, Facebook, Twitter, or blogs. Users can send notes or messages to their friends or to other travelers, and they can share their personal travel diaries with them. Furthermore, they can consult other travelers' or friends' diaries and can ask them for information and tips. These social ties will serve as an additional incentive to motivate behavior change.

6.4.3 Improve Learning

The central objective of the Travel Machine is to trigger a learning process that is both informative and reflective and that combines textual and visual elements. Thus, the advisor provides a range of practical (emergency numbers, addresses, etc.), current (news, events, weather, etc.), general (history, religion, politics, currency, etc.), and more culture-specific (values, symbols, rituals, dress code, etc.) information, but it also proposes a range of activities that are supposed to help travelers themselves discover cultural particularities and characteristics of a destination. The idea is to support, assist, and stimulate travelers during their trips, in order to make them observe certain things and to make them reflect actively about their experiences.

We also sought to make the education process both informative and entertaining. Therefore, we proposed three different games that refer to vocabulary knowledge of the respective foreign language, to destination-related general and more culture-specific information, as well as to the challenges posed by the Travel Advisor. Through playing games featuring educational information, users will be able to increase their intercultural knowledge and understanding, without getting bored.

6.5 Metaphors and Mental Model

Based on the UCD UX approach, on the results of the market research, and on the comparison studies, as well as on the persuasion and motivation theories of Fogg and Maslow, our main UX design objectives for the Travel Machine were to ease, enhance, and enrich the experiences of people going abroad; to foster a deeper-going process of self-development, of intercultural understanding and learning; and to thus contribute to a short- and long-term change of typical travel behavior via a shift *from leisure to cultural tourism.*

Therefore, the focal point of the Travel Machine, as opposed to the majority of existing mobile applications in the field of travel and tourism, is the traveler, not on the destination itself, which will nevertheless play an important role in the concrete implementation and conception of the software. The Travel Machine is intended especially for travelers aged up to 50 years, from higher to average economic and educational demographics, who have a general interest in and openness towards foreign countries, cultures, and people.

To put it simply, the Travel Machine aims to answer the following two critical questions:

- How can information design/visualization present persuasive information to promote a personally enriching, profound, as well as educative short- and long-term travel behavior change?
- How can mobile technology assist in presenting persuasive information and promote behavior change of medium- to high-income and educated people?

The Travel Machine contains several key components. The many detailed functions are grouped into five sets or "tabs": a dashboard (Tab 1), an individual travel diary (Tab 2), a mobile Travel Advisor (Tab 3), a tab related to social comparisons and social interactions (Fellows, Tab 4), and a section dedicated to games (Challenge, Tab 5). The sections below explain each tab.

6.5.1   Tab 1: Dashboard

The Travel Machine design concept contains a dashboard, which can be considered as the "homepage" of this mobile application. The dashboard contains the most important information travelers may need at any time during their trips. Therefore, potential elements of the dashboard are the following:

- Current time and date: Users can set two clocks, one displaying local time and the other one indicating the current time of a home city or country.
- Current weather and weather forecasts.
- Local, regional, and national news.
- Events of the day: By clicking on an "Events" symbol, users will see the most significant events taking place in a current location (and surroundings) on the same day, as well as during the following two days.
- Current location with possibility to show a large-scale map of the general scene.

This map will show, also, the main attraction points close to users, and it will indicate whether and where there are other travelers around using the same app. Furthermore, "dangerous" neighborhoods, streets, and places will be highlighted in the map. Less-experienced travelers might be less prudent and careful; therefore, they may be provided with additional "security tips" (e.g., carefully watch one's purse or wallet when on public transportation, not to walk around alone after 9 pm, etc.).

A good example of a mobile application offering highly informative and detailed maps is the National Park Maps HD app, which is provided by *National Geographic* (http://www.nationalgeographic.com/mobile/apps/national-park-maps/). In addition to showing park visitors' current exact position, the maps also show what is around them, they guide them to chosen locations, and make further suggestions about where they may want to go.

- Remaining travel budget: Travelers may set up a preset travel budget before going on a trip. Similar to a fuel gauge, the dashboard will keep track of expenditures and will show at any time how much money has been spent and how much is still left for the respective day, week, or month or for the entire trip.
- Time budget: If travelers have already decided about certain activities at certain times in their travel agenda, which will be integrated into their Travel Diary (see Tab 2 below), the dashboard will indicate how much free time is left until a next appointment.
- "Fuel gauge" for social relations and commitments: Before the trip, travelers can determine how many postcards, emails, or social media updates they will have to send and how many presents they will have to buy.

The dashboard then indicates the travelers' "social progress," i.e., how many postcards have already been mailed or how many souvenirs have been purchased and for whom. This capability might at the same time serve as an incentive for travelers to improve their social skills and their current social level. The dashboard will regularly give "gift notifications" in order to remind users of their social tasks.

All of these elements are optional for Travel Machine users, which means that users can personalize and customize the dashboard according to their needs and interests, by deciding which of the abovementioned features to show and which ones to hide.

6.5.2   Tab 2: Diary

A central element of the Travel Machine is the integration of a personal, customized Travel Diary, which might be considered as the modern and technologically advanced realization of the small Moleskin book of Hemingway's days in the early twentieth century. It must be stressed, however, that not only Hemingway and other intellectuals enjoyed meticulously recording their travel observations and experiences but also ordinary travelers frequently made an effort to accurately write down what they saw, heard, felt, and thought during their trips, as is shown in the following authentic travel records of an ordinary traveler in 1975 (Fig. 6.1).

Travelers of the twenty-first century should be able to document digitally their travels and to keep daily records of their personal travel itineraries, in the form of texts, pictures, audio recordings, and/or video recordings, thus, to track the continuous progress of their trip experiences. Notwithstanding the strong criticism and fears of data protection specialists, this idea of providing a highly personalized

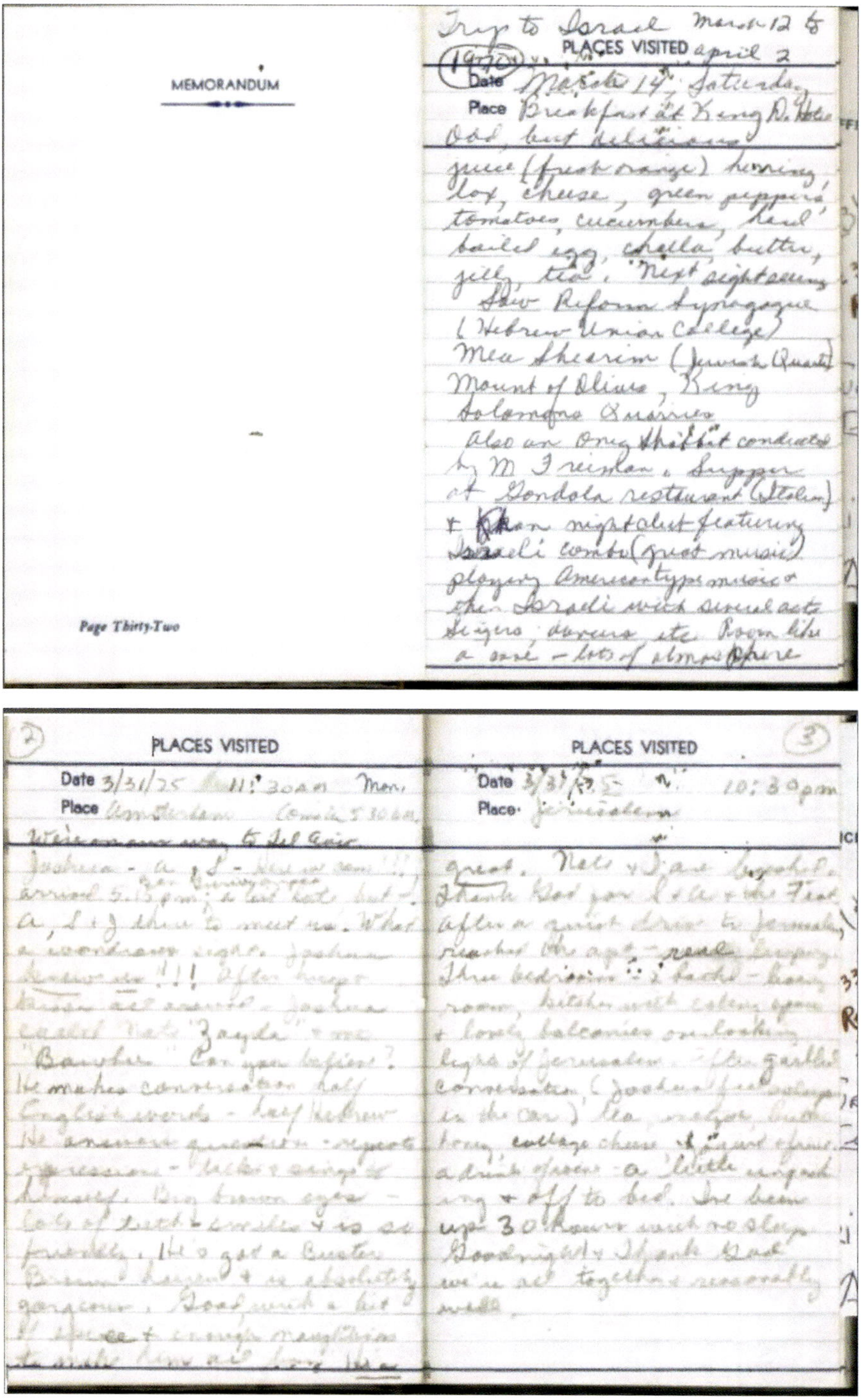

Fig. 6.1 Extracts from an actual travel diary of an "ordinary person," from the travels of Libbie Burstein Marcus, 1975

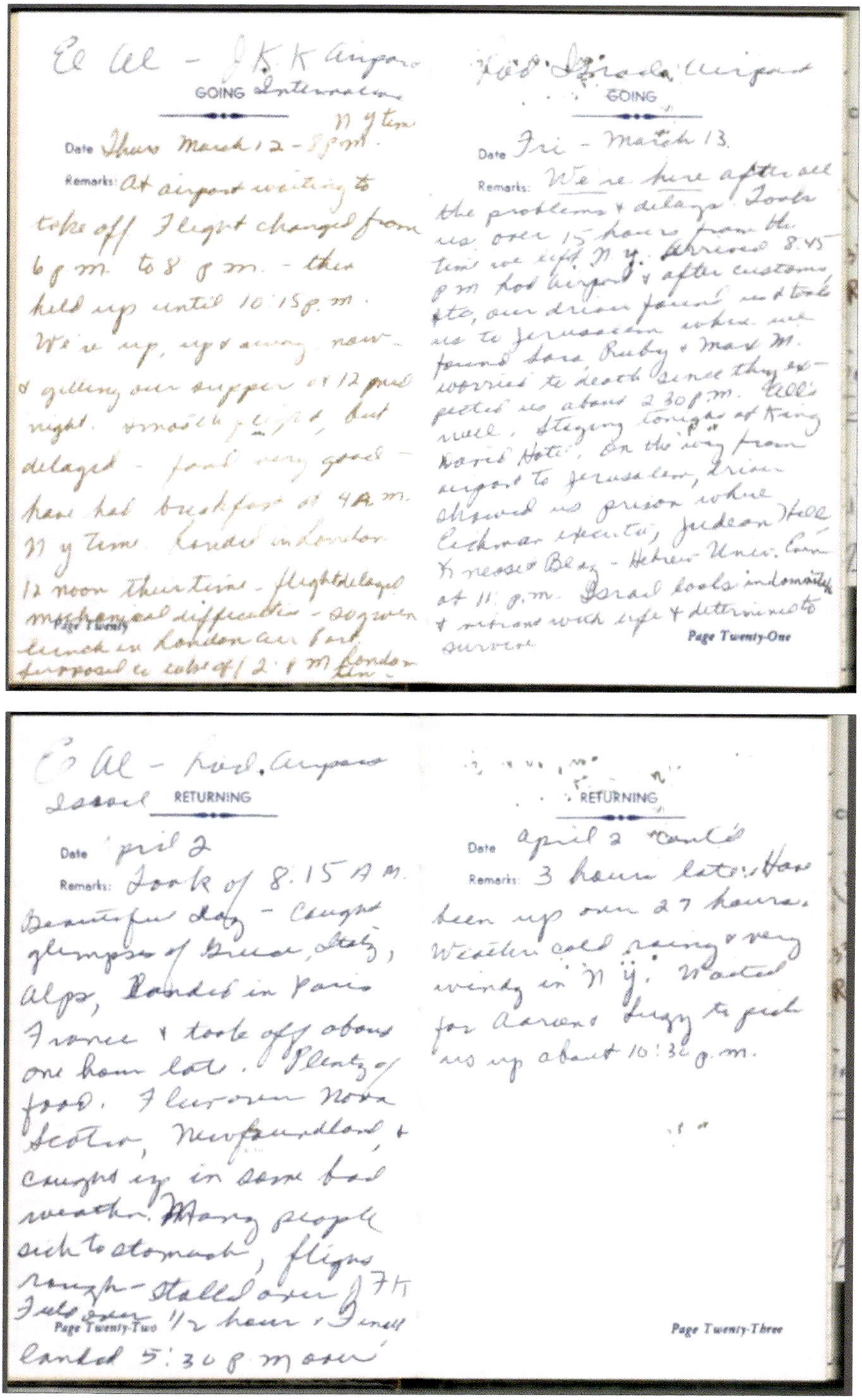
El Al - J.F.K Airport International N.Y. time

GOING

Date Thurs March 12 - 8 p.m.

Remarks: At airport waiting to take off. Flight changed from 6 p.m. to 8 p.m. - then held up until 10:15 p.m. We're up, up & away now - & getting our supper at 12 mid night. Smooth flight, but delayed - food very good - have had breakfast at 4 A.m. N.Y. time. Landed in London 12 noon their time - flight delayed mechanical difficulties - so given lunch in London air port. Supposed to take off 2 p.m. London time -

Page Twenty

Lod Israel Airport

GOING

Date Fri - March 13.

Remarks: We're here after all the problems & delays. Took us over 15 hours from the time we left N.Y. Arrived 8:45 p.m. Lod Airport & after customs etc, our driver found us & took us to Jerusalem where we found Sara Ruby & Max M. worried to death since they expected us about 2:30 p.m. All's well. Staying tonight at King David Hotel. On the way from airport to Jerusalem, driver showed us prison where Eichman executed, Judean Hills Knesset Bldg - Hebrew Univ. Came at 11 p.m. Israel looks indomitable & vibrant with life & determined to survive

Page Twenty-One

El Al - Lod Airport Israel

RETURNING

Date April 2

Remarks: Took off 8:15 A.M. Beautiful day - Caught glimpses of Greece, Italy, Alps, Landed in Paris France & took off about one hour late. Plenty of food. Flew over Nova Scotia, Newfoundland, & caught up in some bad weather. Many people sick to stomach, flight rough - Stalled over JFK field over 1/2 hour & finally landed 5:30 p.m. our

Page Twenty-Two

RETURNING

Date April 2 cont'd

Remarks: 3 hours late. Have been up over 27 hours. Weather cold rainy & very windy in N.Y. Waited for Aaron's Iggy to pick us up about 10:30 p.m.

Page Twenty-Three

Fig. 6.1 (continued)

chronicle of users' individual experiences is also, at least in part, behind Facebook's profile concept of the timeline.

Specific features of such a Travel Machine Travel Diary could be the following:

- A virtual notebook with pages for notes and ideas dedicated to every day of the journey. This notebook will enable travelers to keep track of their itinerary with some automatic data fill-in via GPS sources and to dedicate special records to memorable incidents and encounters during the trip. Moreover, the Travel Machine users will have the possibility to personalize and customize their diary, i.e., to choose the background color and picture of the pages, as well as the font and the information to be shown on top of each entry (date, place, mood, budget and daily expenditure, time spent walking, number of steps taken, etc.). In an online Travel Machine store, the app users may also pay for the acquisition of more sophisticated backgrounds or fonts. We suggest that an information visualization summary might enable users to see a thumbs-up/thumbs-down or happy/sad summary of all their travel days based on summaries of mood indicators to help travelers get a sense of the emotional tone during or after their journeys.
- A personalized travel photo album, in which travelers can comment on single photographs in written or oral form, and where they can opt for a "photo of the day." The pictures will be displayed in the form of a gallery on the diary page of the respective day, on which they were taken, and they will be arranged in a long stripe on the right margin. By clicking on them, they become enlarged; by dragging them to the right, they are deleted. These entries might be entered automatically or manually into travel blogs, depending on the users' settings.
- An individual travel vocabulary that not only contains a dictionary with basic words and sentences with which one should be familiar but that gives the possibility to travelers to add terms or phrases that they have learned or accidentally have picked up during their trips, again to increase awareness of the learning process and progress they are undergoing. Also short vocabulary tests in the form of quizzes may contribute to this form of feedback (see Tab 5 below). The dictionary will be organized in five sections: small talk, food, direction, culture, and mixed.

Firstly, such a personal travel diary can be both useful and motivating for travelers while they are on a trip: It makes them aware of the steady progress of their travel itinerary, it makes them recall the moments and things they already experienced, and it might also make them more attentive and eager to discover new places of a destination, in order to further enrich the daily records.

Secondly, the value of the travel diary also concerns the after-trip phase. Instead of having a series of pictures that could have been taken by anyone, and the exact content of which is often even unknown to the travelers themselves, the diary would represent a unique and highly individual record of travel souvenirs, containing little stories about personal encounters with locals; commented photographs of impressing objects, people, or views; vocal comments about allegedly curious observations and experiences; a collection of names for dishes one had never heard of and one had never tasted before; and so on and so forth. Having a look at such personal travel notes, and reexperiencing in this way the past travel route, is without doubt more motivating, enriching, and memorable for travelers than going through a rather monotonous succession of photographs.

Overall, the conception of an individual completely customized travel diary might thus stimulate intercultural interest, openness, and learning during the trip, and it might, also, enable travelers to take long-lasting, deeper-going memories out of their experiences abroad. At the same time, such a form of travel documentation represents a particular combination of cultural and digital storytelling. The Travel Machine will help users to create and tell stories about people (a kind of travel-oriented Story Machine, as described earlier in Chapter 4), places, incidents, events, etc. Such intimate and personal tales are usually the most enriching experiences when going abroad, more than merely looking at attractions and sights (Robbins 2012). This aspect is closely linked to the social dimension, by which the application will be fundamentally enriched and which will be dealt with more in detail below (see Tab 4).

In addition, the travel photo album deserves special attention, given that it contains enormous potential for enhancing and enriching travelers' experiences during and, also, after their trip. By combining the photo album with, for instance, tag and face recognition functions, the application can show to travelers who and what is around them at a certain moment in time. By attaching a URL to a picture, they will just have to click on it in order to get more information and further details about what is in the picture. Moreover, by using geo-tags, the application could indicate to users' specific locations, buildings, or cities that appeared in famous movie scenes. After photographing and geo-tagging, for example, the Trevi Fountain in Rome, travelers would have the possibility to directly watch on their mobile phone a video clip of the famous scene of the movie "La Dolce Vita." As far as pictures of food and drinks are concerned, a URL link could directly refer users to the respective recipes.

Taking into account possible future improvements, and technical developments in the field of photography, such as, for instance, the potential arrival of gigapixel cameras (Naik 2012), it becomes clear that more and more "hidden details" will be made visible in travelers' photographs, which will constitute both a huge challenge and at the same time a great opportunity for enriching trip experiences. At the same time, the fact that some travelers have even started to hire professional photographers to record their journeys (Petersen 2012) clearly illustrates the huge importance of this technology for creating long-lasting travel memories.

In addition to documenting the *during-trip phase*, there is also a need to integrate the *before-trip phase* of information collection and schedule planning. For this reason, Travel Machine users will have the possibility to insert into their diary a provisional schedule or calendar of their trip, writing down the main attractions and activities they are planning to integrate into their travel itinerary. Since most of the travelers may carry out their pre-trip Web research via a desktop computer, by consulting pages such as Google (Maps), TripAdvisor, Expedia, or Lonely Planet, they will be able to gather and organize all this information on the Travel Machine Web portal. They can then transfer the created travel schedule to their mobile phone and insert the schedule into the diary. In this way, users can also set on their phone the time frame of their trip.

In addition to planning and scheduling a trip itinerary, the portal may also include possibilities to find appropriate flights and hotels, which again is part of a before-trip phase. The integration of travel search sites, such as *Hipmunk* (http://www.hipmunk.com/), into the Web portal, will assist travelers in choosing appropriate accommodation and transportation options.

In order again to enhance the cultural dimension of the trip, the Web portal will suggest that users not only make use of the abovementioned, most common, and popular travel information sites, but it will recommend more specialized culture- and history-related Webpages for the respective travel destinations (see, for example, http://www.chinaculturecenter.org/ for China).

Of course, Travel Machine users will be able later, while on the trip, to constantly modify and update their provisional plans directly via mobile phone, i.e., to eliminate, change, or add elements (regarding places to go, attractions to see, dates, itineraries, etc.).

6.5.3 Tab 3: Advisor

Concerning the creation of a personal travel diary, one has to keep in mind that not all travelers may possess the background knowledge and the creativity of an Ernest Hemingway. Therefore, Travel Machine users will be assisted and accompanied during their trip by a mobile Travel Advisor. It is important to mention that this Travel Advisor will be essentially different from the typical push services, which numerous applications already offer (see, e.g., (Coelho/Dias, 2011)) and which are based on both the current location of travelers and on the preferences they have expressed in advance. Such services belong to the field of so-called personalized hypermedia. Based on users' characteristics, needs, and preferences, on usage behaviors, and on the respective usage environment, they provide personalized and tailor-made offers and recommendations.

These services literally "spoil" travelers by bringing them to places that they will most probably like and that are consistent with their common preferences and with their expectations. The recommendations can be a shop, where they will get discounts, a bar that is closest to their current position, a restaurant that corresponds to their usual tastes and predilections, the nearest McDonald's, etc. However, in order to really immerse oneself in another culture and in order to learn something about that culture, it is necessary, at least to a certain extent, to leave one's own cultural environment and to go a bit off the familiar, most convenient path. That means that travelers should be induced to step out of their personal comfort zone, to get rid of usual habits and behaviors, and to leave behind part of their "cultural baggage" in the course of the trip.

Exactly this behavior change is the main goal of the Travel Advisor function. By suggesting specific activities, by giving information, instructions, incentives, goals, and challenges, the Travel Advisor tries to get travelers further, to enhance and enrich their intercultural learning process, and to assist them in making the best out of their trip. This approach is substantially different from that of a traditional mobile travel guide: The Travel Advisor asks/suggests to users to do certain things, to discover specific places, and to find out something specific about another culture or country. The Travel Advisor gives tasks to travelers and tries to stimulate their intercultural interest and openness. In order to change Travel Machine users' behavior, the Travel Advisor does not spoil users but tries to push, to motivate, and to challenge them, by giving different inputs in the form of ideas, suggestions, and tasks. This approach of challenging application users already has been implemented successfully in other areas, for example, in the field of health (Kim 2012).

Thus, the Travel Advisor proposes to Travel Machine users a range of activities, which can be related in many different ways to the respective location's culture, religion, literature, history, culinary traditions, etc., such as the following:

- "You have just arrived in X. These are the first 10 things you should do during your first day."
- "Go to a shop, and buy something you have never eaten before."
- "Go to restaurant X, and order something for which you have no idea of what it actually is."
- "Go to this place, and ask a local in his language about the closest local bar. Go there, and ask for traditional local drinks."
- "Go to this museum to see the painting of X, and find out why it is important for the local culture."
- "Go to the fruit and vegetables stall, and interview the vendor about from where he gets the fruits and vegetables, about what his preferred/less-preferred tourists are, where he was born and raised, *etc.*"
- "Take a local bus to get to point X. From there go to point X by foot."
- "Go along this road, and look at the people: What is different from your country?"
- "Ask someone to write your name in the local-language characters."
- "These are your five daily challenges: *etc.*"

By looking at these examples, it becomes clear that the Travel Advisor is not trying necessarily to push users towards shopping or eating in a specific location but towards meeting and getting in touch with the people and with the culture of the respective place and towards doing and observing specific things that will help them understand and experience cultures different from their own. The Travel Advisor teaches travelers how to be open for a foreign culture and how to learn more about it. By presenting the suggestions in the form of goals or challenges, the Travel Advisor gives *persuasive incentives* to users and stimulates their need of self-accomplishment. The Travel Advisor can also provide travelers with regular feedback about their progress and can thus further enhance their motivation.

There are also several ways by which the Travel Advisor can be aligned with the above-described dashboard. If the dashboard indicates a certain amount of free time, travelers may ask the advisor to make suggestions on how to spend these free hours, on where to go, what to see, what to try, etc. Furthermore, by linking the

Travel Advisor to the social "fuel gauge" on the dashboard, the advisor can also act as a tutor for social relations, by suggesting to travelers to buy the still missing presents or to send the lacking postcards.

For the explicit configuration of activities proposed by the Travel Advisor, it will be helpful to take into consideration the main theories, concepts, and cultural dimensions of cultural theorists, for example, Hofstede 2012. Even though there are without doubt several limitations, and debatable points in Hofstede's research, his work nevertheless represents a helpful instrument when it comes to the analysis and comparison of different cultures. Therefore, the consideration of Hofstede's five main dimensions, which refer to high/low power distance, individualism/collectivism, masculinity/femininity, high/low uncertainty avoidance, and long-term/short-term time orientation, could be very useful for the creation of activities to be suggested by the Travel Advisor. If one thinks, for example, about American travelers going to Japan, effective ways of making them experience cultural differences could include the following:

- Help them to participate in a group excursion together with other Japanese people (US individualism vs. Japanese collectivism).
- Help them to observe Japanese behavior towards officials and authorities (US low *vs.* Japanese high power distance).
- Help them to ask a business woman whether a high job position is something common for Japanese women and how much Japanese people usually work (US average *vs.* Japanese high degree of masculinity).
- Help them to discover Japanese rituals and important etiquette (US low *vs.* Japanese high need for uncertainty avoidance).
- Help them to discover basic theories of Confucius and their importance for the Asian world (US short-term *vs.* Japanese long-term time orientation).

Also, if we assume the opposite situation, i.e., Japanese travelers going to the United States, Hofstede's dimensions can be a helpful instrument to generate activities, which could be proposed by the Travel Advisor and which could make Japanese travelers understand fundamental cultural differences between their country and the United States:

- Help them to take a bus, and suggest to them to randomly get off at the 5th bus stop (US low vs. Japanese high uncertainty avoidance).
- Help them to observe US behavior towards officials and authorities (US low *vs.* Japanese high power distance).
- Help them to understand the importance of traditions and religion in US society (US short-term *vs.* Japanese long-term time orientation).

We mention, again, that the Travel Advisor must be considered distinct from a typical travel guide, in that the Travel Advisor only gives inputs and tries to motivate travelers to discover things by themselves and to then tell or show them to the uses, ideally via the diary. Unlike a typical guide that provides perfectly prepared and "precooked" information, this form of coaching enables an education and learning process that is based, also, on reflective activity. Users are motivated to observe and to explore something; they get assistance in this process, but they nevertheless need to individually reflect on and make a meaning out of their experiences. It is an education and learning process that goes beyond the solely textual dimension, but that is both textual and visual. The Travel Advisor does not tell travelers "go there and you will find this, do this if you like this, *etc.*" The focus is much more on the intercultural and personal experiences of the Travel Machine users, who get guidance, but who are discovering the place by themselves, in a more independent way.

In summary, the main function of the Travel Advisor is to ease, stimulate, and enhance the travelers' process of intercultural learning and to assist them in gaining more profound, deeper travel experiences. The Travel Advisor helps them to become "good" and open-minded, *active, more independent*, and *wise travelers* and to develop an interesting travel history. By suggesting to Travel Machine users that they engage local means of transport or that they more frequently go by foot, the Travel Advisor can also motivate travelers to travel in a more *sustainable* way with a low carbon footprint. Similarly, the Travel Advisor may recommend to users to eat at a "green" restaurant or to spend the night in a sustainability-labeled hotel.

Travelers, in turn, shall document their travel history in the above-described diary. Also in this composition process, the advisor can have an assisting function by asking users for short daily summaries:

- "What was the most interesting/beautiful/surprising thing you have seen today?"
- "What was the most beautiful photograph you took?"
- "What differences in people's behavior/reactions/manners/values, *etc.*, did you notice?"

Of course, the Travel Advisor should nevertheless also provide users with practical, location-related information and advices (e.g., maps, weather and news updates, time zones, local holidays, daily rhythm (opening hours, business hours, breakfast/lunch/dinner hours), public transportation, currency, prices, tips, security issues, urgency numbers of police/hospital/doctors/pharmacies, contact data of local DMOs or tourist information centers, addresses of banks/post offices/grocery stores, passport, and customs regulations) as well as with basic knowledge about the trip destination (e.g., history, facts and figures (number of inhabitants, *etc.*), spoken language(s), climate, national anthem and flag, traditions, traditional and contemporary music, stars, sports, politics, and religion). This background information should also reach a deeper-going cultural dimension, by dealing with issues such as culture-specific gestures and symbols (Wagner/Armstrong, 2003), dress codes, appreciated versus disliked values, attitudes, and behaviors, perception of foreigners, etc. Also in this case, the culture dimensions of cultural theorists such as Hofstede's could constitute very useful points of reference.

The following extracts of the already cited travel diary of 1975 are an interesting source for the conception of the Travel Advisor, insofar as they show important things that someone thought travelers ought to know back then when going abroad (Fig. 6.2).

AMERICAN CUSTOMS REGULATIONS

Upon your arrival in New York your baggage will be subject to the same inspection on landing as on landing abroad. American citizenship does not permit you to bring dutiable goods into the country without paying duty.

A blank will be furnished you aboard the steamer before landing. This must be filled out, listing in detail every article you obtained abroad which you are bringing home. The list is then given to the ship's purser.

This list is called your "declaration" and should include all wearing apparel, jewelry and other articles, whether worn or not, carried on your person, in your clothing, or in your baggage. These items must give their cost or value abroad and whether they were bought or given to you. Also jewelry and wearing apparel, taken out of the United States, and remodeled abroad, must be listed with the cost of remodeling.

MISCELLANEOUS

Travel light. Secure a deck chair when obtaining ticket and arrange that the deck chair is placed on the starboard side of the promenade deck when sailing East, and port side when sailing West. Thus you will be facing South all the time. Since there are entertainments on board the ship, it is advisable to be prepared for such occasions, especially when fancy dress balls and the like are included.

Page Ten

GENERAL INFORMATION FOR AIR TRAVEL

BAGGAGE—Baggage includes all luggage, parcels, and packages carried in the cabin or checked in the cargo space. Overcoats, umbrellas, ladies' handbags, or ladies' pocketbooks, and small cameras carried by the passenger will not be weighed as "baggage." Airline may limit the weight, size and type of baggage.

BAGGAGE ALLOWANCE—IN THE U. S. A.—40 lbs. carried free on each first class or coach ticket except when travel is part of an international trip by air. INTERNATIONAL FIRST CLASS—66 lbs. (30 kilograms) carried free from each full or half fare first class ticket, including portions of travel in the U. S. A. INTERNATIONAL ECONOMY CLASS—44 lbs. (20 kilograms) carried free for each full or half fare economy class ticket, including portions of travel in the U. S. A. CHILDREN—Children paying half fare are granted free baggage allowance on the same basis as a passenger paying the full adult fare. Children traveling free or at the 10% fare receive no free baggage allowance.

BAGGAGE—EXCESS WEIGHT CHARGES—Baggage weight in excess of the above free allowances shall be charged for as follows:
IN THE U. S. A.—Baggage over 40 lbs. is charged for at one half of 1% of the first class adult one way fare per pound.
INTERNATIONAL—Excess baggage is charged for at 1% of the first class adult one way fare per kilogram (2.2 lbs.).

BAGGAGE—EXCESS VALUE—Airlines liability for baggage is limited to its actual value, but not more than $250 per passenger on trips in the U. S. A. or $16.50 per kilogram for checked baggage and $330 per passenger for unchecked baggage on international trips, unless a higher valuation is stated before the flight and a charge of 10¢ on trips in the U. S. A. or 15¢ on international trips is paid for each extra $100 or fraction thereof claimed.

Page Eleven

LEGAL OR PUBLIC HOLIDAYS
IN THE UNITED STATES

The Chief legal or public holidays are:

Jan. 1—New Year's Day (all the States and Territories).

Jan. 20—Inauguration Day beginning 1937 and every fourth year thereafter (in the District of Columbia only).

Feb. 12—Lincoln's Birthday (Ark., Cal., Col., Conn., Del., Ill., Ind., Iowa, Kan., Ken., La., Md., Mass., Mich., Neb., Nev., N.J., N.Y., N.D., Ohio, Ore., Pa., S.D., Tenn., Tex., Utah, Vt., Wash., W. Va., Wis., Wyo., and the Territories).

Feb. 22—Washington's Birthday (all the States and Territories).

Good Friday—(Conn., Fla., Ill., Ind., La., Minn., N.J., N.D., Pa., S.C., Tenn., C. Z., Hawaii, Puerto Rico & Virgin Islands).

May 30—Memorial or Decoration Day (All States and Territories except Ala., Ga., La., Miss., N.C., S.C.)

July 4—Independence Day (all the States and Territories).

Sept. (1st Monday)—Labor Day (all the States and Territories).

Oct. 12—Columbus Day (Ariz., Ark., Cal., Col., Conn., Del., Fla., Ga., Ill., Ind., Kan., Ken., La., Md., Mass., Mich., Minn., Mo., Mon., Neb., Nev., N.H., N.J., N.Y., N.M., N.D., Ohio, Okla., Ore., Pa., R.I., Tex., Utah, Vt., Va., Wash., W. Va., Wis., Wyo., P.R., Ala-Fraternal Day).

Nov. (1st Tue. after 1st Monday)—General Election Day—(all States except Ala., Conn., D.C., Ga., Kan., Ken., Mo., Mass. Miss., Neb., N.M., Utah, Vt.) Observed only when General or President Elections are held.

Nov. 11—Veterans Day (all the States and Territories).

Nov. (4th Thursday)—Thanksgiving Day (all the States and Territories).

Dec. 25—Christmas Day (all the States and Territories).

Page Fourteen

NATIONAL PARKS OF THE UNITED STATES

Following are the National Parks including date established, location, acreage and outstanding characteristics.
Acadia (1919, Maine, 28,382)—group of granite mountains on Mount Desert Island.
Big Bend (1944, Texas, 691,339).
Bryce Canyon (1928, Utah, 36,010)—Eroded pinnacles, vivid colors.
Carlsbad Caverns (1930, New Mexico, 45,527)—Beautifully decorated limestone caverns.
Crater Lake (1902, Oregon, 160,290)—Extraordinary blue lake in crater of extinct volcano. Sides 1000 feet high.
Glacier (1910, Montana, 997,487)—Grand mountain scenery, 200 lakes, 60 glaciers, tremendous precipices.
Grand Canyon (1919, Arizona, 645,136)—World's greatest example of erosion. Mile deep gorge, 4 to 18 miles wide. Rare coloring.
Grand Teton (1929, Wyoming, 94,893)—Spectacular Teton Mountains.
Great Smoky Mountains (1930, No. Car. and Tenn., 460,882)—highest mountain range in the East. Splendid forests.
Hawaii (1916, Hawaii, 173,405)—Volcanic areas.
Hot Springs (1921, Arkansas, 1,019)—Noted for healing properties.
Isle Royale (1940, Michigan, 133,839)—Forested Island.
Kings Canyon (1940, California, 452,905) Big trees, mountains.
Lassen Volcanic (1916, California, 103,269)—Only U. S. volcano.
Mammoth Cave (1936, Kentucky, 50,548)—150 miles of caverns. Limestone formations. Underground river.
Mesa Verde (1906, Colorado, 51,018)—Cliff dwellings.
Mt. McKinley (1917, Alaska, 1,939,199)—North America's highest mountain.
Mt. Rainier (1899, Washington, 241,525)—28 glaciers; 40 sq. miles glaciers; 50 to 500 ft. thick.
Olympic (1938, Washington, 848,212)—Forests; glaciers.
Platt (1906, Oklahoma, 912)—Cold mineral springs.
Rocky Mountain (1915, Colorado, 252,626)—Heart of the Rockies; Peaks 11,000 to 14,255 ft. altitude; Glacial records.
Sequoia (1890, California, 385,100)—Scores of giant trees 20 to 30 ft. in diameter; Thousands over 10 ft.; Mt. Whitney.

Page Fifteen

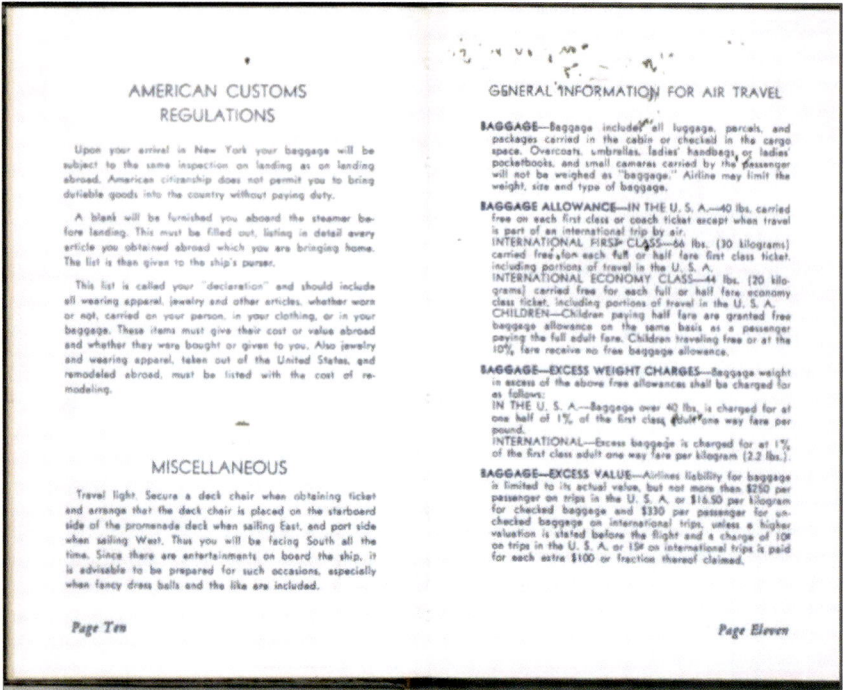

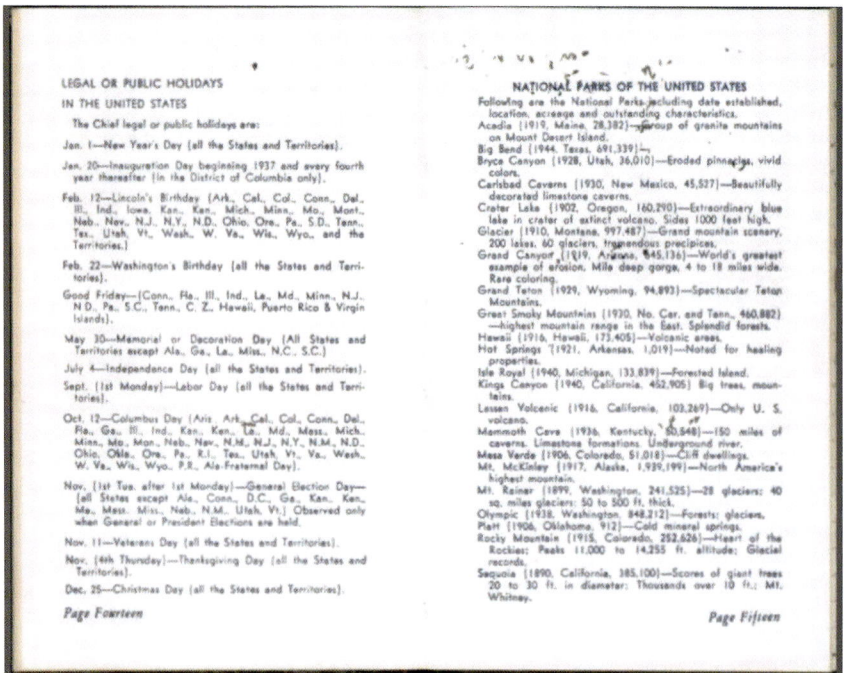

Fig. 6.2 Extracts of an authentic travel diary of an "ordinary person" with advice and recommendations, 1975. Image credit: Aaron Marcus

INFORMATION & SUGGESTIONS

Travellers' checks can be obtained through your bank or travel agency for a fee of about $1.00 per $100, in denominations of $10 to $100. These for the most part should be in small denominations so that when cashing checks in one country, you will not have more currency than needed in that country. If forced to exchange local currency in passing from one country to another, you will lose by such an exchange. On each check there are two places to sign. One is signed upon receipt of the checks. The other is signed in the presence of the person who will cash it. Should your checks be lost or stolen, notify the nearest branch of the organization which issued them. The value of the lost checks eventually will be refunded. Banks or travel bureaus will cash your checks at nearly the current rate of exchange, while hotels or business establishments will usually cash them at a lower rate of exchange. It is well to remember this and try to arrange to cash your checks during banking hours.

Letters of credit are certificates that you have deposited a certain amount of money with the issuing bank or company. To draw money on a letter of credit it must be presented, with identifying papers, to the foreign correspondent of your bank. After you have signed the draft and received your money, the bank will make a memorandum of the amount drawn on your letter of credit.

For the ordinary tourist, however, Travellers' checks are more convenient than letters of credit.

When passing from one country to another, it is always best to be provided beforehand with a small amount of money of the country into which you are going, to cover tips and first expenses after arriving. However, regulations of each country vary on the total amount of its currency you may bring in with you.

BAGGAGE

Ship all heavy baggage to arrive at the steamship piers at least twenty-four hours before sailing. Always make sure

Page Four

INFORMATION & SUGGESTIONS

either by personal observation or proper claim check, that all baggage is going with you. Baggage insurance is strongly recommended. From the steamship company you can obtain labels which should be pasted on all your baggage. The trunks you will not need on the voyage are marked "Hold." Those needed during the voyage are marked "Wanted," and will be held for you in the ship's baggage room. Your steamer trunk and hand baggage will accompany you to the stateroom.

SEASICKNESS

Should you feel yourself becoming ill, keep moving or keep busy in the fresh air, and this will often relieve the condition. The ship's doctor can be called if necessary, although you may secure reliable relief from your cabin steward or stewardess. Rough seas increase the number of seasick persons on board ship, although rough weather and storms occur most frequently during the winter months.

STORMS AT SEA

Storms at sea are apt to create some nervousness on the part of inexperienced travellers, but do not cause any concern to the ship's officers or personnel. Modern ships are designed and constructed to withstand the worst storms, and the officers are especially trained in the efficient handling of the ship under all conditions. In certain stretches of ocean, you are more likely to experience rough seas than in others. The sea area around Cape Hatteras, for instance, is noted for its gales and is invariably rough. However, with favorable weather you should experience a calm, pleasant crossing.

DECK SPORTS

There are many popular deck sports which you may enjoy while on board ship. If you are to enjoy these games, you must enter into them without feeling that it is necessary to have been formally introduced to the participants. Ship's passengers are in a sense "one big family."

Page Five

INFORMATION & SUGGESTIONS

SHIP'S POOL

The "pool" on a steamer usually refers to the game of chance played on the ship's daily run. Those desiring to participate are asked to draw from one of ten numbers at so much each, the winner being the holder of the number that corresponds to the last figure of the day's run in miles.

AVOID PROFESSIONAL GAMBLERS

On many steamers, notices are posted warning passengers against professional gamblers, yet in spite of these, every once in a while there is evidence that all travellers do not heed this warning. The smoking room or the lounge affords ample opportunity for all who wish to play cards.

FOREIGN HOTELS

When engaging rooms at hotels where there are no printed rates, it is always best to have a definite understanding as to the price per day or week. In Southern Europe when travellers do not bargain for their rooms, they may be overcharged. As a rule, the upper floor rates are more reasonable in price than those below, and are often more comfortable due to less street noises and more ventilation. Avoid extras if possible, especially in food, for, unlike America, exorbitant prices are often charged for them.

Except for the larger hotels in the capital cities, there are very few rooms with bath. There is usually a bath on each floor. Arrangements may be made at the desk for its use by payment of a slight additional charge. Soap is seldom supplied, and it is best to carry your own.

Page Six

CHANGING THE CLOCK

Between New York and London there is a difference in time of five hours, and as the sun rises in the East, as we say, when the ship is going eastward, she meets sunlight earlier each day and thus gains time. Exactly how much is computed each day at noon, and the ship's clocks are immediately set at the correct time for that longitude. On a vessel which makes the crossing in five days the clocks will be set ahead each day approximately an hour; on slower ships, of course, less. Going westward the clock is set back daily in similar fashion.

DIFFERENCE IN TIME

When it is noon in New York, the time in various cities and parts of the world is as follows:

Amsterdam (Holland)	5.20 p.m.
Berlin (Germany)	5.54 p.m.
Brussels (Belgium)	5.00 p.m.
Calcutta (India)	10.50 p.m.
Cape Town (South Africa)	5.50 p.m.
Chicago	11.00 a.m.
Copenhagen (Denmark)	5.50 p.m.
Cherbourg (France)	5.00 p.m.
London (England)	5.00 p.m.
Madrid (Spain)	4.45 p.m.
Manila (P. I.)	*1.00 a.m.
Melbourne (Australia)	*2.40 p.m.
Paris (France)	5.09 p.m.
Pekin (China)	*1.00 a.m.
Leningrad (Russia)	7.01 p.m.
Rome (Italy)	5.50 p.m.
San Francisco	9.00 a.m.
Stockholm (Sweden)	6.13 p.m.
Switzerland	6.00 p.m.
Vienna (Austria)	6.06 p.m.
Yokohama (Japan)	*2.00 a.m.

*Next day

Page Seven

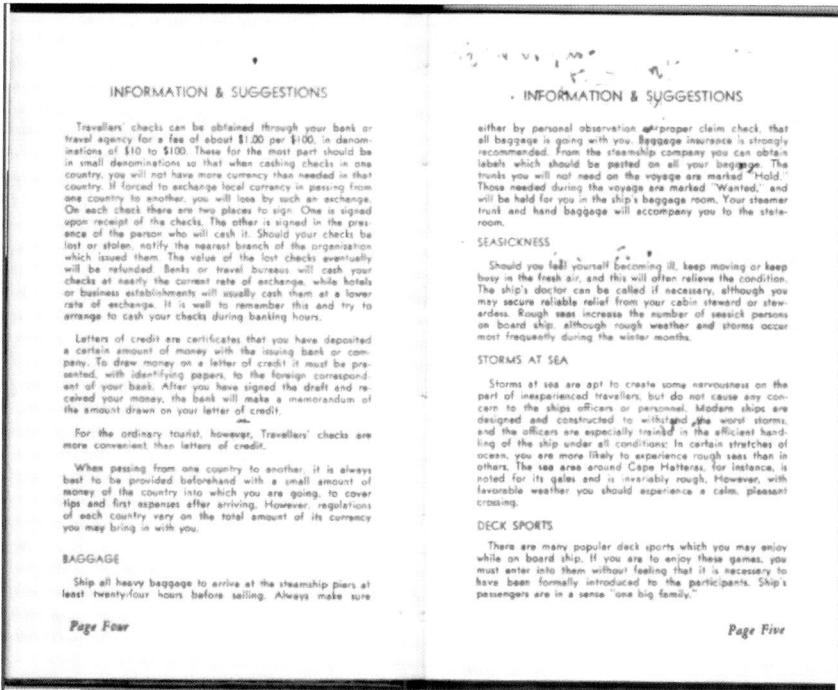

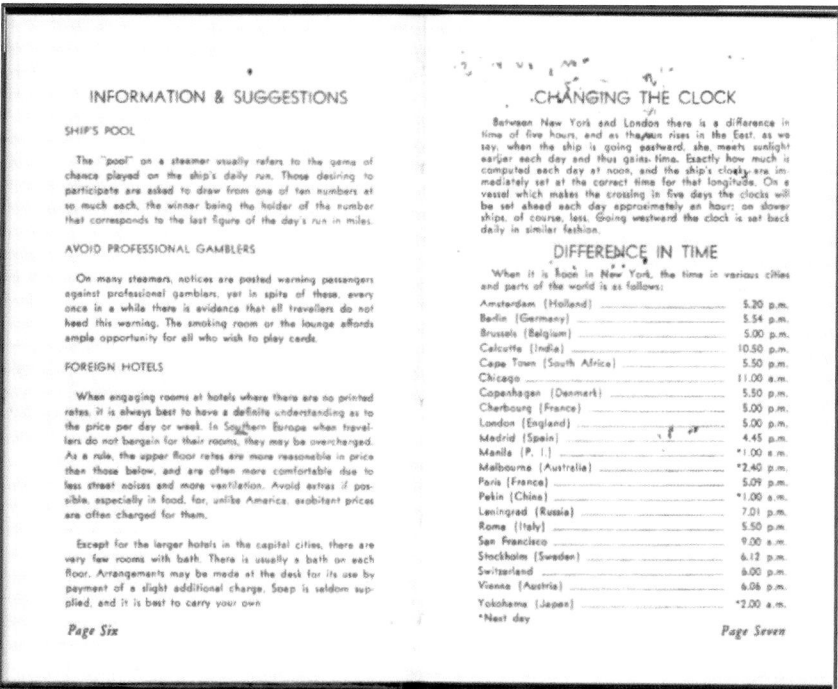

Fig. 6.2 (continued)

6.5.4   Tab 4: Fellows

The above-described features and functions of the Travel Machine could be significantly enhanced in their persuasive and motivational power by incorporating also a social dimension into the application, i.e., by providing possibilities for social interaction, comparison, and competition. All functions dedicated to these aspects appear under a tab provisionally labeled "Fellows."

It is an acknowledged and well-proven fact in the field of persuasion theory (Fogg 2003) that this social component can become a superior incentive for behavior change of individuals. By leveraging the features of successful social networks such as Facebook, by creating specific forums, or a specific Web portal for the Travel Machine, several means of social interaction and exchange seem likely:

- Travelers can share their diary with friends or with other travelers at the same or also at other places. Taking up again the idea of cultural and digital storytelling, it becomes clear that one of the main motivations to tell a story is also to have someone listening to it, so that travelers will have an interest in sharing their diary and their experiences, in receiving feedback and comments from their diary followers or "subscribers," and as well as in giving recommendations to others. By linking the travel diaries of Travel Machine users, there is also the possibility to match travelers with similar interests and itineraries. A traveler may be interested to know whether there are some fellow countrymen around with whom he or she could meet and maybe have lunch or dinner together. At the same time, this function could be a means to match travelers with local people, so that they get the occasion to speak the local language and to learn about local culture and customs via first-hand reports. Looking further, such a function could also be enhanced and further refined in the future, for example, by having an airline, bus, or taxi operator use it as an efficient instrument for *social seating* (SeatID 2012).
- At the same time, a social platform for exchanges and interactions could give Travel Machine users the possibility to look at others' travel diaries, to follow or "subscribe" to them. Thus, they can see which places other travelers visited, which curious and interesting things they discovered, which photographs they uploaded, and so on and so forth. Users might even choose for themselves a sort of model or mentor, and they might then adapt their own travel itinerary accordingly. Moreover, users can also just collect information about a place during the before-trip phase by analyzing the travel diaries of their predecessors. Questions guiding them may be those such as the following:
  - "What has been the most photographed object at that place?"
  - "What has been the most popular thing? What did people like most?"
  - "What was the most preferred local food?"
- The attribution of points and virtual rewards in the below-described games section (see Tab 5) can then trigger social competition among application users and can further stimulate their motivation to discover the foreign culture and to proceed in their learning process. The more points and rewards travelers accumulate, the higher they will rank in the list of a destination's experts, and the more

likely it is that other travelers will ask them for information and hints. Travel Machine users will be motivated even more if they can communicate their current expert status on their preferred social media, e.g., Facebook profile, or via Twitter.

- Another form of competition could be to let users decide daily about the "most beautiful photograph" that has been uploaded to a travel diary, which would again bring points to the traveler who took the acclaimed picture.
- The possibility of social interaction can also serve as a way to animate discussions among travelers. These exchanges would represent another form of reflective activity, and they would thus enhance the individual learning process. In this case, reflection would occur within a community and through interaction with others.

6.5.5 Tab 5: Challenge

The travelers' knowledge and their learning progress will then be tested in the form of games or quizzes, which will appear under the tab "Challenge." This form of knowledge-testing and feedback-giving will serve as a major incentive. Challenges will have a stimulating effect on users and will increase their motivation to become familiar with the places they are visiting. Especially the provision for virtual rewards, in the form of titles (e.g., "X is now a four-star Japan expert!") or medals (e.g., bronze, silver, and gold), represents a strong incentive and increases use frequency. Moreover, real rewards can be given to successful Travel Machine users. Having achieved a certain level of knowledge about a culture or a country, users could be rewarded with an eBook, with free subscription to an eMagazine, with travel air miles, or with a tangible gift, such as a coffee mug. Companies and corporations working with the Travel Machine may even offer coupons or price reductions for the future. An airline may provide air passengers with a 5 % discount for the next flight, for example, once they have achieved a certain level of knowledge about a particular country or culture.

The integration of games will, at the same time, add a component of fun and play to the application and thus will make the education process not only informative and reflective but also entertaining.

Apart from a knowledge-testing game (Game 1), the Travel Machine will contain a vocabulary quiz (Game 2). Furthermore, a game in accordance with the Travel Advisor's instructions and suggestions (Game 3) will increase the travelers' motivation to accomplish the presented challenges. For example, users will get points for uploading a photograph of the dish the Travel Advisor suggested them to try. Or the Travel Advisor might have said to them: "Take a local bus to get to point X, and take a picture of X."; "Go to point X by foot, and interview a local person there." Travelers will get points and virtual rewards for doing so and for uploading a photograph or video as a proof. Also, Games 1 and 2 will reward users with points for correct answers to knowledge- and vocabulary-related questions. Different point levels will then correspond to specific expert status levels. Moreover, players can use their points in the "Point Store" and exchange them for real rewards.

Such challenges, as well as the prospect of rewards and of positive feedback, will motivate application users to become active, wiser, and more sustainable travelers.

6.5.6 Further Possible Implementations

One might think, also, about implementing the above-described ideas of social competition and comparison in other contexts. In the framework of a school excursion, for example, students could be asked to document their trip experiences and impressions with the help of their digital diary, and they would get suggestions and stimuli via the Travel Advisor. The teacher could award a prize to the best diary, and the teacher could even use the students' records in order to give final grades.

Moreover, a special version of the Travel Machine could be conceived for group use, e.g., for groups of tourists doing an organized package tour to a certain destination. By further enhancing the above-described features linked to social interaction and exchange, the Travel Machine and its main functions could significantly enrich the overall trip experience, as well as the after-trip phase of sharing and reviewing, of a group of several travelers going abroad together.

Another possible application of the diary and of the advisor functions could be for use with large events, such as the EXPO2015 in Milan (http://www.expo2015.org/), where again the mobile application could be a suitable means of documenting, easing, and enriching the visitors' experiences. The central EXPO2015 themes of food and sustainability, which are expressed in the slogan "Feeding the planet. Energy for life," undoubtedly lend themselves for the integration of such a technology. Also in this case, the application would be an excellent basis for a competition among visitors (e.g., What was the best travel/visitor diary, what was the best photograph that has been uploaded, *etc.*?). Such an extension of the Travel Machine would allow, at the same time, organizers to be better informed about visitors' preferences and itineraries. Local and regional direct marketing organizations could benefit as well from this potential extension of the Travel Machine.

Of course, in addition to its specifically traveler-centered functions and data, the Travel Machine could still be equipped with more "traditional" mobile application services, such as booking, attractions, and sights, accommodation, transportation, restaurant, and shopping databases or other location-based services (LBS). Theoretically, also the abovementioned push services could be integrated for commercial objectives, so that the Travel Advisor might tell travelers "Go to the specific restaurant X (based on paid-for recommendations), and eat X there" and the users could then get points or discounts for doing so. However, such functionalities would not be in line with the main focus and objectives that underlie the Travel Machine concept.

6.6 Personas

As mentioned in Chap. 1, personas are characterizations of primary user types and are intended to capture essentials of their demographics, contexts of use, behaviors, motivations, and subsequent impacts on design solutions. Personas are also called user profiles. Typically, UI development teams define one to nine primary personas.

For the Travel Machine, we determined the following personas by analyzing available data.

6.6.1 Persona 1: Michael Johnson

- 22-year-old Caucasian male.
- Undergraduate student, from Orlando, Florida (USA).
- English as native language.
- Familiar with ICTs and smartphones.
- Already uses some mobile applications.
- Wants to do a graduation trip for himself.
- Persona image credit: AM+A.

6.6.1.1 Textual Summary

Michael is an easy-going and open-minded person, who has no problems asking questions to strangers. He also likes to be adventurous. His mother tongue is English, so usually he does not encounter problems when going abroad, but still he would like to gain some additional knowledge of other languages. He is very familiar with ICTs, so that he can book a hotel or a flight by himself, and he knows how to acquire online general, basic information about a destination before going there. Also, Michael plays games, enjoys online competitions with his friends, and uses online social networks. He especially likes the tagging features and makes frequent use of them.

Michael has already done some city trips in the past, and he has visited friends in other countries. But he would not define himself as an experienced traveler. Michael now wants to travel to Tokyo for 1 week, with his best friend Marco. However, he does not like the idea of traveling in a group, of buying a packaged tour, and of following the instructions of a guide. Being guided and directed by others is too boring for him, and he prefers traveling in a more independent and self-directed way. Therefore, he wants to create his own travel plan carefully, based on the limited budget he has as a student. In general, Michael is quite interested in other cultures and places. For this reason, he would like to learn more about the Japanese history, traditions, cultural idiosyncrasies, etc. Nevertheless, he usually finds some difficulty in motivating himself to get into another culture; he has not found the right approach so far. Just reading books or online articles easily bores him. Concerning the planned trip to Tokyo, Michael does not necessarily want to visit all the "must-see" places of the city or limit his activities to the list of sights and attractions common tourists go to see. Instead, he is interested in feeling and living the place, so that he can understand better the local people and culture.

6.6.1.2 Design Implications Summary

Objectives

- Gain some additional knowledge of another language.
- Learn more about history, traditions, and cultural idiosyncrasies of another country.
- Travel in an independent way with a limited budget.
- Stay connected to friends via social networks and share travel experiences remotely.
- Have fun with games and online competitions among friends and with other travelers.

Context

- Travel as an undergraduate student during holidays.

Behavior

- Regular use of smartphone and mobile applications.
- Past travels abroad to visit friends in other countries.
- Independent online booking of hotels and flights.

- Frequent use of social networks (especially Facebook).
- Regular participation in games and online competitions with friends.

Design Implications/Objectives

- Enable travelers to plan their independent, personalized travel schedule in a convenient, effective, and efficient manner (Web portal).
- Help travelers keep track of their expenses and have better control of their travel budget (dashboard).
- Enhance users' travel experiences abroad by sharing information and interacting with other travelers or locals (diary).
- Enrich travelers' knowledge of local cultures and languages (vocabulary and knowledge quizzes).
- Share travel experiences instantly with friends and tag friends in the photographs (diary, social interaction, and social networks).

6.6.2   Persona 2: Sara Robertson

- 26-year-old Caucasian female.
- Graduate student, from Chicago, Illinois (USA).
- English as native language and familiarity with other languages.
- Familiar with ICTs and smartphones.
- Already uses interesting mobile applications occasionally.
- Wants to become a social worker.
- Persona image credit: AM+A.

6.6.2.1   Textual Summary

Sara is open-minded and sociable, she likes getting to know and having conversations with new people, but she sometimes is shy and unsure at the beginning. She likes painting and writing little poems in her free time, and she also keeps a diary. In general, Sara would define herself as a rather individualistic person; however, she still likes interacting with her friends via social networks to exchange information and experiences and to always be up to date about what they are doing. Apart from English, which is her native language, she is fluent in several other languages, e.g., French, Spanish, and German. Sara has a smartphone, although she does not make regular use of mobile applications. She uses them only occasionally, if she happens to come across an interesting application.

Sara is very interested in other cultures and languages, in learning, and discovering new things and new people. She has already had several travel experiences, also outside the United States. However, due to her studies and her budget constraints, she is still in the initial phase of her "travel career." She is now planning to make a trip for 2 weeks to Japan, in order to discover Tokyo and several other places nearby. Since Sara speaks several languages, but not Japanese, and since she has a lot of intercultural knowledge and is familiar with ICTs to inform herself before the trip, it is no problem for her to travel by herself. More importantly, Sara prefers doing things with specific objectives or goals in her mind. She really enjoys the process of planning a trip and then completing the objectives set by herself. Given that she has already read much about the country and has gained some basic knowledge about its history and cultural specificities, she would like now to experience the culture directly, to meet and ask questions to locals, to taste local food, etc., in order to gain more understanding of the Japanese culture and to broaden her personal horizon.

6.6.2.2   Design Implications Summary

Objectives

- Make an individual travel plan, and complete set goals and challenges.
- Keep a personalized and convenient travel diary on the smartphone.
- Learn and discover new things and new people.
- Gain deeper understanding of a foreign culture and broaden the personal horizon.
- Interact with friends via social networks and always be up to date about what they are doing.

Context

- Former student, who has recently completed the master's degree.

Behavior

- Past travels abroad, also outside the United States.
- Basic knowledge about history and cultural specificities, plus additional information gathering before going to another country.
- Frequent use of social networks (especially Facebook).
- Preference for activities with a determinate meaning, objectives, and aims.

- Regular smartphone use.
- Occasional use of interesting mobile applications.

Design Implications/Objectives

- Help travelers conceive a trip schedule, and set some goals as a personal challenge (Web portal).
- Enable travelers to record their travel experiences in a convenient manner (diary).
- Guide users in the planning of their trips, by giving them access to other travelers' experiences and tips, written down in their individual travel diaries (social interaction and social networks).
- Enrich users' knowledge of local cultures and languages (vocabulary and knowledge quizzes).
- Share travel experiences instantly with friends and other potential travelers; tell stories about trips (social interaction and social networks).
- Give possibility to search for travel partners and to make new friends with like-minded people (social interaction and diary sharing).

6.6.3   Persona 3: Richard Harris

- 31-year-old Caucasian male.
- Business professional, from New York City, New York (USA).
- English as native language.
- Familiar with ICTs, has a smartphone, and likes informative applications.
- Does frequent business trips.
- Persona image credit: AM+A.

6.6.3.1   Textual Summary

Richard is president of a renowned international trade company. He is not married, and, due to his job, he often has to travel around the world in order to attend business conferences. Richard is a sociable person, even though he rarely makes use of social networks. He likes purchasing lovely presents for his friends when he travels abroad. Richard uses a smartphone and numerous applications. He is particularly interested in helpful and informative mobile applications, especially apps that can educate him and that provide him with useful information, such as the stock app. Next week, he will have a meeting in Beijing, and he will stay there for 7 days in total. It will be his first time in China.

Usually, Richard would prepare his business trips by acquiring some basic background knowledge on the country and its culture. However, he thinks that the best way to really understand other traditions and cultures is to talk to local people and to visit local places, especially those that might be less popular and less well known. At present, Richard would like to know more about Chinese traditions, customs, and the overall culture, which are quite different from those of the United States. For example, he wants to know how to be polite and how to behave well, or how to more easily make friends with locals during his stay. Additionally, Richard hopes to enjoy his free time after hard and long working days. Since he will be free of commitments only in the evening, or on weekends, he has only limited time for leisure activities. Nevertheless, Richard does not want to visit only the "must-see" sights and places. Instead, he wants to figure out some special and less well-known locations to go, and some special things to do, so that he can directly get in touch with local people on the place and better understand the cultural idiosyncrasies of the city he is visiting.

6.6.3.2   Design Implications Summary

Objectives

- Understand the traditions, customs, and culture of another country, in order to be polite and to act in an adequate manner.
- Get to know new people and make friends with locals.
- Gain additional knowledge of a foreign language and culture.
- Travel in an independent way with a flexible and personalized schedule.
- Remember buying gifts for family members and friends during the trip.
- Conveniently record and share travel experiences with friends.

Context

- President of a renowned international trade company.

Behavior

- Due to job frequent business travels abroad.
- Acquisition of general knowledge about traditions, customs, and culture of other countries in before-trip phase.
- Rare use of social networks, but interest in sharing experiences with friends.

- Preference for doing things that pay off, instead of killing time.
- Regular smartphone use.
- Regular use of helpful and informative mobile applications.

Design Implications/Objectives

- Significantly ease travel preparation phase; enable time saving for searching information and for planning the trip in advance (Web portal).
- Help travelers discover a city in depth with useful suggestions and tips (advisor).
- Enable convenient recording and sharing of users' travel experiences via smartphone (diary, social interaction, and social networks).
- Enrich travelers' knowledge of local languages and cultures (vocabulary and knowledge quizzes).
- Match potentially like-minded travelers (social interaction and social networks).
- Remind people of buying gifts (gift notification in dashboard).

6.6.4   Persona 4: Mike Fenning

- 26-year-old African-American male.
- Professional photographer, from Minneapolis, Minnesota (USA).
- English as native language and fluency in several other languages.
- Familiar with ICTs, has a smartphone, and likes trying new applications.
- Travels around the world and enjoys taking professional pictures.
- Persona image credit: AM+A.

6.6.4.1   Textual Summary

Mike works for a travel magazine, and he often travels abroad in order to take pictures. From time to time, he also writes small articles for the magazine about his travel experiences. Therefore, it is important for Mike to remember and to be aware of the meaning and information contained in every single photograph. Mike loves his work, because he can visit interesting places all around the world and meet all kinds of different people. He is interested in foreign languages and cultures, too. Mike always travels by himself, because in this way it is up to him to decide on his preferred itinerary and schedule, and he has enough time to look for beautiful sceneries and to take pictures as he wishes. As a young man, Mike is passionate about new technology. He owns an iPhone and likes trying out new mobile applications.

Mike has many friends, who share his interests and passions. Some of them are also photographers. Traveling is a popular discussion topic among Mike and his friends. They often tell and exchange stories and recommend interesting, special places to each other. This time, Mike will be going to Paris to take pictures and to write a short article about his experiences there. Paris is a major city that Mike himself and most of his friends have visited already. However, this time Mike wants to visit some unique, "hidden" places or events, which represent French culture and lifestyle, not only famous tourist attractions every visitor would go to. He wants to create special memories and extraordinary photographic works during this journey.

6.6.4.2   Design Implications Summary

Objectives

- Search for new places abroad and take nice pictures there.
- Deeply explore a city to understand local culture, spirit, and stories.
- Record travel experiences in the form of words, photographs, audio, and video recordings.
- Tell and exchange travel stories among friends, and talk about and recommend good places to each other.
- Gain new, profound knowledge about people and culture of a destination.

Context

- Job as professional photographer for travel magazine.
- Frequent, rather short travels for job purposes.
- Numerous travel lovers as friends.

Behavior

- Regular use of iPhone.
- Regular use of social networking sites (especially Facebook and Twitter).

Design Implications/Objectives

- Enable travelers to create their independent, personalized travel schedule in an effective and efficient manner (Web portal).
- Help users discover a city's spirit and culture, while being guided by useful suggestions and challenges (advisor).
- Enable users to record their travel experiences in a convenient manner (diary).
- Help travelers record experiences and views with mobile cameras, to figure out "hidden" details and information behind the photographs (diary photo album).
- Share travel experiences instantly with friends and other potential travelers; tell stories about trips (diary, social interaction, and social networks).
- Enrich travelers' knowledge of local cultures and languages (vocabulary and knowledge quizzes).

6.6.5   George Shen and Julia Lin

- 67 and 65, Asian retirees, from Suzhou, China.
- Chinese as native language and familiar with English.
- Know ICTs, have smartphones, and use simple applications.
- Travel to San Francisco to see their daughter Lisa.
- Persona image credit: AM+A.

6.6.5.1   Textual Summary

George and Julia have been married for 40 years already. Their only daughter, Lisa Shen, moved to San Francisco 7 years ago, and she has been working there since then. Given that Lisa lives overseas, George and Julia use their smartphones and some mobile applications to keep in touch with their daughter. George and Julia are both well-educated, and they read a lot. Therefore, they have already acquired some basic historical and cultural knowledge. They have some familiarity with ICTs, but nevertheless they are still a bit suspicious of new technologies. However, they would like to learn more to become more skilled in using them, since they are at the same time fascinated by this new world of ICTs and by the numerous possibilities it offers.

Recently, George and Julia have retired, and they are planning now a trip to San Francisco to visit Lisa. Though the couple lacks rich experience of traveling abroad, they know how to interact with people of different cultures, and they enjoy traveling by themselves. Since Lisa is busy and will have only little time to accompany her parents, she uses the Travel Machine to make a trip plan for them on the Web portal and then transfers the schedule to their smartphones. George and Julia have already visited San Francisco during a short 3-day trip about 5 years ago. Thus, they have seen at least the most important and most famous places of the city. However, they feel that they have not fully explored some places and that they have not become familiar with the city's idiosyncrasies and with its characteristic culture. They actually do not even remember much of the trip, and when they look at the photographs they took back then, very often they could not even tell what exactly they represent. It is only a series of pictures, which don't have much life in them; they lack a more personal touch. Also, since they only rarely travel abroad, George and Julia are more curious about different cultures and lifestyles outside China, and they are interested in something new and unknown. In addition, as seniors they are more prone to get tired while traveling. Therefore, they want a relaxing and enjoyable trip and at the same time experience a city and its local spirit, for example, by going to local restaurants, by eating authentic and traditional food, by talking to locals in cafés, etc.

6.6.5.2   Design Implications Summary

Objectives

- Record travel experiences in a structured and clearly arranged way, thereby enabling users to take more intensive and long-lasting memories out of their trip.
- Suggest a new and innovative approach to the discovery of an unfamiliar place.
- Plan a relaxing and enjoyable journey to discover new things and new people, to learn something so far unknown.
- Learn more about the history, traditions, and cultural idiosyncrasies of other countries and cities.
- Record and share travel experiences with friends and family members in a convenient manner.

Context

- Retirees visiting a family member abroad.

Behavior

- Limited experience of traveling abroad.
- Some basic historical and cultural knowledge.
- Rare usage of social networks, but interest in sharing travel experiences with family members and close friends.
- Regular smartphone use.
- Regular use of simple mobile applications.

Design Implications/Objectives

- Enable travelers to find out a new and innovative approach to not well-known places (advisor).
- Help users record travel experiences in a structured and clearly arranged way; help them take more intensive and long-lasting memories out of the trip (diary).
- Guide travelers in planning their trip schedule, and assist them in discovering a city in depth with useful suggestions (Web portal; advisor).
- Enable convenient sharing of users' travel experiences via smartphones (diary and social interaction).
- Enable travelers to bridge waiting time and to enrich their knowledge of local languages and cultures (vocabulary and knowledge quizzes).

6.7 Use Scenarios

As explained in Chap. 1, use scenarios are a UI development technique whereby, in the development of a prototype to simulate the major characteristics of a software product, a story of usage is written to determine what behavior will be simulated. A scenario is essentially a sequence of task flows with actual content provided, such as the subject user's demographics and goals, the details of the information being worked with, *etc.* For the Travel Machine, we developed several use scenarios as explained below.

6.7.1 General Use Scenario

The following use scenario topics are drawn from the five preceding personas. Some specific examples might be relevant only to users of a particular age group, education level, gender, or culture. Note the general usage of the terms "objective" and "goal." An objective is a general sought-after target circumstance. A goal is more specific and is usually qualified by concrete, verifiable conditions of time, quantity, *etc.*

Dashboard

- Personalize and customize the users' dashboard according to their needs and interests.
- Track important travel information very quickly, e.g., current location, time, weather and weather forecasts, local news and events, security tips, budget notification, etc.

Diary

- Before-trip phase: Collect online information and determine a travel schedule on the Travel Machine's Web portal; then transfer the created travel plan to the mobile phone, and insert it into the diary.
- Record and keep track of daily travel experiences in a customized, personal travel diary in the form of texts and comments, photographs, audio, and video recordings; attach textual or audio comments to photographs or silent videos.
- Review diary entries both during the trip and after the travel experience, in order to become more aware of, more deeply involved with, and more proud of the personal travel progress and chronicle; get motivation for further travel discoveries and experiences by looking at the personal travel timeline.
- Transform the data of the diary from the smartphone to a PC.
- Photo album: Combine photo album with advanced technical functions, such as tagging and face recognition; take pictures with the smartphone and use the Travel Machine to figure out "hidden information" behind the pictures; accede to additional information via URL tags.

Advisor

- Have access to see and read basic country-/city-related and current information (news, current events, weather, etc.), as well as practical advice and tips (currency, local holidays, daily rhythm, useful addresses and numbers, safety issues, etc.).
- Have access to see and read deeper-going, culture-related information (dress code, symbols, gestures, etc.).
- Increase level of knowledge and expertise about a culture/destination by reading tips and following suggestions/challenges provided by the Travel Advisor.
- Discover culture- and location-specific objects, food, traditions, etc., with the assistance, challenges, and proposals of the advisor.
- Learn about the destination by adopting a new, unusual, and innovative approach in discovering it.
- Get in touch with locals and location-specific culture, by accomplishing the challenges and goals set by the Travel Advisor.

Challenge

- Increase level of knowledge and expertise about a culture/destination by learning in a playful, entertaining way, doing games (vocabulary, destination-related information, etc.).
- Get virtual rewards (titles, "virtual" medals) and real prizes (e.g., ebook, emagazine, or coupon) as incentives for each new level of expertise.

- Via social media (Facebook, Twitter, fora, etc.): Communicate expert status/achievements, and compete with other travelers and friends on the degree of expertise.

Fellows

- Compare and share travel diary with friends, family members, and other travelers worldwide, via social networks and via upload on Travel Machine Web portal.
- Consult other travelers' diaries to get ideas and inspirations out of them, to learn about popular and interesting activities/things/locations in a destination, and to orient travel behavior towards a model/mentor.
- Connect with travelers at the same destination or with similar travel interests.
- Interact and communicate/chat with friends and other travelers, and read/react to messages, suggestions and comments, emails, or other communications about a destination/about diary entries, *etc.*
- Exchange and share information/photographs/videos, etc., via social media and Travel Machine Web portal; ask for advice.

Future Extension

- Record and enhance school excursion experiences.
- Group use (e.g., organized package tour): Enhance group experience, e.g., by interconnecting all group members and their diaries, by tracking the position of fellow travelers, etc.
- Benefit from location-based services (LBS) via GPS.

6.7.2   Personas' Use Scenarios

6.7.2.1   Michael's Use Scenario

Michael is 22 years old, and he likes to be independent at his age. Therefore, he does not enjoy following a pre-planned guide, but instead he hopes to discover the destination by himself. So he can use the Travel Advisor to read related news and information as well as the travel diaries of others to get hints, suggestions, and ideas for his own travel schedule. In addition, Michael makes use of the Travel Machine to control his limited travel budget as a student.

Michael is adventurous and curious about new and unknown things. He uses the Travel Machine to search for other travelers' recommendations of local attractions, cultural particularities and differences, new food, etc., which he didn't know in the past.

Michael also likes using his smartphone to take photographs. The Travel Machine allows him to store and order his pictures in a personal travel diary. Since he is not very familiar with foreign cultures, and not a highly experienced traveler, he would otherwise not remember the content of all the pictures he took, the names of the places he visited, etc. The diary will be his personal travel chronicle, where he can trace his travel experiences, see daily the travel progress he has made, and reexperience the trip with his friends even in the after-trip phase.

Michael will share his experiences (recorded in the diary) with his numerous friends and family members via social networks, as well as via the Travel Machine Web portal. Since he will be able not only to share but also to compete with his friends and other travelers in his discovery and knowledge of the new destination, he will be even more motivated to become an expert of the place.

Given that Michael likes games, he will enjoy learning something about Tokyo in an entertaining and playful way, by doing vocabulary and knowledge quizzes on his smartphone. Michael does not like reading too much information, and he becomes bored easily when going through, for example, a guide, so that such a playful way of information transmission will be more effective and efficient in his case.

Michael's Behavior Changes

- In the past, before doing a trip, Michael used to ask advice from friends who had already been to the respective place. Or he had to spend plenty of time searching for information about a destination and developing a schedule. Now he can save a lot of time and energy by reviewing others' travel diaries and comments, which will influence his own travel plan. Moreover, it is now easier for Michael to keep track of his travel expenses and of his remaining budget.
- Though Michael is adventurous and likes new things, he is usually a bit lazy in discovering unfamiliar places and cultures. However, the Travel Machine now challenges and pushes him towards discovering local attractions, cultural particularities and differences, new food, etc. Especially the game tab and the attribution of rewards stimulate Michael's interest additionally.
- When looking at his travel photographs, in the past, Michael often could not remember the name of certain places, sceneries, attractions, etc., or the stories behind determinate monuments or buildings. Already at the end of a trip, it was usually not easy for him to remember the content and meaning of the pictures taken on the first day. Now he can overcome this problem by easily recording all his experiences in the travel diary and by commenting and describing his pictures and videos. Thus, he will be able to access and to make sense of his travel records more easily.
- Michael used to share his travel experiences occasionally with his friends and family members via social networks (Facebook, Twitter, Skype), usually only via short messages or single pictures. Now, he can make his friends and family members reexperience his trip by giving them access to his diary. Moreover, he will not only share experiences but also compete with his friends and with other travelers in his discovery and knowledge of the new destination. He is thus even more motivated to become a destination expert.
- Michael particularly appreciates the gaming functions of the Travel Machine. Now he can gain information and learn something in a playful, efficient, and effective way, instead of just reading information online or in books, which easily bored him before.

6.7.2.2   Sara's Use Scenario

Sara already knows a lot about the country, but she likes using the Travel Machine's social networking functions to ask questions to travelers that have been to Japan before and also to communicate with local people there. Furthermore, she enjoys looking at other travelers' or friends' travel diaries and taking new ideas and inspirations out of them. She may even find a sort of mentor or model traveler, on whose diary she will base her own trip itinerary.

Sara likes to do things with a certain meaning and with a certain aim. Concerning the games offered by the Travel Machine, she might therefore not be very interested in competing with others. Instead, she prefers meeting personal challenges for herself (advisor) and increasing her Japanese vocabulary (vocabulary game), also by adding new words that she picks up during the trip. She also enjoys playing the knowledge game, which gives her the possibility to acquire more knowledge about Tokyo and its surroundings.

The idea of tracing her trip in a mobile travel diary perfectly suits Sara's idea of traveling and her character disposition: Already in daily life, Sara writes a diary, and the concept of a travel diary gives her the possibility to collect and organize her travel experiences in a very personal and individual way. She can give free rein to her creativity, enrich the photographs she takes by short textual and visual comments, and register spontaneous ideas and impressions. In the after-trip phase, Sara will then be able to reexperience her trip to Japan and to further increase her knowledge based on the travel records in her diary.

Sara likes to get to know new people, but she is still a little bit unsure and shy. Therefore, she would like to have at least some points of reference and some guidance or assistance during her first trip to Japan. This role will be taken over by the Travel Advisor of her Travel Machine application. The advisor will help her to overcome her initial reticence and shyness and will push her towards seeing particular places, interacting with locals, and discovering locations or objects that maybe even she had not heard of before.

Sara likes to interact with her friends via social networks to exchange information and experiences and to always be up to date about what they are doing. She can easily stay in touch with her friends via the Travel Machine, in an even more convenient way. Sara will love sharing her diary with her family members and friends on the Travel Machine Web portal and on social networking sites like Facebook or Twitter.

Sara's Behavior Changes

- Sara always spent plenty of time preparing and planning her trips. Now she saves much time since she uses the Travel Machine Web portal and its social networking functions to get information and advice. She starts to find out how easy it is to plan a good trip with the help of other travelers' experiences and recommendations.
- Sara used to feel bored when she was waiting in line for a scenery spot or attraction, since she doesn't like playing games just to kill time. However, Sara now enjoys

playing the vocabulary game of the Travel Machine, since it enables her to learn something and to have fun at the same time. She is also keen on challenging herself to get to a higher expert level.

- Sara already keeps a diary in her everyday life, because she likes recording everything and keeping her memories in a certain order. However, in the past, it was not very convenient for her to keep a paper diary during her journeys. Now, Sara uses the travel diary in order to record vivid trip experiences in variegated formats and forms. She now even prefers "writing" a digital diary on her smartphone, instead of using the more traditional paper version.
- Sara is sometimes shy and unconfident in establishing social contacts, especially with foreigners. Yet, now she no longer feels unsure about how to interact with them, and she is more open-minded even towards people she doesn't know. She has already read about other travelers' experiences and hints, and she may even have had contact with foreigners all around the world via the Travel Machine Web portal.
- Sara used to share her travel experiences instantly with her friends via social networks (Facebook, Twitter), usually only in the form of short messages or single photographs. With the Travel Machine, she can share her whole diary (including texts, photos, audio, and video recordings) with them and make them really participate in her experiences abroad.

6.7.2.3   Richard's Use Scenario

As the president of a renowned trade company, Richard works in a fast-paced business environment. With the assistance of the Travel Machine, Richard can read other travelers' diaries and learn from their experiences. Thus, he can save plenty of time in the planning phase of his trips, and it is much easier for him to arrange a proper itinerary within his tight schedule. The Travel Machine is helpful for his busy, on-the-go lifestyle.

Richard also uses the Travel Machine to discover the city in depth with a more flexible schedule. He gets necessary and useful hints, as well as suggestions and ideas from the Travel Advisor. He likes to follow these advices in order to push himself towards discovering local attractions, cultural particularities, and differences he may have paid less attention to in the past.

Richard wants to know the customs and etiquette concerning social contacts, so that he can behave in an appropriate and socially accepted manner. Moreover, he uses the Travel Machine to communicate with local people and with other travelers, and he consults them for specific questions he has. He also gains useful competences through the knowledge and vocabulary quizzes of the application.

Richard takes all of his photographs with his smartphone, and he stores them in his personal travel diary. Though he rarely makes use of social networks, he would like to show and share his travel experiences with his co-workers, closest friends, and family members.

Richard is heavily involved into all businesses of his company, so he desires to
have a good and agreeable time after work. He uses the Travel Machine to search
for recommended, enjoyable places, where he can relax and experience the culture
abroad, instead of limiting himself to the normal recommendation list of "must-see"
spots.

Richard's Behavior Changes

- Richard used to feel tired of business trips, since he did not have enough time to
  plan his free time after work properly, and since he was bored of just rushing to
  some famous spots all other visitors and tourists would go to as well. Now he can
  enjoy more his business conferences all around the world. With the Travel
  Machine, it takes him only a few minutes to get an idea of where to go out at
  night, even if he hasn't prepared anything in advance. And he always has a good
  time when going to the places other travelers "tested" and recommended before.
- In the past, Richard felt that it was not easy to discover the local, unique features
  in an unfamiliar city, even if he could ask advice from his local co-workers. Now,
  via other people's travel diaries, he can use the Travel Machine to get hints and
  ideas in order to discover the city in depth. He gets to learn much by visiting
  those recommended places.
- Richard used to search and read some information about the nation and culture
  of other countries before he went there. Nevertheless, he had to realize that often
  he was not adequately prepared for the customs and etiquette; his intercultural
  knowledge was not accurate enough or sometimes outdated. Now, he can more
  easily get access to local people or to travelers who have been at a certain place
  before, and he can get to know the traditions, usages, and the overall culture of
  the respective place in a more comprehensive and a more accurate way.
- Due to a certain gap between different cultures, Richard sometimes had difficulty
  in finding a common conversation topic with his foreign co-workers. Now, showing his travel diary or the diary records of former travelers is a convenient and
  pleasant way for Richard to start and manage conversations and intercultural
  communication.
- In the past, Richard used to stay in his hotel after work, since he didn't want to
  take the trouble to visit overcrowded famous spots. Now, he often rather randomly
  picks some recommended places, or things that attract him, and he just goes and
  tries. Most of the times, he is amazed by what he experiences, especially
  in locations which are usually not considered being "must-see" spots.

6.7.2.4   Mike's Use Scenario

Mike wants to know the interesting spots and locations all around the destination.
With the Travel Machine, he can get to know recommended places, which might be
of interest for him.

Mike also uses the Travel Machine to record his travel experiences. When he gets
to a new place, he writes down his daily experiences in a personal travel diary.

The diary also includes all the photographs he takes; he is not interested in publishing or sharing them in the first place, but he enjoys keeping them for his personal and individual memory. In addition, he integrates into his diary multimedia sources like videos and audio comments.

Mike wants to share some of his travel experiences with his closest friends and with potential future travelers. When he comes across beautiful and fascinating places or destinations, which might not be well known to others, he wants to communicate these places to other interested travelers. At the same time, he is sometimes a little bit disappointed by certain allegedly "must-see" spots, and he wants to share also these negative experiences (i.e., also cautions or warnings) with friends and other travelers. The Travel Machine is ideally designed to satisfy such needs.

Moreover, Mike enjoys playing games in the Travel Machine, in order to check for himself, how familiar he is with the culture, and maybe with some vocabulary of the foreign language. A high score will entitle him to certain virtual and real rewards, which further stimulates his motivation and interest to play.

Mike is fascinated by the fact that the Travel Machine can figure out "hidden" details and information in his photographs. For instance, by using geo-tags, the Travel Machine can indicate to him specific locations, buildings, or cities that appeared in famous movie scenes in the past. Thus, Mike gets to know many interesting stories behind his pictures, which significantly enriches the content of his photo album, as well as his overall travel experience.

Mike's Behavior Changes

- Mike was frustrated because it was difficult for him to discover new places that would have been different from those recommended by most travel Web sites and visited by masses of people due to their high popularity. Now, he is surprised that the Travel Machine provides him with numerous interesting travel options that he would not have considered before. He enormously enjoys the variegated, enriching experiences he can have.
- Mike needed to write down his feelings and thoughts on paper; otherwise he would not have remembered them later, when composing his travel articles. Still, his notes sometimes were confusing and mixed up. Now, Mike uses the Travel Machine to keep track of travel experiences, and he takes "digital" notes in connection with photographs, as well as with audio and video recordings. It is thus much easier for him to organize his record materials and to later publish them on his travel blog together with pictures.
- Mike shares his thoughts with friends and other like-minded people. It is a great experience for him, because the Travel Machine provides a unique meeting platform, which is designed for telling and exchanging stories among travelers, be they already friends or not.
- Having used the Travel Machine, Mike has been able to get to know many cultures in Africa, Asia, and Europe. The application also helped him take better pictures with culture-related annotations and with little stories about funny or interesting episodes he experienced during his trip.
- Frequently, Mike used to come across interesting and beautiful architectural works, which he didn't know about before. Thus, he had to spend time searching

for information about the respective buildings, their history, their features, etc. Now, he obtains all this background information more quickly via the URL offered by the Travel Machine. He is also inspired by the movie stories behind those architectural works and places, and he wants to write some related articles. The Travel Machine is therefore a highly useful tool for Mike and for his work.

6.7.2.5   George's and Julia's Use Scenario

George and Julia use the travel diary of the Travel Machine to record their travel experiences in a more structured and clearly arranged way and to take more intensive and long-lasting memories out of their trip.

Even if George and Julia already know quite well numerous places they visited in the past, they are very interested to see what other travelers visited and what they liked particularly. They do this by consulting other people's travel diaries, which gives them many new ideas. Furthermore, they both highly appreciate the Travel Advisor function of the Travel Machine, since it helps them go beyond the travel experiences they had in the past.

George and Julia are also happy to get some feedback, and at least some guidance both by the application's advisor and by other travelers, as it was the case during their last trip to San Francisco: They could not speak English fluently, and they were a bit afraid of not being familiar with many intercultural differences. Thus, they needed at least some instructions and proposals based on which they could then go further. With the usage of the advisor of the Travel Machine, they could not only see and observe but also learn and enrich their minds and broaden their horizons by getting to know locals and the local culture.

Usually, George and Sara do not make extensive use of social networks, such as Facebook and Twitter, even if both of them have created a Facebook account. However, they would love to send photographs and impressions of their trips to their siblings or nephews, as well as to their closest friends. By using the Travel Machine, they can quickly share with them what they are currently living and make them participate in their experiences.

Even though George and Julia normally do not make extensive use of electronic devices (PC, mobile phones) for playing games, they find it entertaining to pass the waiting time at attractions, or the time spent on buses and trains by doing together little knowledge and vocabulary quizzes in the Travel Machine.

George's and Julia's Behavior Changes

- Julia always complained about the fact that, with the new era of digital photographs, no one created nice photo albums anymore, and memories were stored in just a few travel folders on the PC. Now, she is pleased that she and George can store their travel experiences in their own travel diary and reexperience the beautiful memories afterwards.
- Concerning their past travel experiences, George and Julia feel that the rhythm and structure of those trips very frequently were quite the same. For this reason, they find highly fascinating the idea of doing a trip with the assistance of a Travel

Advisor: They can explore the same destination in a totally different and more exciting way they have never thought of before.

- Also, by looking at other travelers' diary entries, photographs, and video records, they get new ideas and inspirations for their own travel itinerary.
- George and Julia are not very young anymore, so they no longer enjoy walking around everywhere during the whole day, in order to see as many places as possible. Now, they make use of the Travel Machine, and they figure out a new and innovative approach to their trips. They learn to focus more on the culture, traditions, and spirit of the places they visit, and they can have relaxing and at the same time culturally and personally enriching trips.
- George and Julia used to send their travel photographs to family members via email once they had come back home from their trips. Now, it is more convenient for them to share their experiences instantly via smartphone. Also, their family members get a much better understanding of their travel experiences with the help of the diary, which includes stories, pictures, and highly personalized descriptions.
- George and Julia rarely played digital games on the computer or on the mobile phone. However, they start to have fun when doing the vocabulary quizzes together. When they are bored during a journey, or when they have to wait somewhere, it is a both an enriching and entertaining pastime to play with this simple but interesting application.

6.8   Market Research

In order to have a clearer vision of the target market for the Travel Machine, AM+A conducted a comprehensive market research with potential customers, in collaboration with marketing students of UC Berkeley Extension. One main objective of the research was to find out more about people's general usage of smartphones and mobile applications. The second objective was to get a better understanding of their travel behavior and of the importance of ICTs, especially of smartphones and mobile applications, both before and during a trip.

6.8.1   Secondary Data

In a first step, secondary data was collected in order to get a general overview of people's mobile usage and travel behavior (all data in this section from Figueira et al. (2012). 80 % of the world's population now has a mobile phone, which accounts for about five billion phones. Currently, around 1.08 billion of these mobile devices are smartphones, i.e., approximately 20 %. Current trends seem to indicate, however, that their number will rise considerably during upcoming years. Solely in the first quarter of the fiscal year 2012, the company Apple, for example, sold a total of 37.04 million devices, which corresponds to about 377,900 mobile phones per day.

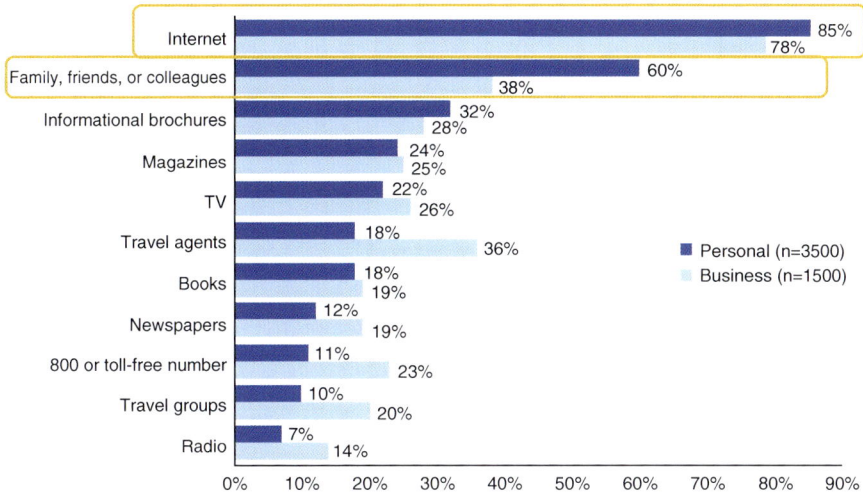

Fig. 6.3 Travel planning sources (Figueira et al. 2012)

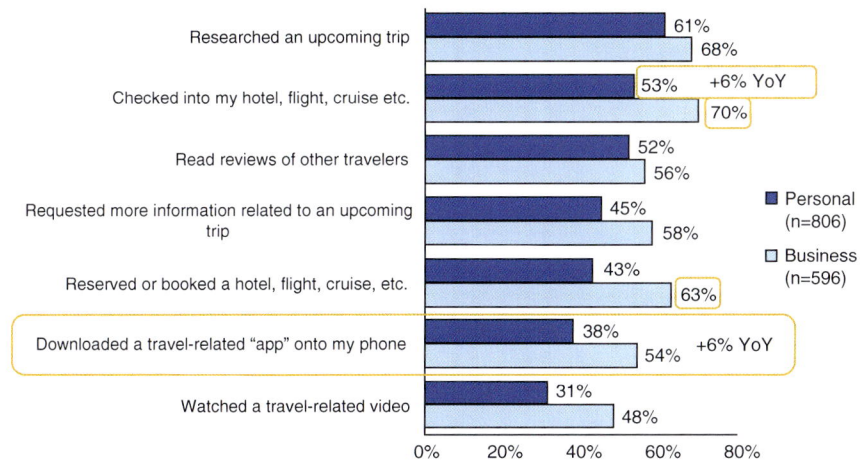

Fig. 6.4 Use of smartphones for travel purposes (Figueira et al. 2012)

Interestingly, not only teenagers and young people make use of smartphones. Most users are between 18 and 44 years old. They use their phones mainly to write text messages (92 %), to go online (84 %), to write emails (76 %), to play games (64 %), and to use downloaded apps (69 %).

As shown in Fig. 6.3, when it comes to travel behavior, the secondary data reveal that people rely primarily on the Internet and on word of mouth as planning sources, both for personal and business travel purposes. As shown in Fig. 6.4, 38 % of the sample for these secondary data stated that they had already downloaded a travel-related app onto their phones. Even 54 % of business travelers made the same statement.

In order to gain further information, qualitative research was carried out in the form of a focus group discussion and in-depth interviews.

6.8.2   Focus Group

The focus group was composed of 1 American (female, 27 years), 1 Korean (female, 26 years), and 1 Spanish (male, 38 years) participants. One interesting finding was the importance of Internet research prior to a trip, to collect necessary information and to decide on a travel itinerary. Moreover, the importance of well-designed maps, which should be accessible also offline, was highlighted. In addition, especially women seem to attribute high importance to other travelers' recommendations and reviews, and they are also interested in sharing their own experiences.

The focus group participants were asked to draw their "ideal mobile travel app." Even though the respondents did not deal with design, specifically, layout issues, their drawings are interesting in terms of desired content and needs. Apart from general information about where to sleep and eat, the provision of maps and routes seems again to be of high relevance. Moreover, user-friendliness and the design of a "clean" mobile application are underlined. Interestingly, also safety and security appear to be an important matter when traveling to foreign places and countries, especially for women.

6.8.3   In-depth Interviews

In-depth interviews were carried out with 7 participants from 6 different countries (America, Brazil, Spain, Chile, Taiwan, and Thailand), 4 male and 3 female, who were aged between 23 and 34. The interviews particularly focused on the information needs and activities of travelers before, during, and after their trips. As shown below, the respondents named numerous subjects in which they were interested, especially regarding where to go, how to move, and what to see. Moreover, an interest in culture and the meeting of new people was mentioned repeatedly. Also, social networking plays a significant role in people's travel experiences, particularly in the during- and after-trip phases (Fig. 6.5).

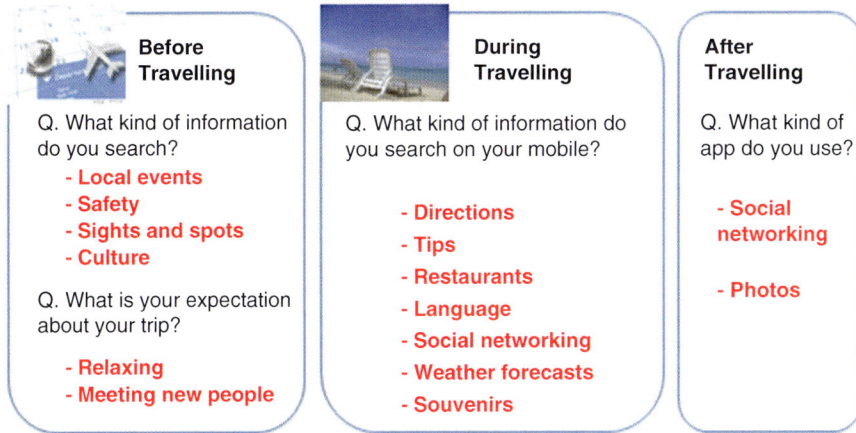

Fig. 6.5 Users' information needs in before-, during-, and after-trip phases (Image credit/data source: Figueira et al. 2012)

Table 6.1 Countries of origin of questionnaire respondents (Figueira et al. 2012)

Nationality	%	Sampling size
Brazilian	27.4	10
Spanish	18.8	7
Japanese	12.0	4
Tawanese	6.8	2
Indian	6.0	2
French	5.1	2
Korean	5.1	2
Chilean	3.4	1
Russian	3.4	2
Turkish	3.4	1
Thailland	2.6	1
Venezuela	2.6	1
Chinese	1.7	1
Belgian	1.7	1
Total	**100.0**	**36**

6.8.4   Survey

In order to enrich the market research with quantitative data, AM+A prepared a questionnaire, which was distributed among 36 UC Berkeley Extension International Diploma Program students from 14 different countries, 53 % male and 47 % female, most of them aged between 20 and 30 years (Table 6.1).

The majority of respondents (15 out of 36) claimed to travel only once per year, regardless of nationality and gender. This fact may be due to limited financial resources for students. Concerning their main information needs before and during

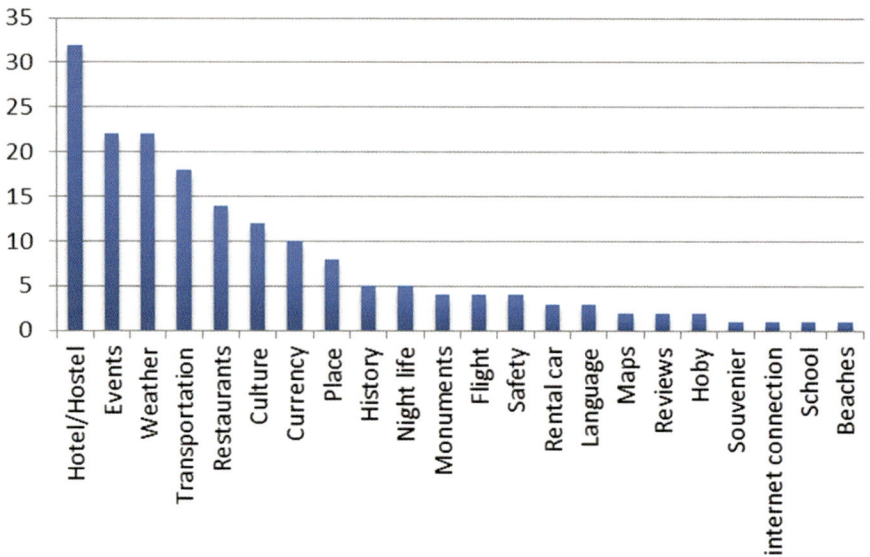

Fig. 6.6   Information needs before traveling (Figueira et al. 2012)

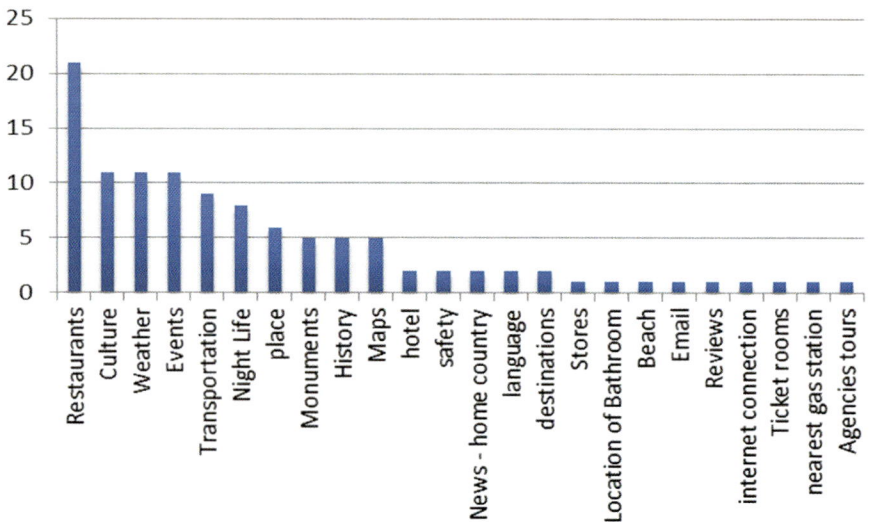

Fig. 6.7   Information needs during traveling (Figueira et al. 2012)

a trip, the major requirements of the questionnaire respondents seem to be quite similar in the two phases; they particularly refer to accommodation, catering, events, transportation, weather, and culture (Figs. 6.6 and 6.7).

When it comes to information research, Google, TripAdvisor, Expedia, and blogs appear to be the most popular and most used sources, as shown in Fig. 6.10 (Fig. 6.8).

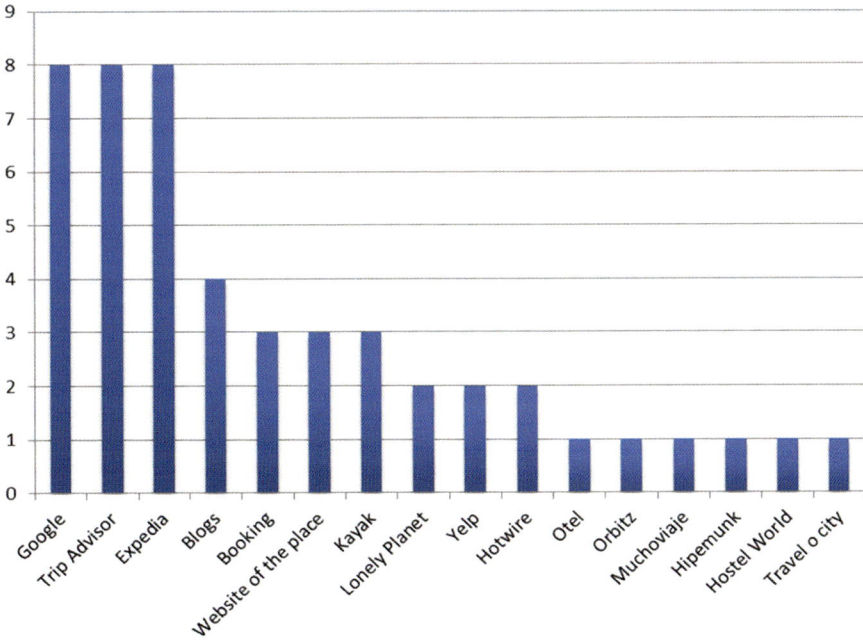

Fig. 6.8 Sources for information research (Figueira et al. 2012)

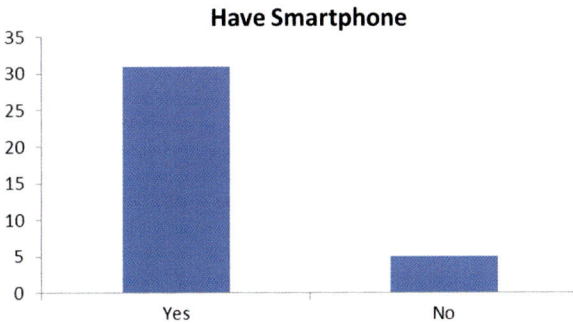

Fig. 6.9 Smartphone distribution among respondents (Figueira et al. 2012)

Concerning respondents' mobile usage behavior, Figures 6.11 and 6.12 indicate that an overwhelming majority already have a smartphone, most of them an iPhone (Figs. 6.9 and 6.10).

Most respondents use their smartphone to make phone calls, to send text messages, to accede to social networks, or to surf on the Internet. Furthermore, almost all of them make usage of mobile applications, especially those related to social networks (Facebook, Twitter, Foursquare, etc.), WhatsApp, Viber, as well as music-, map-, and game-related applications (Figs. 6.11, 6.12 and 6.13).

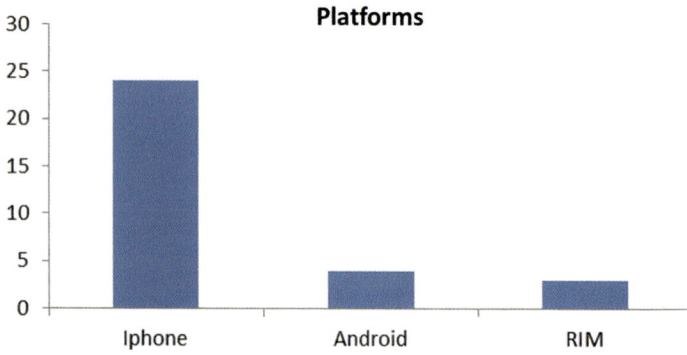

Fig. 6.10 Distribution of smartphone models of respondents (Figueira et al. 2012)

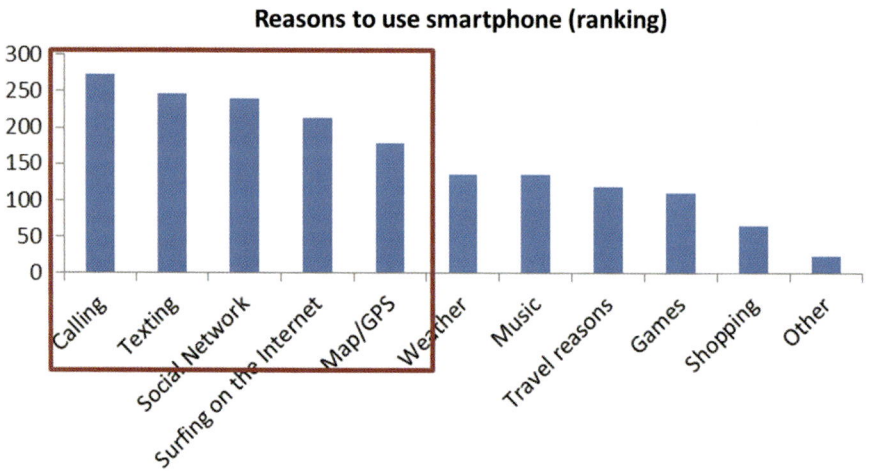

Fig. 6.11 Reasons for smartphone usage (Figueira et al. 2012)

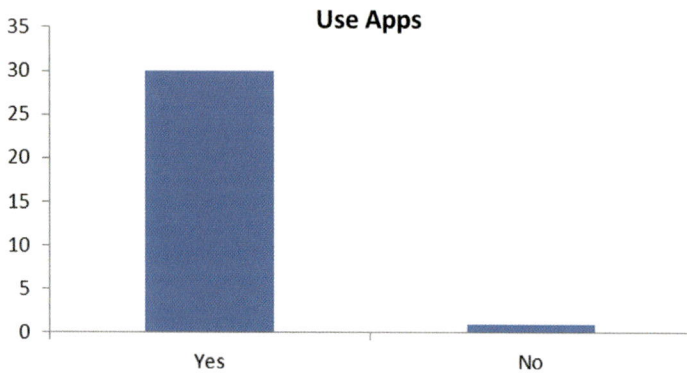

Fig. 6.12 Usage of mobile apps (Figueira et al. 2012)

Apps used frequently

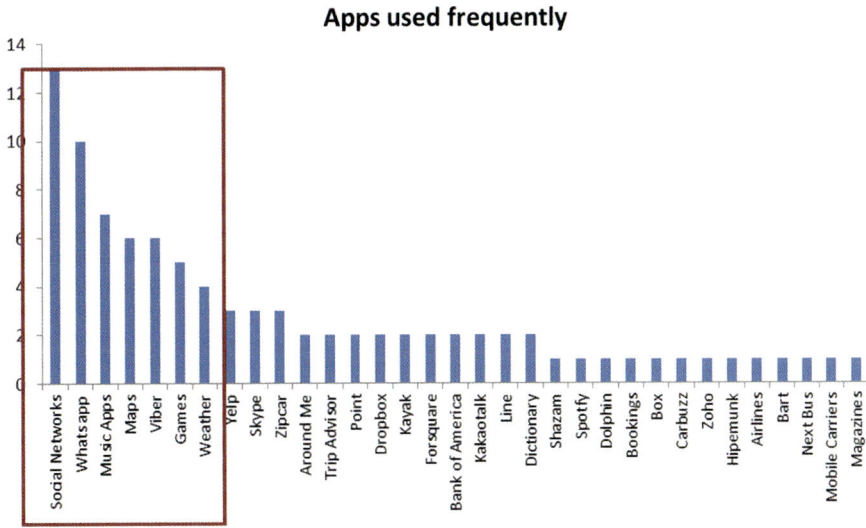

Fig. 6.13   Most popular mobile apps (Figueira et al. 2012)

How much did you pay for this app?

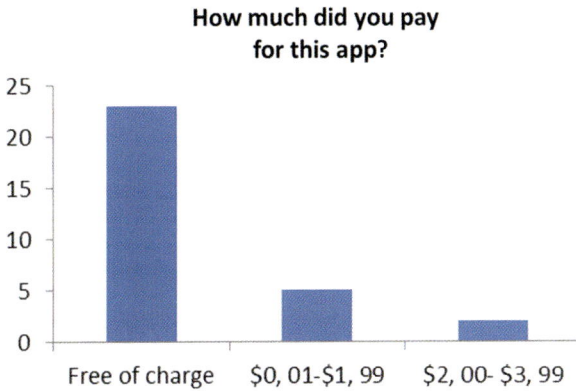

Fig. 6.14   Willingness to pay for mobile apps (Figueira et al. 2012)

Interestingly, most respondents did not seem, however, to be willing to pay money for the mobile applications they download and use (Fig. 6.14).

When it comes more specifically to the use of travel-related mobile applications, more than half of the respondents already made use of them when going to foreign places and countries. They did so mainly to find, book, or rate hotels and restaurants, to get information on their current locations, to obtain transportation schedules and prices, or to inform themselves about local culture and places. Those who do not rely on travel apps when going abroad retrieve most of the necessary information and data from travel-related Web sites (Fig. 6.15 and 6.16).

Fig. 6.15 Usage of
travel-related apps
(Figueira et al. 2012)

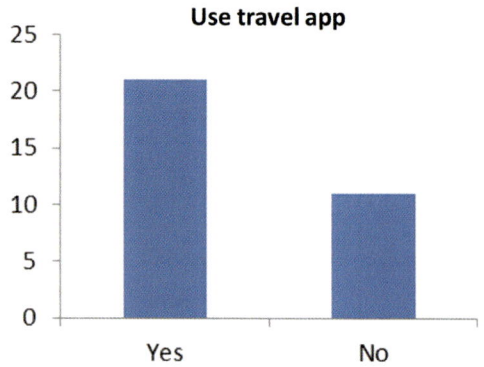

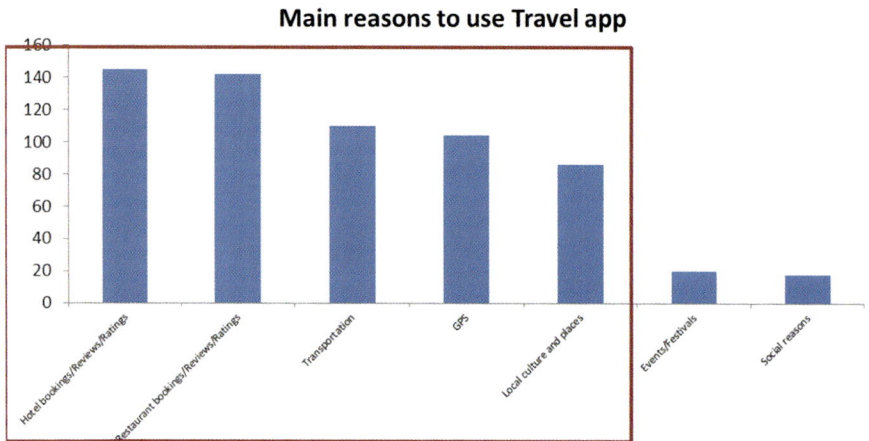

Use travel app

```
25

20

15

10

5

0
            Yes              No
```

Main reasons to use Travel app

```
160
140
120
100
 80
 60
 40
 20
  0
    Hotel bookings/Reviews/Ratings
         Restaurant bookings/Reviews/Ratings
              Transportation
                   GPS
                        Local culture and places
                             Events/Festivals
                                  Social reasons
```

Fig. 6.16 Main reasons to use travel-related apps (Figueira et al. 2012)

When asked about the makeup of an "ideal travel app," the most-mentioned contents were restaurants, events of the day, maps/GPS, hotels, transportation, night life, reviews, and the recommendation of interesting places to visit. Also, it became clear that most potential users would not be willing to pay more than about $2 for a mobile travel application, as shown in Figs. 6.17 and 6.18.

Another interesting finding of the quantitative research is respondents' behavior in relation to reviews and ratings. Even though a large majority of them regularly read and rely on other travelers' reviews, ratings, and suggestions, they are quite reluctant to provide this kind of user-generated content (UGC) themselves (Fig. 6.19 and 6.20).

A further important point that emerged in the course of the market research is the partially difficult or expensive access to Internet when traveling abroad. Even though the number of public hotspots offering free Wi-Fi (parks and squares, tourist attractions, transportation places like airports or train stations) is expected to rise

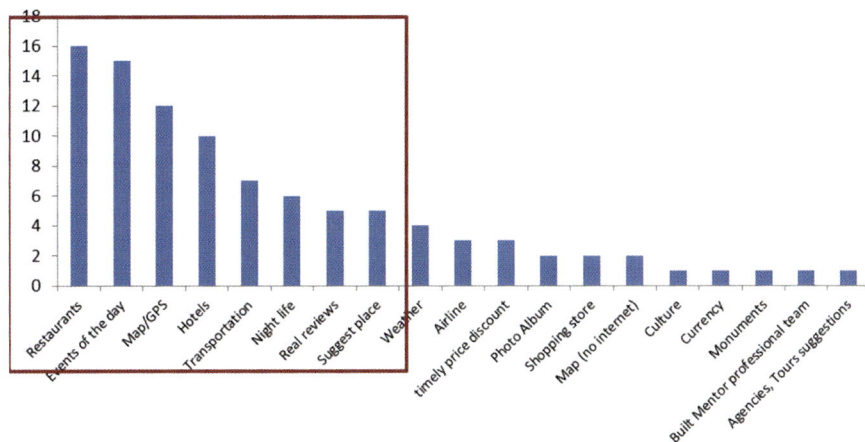

Content Ideal Travel App

Fig. 6.17 Main reasons to use travel-related app (Figueira et al. 2012)

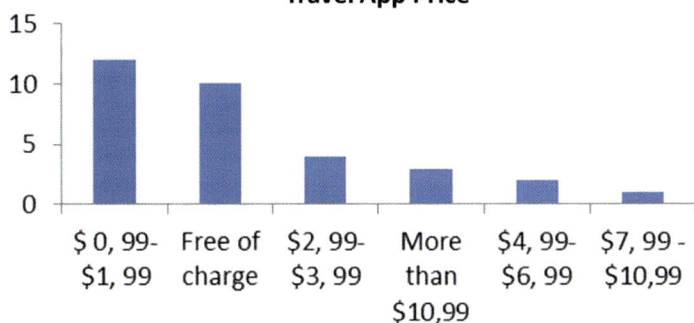

Travel App Price

Fig. 6.18 Willingness to pay for a travel app (Figueira et al. 2012)

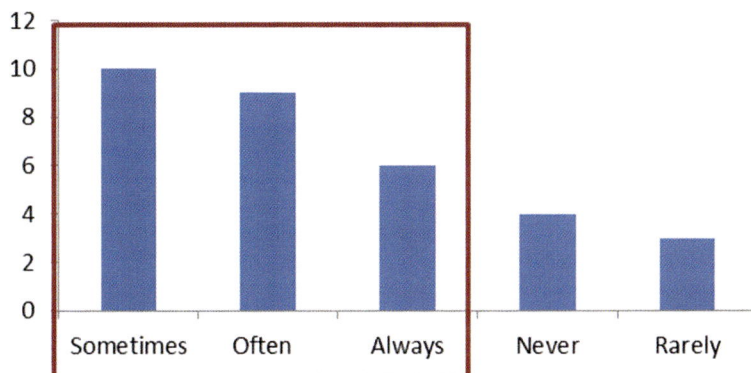

Search for reviews/ratings/suggestions

Fig. 6.19 Search for reviews/ratings/suggestions (Figueira et al. 2012)

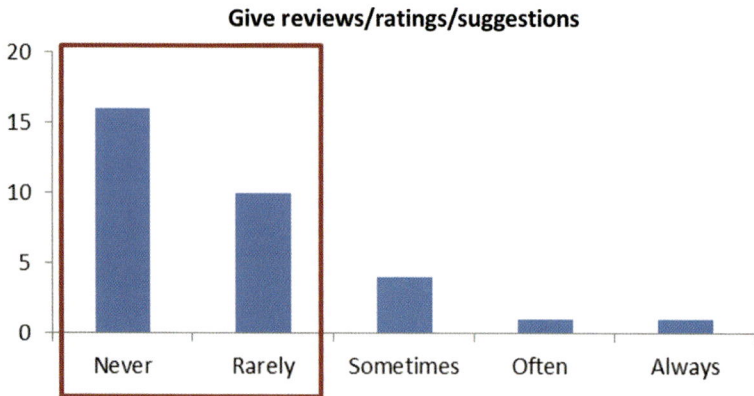

Fig. 6.20   Provision of reviews/ratings/suggestions (Figueira et al. 2012)

from 1.3 million in 2011 to 5.8 million in 2015, with the continuous increase of smartphone users, the cheaper provision of Internet will be an important future topic for telecommunications companies. At the same time, this fact makes it recommendable for a travel application to provide also various offline features, i.e., functions that can be used even without having Internet access.

6.9   Competitive Analysis

Before undertaking conceptual and perceptual (visual) designs of the Travel Machine prototype, comparison studies were carried out. AM+A studied in detail 11 smartphone applications in 2012. Through screen comparison and customer review analyses, AM+A derived these applications' major benefits and drawbacks. This in-depth analysis helped to further develop initial ideas for the Travel Machine's detailed functions, data, information architecture (metaphors, mental model, and navigation), as well as look and feel (appearance and interaction).

6.9.1   TripAdvisor

Who Were They in 2012?
- Founded by Stephen Kaufer in February 2000.
- TripAdvisor is a public company based in Newton, Massachusetts, which provides travel services for users.
- TripAdvisor manages and operates 19 other travel media brands, including Cruise Critic, Flipkey, Holiday Watchdog, SeatGuru, TravelPod, VirtualTourist, etc.

- TripAdvisor is the world's largest travel site. It launched a mobile application for iPhone and Android, providing free versions in different languages, such as English, German, French, Arabic, and Chinese.
- TripAdvisor has gained 4 stars from 26,229 ratings among Apple users. It is also a top developer for Android's mobile applications. It has had 5,000,000 downloads for Android, and 12,000 people liked this application (July 2012) [Google Play App Store].

What Was It in 2012?

- Mobile application that offers a wide variety of travel options and planning features, with seamless links to booking tools.
- Platform that assists customers in gathering travel information with trusted advice from real travelers and in posting reviews or other travel-related content.
- Makes travelers engage in an interactive travel forum.
- An early adopter of user-generated content (UGC).

What Did the App Let Users Do in 2012?

- Search for travel information about hotels, restaurants, things to do, etc.
- Submit reviews of hotels, vacation rentals, attractions, or restaurants, in an open travel community, where other users can view and vote for the most helpful contributions.
- Log in with a Facebook account to create a personal profile, and view other users' reviews.
- Participate in discussions held in the TripAdvisor fora.
- Filter points of interests (POIs) based on distance from users' location and based on ratings/popularity.
- Save interesting attractions to users' personalized "My Saves."

Mobile Strategy

- Service free to all users.
- Provides clear division of functions and offered services.
- Feature "My Saves" helps users keep track of places that they are interested in.
- Interconnected with Facebook, but users cannot view friends' activities via mobile application.
- Has unique "Near Me Now" mobile function (not available on the Web site), providing information on attractions based on GPS.
- TripAdvisor uses Bing Maps for its map system. The iPad version also integrates Google Maps to direct users to selected POIs, based on their current location.

User-Friendliness

- Platform is easy to use and to understand.
- Fast registration; possible also via Facebook account.
- Clear arrangement of different categories. Thus, easy for new users to get started immediately.
- Lack of social networking functions.

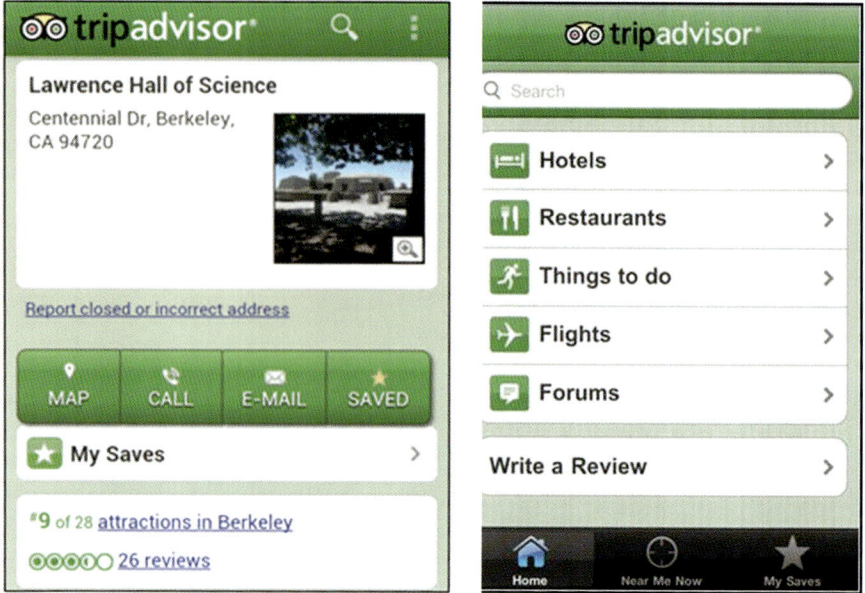

Fig. 6.21 Screenshots of the TripAdvisor smartphone version in 2012 [Images: TripAdvisor, LLC.)

- Lack of user interaction and feedback. Function "vote for useful reviews" is not available on the mobile application, and users cannot filter reviews based on the most "helpful votes."

Visual Analysis

- White background with green base color makes application look very clear and makes it easy to use.
- Services are categorized in a clear and structured way, allowing travelers to easily find necessary information (Figs. 6.21 and 6.22).
- No possibility to display several points of interest (POIs) simultaneously on one map, neither on the Web site nor on the mobile phone. However, the iPad version can support this function by combining Bing Maps and Google Maps. It is thus capable of directing users to selected POIs from their current location.
- Provision of a travel guide with brief information, including place name, address, photograph, and rating. Users can view more by reading others' reviews.

Primary Competitors

- Yahoo! Travel.
- TravelPost.
- Virtual Tourist.

Tips

- TripAdvisor guides users to select POIs, by providing them attraction rankings for specific regions.
- TripAdvisor offers user fora that enable discussions and user interaction.

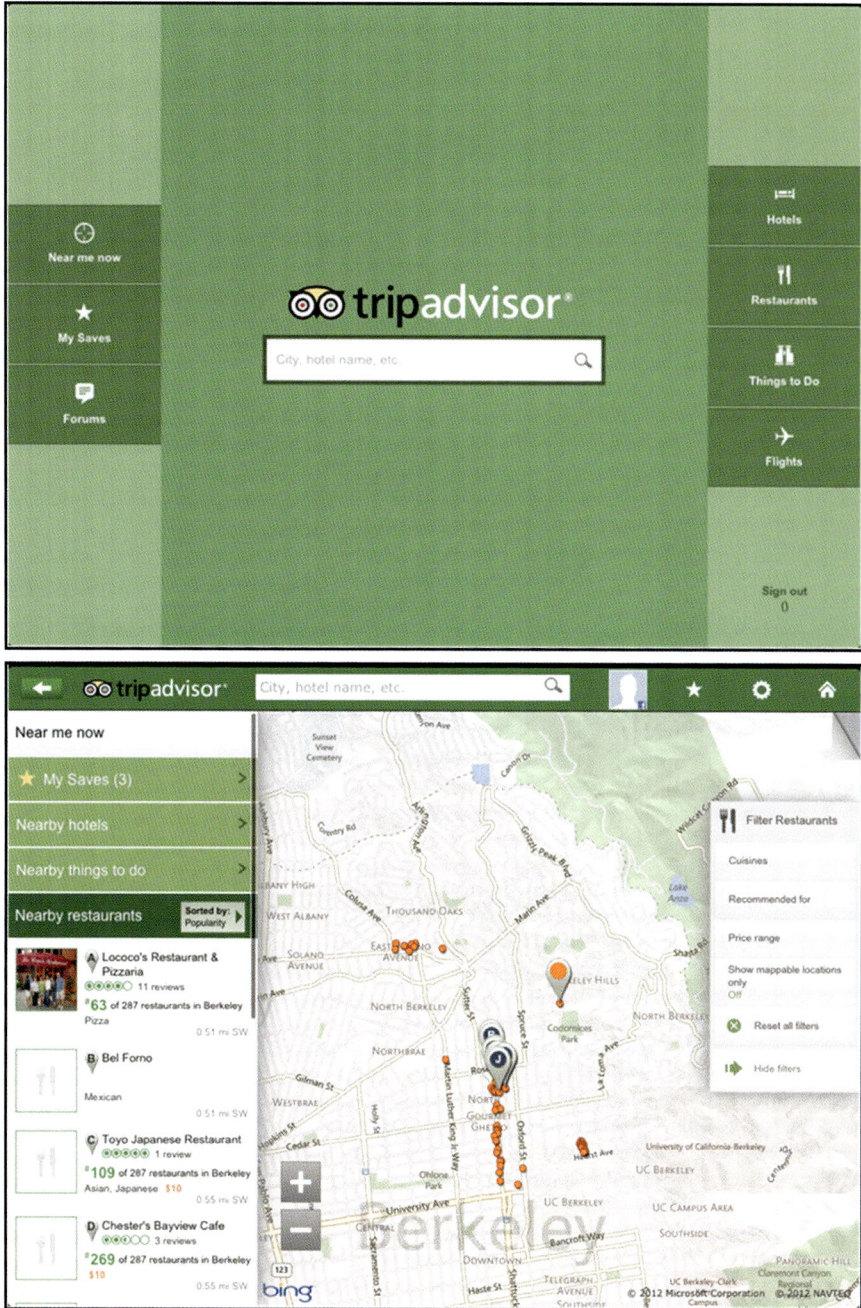

Fig. 6.22   Screenshots of the TripAdvisor from an iPad in 2012 (Images: TripAdvisor, LLC)

Social Networking

- Integrated with Facebook.

SWOT Analysis

Strengths

- "Write a Review," "Near me now," and "My Saves" functions are intuitive and easy to use.
- UGC engages travelers to make use of the platform and to participate in the forum.
- Integration with Facebook gives users more channels to engage and communicate via the platform.
- TripAdvisor has a huge number of users on its Web site. It is very probable that these users will also use the TripAdvisor mobile applications for iPhone and Android.

Weaknesses

- No possibility to filter reviews based on most "helpful votes."
- "Near Me Now" function of the mobile application is excellent, but the mobile map system (for smartphone use) is inadequate.
- Some important features of the Web site are not included in the TripAdvisor mobile application, e.g., view friends' activities and updates, vote on the most helpful reviews, record travel experiences and expectations on the travel map.

Opportunities

- Improve map system with the "Near Me Now" function on the mobile application.
- Add filter function based on "helpful votes."
- Develop more functions linked to social networking.

Threats

- Allowance of anonymous reviews to be posted for any hotel, B&B, inn, or restaurant. This policy partly destroys the credibility of the reviews on TripAdvisor.

6.9.2   Lonely Planet

Who Were They in 2012?

- Founded by Tony and Maureen Wheeler in 1972.
- Lonely Planet is the largest travel guide book and digital media publisher in the world. The company is owned by BBC Worldwide.

What Was Is It in 2012?

- The world's most favorite travel guides developed for iPhone. It sets the standard for guidebook applications with high-quality level of content and organization.
- A collection of recommendations and reviews from expert authors, including popular "must-see" sights and places, information on where to shop, eat, go out, etc.

What Did the App Let Users Do in 2012?

- Download and store city travel guides on the mobile phone.
- Obtain comprehensive information from the travel guide, including basic introduction, map information, neighborhoods, things to do, etc.
- Get the full version of the mobile application, where users can use offline maps with GPS and get additional information from Lonely Planet.

Mobile Strategy

- Lonely Planet aims at catering for all kinds of needs during a trip, in a compact mobile app package.
- Lonely Planet has a free version for all users. However, the full city guide versions cost around $ 6 per destination.
- The free city guide versions offer insufficient and little information compared to the Lonely Planet Web site.
- Mere provision of a travel guide. The application is not connected with any social networks like Facebook or Twitter.

User-Friendliness

- Platform is easy to use and manage.
- No registration needed, fast to get started.
- City guides are well designed and clearly structured, allowing users to easily find necessary information.

Visual Analysis

- Blue background with white-colored titles makes it easy for users to read.
- Large and catchy photographs above clearly arranged categories (e.g., "Hello Barcelona," "Map," "Practicalities," "Neighbourhoods," and "Things to do").
- Differently colored areas on the maps to distinguish different neighborhoods.

Primary Competitors

- TouristEye.
- TripAdvisor.

Tips

- Lonely Planet offered links to related merchandises (e.g., travel books, travel magazines), where users could obtain further information.

Social Networking

- No social networking functions.

SWOT Analysis

Strengths

- Lonely Planet is the largest travel guide book and digital media publisher in the world. Thanks to its own production of books, magazines, and documentaries about travel and tourism, the Lonely Planet mobile application has enormously rich travel guide resources, compared to any other similar mobile application.

- As one of the world's most favorite travel guide, users might be willing to use the mobile app of Lonely Planet, even if they need to pay more than for other travel applications.
- Numerous different functions are wrapped up in an intuitive, easy-to-use mobile application, which does not necessitate an Internet connection. Thus, the app can be used everywhere and at any time, without creating data or roaming fees (full version only).

Weaknesses

- Lonely Planet charges users with $ 5.99 to get the full version of a specific city guide. This price is relatively high on the travel app market.
- The mobile application exclusively offers a travel guide, without any social networking features.

Opportunities

- More functionalities linked to social networking.

Threats

- As more similar travel applications appear, the high cost of purchasing a guide from Lonely Planet may prevent current users from using the mobile application in the future.

6.9.3   TripIt

Who Were They in 2012?

- Founded in Silicon Valley in 2006.
- TripIt's mission is to simplify users' travel experiences. It is the leading mobile travel organizer of Concur.
- TripIt has gained 4 stars from 22,028 ratings among Apple users (http://itunes. apple.com]. It is also an "editor's choice app" of Android's mobile applications.
- TripIt has 1,000,000 downloads for Android, and 7,500 people like this application (July 2012) [Google Play App Store].

What Was It in 2012?

- TripIt is an intelligent mobile travel organizer that makes it easier for users to organize, plan, and share their trips.
- It is a personal travel agent for users, helping them to plan every detail of their trip, from car rental to lodging, to restaurants, *etc.*
- Travelers can forward confirmation emails from anywhere they book to plans@ tripit.com. TripIt "automagically" creates one simple, smart itinerary, which can be accessed via smartphone or anywhere online.

What Did the App Let Users Do in 2012?

- Construct customized itineraries by hand or simply forward purchase confirmation emails to TripIt. TripIt then automatically creates online master itineraries with travel plans and other critical information (maps, directions, weather, etc.).
- Option of booking restaurants, theater tickets, activities, and more, right from the online itinerary.
- Moreover, users themselves can organize and integrate different trip details and components into one online master itinerary. This is possible even if bookings and reservations are made on multiple travel Web sites. They can then print and share itineraries with TripIt.
- Usage of TripIt as a personal travel assistant to keep track of flight status, alternative flights, and frequent traveler points. Users can also check in for flights via TripIt (TripIt Pro version only).
- Upload of additional photographs to trip objects.
- View all personal trips on one single map. This function is, however, not supported by Android mobile phones.

Web Strategy

- Web site looks clean and users-friendly. A big picture of the application interface occupies more than half of the screen.
- Encourages users to sign in and to become members of TripIt.
- Very limited information about the application before users register and sign in for the first time.

Mobile Strategy

- TripIt has a free version with advertisements for all users, and it has a version without ads for $3.99. TripIt also offers a professional version with more enhanced and innovative functions, which requires a payment of $ 49 for an annual subscription.
- Clear division of functions and offered services. Thus, it is easy for new users to get started immediately.
- The mobile application is connected with the Web site, so that changes on the Web portal are automatically synchronized with the mobile application.
- The mobile application has only limited social networking functions. The Web portal allows more social interaction and exchange among users and friends.

User-Friendliness

- Fast registration with an email address. Registration also possible via Google account.
- Platform is easy to understand due to a clearly structured list for the organization and planning of trips.
- Easy and smooth navigation through different sections.
- Social networking functions of the mobile application are limited. For instance, users cannot search for friends or add friends via mobile phone, which is not very convenient. These functions are only offered on the Web portal.
- No demo or trial version. No video tutorials available.

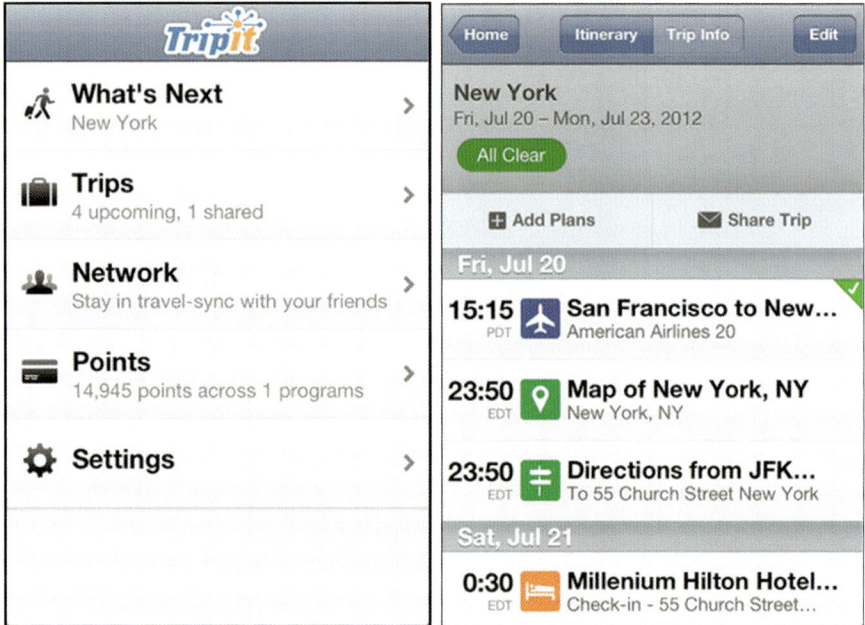

Fig. 6.23 Screenshot of the TripIt mobile app in 2012 (Images: Concur Technologies, Inc.)

Visual Analysis

- Usage of blue and gray as primary colors (Fig. 6.23).
- Buttons and icons seem easy to recognize and understand.
- Usage of bigger fonts than most applications.

Primary Competitors

- Pageonce.

Tips

- None.

Social Networking

- Integrated with Google account only on the mobile application. However, users cannot fully manage their social networks via mobile usage of the app.
- Integrated with various social networks via Web site. Users can sign in via Facebook, Google, Yahoo, etc., and they can connect with their friends on TripIt via email.

SWOT Analysis

Strengths

- Creative idea for organizing trips. Focus on a different phase of the travel experience compared to most travel applications on the market.

- Mobile application is synchronized with the Web portal, which enables users to organize their trip in a convenient manner.
- User-friendly interface and very handy functions.

Weaknesses

- Social networking functions are limited on the mobile application. It is not convenient for users to have to depend on the Internet and the Web portal to fully manage their social networks.

Opportunities

- Application does already very well on iPhones, but it could improve its iPad performance.
- TripIt added several important features in its recently updated version, which includes "photo uploading," "local search and itinerary map," "quickly find directions between trip items," and so on and so forth. These functions, especially the "search for POIs and quickly add to the itinerary" feature, significantly improve the performance of TripIt as a travel-related application.

Threats

- Users are unsatisfied with the advertisements included in the free TripIt version. At the same time, they are not willing to pay $ 3.99 for the more advanced version without ads. This might be a problem in the long term.

6.9.4 TouristEye

Who Were They in 2012?

- Founded by Ariel Camus and Javier Fernandez Escribano in November 2009.
- TouristEye is a privately held company based in Madrid, Spain, which provides travel guide information for its users.
- TouristEye launched its first version on the market in 2010, achieving considerable success and a major impact in the media. In the course of 2010 and 2011, the TouristEye team launched a second version of the product with improvements and clearer focuses.
- TouristEye has been downloaded around 100,000 times to Android phones. 602 people liked this application (July 2012) [Google Play App Store].

What Was It in 2012?

- A smart offline travel guide that helps travelers plan their trips and find and access tourist information by using their browsers or smartphones.
- A social mobile travel guide that enables travelers to share and exchange their experiences, activities, thoughts, opinions, and photographs with friends via a personal trip journal.

What Did the App Let Users Do in 2012?

- Tired of paper travel guides and big maps that are difficult to understand, TouristEye was set up with the objective to create a guide for mobile devices, which can be perfectly tailored to each individual.
- Account: Log in via personal Facebook or Twitter account, to view and share travel experiences with friends.
- Built-in trip planner: Use trip planner to decide what to do each day and to have offline access to information from the mobile device at any place and at any time. Moreover, recommendations are integrated in the trip planner, so that users can see suggested locations and destinations while planning their trip.
- Trip journal: Use the TouristEye trip journal to record one's individual travel itinerary. Edit and share activities, thoughts, opinions, and photographs with the help of this trip journal. The journal also functions without travelers having access to the Internet.
- Travel guide: Access to the TouristEye travel guides for over 60,000 destinations, including monuments, restaurants, hotels, and other trip-related locations. Share interesting travel information and recommendations with friends via social networks.
- Offline feature: Download personalized travel guide to users' mobile phone, so that no Internet connection is needed to access the information.
- Personalized recommendations: Create user "passports." Users are asked to vote on suggested landmarks and attractions, so that their tastes and preferences can be identified. Based on their expressed predilections, users are provided with personally tailored recommendations and suggestions.
- Other: Add new places and destinations that are not included in the current TouristEye database. Filter POIs based on popularity, distance, and recommendations.

Web Strategy

- It is a user-friendly Web site, which provides a quick tour through the page, in order to explain the services in detail, with the help of pictures and texts.
- Web site contains a huge amount of information that can make travel planning more engaging and convenient.
- New users can create an account within seconds or join the community by signing up via their Facebook or Twitter account.
- Web site integrated with several social media, so that users can directly share interesting tours with friends via Facebook, Twitter, email, blogs, Pinterest, etc.

Mobile Strategy

- Use of the "passport" feature to make travelers rate landmarks and attractions, in order to then generate personalized recommendations, based on the individual users' tastes and preferences.
- No instructions. Not easy for new users to start. Some functions are not clearly understandable, due to a lack of explanations or interpretive signs.

- Several features are not well developed yet. Users cannot change or delete a trip, and they do not have the possibility to erase comments in their trip journals.

User-Friendliness

- Lacks necessary instructions, or interpretations for the TouristEye mobile application, e.g., concerning the map system or the "passport" function.
- Provides users with a combination of texts and images. Each location is presented with a name and a thumbnail image, so that users easily find what they need.
- A list of locations is presented on the left side of the screen. It is very well organized and provides users with all details and information they need, without forcing them to navigate through numerous pages.
- Allows users to create new descriptions for locations and destinations and to add them to the existing list of places.

Visual Analysis

- White background with bright green base color (Figs. 6.24 and 6.25).
- Some features of TouristEye are not labeled clearly, both on the Web site and in the mobile application. Various tools are divided into different categories (wish list, attractions, restaurants, and entertainment), which are also distinguished by different colors.
- Important spots are shown on the map, with different colors and signs to distinguish them more easily. However, users need to zoom into the map to recognize the signals.
- The travel guide Web pages are catchy, due to bright colors and thumbnail images for each destination.

Primary Competitors

- Frommer's Travel Guide.
- Triposo.

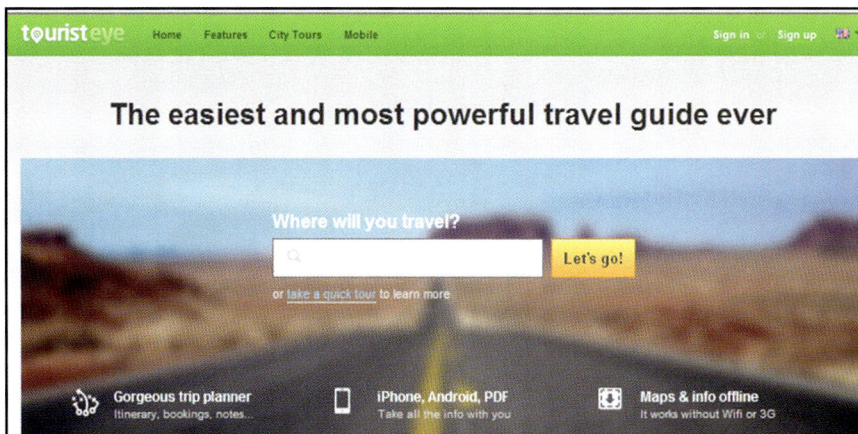

Fig. 6.24 Screenshot of the TouristEye Web site in 2012 (Image: TouristEye, Inc.)

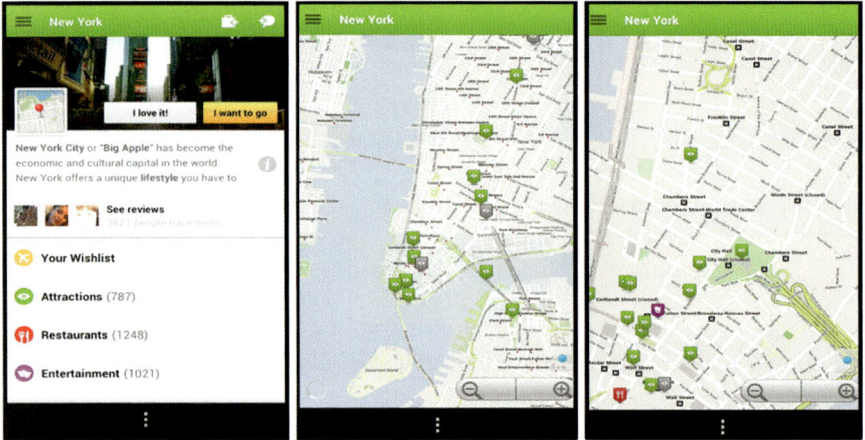

Fig. 6.25 Screenshots of the TouristEye mobile app in 2012 (Images: TouristEye, Inc.)

Tips

- TouristEye guides users by providing personalized recommendations, based on users' trip planning decisions.
- TouristEye encourages travelers to explore places, by indicating how many of their friends have already been there before.

Social Networking

- Integrated with Facebook.

SWOT Analysis

Strengths

- Provision of several good features: Download travel guide to mobile phone and use offline; edit travel journal offline; add new places, which are not yet listed in the destination database; *etc.*
- Well-developed and quite professional travel guide. TouristEye covers a lot of ground, geographically speaking, which might make it more appealing to some travelers.
- Free version to all travelers.
- Facebook and Twitter integrated.

Weaknesses

- System is not stable, several technical problem exist.
- User-friendliness of the user interface needs to be improved.

Opportunities

- Integrate rewards to encourage users to add new places or to write reviews.
- Add functions to track and manage other users' reviews.

Threats

- Charging of fees for usage of "offline" features.
- No comprehensive and overall integration of social networking features (as, e.g., in TripAdvisor); *e.g.*, no direct way to view friends' travel activities via TouristEye.

6.9.5   Wikitude World Browser

Who Were They in 2012?

- Wikitude is a mobile augmented reality (AR) software, which was developed by the Austrian developer company Wikitude GmbH (formerly Mobilizy GmbH) and published in October 2008 as freeware. Wikitude was the first publicly available application that used a location-based approach to AR.
- Wikitude has been downloaded around 1,000,000 times to Android phones, and 3,000 people liked this application (July 2012) [Google Play App Store].

What Was It in 2012?

- A mobile application that allows users to discover their surroundings in a completely new way. The application displays information about the users' surroundings in a mobile camera view.
- Users can create their own world of special places and things and share it with their friends.

What Did the App Let Users Do in 2012?

- Using the camera of users' smartphones to discover and explore what is around them in AR. See places, POIs, and other exciting AR content through the camera's field of vision.
- Create users' individual AR world. The "create place" tool allows users to mark the location of their favorite restaurants or of any other location that is of importance to them.
- Users share their created worlds with their friends on Facebook.

Mobile Strategy

- Interesting and advanced AR technology to attract users.
- Integration of various categories of Web sites and resources, among which users can choose.
- Encouragement of users to build their own "worlds" with objects and places identified by themselves, even if those items are not famous at all.

User-Friendliness

- There are no instructions. It is difficult for new users to get started, especially since AR is a rather new technique, which is integrated in mobile applications.
- Different functional categories are clearly divided and recognizable.

Fig. 6.26   Screenshot of the Wikitude mobile app in 2012 (Image: Wikitude GmbH)

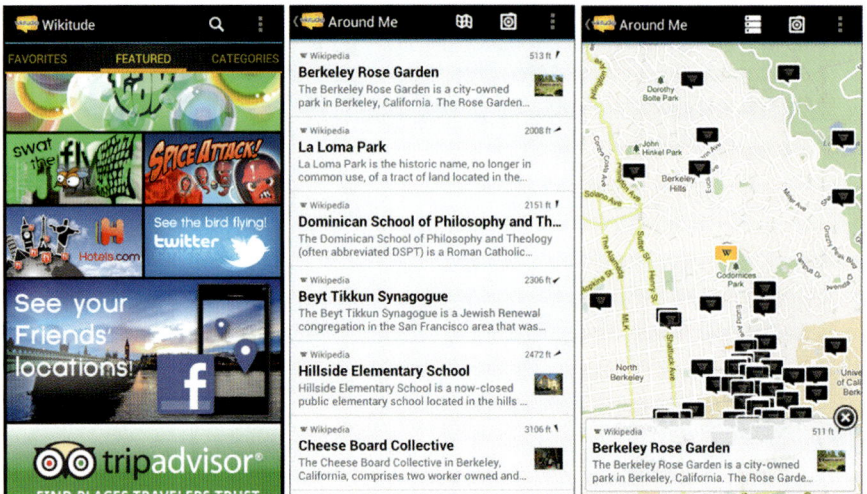

Fig. 6.27   Screenshots of the Wikitude mobile app in 2012 (Images: Wikitude GmbH)

Visual Analysis

- White background with big titles (in yellow) and subtitles (in black) (Figs. 6.26, 6.27 and 6.28).
- The home page of the mobile application is colorful and contains links and pictures of various recommended sites (such as Twitter, Facebook, etc.).
- Wikitude has traditional formats that show places based on their distance or simply on a geographical map. It also has an AR format to show the space around users though their mobile camera.

Primary Competitors

- Layar.

Fig. 6.28 Screenshot of the Wikitude mobile app in 2012 (Image: Wikitude GmbH)

Tips

- Wikitude attempts to guide users, and to help them gain information from various other resources, by providing them with numerous links on the Webpage.

Social Networking

- Integrated with Facebook.

SWOT Analysis

Strengths

- An engaging and immersive experience for travelers.
- Possibility for users to create their own worlds, based on their unique and individual experiences.

Weaknesses

- New technology. Not very easy to learn and to use.

Opportunities

- Only few competitors on the current market.
- Even though the application does not focus on travel use, it has some good features of "discovery" during a trip.

Threats

- Difficulty of fully and smoothly integrating AR techniques into mobile applications.

6.9.6   World Explorer – Travel Guide

Who Were They in 2012?

- World Explorer is a travel guide application developed by AudioGuidia, a company whose main activity is the creation of software for smartphones.
- World Explorer has released different versions (World Explorer – Travel Guide, World Explorer 360 Tour Guide, World Explorer Premium).
- World Explorer has gained 4.5 stars from 193 ratings among Apple users [http://itunes.apple.com]. It has been downloaded around 100,000 times to Android phones, and 899 people liked this application (July 2012) [Google Play App Store].

What Was It in 2012?

- A mobile application, which simply uses a person's current GPS location to spot and to suggest the best surrounding places to visit. Based on this, it provides a travel guide for weekend and longer holiday trips.

What Did the App Let Users Do in 2012?

- Simply open the application, and check surrounding POIs based on users' current GPS location.
- Filter all the spots around based on rating stars, distance (miles), or initial letter.
- Use application to get information about the respective spots and to access to a travel guide.
- Display all surrounding POIs on a map, with different colors according to users' ratings (full version).
- Search for any location worldwide, and visualize all POIs in the respective place (full version).

Mobile Strategy

- Focus on the "Travel Guide" feature, which provides detailed and professional information about numerous POIs.
- Free version of the application is limited to a few good features.
- Simple design and very easy to use.

User-Friendliness

- Simple design enables easy learning and usage: It is enough to click on the app to make it display automatically surrounding POIs, based on users' current GPS location.
- Interface is simple and clear for users to understand and to use, but not very catchy.
- No feedback from users and lack of interactions.

Visual Analysis

- White background with blue base color. User interface is clean and simple.
- Tools are categorized in the form of symbols without words: ratings, miles, A–Z, directions, search for (Figs. 6.29 and 6.30).

Fig. 6.29 Screenshot of
the World Explorer mobile
app (Image: AudioGuidia)

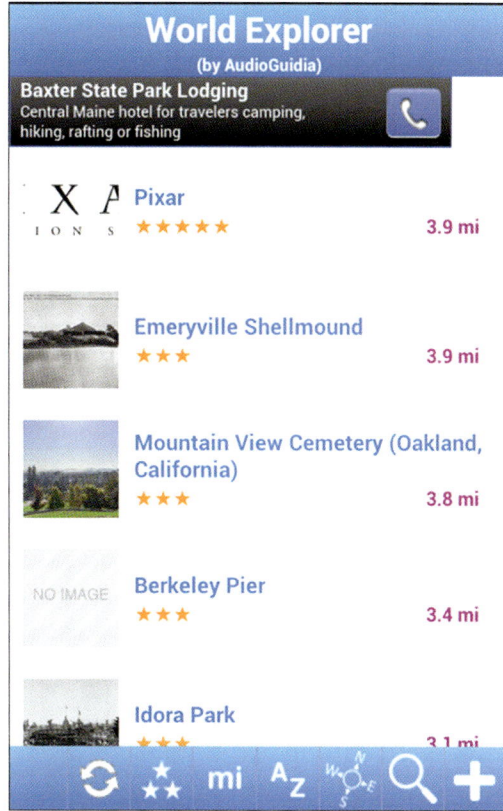

Fig. 6.30 Screenshot of the World Explorer menu tab (Image: AudioGuidia)

Primary Competitors

- Frommer's Travel Guide.
- TouristEye.

Tips

- None.

Social Networking

- Integrated with Facebook.

SWOT Analysis

Strengths

- Simple design and thus very easy to learn and to use.
- Interesting idea, very practical, and convenient for planning a short trip for weekends or also longer holidays.
- Professional travel information, including numerous details compared to common travel guides.

Weaknesses

- Lack of interaction with users and customer feedback.
- Some of the travel information is too long and too formal, which might be boring for some travelers.
- Free version is limited to some good functions. The full version costs $ 4.99.

Opportunities

- As a travel application, World Explorer aims at a niche market, which distinguishes it from the most common and most popular travel applications.
- The app is designed for mobile users who just want to go outside during their free time, not primarily for people planning long trips in detail beforehand.

Threats

- Most of the travel information is not original or unique. Thus, the application can be copied or replaced easily.

6.9.7 World Travel Guide by Triposo

Who Were They in 2012?

- World Travel Guide by Triposo was developed by Triposo and founded by Jon Tirsen, Douwe Osinga, and Richard Osinga.
- World Travel Guide by Triposo is an application that provides a general travel guide for places all around the world. Triposo developed numerous other applications for specific cities, for example, Rome Travel Guide Triposo, Paris Travel Guide Triposo, London Travel Guide Triposo, etc.
- World Travel Guide by Triposo has been downloaded around 50,000 times to Android phones, and 509 people liked this application (July 2012) [Google Play App Store].

What Was It in 2012?

- Free mobile application that provides travel guides for places all over the world. It works also offline.

What Did the App Let Users Do in 2012?

• Download information about a destination, and store it on the mobile phone. Later use it also offline, and look through it like through a travel guide.
• Use the app to check different kinds of information, i.e., the weather during the following 5 days, prices of tickets, information about tours and excursions, addresses of the grocery stores, banks, transports, etc.
• Save interesting places in a personal, customized archive. If, for example, users collect data and information about San Francisco, the respective places are stored in "My San Francisco."
• "Check in" at the spots with texts and commented photographs (similar to Foursquare).
• Filter places by initial letter, popularity, distance, and type. Filter tours by popularity and price.

Mobile Strategy

• Possibility to use the app offline, after having downloaded the necessary information. This feature can save data and costs for travelers, especially when doing international trips.
• Display of the most popular and most famous spots on the top, with large and catchy pictures. Provision of comprehensive and detailed information.
• Provision of fresh and inspiring travel suggestions, based on location, time of day, and weather.

User-Friendliness

• Easy to use due to precise explanations and clearly structured categories.
• Stylish and catchy user interface.
• Travelers cannot rate or rank the places they visited. Thus, users cannot seek advice through reviews and accordingly make better decisions.

Visual Analysis

• Clean and simple user interface with white background.
• Big and appealing pictures of attractions are displayed on top. Different elements are clearly labeled and distinguishable by different colors.
• POIs are shown with different colors and symbols on the map and are thus easily distinguishable (Fig. 6.31).

Primary Competitors

• TouristEye.

Tips

• Triposo guides users to select POIs, by providing recommendations based on popularity.

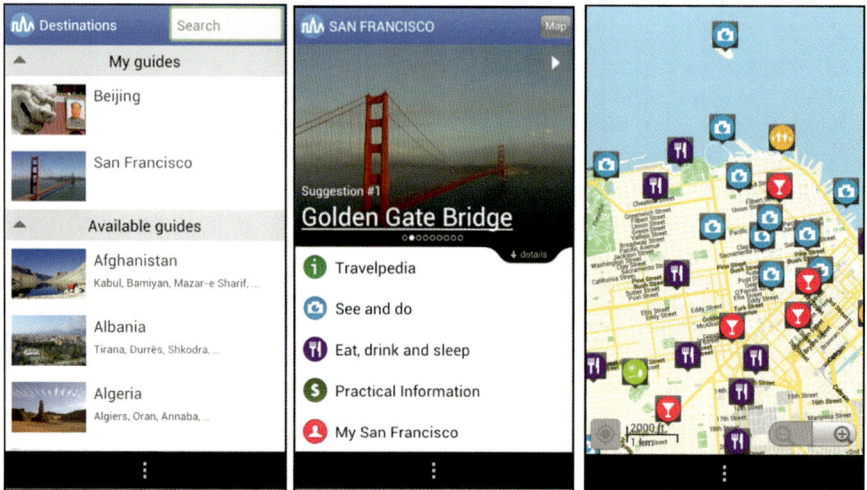

Fig. 6.31   Screenshot of the Triposo mobile app in 2012 (Images:Triposo.com)

- Triposo makes use of "Travelpedia" in the travel guide, in order to educate users about language, history, and culture. It also provides URLs to give more information.

Social Networking

- Integrated with Facebook.

SWOT Analysis

Strengths

- Offline features are very attractive.
- Free version for all travelers.
- Travel guide is very comprehensive compared to user-generated content. There is even a special part dedicated to culture.
- Provision of much practical information that many other travel applications do not contain, such as addresses of grocery stores and banks, etc.

Weaknesses

- No possibility to see other travelers' reviews and ratings.
- Destinations are limited to famous spots, thus, only limited choices for travelers.

Opportunities

- Enhance user interaction features, and make application more customized.

Threats

- Restricted number of places in the database limits number of potential users of the app.

6.9.8   *Sutro World*

Who Were They in 2012?

- Sutro World is one of the mobile travel applications developed by Sutro Media, an app company that publishes real-world guides.
- Sutro Media partners with experts to make it easier and more entertaining for people to explore the world while using their mobile phones. As a new type of app publishing company, Sutro Media amasses the world's largest collection of indie experts, who love sharing what they know.

What Was It in 2012?

- Sutro World is a travel guide app. With more than 375 guides for different destinations, Sutro World is the world's largest publisher of independently authored travel guides.
- Sutro guides cover everything from major urban centers, including top ranked guides to Las Vegas, New York, Rome, or Berlin, to less famous locations and attractions, such as Texas' wildflowers, Basque country cuisine, Chicago architecture, Hawaiian beaches, etc.

What Did the App Let Users Do in 2012?

- Purchase and download travel guides from Sutro Media.
- View maps, recommendations, photographs, and comments, which are all also available offline.
- Connect online to watch videos, to participate in discussions, or to ask questions to authors about specific destinations.

Web Strategy

- There is no specifically developed Web site for Sutro World.

Mobile Strategy

- Free sample of San Francisco is available. Guides for other destinations cost between $ 3 and $ 5.
- Allows users to view information offline, after having downloaded the guide, in order to save data and costs when traveling abroad.

User-Friendliness

- Clearly structured and well-designed user interface.
- Easy to find destinations.
- Free sample guide as a quick tutorial for users.
- Limited customer support (Fig. 6.32).

Visual Analysis

- Usage of blue as basic color.

Fig. 6.32 Screenshots of the Sutro World mobile app (Images: Sutro Media)

- Background is simple, but icons are colorful.
- Contents are easy to browse through and have special filter function.
- Maps seem easy to read.

Primary Competitors

- Lonely Planet.
- Triposo.
- TripAdvisor.

Tips

- Sutro World aims at educating users, by providing them with popular download links of Sutro Media's published travel guides.

Social Networking

- No social networking functions

SWOT Analysis

Strengths

- Each Sutro guide is an authentic, independently authored work, which was created specifically for mobile devices. This means no endless pages of guidebook text, scraped Wikipedia articles, or advertorial content.

- Sutro World has rich and unique resources for travel-related content, thanks to its company Sutro Media.
- Sutro asks a relatively low price for its travel guides, compared, for example, to the similar mobile application of Lonely Planet.

Weaknesses

- Lonely Planet, for example, provides part of the travel guide for free and users can choose if they want to buy the full guide and thus get more information. Sutro, however, only offers one free sample, and, unlike Lonely Planet, users cannot get any information unless they purchase a travel guide. This may prevent potentially interested users from using this mobile application.

Opportunities

- Sutro World encourages users to ask travel-related questions to experts. Sutro amasses the world's largest collection of indie experts, who love sharing their knowledge and experiences. With such unique resources, further opportunities might emerge for Sutro in the future.

Threats

- Most users are not willing to pay between $ 3 and $ 5 for one guide. Those who pay money for their mobile travel guides might rather opt for Lonely Planet's app, given that, despite a higher price, Lonely Planet is more popular and usually better known.

6.9.9   Foursquare

Who Were They in 2012?

- Foursquare Labs, Inc., is a private company based in New York City, New York.
- Foursquare is a location-based, social networking Web site for mobile devices, such as smartphones. Users "check in" at venues. The service was created in 2009, by Dennis Crowley and Naveen Selvadurai.
- Foursquare has gained 4 stars from 75,984 ratings among Apple users for its precedent app; however, the latest version of Foursquare is gaining lower ratings [http://itunes.apple.com]. Foursquare so far has been downloaded around 5,000,000 times to Android phones, and 36,000 people liked this application (July 2012) [Google Play App Store].

What Was It in 2012?

- Social networking platform, which allows users to interact with their environment.
- Tool for users and their friends to make the most out of where they are.

What Did the App Let Users Do in 2012?

- Sign in with a Facebook account, or quickly register via email.
- "Check in" at venues and earn "badges," by using a mobile Web site, text messaging, or a device-specific application.
- Explore surrounding things and places, and interact with the environment via mobile phone.
- View friends' activities, and share own activities with them.
- Create user profiles containing information of friends, taken photographs, written comments and tips, saved lists, badges, etc.
- Easily "save" recommended places to visit them later; mark them as "done" after the visit, and remove them from the list.
- Get travel tips from Foursquare, based on users' previous check-ins. Receive personalized suggestions of "top picks" at other places all around the world.

Mobile Strategy

- Application is free for all users.
- Integration with Facebook.
- Creation of user profiles, for tracking and managing their activities.
- Famous for its "check-in" feature, which records users' current locations with texts and photographs. In addition, each "check-in" rewards users with points and badges.
- The updated version of the mobile application enhances the feature of "exploring" the surroundings and adds more places. The mobile application has stopped being just a game of collecting points and has become instead a new exploration and social interaction machine.

User-Friendliness

- Fast and easy to log in with a Facebook account or register via email.
- The page for check-in is not well designed.
- Maps seem too small.
- Display of all friends' activities; no possibility to filter or select.
- The new design for "explore" is innovative and catchy, but it costs the user time and effort to look into each topic.
- Profile is detailed, and it allows easy tracking of activities.
- "Explore" by different categories, with different symbols displayed on the map.

Fig. 6.33 Screenshot of the Foursquare app menu tile in 2012 (Images: Foursquare)

Fig. 6.34 Screenshot of the Foursquare mobile app (Images: Foursquare)

Visual Analysis

- White background with blue base color and simple labels.
- Clear and appealing interface for users' profiles.
- Display of various kinds of potentially interesting places and things around users' current locations. Users can select the most appealing ones for them to check in (Figs. 6.33 and 6.34).

Primary Competitors

- TripAdvisor.

Tips

- Guide people in selecting POIs, with display of check-in numbers to identify hot spots.
- Encourage users to explore places and things, by rewarding them with badges in the check-in game.
- Encourage users to explore the city, by providing friends' check-in activities on the map.

Social Networking

- Integrated with Facebook.

SWOT Analysis

Strengths

- Specific searches for things like "tiramisu" or "wine list," or adjectives like "romantic" and "delicious," or time-significant keywords like "Friday" or "summer" function perfectly, since tips and lists are indexed.

- Money Savings: Location-based coupons and deals provide users with "Foursquare specials."
- Popularity: Display of check-in numbers helps to identify hot spot locations. If there is a high number of check-ins by users' friends, they have a higher chance of meeting a friend at the respective place.
- Travel tips: Based on users' previous check-ins, Foursquare suggests personalized "top picks" at other places all around the world.
- Social networking features: Enhance relationships by learning which places users may unexpectedly have in common with their Foursquare acquaintances and friends.

Weaknesses

- Interface is not well developed and not user-friendly, especially concerning the mobile application.
- New explore feature needs to be improved.

Opportunities

- With its huge number of users, and its numerous social networking characteristics and functions, Foursquare has considerable advantages over pure travel applications.

Threats

- The search and explore features are weak and not very competitive. Though Foursquare is most famous for its check-in function, other applications already contain similar features.

6.9.10   Trip Journal and VA Journal

Who Were They in 2012?

- Trip Journal was developed by IQAPPS. Only the first trip can be recorded free of charge; subsequent trips require the full version, which has to be paid.
- VA Journal was also developed by IQAPPS. It has a free version for all users with almost the same functions, but with different features.
- Trip Journal is the #1 Google Awarded Travel Application currently available for iPhone, Android, Symbian, and Facebook. In December 2009, it received a $100,000 prize from Google for its innovative concept and design.
- Trip Journal won Android's development challenge. So far, it has been downloaded around 50,000 times to Android phones, and 249 people liked this application (July 2012) [Google Play App Store].

What Was It in 2012?

- An application using latest technology to fulfill a classical demand for tracking, recording, documenting, and sharing users' travels with friends and family.

What Did the App Let Users Do?

- Track, record, and document vacation experiences with text, photographs, videos, etc., including information about distance, time, and geography. Much of this information is automatically documented in the journal.
- Share travel experiences with friends and family via social networks. Impress others with real-time updates from the visited destinations, and let them see photographic proofs of users' latest adventures.
- Manage and edit recorded trips of the past.

Mobile Strategy

- Focus on the travel journal feature, various functions, and features to record and share a personal travel journal.
- Combined and integrated with advanced technology, latest platforms, and popular social communities.
- Well-designed and attractive interface.

User-Friendliness

- Helpful tutorial with detailed information about main features of the app on its official Web site.
- Easy to learn and get started for new users, due to clearly labeled icons and descriptions.
- Possibility to create conveniently a personalized and individual travel journal/diary.
- Catchy and attractive user interface.

Visual Analysis of Trip Journal

- Appealing cover as first intro page. The user interface is in retro style, with a light brown background. The application is designed as a real journal.
- All recorded information can be shown on a map. It is visualized in the form of different symbols (Fig. 6.35).

Visual Analysis of VA Journal

- Light pink background with clearly labeled features in white boxes.
- Interface has an appealing design, with various tools for users to track their experience.
- Automatic tracking of routes on a map (Fig. 6.36).

Primary Competitors

- Travel Diary.
- Travel Blog.

Tips

- Trip Journal encourages users to record and share their travel experiences.

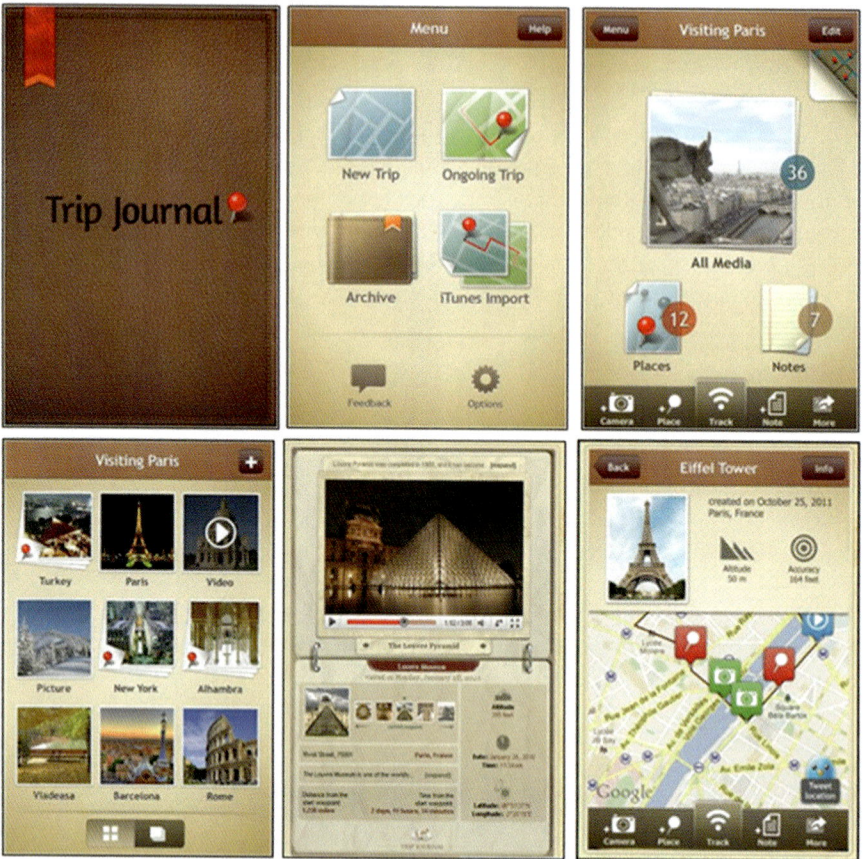

Fig. 6.35 Screenshots of the Travel Journal mobile app (Images: IQAPPS)

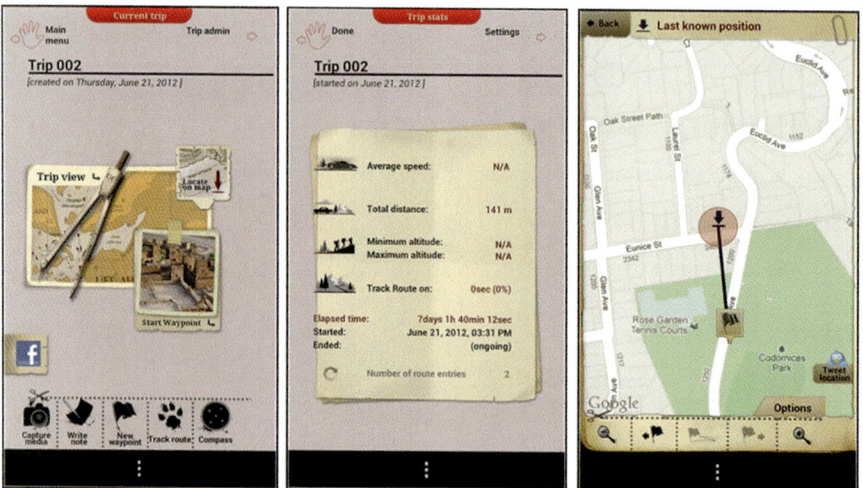

Fig. 6.36 Screenshots of the VA Journal mobile app (Images: IQAPPS)

- Trip Journal guides users in recording their travel experiences in various formats, including texts, photographs, videos, etc.

Social Networking

- Integrated with Facebook.

SWOT Analysis

Strengths

- Clear focus on travel journal functions, with various features that are very professional for a travel diary. With the integration of Google Earth, Trip Journal allows friends and family members to easily see app users' travel routes, waypoints of visited locations, full-screen photographs, videos, comments, and blog entries. Trip statistics such as distance, time, and geography are also documented in real time.
- Trip Journal brings much social interaction, since the app is integrated with popular social networks, as well as with content-sharing portals, including Facebook, Flickr, Picasa, YouTube, and Twitter.
- Best travel journal/diary application so far. Trip Journal contains a well-designed and appealing interface, and it integrates in the app the most recent technology advancements. Most recent platforms are chosen for the app, and a lot of effort is put into developing a top GPS correction algorithm, which makes Trip Journal probably the most accurate GPS tracking solution currently available.

Weaknesses

- The stability of the version needs to be improved. There are still some technical errors.

Opportunities

- Currently, there are hardly any competitors on the market.

Threats

- The full version for an unlimited journal costs $ 4.99. The relatively high price may prevent potential users from making use of this app.

6.9.11   MobilyTrip

Who Were They in 2012?

- MobilyTrip was founded by Benoît Le Ny.
- MobilyTrip is a mobile application for iPhone and iPad, providing free versions in different languages, including English, French, German, and Spanish.
- MobilyTrip is a very recent travel-related mobile application. So far, it has gained 4.5 stars from 123 ratings (July 2012) [http://itunes.apple.com].

What Was It in 2012?

- Mobile application that provides a huge number of free travel guides for destinations all over the world.
- Social network that enables travelers to create, store, and share personalized travel journals with friends and family members.

What Did the App Let Users Do?

- Record a travel journal.
- Current trip: Use MobilyTrip to describe and record one's personal trip, including paths and itineraries, visited places, photographs, and comments. All these data can be recorded easily on the iPhone and do not necessitate Internet connection.
- Past trip: Relive and create journals from past trips, by importing geo-localized pictures to users' iPhones and to the app.
- Classify photographs into different categories, based on the options provided by MobilyTrip.
- Add photographs taken with other cameras to users' online MobilyTrip journal.
- Share a travel journal.
- Synchronize journal with users' Web portal accounts (only with Internet access), to share travel experiences with friends, relatives, or with other travelers. The online interface is appealing and well-designed. It is created automatically.
- Manage privacy settings for the travel journal, by deciding who will be notified about changes and updates.
- Travel guide.
- Download free travel guides, which contain a map and a list of interesting places to visit, illustrated with photographs and detailed descriptions.

Web Strategy

- Fast and easy registration.
- Synchronized with mobile application.
- Interface of online travel journal is automatically created after users' upload. Web interface is different from mobile version.

Mobile Strategy

- Free application with free travel guides.
- Application can be used offline with an embedded map.
- Clean and simple interface.
- Interconnected with Facebook and email account.

User-Friendliness

- Quick tutorial with pictures on the entry page and help bottom to get more detailed instructions.
- Clearly structured platform for users to start. Simple choice among only two categories.
- Fast registration, possible also via Facebook or email account.

Fig. 6.37   Screenshots of the MobilyTrip smartphone version in 2012 (Images: MobilyTrip)

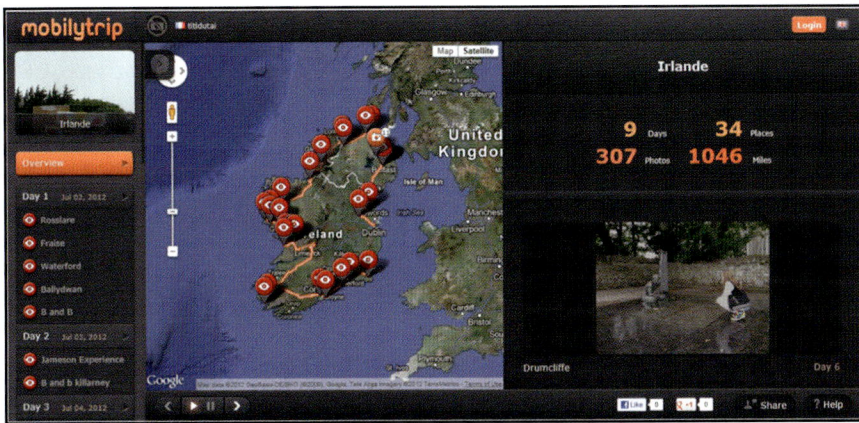

Fig. 6.38   Screenshot of the MobilyTrip web version in 2012 (Image: MobilyTrip)

Visual Analysis

- Appealing photographs flow in the background. Only two categories are displayed on the entry screen. Interface is well-designed and easy to use.
- User interfaces of mobile travel journal and Web portal version are different. Simple and easily manageable interface to classify photographs in the album.
- MobilyTrip separates photographs from text in the travel guide. Photos are visualized in an extra album (Figs. 6.37 and 6.38).

Primary Competitors

- Trip Journal.
- Trip Color.

Tips

- In the travel guides, MobilyTrip recommends POIs to users and filters them based on importance and popularity.

- MobilyTrip provides users with comprehensive background data and information, including history, geography, economy, etc., of a destination.

Social Networking

- Integrated with Facebook and email account.

SWOT Analysis

Strengths

- Mobile application and all its services are free for all users.
- MobilyTrip combines within one travel application the functions of a travel guide and of a travel journal.
- MobilyTrip is integrated with Facebook and private email accounts.
- Users can manage privacy settings easily.
- Interface is well-designed, simple, clean, and user-friendly.
- Once users have uploaded their data to the Web site, the travel journal will be created automatically. Users spend no additional time for organizing their data.

Weaknesses

- Currently, only 125 travel guides are available. Therefore, the number of destinations for travelers to choose among is still very limited.
- The format of the travel guides is very different from standard guide book styles, so that users might not be used to it. Moreover, the text provided by the guide is rather long and quite detailed. Users may get bored quickly.

Opportunities

- MobilyTrip should develop further versions that are also compatible with other systems.
- As mentioned above, MobilyTrip is so far not very well-known as a new travel-related mobile application. However, current users appear to be very satisfied with the application. In the near future, MobilyTrip might become a popular travel application, due to its creative and uncommon approach and its user-friendly interface.

Threats

- The rather uncommon format of MobilyTrip's travel guides may prevent travelers from using them.
- Further noteworthy travel-related mobile applications and therefore potential competitors are Tricolor, Trip Viewer, and Off Exploring.

6.9.12   Results of Competitive Analysis

Results from our investigations of about 15 mobile travel applications including those cited above, AM+A concluded that usable, useful, and appealing user interface (UI) design must include incentives to lead to behavior change. Good travel-oriented mobile phone applications should not only provide users with practical information and tips, but it should encourage and motivate them to immerse themselves into another culture or country, thus provoking a shift from mere leisure to cultural tourism. For this purpose, the Travel Machine should include group comparisons, charts, illustrations, goals, competition, and/or instructions and advice to motivate people to change their behavior. The proposed Travel Machine needs to combine persuasion theory, provide better incentives, and motivate users to achieve short-term and long-term behavior change towards a personally and culturally more enriching travel experience.

Last but not least: The Travel Machine should be fun to use. Well-designed games will serve as an additional appealing incentive to teach and to train. Also, the Travel Machine should allow users to share their experiences with friends, family members, and the world, primarily through Facebook, Twitter, and a Travel Machine Web portal.

Based on these concepts and available research documents, we have proposed and are developing conceptual designs of the multiple functions of the AM+A Travel Machine. Subsequent evaluation will provide feedback by which we can improve the metaphors, mental model, navigation, interaction, and appearance of all functions and data in the Travel Machine's user interface. The resultant improved user experience will move the Travel Machine closer to a commercially viable product/service.

We believe a well-designed Travel Machine will be more usable, useful, and appealing to culture-conscious users, especially those having problems finding the right approach to immerse themselves into and learn about foreign places and cultures. Our objective is to provide a mobile suite of applications that can reliably persuade people to move towards a deeper-going travel habit, in order to allow a shift from leisure to cultural tourism.

6.10   Information Architecture

6.10.1   Machine Information Architecture

In designing the information architecture (IA) for the Travel Machine, AM+A began by examining the IA for past machines, including the Green Machine, the Health Machine, the Money Machine, the Driving Machine, and the Innovation Machine, as discussed in previous chapters. We discovered an overarching model for the IA that permeated throughout each of the past machines. In this model, there are five primary "modules," or branches of the IA, each of which is described below. While we altered slightly the details of each module to fit the needs and requirements of each respective machine, a generalized model was still evident. These modules are described below:

6.10.2   Dashboard

The Dashboard module is similar to a landing page for its respective machine. It is an overview into the status of the user's behavior change. Here, the user gets a view of his/her goals and where she/he stands in achieving those goals.

6.10.3   Process

The Process View module is where the user gets more high-level view of the process and more details regarding each objective and goal. The user sees the progress being made, as well as the next steps in achieving a particular goal.

6.10.4   Social Network

The Social Network module is an integral part of behavior change in modern software. Users engage in focused, subject matter-based connections with friends, family, and/or like-minded people that either share similar goals or wish to support others in achieving behavior-change objectives.

6.10.5   Tips/Advice

The Tips/Advice module provides focused knowledge about a given topic to give users insight into the habits they wish to either get rid of or adopt.

6.10.6   Incentives

The Incentives module presents users with fun and engaging ways to change their behavior. Gamification has proven to be a powerful tool in adopting users to try an application, even with virtual incentives; although in some cases, real incentives could be provided. In addition, a leaderboard allows a user to compare his/her progress with others, tapping in the competitive nature of the human mind to create behavior change.

6.10.7   Travel Machine Information Architecture

AM+A adapted the basic structure of the past machines in creating the Travel Machine information architecture. However, due to particular complexities that were unique to the Travel Machine, we made the following primary changes:

- The respective tabs were attributed different labels: Dashboard, Diary (process), Advisor (tips/advice), Fellows (social network), and Challenge (incentives).
- A different menu navigation system was provided, in order to keep the screens uncluttered. The high-level basic modules are removed from most screen displays, but may be quickly called back to enable users to navigate high-level modules.

6.11 IA Diagram

The following is a diagram of the information architecture for the Travel Machine (Fig. 6.39).

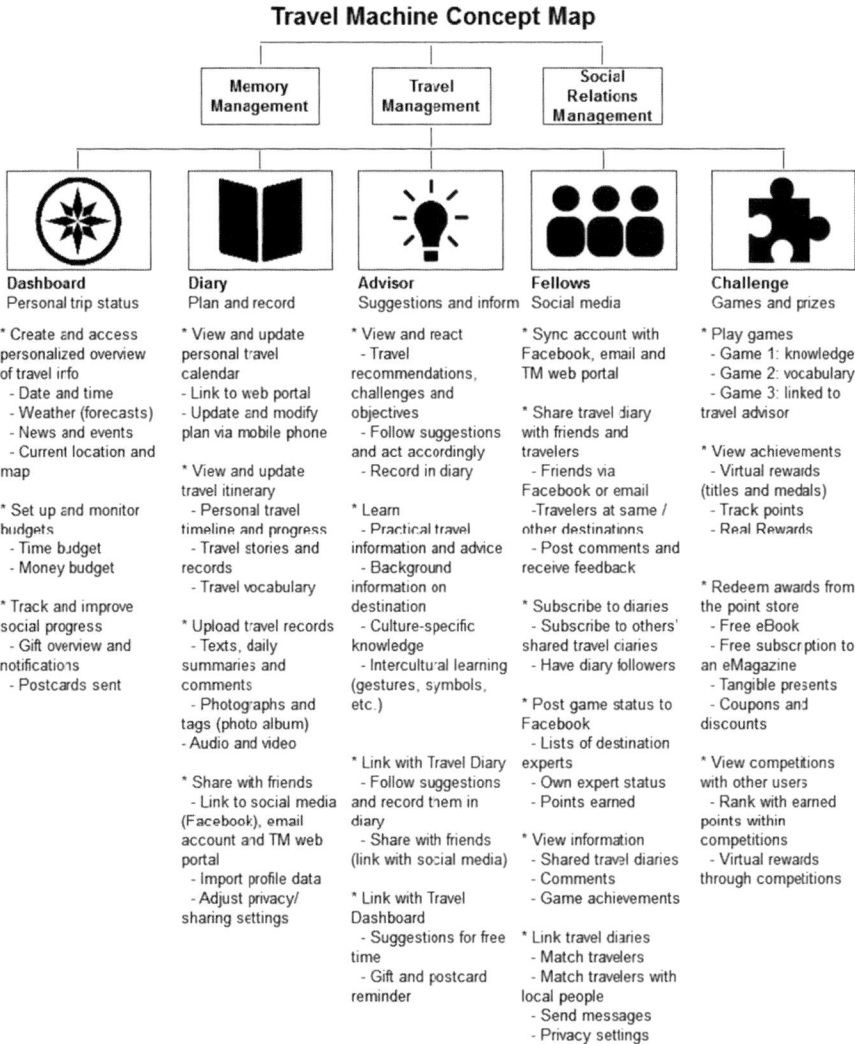

Travel Machine Concept Map

Memory Management | Travel Management | Social Relations Management

Dashboard — Personal trip status

* Create and access personalized overview of travel info
  - Date and time
  - Weather (forecasts)
  - News and events
  - Current location and map

* Set up and monitor budgets
  - Time budget
  - Money budget

* Track and improve social progress
  - Gift overview and notifications
  - Postcards sent

Diary — Plan and record

* View and update personal travel calendar
  - Link to web portal
  - Update and modify plan via mobile phone

* View and update travel itinerary
  - Personal travel timeline and progress
  - Travel stories and records
  - Travel vocabulary

* Upload travel records
  - Texts, daily summaries and comments
  - Photographs and tags (photo album)
  - Audio and video

* Share with friends
  - Link to social media (Facebook), email account and TM web portal
  - Import profile data
  - Adjust privacy/sharing settings

Advisor — Suggestions and inform

* View and react
  - Travel recommendations, challenges and objectives
  - Follow suggestions and act accordingly
  - Record in diary

* Learn
  - Practical travel information and advice
  - Background information on destination
  - Culture-specific knowledge
  - Intercultural learning (gestures, symbols, etc.)

* Link with Travel Diary
  - Follow suggestions and record them in diary
  - Share with friends (link with social media)

* Link with Travel Dashboard
  - Suggestions for free time
  - Gift and postcard reminder

Fellows — Social media

* Sync account with Facebook, email and TM web portal

* Share travel diary with friends and travelers
  - Friends via Facebook or email
  - Travelers at same / other destinations
  - Post comments and receive feedback

* Subscribe to diaries
  - Subscribe to others' shared travel ciaries
  - Have diary followers

* Post game status to Facebook
  - Lists of destination experts
  - Own expert status
  - Points earned

* View information
  - Shared travel diaries
  - Comments
  - Game achievements

* Link travel diaries
  - Match travelers
  - Match travelers with local people
  - Send messages
  - Privacy settings

Challenge — Games and prizes

* Play games
  - Game 1: knowledge
  - Game 2: vocabulary
  - Game 3: linked to travel advisor

* View achievements
  - Virtual rewards (titles and medals)
  - Track points
  - Real Rewards

* Redeem awards from the point store
  - Free eBook
  - Free subscription to an eMagazine
  - Tangible presents
  - Coupons and discounts

* View competitions with other users
  - Rank with earned points within competitions
  - Virtual rewards through competitions

Fig. 6.39 Information architecture of the Travel Machine

6.11.1   Components of Information Architecture Dashboard

Create and Access Personalized Overview of Travel Info

- Date and time displays.
- Weather (forecasts).
- News and events.
- Current location and map.

Set Up and Monitor Budgets

- Time budget.
- Money budget.

Track and Improve Social Progress

- Gift overview and notifications.
- Postcards sent.

6.11.2   Diary

View and Update Personal Travel Calendar

- Create itinerary on Web portal.
- Link to Web portal.
- Update and modify travel plan via mobile phone.

View and Update Travel Itinerary

- Personal travel timeline and progress.
- Travel stories and records.
- Travel vocabulary.

Upload Travel Records

- Texts, daily summaries, and comments.
- Photographs and tags (photo album).
- Audio and video documentation.

Share with Friends

- Link to social media (Facebook), email account, and TM Web portal.
- Import profile data.
- Adjust privacy/sharing settings.

6.11.3   Advisor

View and React

- Travel recommendations, challenges, and objectives.
- Follow suggestions and act accordingly.
- Record in diary.

Learn

- Practical travel information and advice.
- Background information on destination.
- Culture-specific knowledge.
- Intercultural learning (gestures, symbols, etc.).

6.11.4   Link with Diary

- Follow suggestions and record them in a diary.
- Share with friends (link with social media).

Link with Dashboard

- Suggestions for free time.
- Gift and postcard reminder.

6.11.5   Fellows: Sync Account with Facebook, Email, and TM Web Portal

Share Diary with Other Friends and Travelers

- Friends via Facebook or email.
- Travelers at the same or at other destinations.
- Post comments and receive feedback.

Subscribe to Travel Diaries

- Subscribe to others' shared travel diaries.
- Have diary followers/subscribers.

Post Game Statuses to Facebook

- Lists of destination experts.
- Current personal expert status.
- Points earned.

View Other Users' Information

- Shared travel diaries.
- Comments.
- Game achievements.

Link Travel Diaries of Users

- Match travelers with similar interests and itineraries.
- Match travelers with local people.
- Exchange nonpublic messages with Travel Machine friends.
- Set up privacy settings.

Participate in "Most Beautiful Photograph" Activities

- Select "most beautiful photograph" from shared photos.
- Gain points by uploading popular pictures.

6.11.6   Challenge

Play Games

- Game 1: knowledge-testing game.
- Game 2: vocabulary quiz.
- Game 3: game in accordance with the Travel Advisor's instructions and suggestions.

View Achievement

- Virtual rewards (titles and medals).
- In progress.
- Reached.
- Track points.
- Remaining points.
- Total accumulated points.
- Real points.
- Gained.

Redeem Awards from the Point Store

- Free eBook.
- Free subscription to an eMagazine.
- Tangible presents.
- Coupons and discounts.

View Competitions with Other Users

- Rank with earned points within competitions.
- Virtual rewards through competitions.

6.12 Information Visualization

For the development and design of the Travel Machine, numerous ways have been thought of about how to present travelers' progress and itinerary in a useable, useful, and appealing way. The following exemplary maps, charts, and diagrams will therefore be implemented in the Travel Machine:

- Map showing the current geographical position of travelers.
- Map showing POIs around travelers.
- Map showing safe and less safe places/areas within a destination.
- Map tracing a travelers' itinerary.
- Personalized world map displaying a summary of a travelers' trip experiences (cities/countries visited, liked/not liked destinations, wish list for the future, etc.).
- Chart displaying the remaining money budget in "fuel gauge" form (money budget per day/week/whole trip).
- Chart displaying the remaining time budget (per day/whole trip).
- Chart displaying the travelers' social progress (postcards sent, gifts bought).
- Chart displaying the travelers' expert status for a destination (progress, points gained, rewards).
- Diagram showing a travelers' expert status compared to other Travel Machine users, with whom they compete via games.
- Figure summarizing the most significant culture-specific gestures (Fig. 6.40).

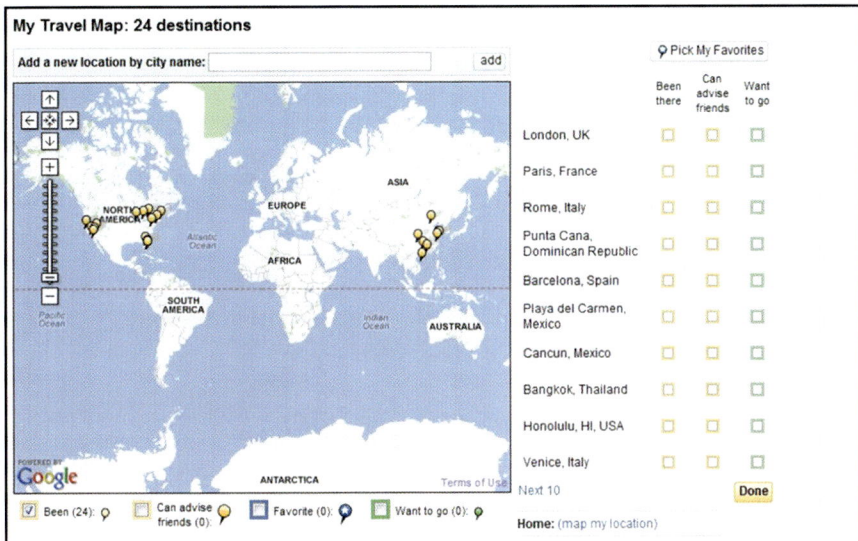

Fig. 6.40 Example of a personalized world map

6.13   Designs

The following subsections discuss and show representative, initial screen designs of the Travel Machine that are based on the information architecture previously described.

6.13.1   Landing Screen

The Landing Screen typifies the main menu and app instruction throughout the Travel Machine. This page displays five major functions with descriptions below (Fig. 6.41).

6.13.2   Diary

The Diary module allows users to record their experiences and to keep a personal diary with texts, pictures, audio, and video recordings (in "Camera"), routes (in "Routes"), and other information (in the drop-down menu on top of the screen).

Fig. 6.41 Screen design of the Travel Machine landing page

Fig. 6.42 Screen designs of the standard diary page

On the right, users can swipe up and down to scroll and view the list of taken photographs. Moreover, users can drag photos away to delete unsatisfying pictures.

In addition, users can customize diary skins and fonts based on their preferences (Figs. 6.42 and 6.43).

6.13.2.1   Diary/Camera

Users can record travel experiences with photographs, audio or video registrations, and geo-tag them in the Camera module of the Diary. In addition, they can import pictures from the gallery of their phone or from other resources, simply by clicking the "clip" button (Fig. 6.44).

6.13.2.2   Diary/Routes

In the Route module, users can automatically track their personal routes via their smartphone's GPS. They can flag places with different colors based on their preferences (green = liked, red = did not like). Users can also share their routes with friends and check shared routes of friends via Facebook or via the Travel Machine portal.

Furthermore, travelers can search for other Travel Machine users near next to their current location (Fig. 6.45).

Fig. 6.43 Screen designs of the customized diary page

Fig. 6.44 Screen designs of the camera page within the diary

Fig. 6.45 Screen design of the route page within the diary (Images: GoogleMaps and Nicola Lecca)

6.13.2.3   Diary/Words

Vocabulary module allows users to add new words to their personal dictionary and to attribute it to a certain category (small talk, food, direction, culture, mixed).

By consulting the Travel Machine dictionary, users can also learn and study words or sentences, according to the above-named five main sections (Fig. 6.46).

6.13.2.4   Diary/Calendar

Calendar module displays users' travel plans with basic and detailed information about the respective locations and about the following events during the month (Fig. 6.47).

6.13.2.5   Diary/Archive

Users can review past trips that they recorded in the Travel Machine, with a brief summary of the respective trip in the archive title (Fig. 6.48).

Fig. 6.46 Screen designs of vocabulary pages within the diary

Fig. 6.47 Screen designs of the calendar page within the diary

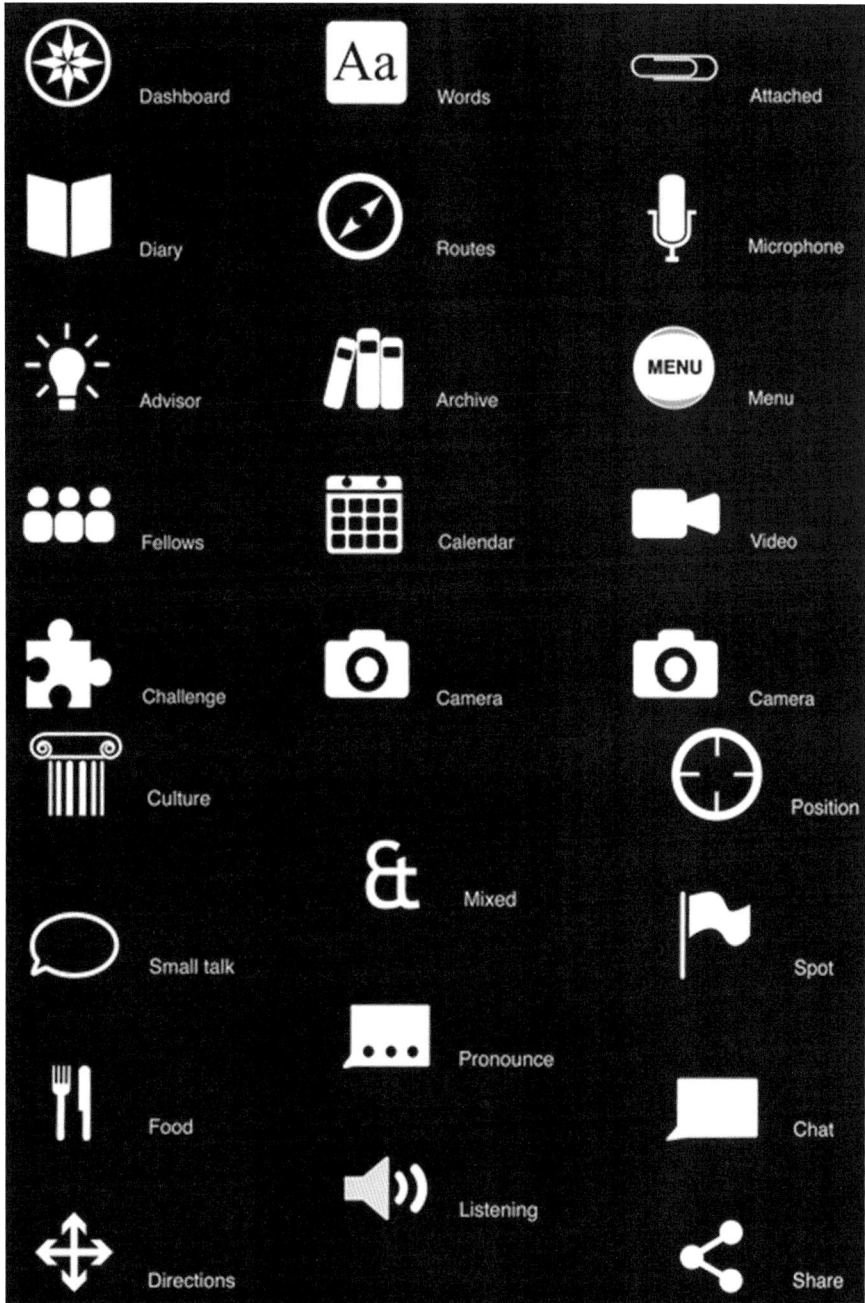

Fig. 6.48 Screen design of the archive page within the diary [Images: Scott Abromowitz, Yuan Peng]

6.13.3   Icon Overview (Fig. 6.49)

Fig. 6.49   Icon overview

6.14   Conclusions

Although the development of each Machine includes, in theory, evaluation of the designs, because of the extended amount of time for previous steps, we were not able to conduct evaluation, usability tests, and to gain feedback that would enable us to design a revised version. This work remains for the future.

In summary: Even though travelers might not come back to travel experiences and records comparable to those of Hemingway's days, we believe the Travel Machine can become an effective means to change people's travel behavior and to contribute to a shift from leisure/business travel to cultural tourism. With the main features of a travel diary and a mobile Travel Advisor, the Travel Machine can assist travelers in the documentation and enrichment of their travel experiences. As stated earlier, the Travel Machine is a traveler- and not destination-centered concept, because it focuses on the intercultural learning process of a person visiting another city, region, or country and on the person willing to absorb and discover its cultural idiosyncrasies. As opposed to the majority of mobile apps in the field of travel and tourism, the Travel Machine is not conceived as a platform for local service providers trying to capture visitors' needs and preferences and eager to attract them with push messages.

The Travel Machine was conceived to intensify users' travel experiences and to assist them during their trip. It incorporates persuasive and motivational elements in order to stimulate users to reflect on and change their travel behavior, arouses their interest for another culture, and increases awareness of it. The Travel Machine is entertaining and fun to use, even though at the same time it provides information and educates travelers, leaving space for reflective activities and personal discovery of another culture. Especially thanks to the Travel Diary, and the possibility of social interaction and comparison, users' travel behavior could be changed, also, in the long term.

Acknowledgements   The author thanks his AM+A associates, designer/analysts:

Mr. Scott Abromowitz, AM+A Designer/Analyst
Ms. Megan Chiou, AM+A Designer/Analyst
Mr. Nichola Lecca, AM+A Designer/Analyst
Ms. Yi Li, AM+A Designer/Analyst
Ms. Yuan Peng, AM+A Designer/Analyst
Ms. Theresa Schieder, AM+A Designer/Analyst (primary Designer/Analyst)

The author also wishes to acknowledge the assistance of these individuals who contributed significantly to this project:

Prof. Lorenzo Cantoni, Università della Svizzera italiana, Lugano (CH), Advisor, and Faculty Advisor to Ms. Schieder
Mr. Bob Steiner, UC Berkeley Extension, International Diploma Program, Marketing Faculty, Advisor
The author acknowledges publications about the Travel Machine subsequent to the AM+A White Paper of 2012, on which this chapter is based:

Marcus, Aaron (2013). "The Travel Machine: Mobile UX Design That Combines Information Design with Persuasion Design." *Proceedings*, Design, User-Experience, and Usability Conference, 20–25 July 2013, Las Vegas, Nevada, pp. 696–704. (Marcus 2013a)
Marcus, Aaron (2013). "The Travel Machine: Combining Information Design/Visualization with Persuasion Design to Change." Chapter 2, in Rizvanoglu, Kerem, and Çetin, Görkem, Editors, Research and Design Innovations for Mobile User Experience, Hershey, Pennsylvania: IGI Global, pp. 23–46. (Marcus 2013b)
Marcus, Aaron (2013). "旅行机-简洁" "The Travel Machine." (in Chinese). *Proceedings*, User Friendly 2014/UXPA China Conference, 21–24 November, 2013, Shanghai, China. Unnumbered pages. (Marcus 2013c)

Further Reading

CrunchBase (2012) TouristEye. Web. http://www.crunchbase.com/company/tourist-eye
CrunchBase (2012) TripIt. Web. http://www.crunchbase.com/company/tripit
Dotopen (2012) TouristEye. Web. http://www.dotopen.com/organizations/view/tourist-eye
Eco U (1970) Semiologie des Messages Visuels. Communication 16:11–52
Eco U (1976) A theory of semiotics. Indiana University Press, Bloomington

Eco U (1985) The semantics of metaphor. Innis (1985), p 262

Facebook (2012) Sutro Media. Web. http://www.facebook.com/sutromedia#!/sutromedia/info

Fineman JA (2012) 10 best free travel apps. US News. http://travel.usnews.com/gallery/10_Best_Free_Travel_Apps/TripIt_1866

Fuchs M, Ricci F, Cantoni L (2012) Information and communication technologies in tourism 2012. Springer, Vienna

Hartshorne C, Weiss P (1933) Collected papers of Charles Sanders Peirce, vol 4: the simplest mathematics, Book 2, Chapters 1–7. Harvard University Press, Cambridge

Hipmunk (2012) About, Web. http://www.hipmunk.com/

Innis RE (1985) Semiotics: an introductory anthology. Indiana University Press, Bloomington

Lakoff G, Johnson M (1980) Metaphors we live by. The University of Chicago Press, Chicago

Law R, Fuchs M, Ricci F (2011) Information and communication technologies in tourism 2011. Springer, Vienna

Levi-Strauss C (2000) Structural anthropology (trans: Claire Jacobson, Brooke Schoepf). Basic Books, New York

MacCannell D (1999) The tourist. A new theory of the leisure class. University of California Press, Thousand Oaks

Marcus A (2002) Information visualization for advanced vehicle displays. Inf Visual 1:95–102

National Geographic (2012) National park maps HD. Web. http://www.nationalgeographic.com/mobile/apps/national-park-maps/

Peirce CS (1933) Existential graphs. In: Hartshorne C, Weiss P (eds) Collected papers of Charles Sanders Peirce, vol 4. Harvard University Press, Cambridge, MA, pp 293–470

Sutro Media (2012) About us. Web. http://www.sutromedia.com/about.html

TripAdvisor (2012) About TripAdvisor. Web. http://www.tripadvisor.com/pages/about_us.html

Tripit (2012) About. Web. http://www.tripit.com/press/about/

TripJournal (2012) About Trip Journal. Web. http://www.trip-journal.com/

TripJournal (2012) Trip journal features. Web. http://www.trip-journal.com/features/features/

Tristram C (2001) The next Mputer interface. Technology Review, pp 53–59

Wikipedia (2012) Foursquare. Web. http://en.wikipedia.org/wiki/Foursquare

Wikipedia (2012) Lonely planet. Web. http://en.wikipedia.org/wiki/Lonely_Planet

Wikipedia (2012) TripAdvisor. Web. http://en.wikipedia.org/wiki/TripAdvisor

Wikipedia (2012) Wikitude. Web. http://en.wikipedia.org/wiki/Wikitude

References

Buhalis D (2003) eTourism: information technology for strategic tourism management. Prentice Hall, Harlow

Cialdini RB (2001a) The science of Persuasion. Sci Am 284:76–81

Cialdini RB ([4]2001b) Influence: science and practice, 4th ed. Allyn and Bacon, Boston

Coelho A, Dias L (2011) A mobile advertising platform for eTourism. In: Law, Fuchs, Ricci (2011), pp 203–214

Figueira R, Kakimoto M, Koken G, Marmo T, Oliveira R (2012) Travel machine market research. Summary report for Prof. Bob Steiner's Marketing Class, University of California/Berkeley International Extension Program, 13 July 2012

Fogg BJ (2003) Persuasive technology: using computers to change what we think and do. Morgan Kaufmann, Amsterdam

Fogg BJ, Eckles D (2007) Mobile persuasion: 20 perspectives on the future of behavior change. Persuasive Technology Lab, Stanford University, Palo Alto

Hartson R, Pyla PS (2012) The UX book. Process and guidelines for ensuring a quality user experience. Elsevier, Waltham

Hofstede G (2012) http://geert-hofstede.com/index.php. Accessed on 12 Feb 2012

Kim R (2012) Kleiner-backed Noom coaches you to better health. GigaOm, 28 June 2012. http://gigaom.com/2012/06/28/kleiner-backed-noom-coaches-you-to-better-health/

Löfgren O (2002) On holiday: a history of vacationing. University of California Press, Berkeley

Marcus A (2011) Health machine. Inf Des J 19(1):69–89

Marcus A (2012a) The money machine. Helping Baby Boomers Retire. User Experience, vol 11:2, 2nd Quarter 2012, pp 24–27

Marcus A (2012b) The story machine: a mobile app to change family story-sharing behavior. In: CHI 2012, Austin, TX, USA, 5–10 May 2012

Marcus A (2013a) The travel machine: mobile UX design that combines information design with Persuasion design. In: Proceedings, design, user-experience, and usability conference, Las Vegas, Nevada, 20–25 July 2013, pp 696–704

Marcus, A (2013b) The travel machine: combining information design/visualization with Persuasion design to change, Chapter 2. In: Rizvanoglu K, Çetin G (ed) Research and design innovations for mobile user experience. IGI Global, Hershey, pp 23–46

Marcus A (2013c) 旅行机-简洁" "The travel machine" (in Chinese). In: Proceedings, user friendly 2014/UXPA China conference, Shanghai, China, 21–24 November, 2013. Unnumbered pages

Marcus A, Jean J (2009) The green machine: going green at home. Inf Des J 17(3):233–243

Maslow AH (1943) A theory of human motivation. Psychol Rev 50:370–396

Maslow AH ([11]2006) Motivazione e personalità. Ed. 11. Armando, Roma

Naik G (2012) Next cameras come into view. Wall St J 21:2012

Petersen A (2012) Don't forget to pack a photographer. Wall St J 3:2012

Robbins L (2012) Follow that tourist. The New York Times, July 21, 2012

SeatID (2012) About. http://www.seatid.com/about/

UNWTO (2012) International tourism to reach one billion in 2012. http://media.unwto.org/en/press-release/2012-01-16/international-tourism-reach-one-billion-2012 (11.02.2012)

Wagner M, Armstrong N (2003) Field guide to gestures: how to identify and interpret virtually every gesture known to man. Quirk Books

Wang D, Xiang Z (2012) The new landscape of travel: a comprehensive analysis of smartphone apps. In: Fuchs, Ricci, Cantoni (2012), pp 308–319

WTTC (2012) Economic impact research. http://www.wttc.org/research/economic-impact-research/ (11.02.2012)

Mobile/Website Travel Applications URLs

The following are relevant to this Machine's development. (Retrieval date for all URLs: 10 July, 2012)

http://abcnews.go.com/blogs/lifestyle/2012/06/airline-passengers-pick-seats-based-on-mood/
http://abcnews.go.com/Travel/social-media-start-ups-bring-compatible-travelers/story?id=16446970#.T9ExfOJYu-t
http://itunes.apple.com/us/app/foursquare/id306934924?mt=8
http://itunes.apple.com/us/app/lonely-planet-travel-guides/id317165182?mt=8
http://itunes.apple.com/us/app/sutro-world/id446206531?mt=8
http://itunes.apple.com/us/app/touristeye-travel-guide-offline/id363369132?mt=8
http://itunes.apple.com/us/app/trip-journal/id341585937?mt=8
http://itunes.apple.com/us/app/tripadvisor-hotels-flights/id284876795?mt=8
http://itunes.apple.com/us/app/tripit-travel-organizer-free/id311035142?mt=8
http://itunes.apple.com/us/app/triposo/id467053028?mt=8
http://itunes.apple.com/us/app/wikitude-augmented-reality/id329731243?mt=8
http://itunes.apple.com/us/app/worldwide-guide-travel-journal/id431464056?mt=8
http://www.planely.com/

http://www.satisfly.com/

http://www.tripadvisor.com/

http://www.youtube.com/watch?v=eL2lWn7oup4

http://www.youtube.com/watch?v=xvB1CVP1Rpw

http://www.booking.com/searchresults.html?src=city&city=-1456928&error_url=http%3A%2F%2Fwww.booking.com%2Fcity%2Ffr%2Fparis.en-us.html%3Faid%3D306395%3Blabel%3Dparis-Pyq81IkhuaRNJLokBM59TwS12915459989%253Apl%253Ata%253Ap11500%253Ap2%253Aac%253Aap1t1%253Aneg%3Bsid%3D7611578cb396677be6ae117cb80ffc34%3Bdcid%3D1%3Binac%3D0%3B&aid=306395&dcid=1&label=paris-Pyq81IkhuaRNJLokBM59TwS12915459989%3Apl%3Ata%3Ap11500%3Ap2%3Aac%3Aap1t1%3Aneg&sid=7611578cb396677be6ae117cb80ffc34&checkin_monthday=5&checkin_year_month=2012-9&checkout_monthday=6&checkout_year_month=2012-9&group_adults=2&group_children=0 (Picture credit "Bel Canto").

https://play.google.com/store/apps/details?id=com.audioguidia.worldexplorer&feature=search_result#?t=W251bGwsMSwxLDEsImNvbS5hdWRpb2d1aWRpYS53b3JsZGV4cGxvcmVyIl0.

https://play.google.com/store/apps/details?id=com.iqapps.mobile.tripjournallite&feature=search_result#?t=W251bGwsMSwxLDEsImNvbS5pcWFwcHMubW9iaWxlLnRyaXBqb3VybmFsbGl0ZSJd

https://play.google.com/store/apps/details?id=com.iqapps.mobile.tripjournalvirgin&feature=search_result#?t=W251bGwsMSwxLDEsImNvbS5pcWFwcHMubW9iaWxlLnRyaXBqb3VybmFsdmlyZ2luIl0

https://play.google.com/store/apps/details?id=com.joelapenna.foursquared&feature=search_result#?t=W251bGwsMSwxLDEsImNvbS5qb2VsYXBlbm5hLmZvdXJzcXVhcmVkIl0

https://play.google.com/store/apps/details?id=com.touristeye&feature=search_result#?t=W251bGwsMSwxLDEsImNvbS50b3VyaXN0ZXllIl0

https://play.google.com/store/apps/details?id=com.tripadvisor.tripadvisor&feature=nav_result#?t=W251bGwsMSwyLDNd

https://play.google.com/store/apps/details?id=com.tripit#?t=W251bGwsMSwxLDIxMiwiY29tLnRyaXBpdCJd

https://play.google.com/store/apps/details?id=com.triposo.droidguide.world&feature=search_result#?t=W251bGwsMSwxLDEsImNvbS50cmlwb3NvLmRyb2lkZ3VpZGUud29ybGQiXQ

https://play.google.com/store/apps/details?id=com.wikitude&feature=search_result#?t=W251bGwsMSwxLDEsImNvbS53aWtpdHVkZSJd

https://plus.google.com/111890356102641093073/about?gl=US&hl=en-US (Picture credit "Cécilia")

Chapter 7
The Innovation Machine: Combining Information Design/Visualization with Persuasion Design to Change People's Innovation Behavior

7.1 Introduction

As described in earlier chapters, Aaron Marcus and Associates (AM+A) has explored the intersection of information design and persuasion theory in designing a series of "Machines," concept prototypes of mobile applications that change human behavior in various capacities. The Green Machine *2009* (Jean and Marcus 2009; Marcus and Jean 2009) aimed to instill better sustainable practices, the Health Machine (Marcus 2010) sought to encourage users to lead healthier lives, the Money Machine (Marcus 2012a) persuaded users to manage their wealth better, and the Story Machine (Marcus 2012b) allowed users to connect with their family history and community. The Innovation Machine is the fifth installment in this series of concept designs.

Innovation is not a single "Eureka" moment. It is not a process that occurs by chance, nor is it a one-person effort. Rather, innovation is a structured process of idea generation and collaboration that is key to industry success. Thus, understanding how to manage and harness the innovation process is imperative.

AM+A has created the Innovation Machine, a concept prototype design, which is intended to guide users through the innovation process more efficiently and effectively. The Innovation Machine is a mobile suite of functions and data, tools and content, best practices, and templates, which are tailored to help users brainstorm for ideas, manage their ideas and projects, find tips and advice regarding the innovation process, and ultimately to make progress implementing their ideas as the context permits/requires. Within this virtual space, users can develop new products/services or improve upon existing products/services. Within this virtual space, innovation communities develop within and between corporations, communities, and individuals, and ideas grow to become more than just passing thoughts.

Current applications and services tend to be specialized and seldom address all aspects of the innovation process. Moreover, after reviewing existing mobile

A. Marcus, *Mobile Persuasion Design*, Human–Computer Interaction Series, DOI 10.1007/978-1-4471-4324-6_7

applications, the author's team found opportunities for further adaptations and improvements to better serve users' needs better. Finally, existing products seem not engaging enough and do not provide an overall "persuasion path" to change users' short-term and long-term behavior. Such a path is essential for leading users to improved innovation habits.

The Innovation Machine is intended for use by people from any company/organization, country, and culture. The application is intended to assist in the complete innovation process, from idea generation to development and collaboration to project implementation. The Innovation Machine is intended to be a platform that facilitates open discussion about ideas and innovations among individuals and across industries.

The Innovation Machine's objectives are the following:

- Combine information design and visualization with persuasion design.
- Persuade users to adapt their lifestyles to include better understanding of generating ideas, discussing their ideas, and moving ideas into projects.

AM+A applied user-centered design along with persuasive techniques to make the Innovation Machine highly usable and to increase the likelihood of success in assisting users to adopt new innovation behaviors.

7.2   Innovation Theory

How does one innovate? Is it just about creativity? Is getting an idea enough? Innovation has been defined differently among different scholars. In *Innovation*, Curtis R. Carlson and William W. Wilmot [Curtis and Wilmot, p. 6] define innovation as "a process of creating and delivering new customer value in the marketplace," while the Center of Innovation Studies (Chesbrough 2011) sees it as "a process that transforms an idea into commercial value." A more systematic definition includes:

- Introduction and commercial sale of a new or improved product.
- Introduction and commercial use of a new method of production.
- Introduction of a new form of business organization.
- New use of existing products.
- New markets for existing products.
- New distribution channels.

Curtis Carlson believes that, contrary to popular belief, innovation is not about an individual happening upon a "Eureka!" moment. For Carlson, innovation should be treated as a process that can be systematically understood and taught to individuals and teams and ultimately become part of a company's "DNA." The Innovation Machine hopes to infuse this philosophy in a mobile application to help individuals, teams, and companies innovate using a structured process.

7.2.1 Open Innovation and Co-creation

AM+A explored two models of innovation, "open innovation," and "co-creation." According to Henry Chesbrough, open innovation is "a paradigm that assumes that firms can and should use external ideas as well as internal ideas… as the firms look to advance their technology" (Chesbrough 2005, p. xxiv). In this model, firms (*i.e.*, companies or organization) use internal and external sources of knowledge to turn new ideas into commercial products and services. Firms employing the open innovation model attempt to draw innovations from their own employees, customers, and competitors. This approach contrasts with the more traditional model of "closed innovation," in which companies are more self-sufficient with their innovation by controlling the creation of ideas via their own research and development (R+D) departments. This older model was a direct result of a lack of commercial application of scientific and academic research in the early twentieth century. Firms opted to create R+D departments that specialized in creating innovations for their respective industries. With the open innovation model, however, ideas and new innovations flow freely between universities, companies, and people.

A similar model, co-creation, allows actual end users to take part in creating new products and services. This model relies heavily on the idea that the value of a product increases dramatically when the users and the creators are jointly able to create and customize the product to fit their needs. According to Chesbrough, the paradigm shift from leaving the customers out of the innovation process to letting the customers actually be a part of the innovation process has worked for companies like Intuit, Threadless.com, and Lego. Co-creation can revive failing businesses, unleash new markets, and provide far more meaningful experience to the customers.

Both open innovation and co-creation share a philosophy that active collaboration among groups and people (including the end users) is an effective and efficient way to innovate. The Innovation Machine hopes to utilize these ideas by creating collaborative innovation communities. Rather than developing ideas within a closed network, ideas are shared, validated, and discussed and, therefore, are more likely to become new products and services.

7.2.2 Innovation Process

The author's team researched different approaches to innovation. They wanted to present a streamlined innovation process that was both clear and comprehensive, suitable for users of varying backgrounds and expertise. One compelling approach was a concept map of innovation developed by the Alberta College of Art and Design (ACAD), Dubberly Design Office (DDO), Paul Pangaro, and Nathan Felde and produced by the Dubberly Design Office (2007) (see Fig. 7.1).

Fig. 7.1 Concept map of the Innovation process (Image: Dubberly Design Office (DDO))

In studying this innovation concept map, AM+A noted similarities between the innovation process and the design process. Thus, AM+A also referenced its own design approach as well as an approach that was developed at the Hasso Plattner Institute of Design at Stanford (the so-called d.school, http://dschool.stanford.edu), which details five modes in the so-called design-thinking process:

- Empathize.
- Define.
- Ideate.
- Prototype.
- Test.

AM+A utilized different aspects of these two approaches together with AM+A's own thirty-year user-experience development process (as described in Chap. 1) to generate a three-part, 11-step process of innovation. (Note: In some ways, this entire plan is similar to the user-centered user-experience design/development process described in Chap. 1, but here it is restated as a step towards the Innovation Machine subject matter and context and the development of a user-centered innovation process, as mentioned below.)

7.2.3 Part 1: Generate Ideas

The first step to any new innovation is to generate ideas. The question is, what is the best way to generate ideas?

Step 1: Observe the Status Quo

According to some proponents of design thinking, empathy is the foundation of a human-centered design process. Thus, in order to truly understand people and their needs, one must first observe the status quo. By examining the problems and inconveniences that exist in society, one can better understand the context in which innovation must occur. AM+A would add that culture observation is part of this contextual and ethnographic research and analysis.

Step 2: Identify Opportunities
After observing the status quo, one has a better understanding of users' needs and how those needs are currently being addressed. Through this understanding, one can better identify possible opportunities or areas of improvement.

7.2.4   Part 2: Manage Discussions

Once possible areas for development have been identified, there are a number of essential questions that must be discussed to determine whether the idea is viable. In the case of a new product, who will benefit from this product? In what ways will this new product be used? What else could potential customers use? To answer these questions, one must engage in a number of discussions.

Step 3: Develop and Discuss User Personas
Who is the user? User personas are characterizations of primary user types that are intended to capture essentials of their demographics, contexts of use, behaviors, and motivations and their impact on design solutions. One can better understand and empathize with possible users by developing and discussing possible user personas. Also, through discussing these user personas with a larger innovation community, one can check assumptions and gain feedback.

Step 4: Develop and Discuss Use Scenarios
How and where will the products be used? One can better understand the particular contexts in which the user personas would use the products through use scenarios. Thinking about use scenarios is useful not only in designing the product but also in evaluating and validating the design. Use scenarios can also be opened up for discussion with communities to check assumptions and gain feedback.

Step 5: Develop and Discuss Market Analysis/Competitors
What else could potential users use? With any great idea, there are at least five people in the world with a similar idea. In designing any new product, it is important to understand possible competitors and perform a market analysis in order to find ways to distinguish oneself from competitors.

Step 6: Develop and Discuss Vision/Mission/Objectives/Goals
What do you want to create? In designing a new product, it is important to develop a clear vision, mission, and an explicit set of goals and objectives, so that all stakeholders and collaborators are in alignment.

7.2.5 Part 3: Implement Project

After user personas and use scenarios have been developed, and after the vision, mission, goals, and objectives have been outlined, one can begin the implementation.

Step 7: Find a Team/Collaborators
After a fair amount of discussion with a social network, one might also identify a need for potential collaborators to fill certain roles. To guide users in forming effective teams, the Innovation Machine provides team templates, influenced by Tom Kelley's *The Ten Faces of Innovation* (Kelley and Littman 2005). The four key roles in any innovative team are the learner, the organizer, the designer, and the builder. The learner (researcher or analyst) is a team member that can collect information on the user, the problem, and how the new product or service can change the world. The organizer (planner, accountant, marketer, or sales) understands how to budget resources and move ideas forward. The designer (industrial, UX, and visual designer) might be needed to make the product more usable, useful, and appealing, using user-centered design methods. The builder (software/hardware engineer) takes insights from the previous team members to implement the product.

Step 8: Design a Prototype
After finding a well-rounded team, one can design a prototype or a preliminary model that simulates the main idea of the product. Rough prototypes are a good way of evaluating an idea before investing significant resources in a finished product or service.

Step 9: Test Your Prototype
From designing, creating, and evaluating this prototype, one can discover errors and design flaws. One can present the prototype to users that match the target user base to gather feedback.

Step 10: Refine and Edit Ideas
After testing a prototype, one understands how best to refine and edit one's ideas, objective, and prototype. Designing any new product or solution means running through several iterations of prototypes. With each iteration, flaws are found and features are added until a version comes that tests success with most users. Often three cycles are required.

Step 11: Finalize and Communicate the Ideas
After cycling through several iterations of the user-centered design process for new products and services, the project may be finalized by assembling white papers or other documents that describe the journey from idea to design iterations, as well as presentations to pitch the idea to potential collaborators.

While this 11-step process does not guarantee innovative breakthroughs, it provides a thoughtful, researched structure for those who may be unfamiliar with a user-centered innovation process. The Innovation Machine aims to ensure that no steps are overlooked and to stress the importance of people and community in two ways:

- Thinking about the user of a product.
- Collaborating with others to discuss and refine a potential project.

7.2.6   Objectives and Goals

AM+A's research of various models of innovation shows that any employee, user, or customer can be an innovator. In other words, anyone is capable of contributing to innovations. Alternately, any employee, user, or customer can harness the innovation process.

With this in mind, AM+A created primary objectives for our design of the Innovation Machine. We may define "objectives" to mean particular functions that we want to enable people (which wish to develop better habits in innovation) to perform. In order to design an application that achieved our objectives, we assembled a list of requisite steps for the development of this application that we have labeled "goals." Examples of both follow:

7.2.7   Objectives

Enable users to:

- Collaborate remotely.
- Manage projects, check progress, and brainstorm.
- Find like-minded people, connect, and create teams.
- Find and share ideas, inspiration, feedback, and problems.
- Be advised of relevant news, trends, ideas, competitors, *etc.*
- Enjoy a fun-filled and, therefore, rewarding user experience.

7.2.8   Goals

- Develop a process that shows the complete innovation process in terms of steps, timeframes, dependencies, personnel, physical assets (equipment and spaces), and other constraints.
- Develop a dashboard that shows summary indicators and provides brief controls for the innovation process.
- Develop social networking paradigms or components or techniques and resources.

- Develop tips/advice: appropriate tags, taxonomies, categories, and content for tips/advice.
- Develop an open environment for people to discuss new ideas, a platform which facilitates open innovation and co-creation with the stake holders and customers.
- Develop incentives, such as games and challenges for the users with respect to creativity, ideas, and innovations.

7.3 User-Centered Design

As noted in Chap. 1, the user-centered design (UCD) approach links the process of developing software, hardware, and user interface (UI) to the people who will use a product/service. UCD processes focus on users throughout the development of a product or service. The UCD process comprises the following tasks, which sometimes occur iteratively. The entire approach mirrors or is similar to the 11-step innovation process previously described above:

- *Plan*: Determine strategy, tactics, likely markets, stakeholders, platforms, tools, and processes.
- *Research*: Gather and examine relevant documents and stakeholder statements.
- *Analyze*: Identify the target market, typical users of the product, personas (characteristic users), use scenarios, and competitive products.
- *Design*: Determine general and specific design solutions, from simple concept maps, information architecture (conceptual structure or metaphors, mental models, and navigation), wireframes, look and feel (appearance and interaction details), screen sketches, and detailed screens and prototypes.
- *Implement*: Script or code specific working prototypes or partial "alpha" prototypes of working versions.
- *Evaluate*: Evaluate users, target markets, competition, and the design solutions, conduct field surveys, and test the initial and later designs with the target markets.
- *Document*: Draft white papers, user-interface guidelines, specifications, and other summary documents, including marketing presentations.

AM+A carried out all of these tasks in the development of the Innovation Machine concept design except for implementing working versions.

7.4 Personas

As noted in Chap. 1, personas are characterizations of primary user types and are intended to capture essentials of their demographics, contexts of use, behaviors, and motivations and their impact on design solutions. Personas are also called user profiles. Typically, UI development teams define one to nine primary personas. For the Innovation Machine, the following personas were determined by analyzing available data.

7.4.1   Persona 1: John McArthur

- Age: 32.
- Race: Caucasian American.
- Occupation: Entrepreneur.
- Tech Usage: Intermediate. Proficient with iPhone, Android, Mac.
- Location: Palo Alto, California.
- Persona image credit: AM+A.

7.4.1.1   Textual Summary

John is the co-founder and President of Mikaya Games, a 22-month-old Palo Alto-based start-up company that creates educational games. The company has have released four games for the iPhone platform and three for the Android, and it is trying to roll out its first history game in the next month.

What started as a small team of four students from Stanford is now a Y-Combinator (YC)-funded company with 13 members, including three interns. Originally, they worked out of their apartment and connected over Skype, but currently, they work from an office that was set up by one of the venture-capital firms.

However, John is often traveling around the United States, meeting with potential investors. He uses Yammer to keep up with his team and emails for formal communication, but he would like a better way to track his team's progress while away. He is always looking for new ideas for games and applications and has to keep himself updated with the market, innovations, and investments. Now that his company is growing, he needs a way to be connected to all of them and facilitate ideation among his other co-founders and colleagues.

7.4.1.2 Design Implication Summary

Note: The following subsections summarize the design implications for this above persona and the ones that follow. As with all of the Machines, these resulted from internal discussions and debate, not from any software-supported, or automatic, or mathematical process, but due to time and budget constraints. This limitation is often a real-life factor of such development.

Objectives

- Stay connected to the team remotely (including progress and deadlines).
- Keep up with the market and other news.
- Find means to brainstorm and take notes.
- Look out for possible stakeholders (investors, funders, and other people to recruit).
- Get and store ideas from users and team.

Context

- Connected to a large global network.
- Not a project manager but likes to keep himself updated with the projects' progress.

Behavior

- Travels around the United States frequently.
- Uses Yammer with teammates.
- Uses Skype and email to communicate.
- Owns an iPhone, Samsung tablet, and Mac Pro.

7.4.1.3 Design Implications

Innovation Machine must help user:

- Stay updated with his team's progress.
- Jot down his ideas.
- Brainstorm with team and take notes side by side.
- Communicate instantly with teammates.
- Remain updated with what similar companies are doing and share his company's progress with the world.

7.4.2   Persona 2: Keas Durden

- Age: 27.
- Race: Caucasian American.
- Occupation: Manager of Starbucks.
- Tech Usage:
- Location: Portland, Oregon.
- Persona image credit: AM+A.

7.4.2.1   Textual Summary

Keas is entering his third year as a manager for a Starbucks in Portland, Oregon. Ever since college, he has been passionate about new technology and technology trends. He spends his breaks reading tech blogs like TechCrunch and is always looking for new and novel iPad apps to rate and review on his own blog, which has 555 followers. One of his other sources of income is advertisements from his Youtube channel where he has reviewed hundreds of iPhone and iPad apps and accumulated 2000+ subscribers. He is looking for a new medium to broadcast his app reviews, increase his audience, gain feedback, and offer his wisdom from having reviewed hundreds of apps.

7.4.2.2   Design Implications Summary

Objectives

- Broadcast his app review blog, increase his audience/subscribers.
- Gain feedback for his app ideas.

- Connect to a network of technology innovators.

Context

- Lives in Portland, Oregon.
- Passionate about technology.
- Owns iPhone, iPad, and Android tablet.

Behavior

- Runs his own mobile app review blog.

Design Implications

- Design Innovation Machine to connect developers with end users.
- Connect people with interesting ideas to those looking for ideas.

7.4.3   Persona 3: Lisa Robertson

- Age: 36.
- Race: African-American.
- Occupation: Head of Operations for a national home-furnishing retailer.
- Tech Usage: Owns a smartphone, proficient with various applications and software, uses technology both personally and professionally.
- Location: New York, New York.
- Persona image credit: AM+A.

7.4.3.1   Textual Summary

Lisa Robertson is a 36-year-old African-American female who heads the operations for a national home-furnishings retailer. She has been with this company for the past seven years, over which she has risen from a store clerk to a project manager. Currently, Lisa manages a group of approximately 70 mid- and entry-level employees. Over the past year, Lisa has noticed that her group has been missing deadlines. Lisa communicates primarily to her middle managers, and from their input, she gathers that there are a number of inefficiencies within the system. She is unsure of exactly what needs to be changed and is interested in getting feedback from her employees. She is overwhelmed by the number of projects she needs to juggle. While she uses Microsoft Project to manage her projects, she is often traveling and is seldom at her computer and feels that the software is minimally useful for her busy, on-the-go lifestyle.

7.4.3.2   Design Implications Summary

Objectives

- Manage multiple projects from flexible locations.
- Collect feedback from employees.

Context

- Has worked within a national corporation for 7 years.
- Has a busy schedule, seldom near a computer.
- Manages a group of seventy employees.

Behavior

- Owns a smartphone.
- Has accounts with various personal networking platforms.
- Uses computer-based project management software.
- Communicates with her employees primarily through email and phone.

Design Implications

- Have a mobile project management platform.
- Have a collaborative, feedback system.
- Be able to interact within a selected network.

7.4.4   Persona 4: Pratika Jain

- Age: 27.
- Race: Indian.
- Occupation: Entry-level customer-service employee for a national telecommunications company.
- Tech Usage: Owns a smartphone, proficient with various technology platforms, but uses them for personal reasons.
- Location: Yonkers, New York.
- Persona image credit: AM+A.

7.4.4.1   Textual Summary

Pratika Jain is a 27-year-old entry-level customer-service employee at a national telecommunications firm. She just started at the company 2 months ago and had a difficult time learning how to deal with dissatisfied customers. Being new to the company, she sees plenty of opportunities for change. However, her job is very solitary and there aren't very many opportunities to exchange ideas. In addition, Pratika just moved to the United States about two years ago. She still feels like an outsider and is often hesitant to express herself for fear of saying something culturally inappropriate. While Pratika owns an iPhone, she uses it primarily to make calls and stay in touch with family overseas. The only networking platform she uses is Facebook, which she uses to stay in touch with family as well.

7.4.4.2   Design Implications Summary

Objectives

- Express new ideas and opinions in a culturally "safe" manner.
- Connect with people outside of her immediate contact circle.

Context

- A new, entry-level employee within a large corporation.
- Her job doesn't involve much physical interaction with other employees.
- Grew up in India, unfamiliar with US customs.

Behavior

- Owns a smartphone, but for personal use.

Design Implications

- Use a guided, simple user interface.
- Include a collaboration platform.
- Be able to interact within a selected network.

7.4.5   Persona 5: Tyler Lee

- Age: 28.
- Race: Asian-American.

- Occupation: Software Engineer at Amdocs.
- Tech Usage: Expert. Proficient with iPhone, iPad, Linux System.
- Location: San Francisco, California.
- Persona image credit: AM+A.

7.4.5.1   Textual Summary

Born and raised in Miami, Florida, Tyler has been programming since high school. He graduated as an engineer from Stanford and worked as a programmer in Oracle for a year. He then got his Masters at University of California at Berkeley and has since been working for Amdocs, a San Francisco high-tech company, for the last three years. While Tyler enjoys the financial security of his corporate job, he is excited by the idea of making a name for himself and starting a company. However, while he is an expert in C, Java, and Python, he doesn't consider himself very creative. He is looking for a way to find people with ideas to collaborate with for exciting side projects and opportunities.

7.4.5.2   Design Implications Summary

Objectives

- Collaborate with people with interesting ideas.
- Help people using his expertise in computer languages.
- Work on new, promising projects.

Context

- Working in a big software company for years.
- Proficient with computer architecture and computer languages.
- Has a 9–5 schedule on weekdays. Leaves a lot of free time on Sundays and Saturdays (when not working).

Behavior

- Owns an iPhone, HP Tablet.
- Collaborates with other engineers, managers, and executives via email and messenger.

Design Implications

- Have a social networking platform that is connected globally.
- Access feed of major technology news with a discussion platform.
- Able to connect to a project-team network and get updates from the specified network.

7.5   Use Scenarios

7.5.1   *Definition*

As explained in Chap. 1, use scenarios are a user-centered UI development technique. In the development of an initial concept prototype, a use scenario helps determine what behavior to simulate. A scenario is essentially a sequence of task flows with actual content provided, such as the users' demographics and goals, the details of the information being exchanged, *etc.*

7.5.2   *General Use Scenario*

Based on internal AM+A discussion, the following use scenario topics emerged, which were drawn from the five preceding personas. However, some specific examples might be relevant only to a particular age group, education level, gender, or culture. As noted in the Glossary of Chap. 1, please note the general usage of the terms "objective" and "goal." An objective is a general sought-after target circumstance. A goal is more specific and is usually qualified by concrete, verifiable conditions of time, quantity, *etc.*

7.5.3   *Idea Creation/Development*

- View and track the discussion surrounding ideas users have posted, commented on, or are subscribed to in the ideation board.
- Broadcast problems, reviews, and ideas on an open discussion board.
- Evaluate and validate other users' ideas and problems.
- Filter through discussions based on popularity and topic (tags).
- Reward users with points for helpful and insightful posts.
- Collaborate and brainstorm with others remotely using a virtual whiteboard tool.

Idea/Project Management

- View and track the discussion surrounding ideas you''ve posted or commented on in the ideation board.
- Visualize and track the progress of your ongoing projects.
- Create and monitor goals for your projects.
- Set deadlines for user and/or project-team members.
- Receive deadline alerts.
- View team and contact team members.
- Collaborate with team and hold brainstorming sessions and whiteboard sessions.

Network

- Create a public profile with contact information.
- Exhibit skills (tools, platforms), experiences (markets, user communities), and interests.
- Find and filter through potential stakeholders or team members.
- Assemble project teams.
- Connect to a broader network of innovators.
- Connect via messaging, emailing, or calling.
- Share project's progress with a set of people or globally.
- Communicate, connect, and work within a specified network according to selected factors (project team, same interests, global network).

Learning

- Receive daily tips and advice via push notifications.
- View tips for brainstorming more efficiently and effectively while using the virtual whiteboard tool.
- Receive relevant articles, blog posts, and resources regarding strategies, trends, competitors, *etc.*
- Filter through the articles using tags, searches, and folders.
- Share researches, articles, *etc.*
- Link to external resources (Fastco, Core77, *etc.*).

Gamification

- Gain digital achievements (*i.e.*, badges) for making deadlines and completing goals.
- Take on challenges presented by the game to brainstorm a potential solution to a global problem.
- Reward users for particularly helpful or insightful posts with points that are redeemable for reputation status and titles.

7.5.4 Persona Use Scenarios

7.5.4.1 John's Use Scenario and Behavior Changes

As the co-founder of a 13-person start-up, Mikaya Games, John works in a fast-paced environment and is heavily involved of all aspects of his company. Having success with Math and Science games, John thinks that maybe the company should branch out to Social Sciences. He is interested in seeing what his team thinks of this idea, so John uses the Innovation Machine to get a sense of his team's sentiments.

John is able to use the Innovation Machine, not only as an information gathering platform, but he is also able to use it to brainstorm privately as well as collaboratively with his team. During these brainstorms, his team can record their session both visually and audibly, and after the brainstorm, John can share the brainstorm transcript with his team. John can also create challenges for his team; the most recent challenge was to brainstorm five new game concepts in five minutes. In fact, their most successful game, Math Busters, was a result of one of his challenges.

John works with a small team, and he likes to keep himself updated with what everyone is doing. With the Innovation Machine, John is able to quickly overview his team's progress on his Track page. Once, John saw that his designer had finished a project ahead of schedule, and he rewarded her with a badge. His company has built upon the badge system, so whoever gathers the most badges each week gets a free lunch.

John also relies on the Innovation Machine to keep him updated with the latest game trends and educational technology news. He also uses the Innovation Machine to receive tips on how to grow a sustainable start-up and how to maintain a culture of innovation. He just read an interesting article about "writable walls." John sent his company the article via the Innovation Machine, and now, Mikaya Games is considering implementing the idea in their office.

John's Behavior Changes

- In the past, John could only check his team's progress by sorting through the numerous emails in his inbox. At times an email would get lost, and John would have to call for an update. John can now easily see the progress of his company on one screen.
- It used to be difficult for John to brainstorm while on the road. John can now hold brainstorming sessions at airport or even in his car.
- During former brainstorming sessions, so many ideas were generated, but once the session was over, people would forget about the ideas. Now, it is easier for John to track and encourage discussion on ideas. Anyone can comment and critique at anytime.
- John used to catch up with the news on start-ups, innovations, funds, and investors by looking through numerous tech blogs and news outlets. Now, John can customize his news feed to aggregate all relevant news. He can share interesting articles with his team.

7.5.4.2   Keas' Use Scenario

Keas is avid technology user who spends most of his time either reading tech blogs or filming application reviews. When he is at work managing his local Starbucks, Keas is likely playing with one of his many tech gadgets. However, while Keas is passionate about new tech, he doesn't have very many like-minded friends or coworkers. Keas is able to use the Innovation Machine as an outlet for his passion. He uses the Connect module to post reviews or comments on latest trends. So many people find his reviews helpful that Keas has obtained Guru status on the Innovation Machine. He uses his newfound fame to publicize his own app review blog. Since using the Innovation Machine, viewership of his video blog has increased threefold. Despite not having the tech background of other Innovation Machine users, Keas' impressive status and badge collection validates him as someone passionate about innovative technology.

From all of his application reviews, Keas has accumulated a number of ideas for new applications, which he posts on the Connect Discussion Board to be evaluated by end users, engineers, and designers in the tech industry. One engineer has been particularly valuable in helping Keas refine his idea. After weeks of discussion through the Innovation Machine, Keas finally learned that the engineer, Chris, lived in Portland as well. They are getting drinks next week.

Keas' Behavior Changes

- Keas uses the Innovation Machine to broadcast his newest reviews to a relevant network and to publicize his own video blog.
- Having used and reviewed several hundred applications, Keas has a few ideas for apps and is able to use the community within the Innovation Machine to evaluate his ideas.
- Keas now connects with like-minded tech-passionate folk.

7.5.4.3   Lisa's Use Scenario

As the Head of Operations for a home-furnishing company, Lisa is constantly flying between factories and suppliers across Asia. When she is waiting in the airport, Lisa uses the Innovation Machine to manage her projects and connect with her employees. Through the application, Lisa is able to see a concise visual summary of all of her current projects. Through this visual summary, Lisa can see how the different components within her projects are developing within an overarching timeline. Employees can update their work on the Innovation Machine, so Lisa receives timely updates on the progress of individual project members. Lisa can set specific deadlines for herself and employees. As deadlines grow closer, Lisa is notified by alerts and can send reminders.

Given her schedule, Lisa has difficulty keeping up to date with the recent overturn of employees within her department. Through the Innovation Machine, Lisa can visit the individual profiles of people within her department. Not only this, Lisa can use the application to pose questions and collect feedback via the Connect Discussion Board. She uses this platform to challenge her team to brainstorm ways to streamline their reporting processes. To encourage her employees to participate, Lisa rewards her employees with virtual points and titles, and at the annual company retreat, she recognizes those with the most points and titles.

Lisa also uses the vast number of resources within the Learn module to learn more about her competitors as well as new management practices. In the past, Lisa didn't really know where to get reliable information, so she would only refer to one or two publications. Now, Lisa can quickly scan headlines and news with the Innovation Machine's comprehensive database.

Lisa's Behavior Changes

- Lisa used to only communicate with her middle managers, but now she can ask for feedback from all levels within her organization. Lisa also uses as a way to reward and recognize employees.

- She used to lose track of her projects because she was always traveling and seldom near a computer with Wi-Fi access. Now, she can track her projects via her mobile, giving instant feedback, and sending notifications and reminders.
- Lisa can read about best project management practices and news of other home-furnishing competitors from a variety of notable resources within the Innovation Machines learning module.

7.5.4.4   Pratika's Use Scenario

As a customer-service employee, Pratika spends most of her workday with her headset and the voice on the other end of the line. As a result, she used to feel disconnected with the rest of her company. Pratika uses the Innovation Machine to learn more about the company, the people within the company, and their new initiatives. She enjoys posting on the Idea Wall and giving feedback to other people's ideas. Since English isn't her first language and since she is shy by nature, Pratika likes this platform much better than the large company meetings she typically attends.

After her own rocky transition as a new employee, Pratika posted an idea about creating a new employee buddy system. Another new employee in the Marketing department saw Pratika's idea and commented that she felt it would have helped her transition into the company as well. Pratika was heartened that she wasn't alone in her sentiments and contacted this other employee through the contact information posted in her profile. They arranged a time to meet, and together, they decided to develop this idea further. They posted a more complete vision on the idea wall, and the head of human resources saw the idea and decided to institute the program. As a result, both Pratika and her new friend will get an extra-large bonus at the end of the year.

Pratika's Behavior Changes

- As a new employee with a different cultural background, Pratika used to be afraid to speak up within her company. However, now, Pratika feels able to share her comments freely. She can post her ideas regarding new employee training initiatives.
- As a customer-service operator, Pratika used to have very little contact with other members within the company. Now, Pratika is connected to the rest of the company and can seek feedback and advice on how to adjust as a new employee.

7.5.4.5   Tyler's Use Scenario

Tyler has been a software engineer for at Oracle for a few years. While he enjoys his job, he recognizes that it is a large corporation which has a tendency to move more slowly. As such, he feels that it's difficult to stay up to date with all the activity within the Silicon Valley. With Innovation Machine, Tyler is able to subscribe to interesting ideas and discussions he sees on the Connect Discussion Board.

Through the Connect Discussion Board, Tyler has also learned about an interesting, new photo-sharing start-up that had just received two million from a VC firm.

Upon looking through user profiles, Tyler realized that one of the start-up founders was his old classmate from college. Using the messaging functions on the Innovation Machine, Tyler contacted his former classmate. It turned out that the start-up team was looking for someone with technical experience; so after a few brainstorming session, the start-up offered Tyler a spot on core team. After accepting the position, Tyler quit his boring, albeit cushy, corporate job and is loving the excitement and challenges that come with start-up life.

Tyler's Behavior Changes

- Tyler was looking for people with promising ideas to collaborate with but didn't have a platform he could use to look out for such opportunities outside his workspace.
- He now uses the Connect Discussion Board on the Innovation Machine to view ideas and discussion from people across the world.

7.6 Competitive Analysis

7.6.1 Comparison Studies

Before undertaking conceptual and perceptual (visual) designs of the Innovation Machine prototype, AM+A studied approximately 10 innovation management Web sites and smartphone applications. Through screen comparison analysis and customer review analysis, AM+A derived these applications' major benefits and drawbacks. This in-depth analysis helped further develop initial ideas for the Innovation Machine's detailed functions, data, information architecture (metaphors, mental model, and navigation), and look and feel (appearance and interaction). These studies were conducted in 2012.

7.6.2 Idea Generation and Management Web Sites and/or Associated Mobile Applications

7.6.2.1 Spigit

Who Were They in 2012?

- Founded by Paul Pluschkell in 2007.
- Spigit is a technology company based in Pleasanton, California, that provides a software as a service (SaaS) platform for enterprise innovation management.
- Customers include big companies like AT&T, Pfizer, Cisco, and Southwest airlines.
- It has been recognized with industry awards, including InformationWeek Startup 50 2009.

What Was It in 2012?

- Platform which lets employees, customers, partners, and fans collaborate to tackle business objectives with interactive and engaging tools to form a community to share ideas.
- Integrate your idea community with Facebook, Yammer, SharePoint, and Jive.
- Engage users in a social environment to submit ideas and gather feedback.
- Allow users to discuss ideas and pitch new ideas through contests.

What Did the App Let Users Do in 2012?

- Submit ideas in an open innovation community where other users can vote on the best ones. Enhance their submitted ideas with documents, video, and images.
- Subscribe to submitted ideas and share them with friends.
- Create profile and view other users' activity and ideas.
- Be rewarded with virtual currency for participation in contests and activity, which can be exchanged for rewards provided by the company, such as vacation days, special lunches, iPads, *etc.*
- View top 10 ideas and top 10 innovators in leaderboard.
- Filter ideas based on date posted, category, rank, *etc.*
- Compete in idea generation contests held by companies using Spigit's "Faceoff" application by submitting idea responses, which are then voted on.
- Buy, sell, trade, and invest in ideas within Spigit's Idea Market.
- Submit tasks into an activity feed.
- Manage community with an administrative dashboard.
- Integrate Spigit activity with Yammer, Facebook, Twitter, *etc.*

Web Strategy

- Web site looks user-friendly and provides a quick tour about their service and what they aim for.
- Quick tours in form of videos and slideshows for Web series they provide for all the platforms.
- It doesn't make it clear where the company should start off and how they proceed.
- A demo has to be requested by providing the information about customer's company.

Mobile Strategy

- Spigit is interconnected with Yammer, so the user can get the activity feed on Yammer mobile application, but there is no native app available.

User-friendliness

- Platform is easy to use and understand.
- Video tutorials available for new users to get started.
- No demo or trial period.
- The tutorials fail to show the process of idea generation through Spigit.

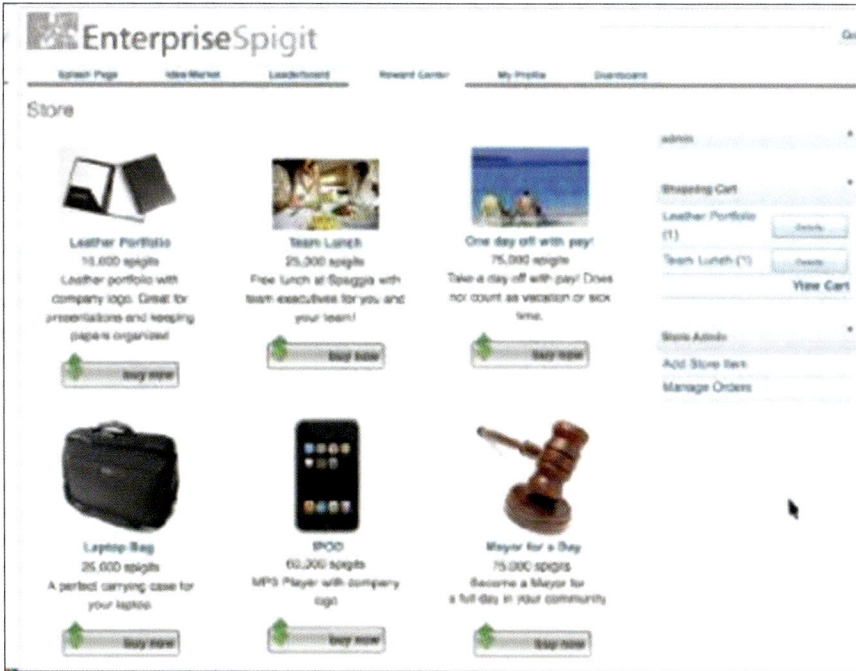

Fig. 7.2 Example screenshots of SpigitEngage and Enterprise Spigit Web sites in 2012 (Image: Mindjet)

Visual Analysis

- Facebook integration follows conventional navigation system which is easy to understand and which also gives a "Facebook" feel to it (Fig. 7.2).
- Spigit for SharePoint uses white and green basic colors making it look very clean and easy to use.
- Use of data visualization in dashboard to give general idea.

Primary Competitors

- Brightidea Innovation Suite.
- IdeaScale.
- Crowdcast.

Tips

- No tips and advice provided to the user for better idea generation.

Social Networking

- Spigit provides integration with numerous other platforms such as Facebook, YouTube, Twitter, Blogs, and Yammer.
- Users can setup their profile and view other users' profiles, which contain their comments and submitted ideas.

7.6.3   SWOT Analysis

Strengths

- Idea discussion forum, voting system, and idea management interface are intuitive and easy to use.
- Spigit's "Faceoff" contests engage users to use the platform and make the application fun.
- Integration with Facebook and Yammer gives users more channels to engage with the platform.

Weaknesses

- Seems to be a platform and not software.
- No mobile version (besides integration with Yammer).
- No project management.
- Guides users through generating ideas and collaborating on them, but does not guide them through the remainder of the innovation cycle: moving ideas to proposals and projects.
- Task management and idea life cycle are complicated and difficult.
- No tips and advices for individual users on how to better generate ideas, but only provides the tools to manage them.
- Helps companies generate new ideas, but not individuals.

Opportunities

- Application for mobile phones and tablets without having to integrate with Yammer.
- More functionality with social networking and creating connections, groups, and teams.
- Implement project management functions to guide users past ideation and into implementation.

Threats

- Users are not able to categorize ideas.

7.6.3.1   IdeaScale

Who Were They in 2012?

- Founded in 2008, when Barack Obama selected IdeaScale to create citizen feedback platforms for 23 different federal agencies.
- IdeaScale provides a service that crowd source ideas for specific companies or organizations from their customers or employees.
- Made *Inc.* magazine's list of the fastest-growing private companies, ranking 172nd overall and 25th among business-service providers.

- Recognized by *Puget Sound Business Journal* as one of the 50 fastest-growing private companies in Washington State.

What Was It in 2012?

- Social engagement platform to collect bugs, ideas, feedback, and conversations about a brand.
- Companies integrate their Web sites with the platform to collect data from their customers.
- Used by some of the major companies like Starbucks, Intel, Xerox, SAP, and Microsoft.
- Has been a part of Barack Obama's "Change" movement.

What Did the App Let Users Do in 2012?

User

Submit ideas for a specific brand on their respective IdeaScale feedback Web sites.

- Comment, up-vote/down-vote, and follow ideas submitted by other users.
- Create and manage his/her profiles and view other users' profiles, as well as their submitted ideas and comments.
- Filter ideas based on popularity, ranking, and time.
- Earn points based on activity, and view leaderboard rankings.

Companies

- Customize the look of the feedback Web site.
- Developers can integrate IdeaScale platform into their iPhone applications, which allows them to submit, view, comment, and vote on bugs and features.
- Make unlimited campaigns and add administrators and moderators for the campaign.
- Moderate ideas by setting them as active, off topic, closed, completed, in progress, *etc*.
- View statistics of the ideas and users.
- Restrict access to community to only a specific IP address range.
- Customize the URL for IdeaScale platform to user's domain.

Web Strategy

- User-friendly Web site which provides a 2.5 min video explaining the service.
- User can start off in seconds for free.
- Web site gives helpful and impressive examples of the IdeaScale in use.
- Sign from the Web site to start working.

User-friendliness

- Easy to follow and understand user interface.
- Allows customization of the platform.
- Provides customer service through live chat and blog.

Web Site Visual Analysis

- Customizable according to the need of the company or brand.
- White background and gray text make it visually neat.
- Follows the conventional horizontal navigation (Fig. 7.3).

Mobile Strategy

- Has iOS and Android plugins.
- While developing an application, developers can add the code given by IdeaScale to implement the platform in their applications.

Mobile Visual Analysis

- Follows iOS and Android design guidelines.
- Seems to be well designed, with clear typography and screen layout (Fig. 7.4).

Primary Competitors

- e-Tipi, Kindling, Brightidea Innovation Platform, IdeaBox.

Social Network Presence

- 2000+ followers on Facebook.
- 2800+ followers on twitter and active.
- Active blog.

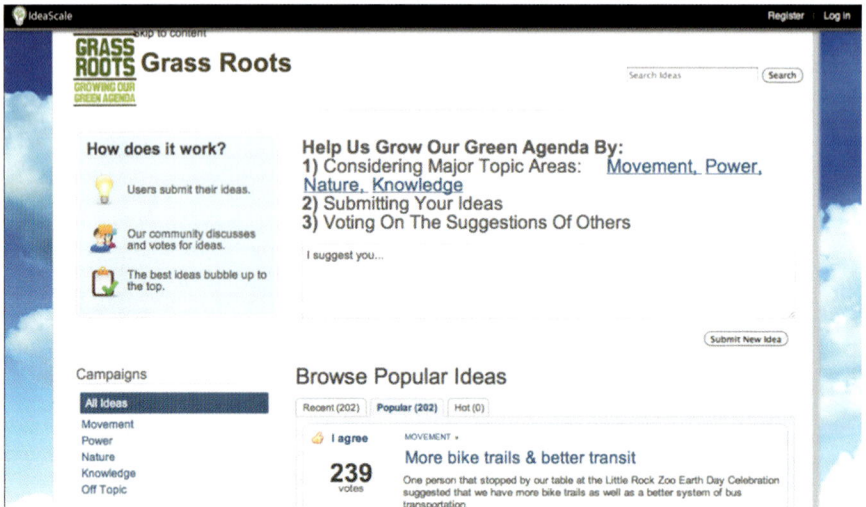

Fig. 7.3 Example of IdeaScale Web site screen in 2012 (Image: IdeaScale, LCC.)

Fig. 7.4 Example of
IdeaScale mobile screen in
2012 (Image: IdeaScale,
LCC.)

7.6.4 SWOT Analysis

Strengths

- Has a free version available.
- Paid versions are affordable and not too expensive.
- Implementation is easy and can be carried out in seconds.
- User-friendly interface.
- Provides a simple, clean URL; for the feedback platform.
- Implementation in iPhone application with an IdeaScale API.
- Idea moderation.
- Facebook integration.

Weaknesses

- Poor idea management.
- No task management.
- It's a brand initiative and not a customer initiative and, therefore, does not guide users to implement their ideas.
- Not for a getting ideas within a company.
- Does not show the total votes given.

Opportunities

- Change idea to innovation generation.
- Add Contests and gamification to engage users.
- Idea generation within an organization.

Threats

- No way to implement the platform within a Web site; creates a separate URL for the platform.
- Not useful for idea generation and solution generation within an organization.

7.6.4.1   Brightidea Innovation Suite

Who Were They in 2012?

- Innovation platform first launched in 2005 by Brightidea, an SF-based company.
- First founded in 1999 to help companies streamline the innovation process and improve their efficiency of moving ideas to implementation.
- Partnered with over 300 companies that range from small companies to bigger ones such as GE, Adobe, Kraft, HP, Cisco, UBS, Bayer, Department of State, and British Telecom (BT).

What Was It in 2012?

- Cloud-based Web platform, integrations with SharePoint, Yammer, Facebook, and more.
- Innovation Suite includes three separate, but connected, pieces of enterprise software: Webstorm, Switchboard, and Pipeline.
- Webstorm: Create and manage a community where people submit, discuss, and vote on ideas.
- Switchboard: Combine ideas to create proposals, form teams, and develop questionnaires and scorecards to evaluate proposals.
- Pipeline: Create projects and manage a project team by assigning tasks and defining a workflow.

What Did the App Let Users Do in 2012?

Webstorm

- Create idea communities with customizable user interfaces.
- Manage users within idea communities.
- Create profiles and network with others.
- Submit ideas into an activity feed, and attach documents and media.
- Comment on ideas, and vote on ideas so that the highly ranked ones bubble to the top of the feed.
- Increase user reputation with increased activity.

Switchboard

- Merge ideas into proposals.
- Set up teams, and track team activity.
- Collaborate with teams in "private collaboration room."
- Evaluate proposals by creating questionnaires and scorecards.

Pipeline

- Create and manage new projects.
- Set up and manage project teams.
- Define workflow and set up milestones.
- Assign action items to team members.

- Create project fan pages.
- Use metrics to assemble financial data.

Web Strategy

- Innovation Suite is cloud-based "Facebook style" software.
- A multitude of features available to customize the user interface.
- New "Webstorms" can be created with a few clicks.
- Users can connect with others through creation of teams and track each other's activity.
- Users within one network present their ideas and completed tasks and may assign each other tasks.
- Facebook integration allows for users to post their task activity and create fan pages for their projects to drive viral success.
- Plugin for Web sites to collect ideas more easily.

Mobile Strategy

- Brightidea Mobile is modeled after its Webstorm product.
- Users can view post ideas to the activity feed in real time, discuss and collaborate on ideas, and attach photos and files.

User-friendliness

- Demo available only upon request.
- 24/7 phone support for customers.
- Innovation community interface and email templates are customizable.
- Follows conventions for social networking (profiles, commenting, voting) for an eased transition for new users.
- Set up wizard for getting started with creating innovation communities and proposal scorecards.

Visual Analysis

- User interface is modern and sleek, and the theme is configurable to the user's preference (Fig. 7.5).
- Various Web elements are clearly labeled.
- User interface is very busy and feature laden, but accessible to people who are familiar with similar products (like Facebook).

  - User interface for mobile is very simple and seems insufficiently exciting.
  - Gray background and large amounts of negative space may make it seem elementary, in terms of mobile user-interface design (Fig. 7.6).

Primary Competitors

- Spigit, Jive.

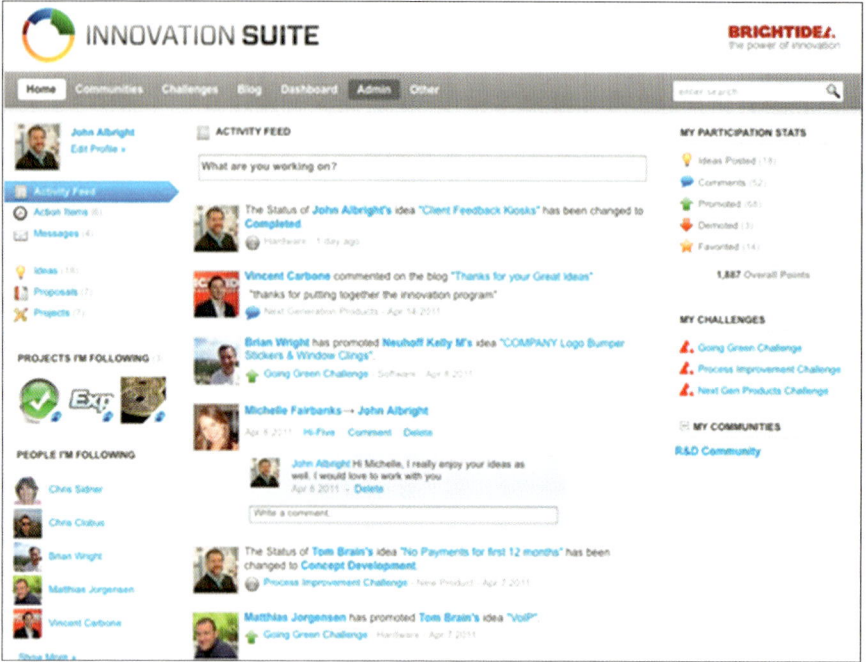

Fig. 7.5 Example screenshot of the Innovation Suite Web site in 2012 (Image: courtesy of Brightidea, Inc.)

Fig. 7.6 Example screenshots of Innovation Suite mobile screens in 2012 (Image: courtesy of BrightIdea, Inc.)

Social Networking

- Users create profiles and can create project teams with others.
- Project managers can assign tasks to members within a team.
- Webstorm relies heavily on the idea of crowdsourcing innovation by having communities collectively decide which ideas are best.

7.6.5   SWOT Analysis

Strengths

- Impressive reputation, with 300+ customers.
- Experience: Brightidea was the first to create an online innovation platform.
- Navigation and interface skins are completely configurable.
- Provides a platform for the whole innovation process and not just idea generation.
- 24/7 support.

Weaknesses

- Massive amount of features, difficult to know where to start.
- Demo available only upon request.
- Little social media exposure and customer reviews are scarce.
- Innovation process spread out thin between three separate pieces of software.
- No persuasion theory to change people's behaviors about innovation.
- No brainstorming tool.

Opportunities

- Incorporate user rankings and gamification to make software fun to use.
- Incentivize users to move through the innovation process, rather than just provide the tools.

Threats

- Security and legal issues with putting ideas on the Web for others to see.

7.6.6   Mind-Mapping and Brainstorming Applications

7.6.6.1   Mindo

What Was It in 2012?

- Mindo is a mind-mapping tool for iPads, helping users organize their thoughts into a 2D space for brainstorming and note-taking.

- 3+ rating on iTunes application store.
- Available in English and Chinese.

What Did the App Let Users Do?

- Organize thoughts, ideas, and information in form of 2D visuals.
- Create Task Lists, Brainstorms, Concept Maps, and meeting notes.
- Export png and pdf versions of maps.
- Synchronize with Dropbox and iTunes.
- Drag and drop text boxes (topics and subtopics) and connections.
- Customize elements of the map using various colors, shapes, and fonts.
- Expand and collapse branches in the map.
- Add comments, tasks, pictures, and icons.
- Share maps via Picasa and email.
- Add boundaries and relationships.

Web Strategy

- Small Web site which gives link to the iPad application.
- Web site also provides a three-minute video session and screenshots of the product.

Mobile Strategy

- iPad application available for 7.99.

User-Friendliness

- Intuitive and easy-to-understand interface.
- Customization of the branches and text boxes.

Visual Analysis

- White background with big topics (in blue) and subtopics (in white).
- Customization to make the map colorful and interesting.

Tips

- No tips or advice.

Social Networking

- Not active on Facebook, Twitter, or any other social networking platform.
- No blog.

Primary Competitors

- iThought (iPhone).
- bubbl.us (Web).

7.6.7   SWOT Analysis

Strengths

- Available for iPads.
- Easy to use.
- Integration with Dropbox, Picasa, and iTunes.
- Export maps as png and pdf.

Weaknesses

- More of an idea mapping tool than a brainstorming tool.
- Doesn't allow remote collaboration with other users.
- Sharing through email and picasa only.
- Mapping only horizontally.
- Not integrated with a project management system.

Opportunities

- Allow networking.
- Syncing of the maps within the network.
- Addition of task management.
- Live mapping within network.
- Application for iPhone.
- Doodling.

Threats

- No networking.
- No description to the maps.

7.6.7.1   Comapping

What Is It in 2012?

- Online mind-mapping software to manage and share information.
- Offline software to make mind maps and share them (when online).

What Did the App Let Users Do in 2012?

- Collaborate with others in creating mind maps.
- Utilize templates for mind mapping, such as agenda, brainstorming, customer database, note-taking, and project management.
- Create new topics and subtopics, make connections, and focus and zoom on topics.
- Comment and add files to mind maps.
- Invite others to collaborate via email.
- Edit maps by all the users at the same time and chat with them.

- Assign tasks to users by providing deadlines, priority levels, and dates.
- Revert to older versions of the mind map.
- Manage, print, and download maps as pdf, svg, rtf, *etc.*
- Offline software which can then be synced online.

Web Strategy

- Web site homepage provides an introduction video.
- Allows a demo without signing in.
- Web site shows how the product can be used for various needs (business and education).
- Allows download of offline version as a software.

Mobile Strategy

- Not available for mobile phones.

User-friendliness

- Demo shows step-by-step tutorial.
- Easy to understand and use interface.
- Provides a free version of the service.

Visual Analysis

- White background with horizontal tools on the top.
- Tools categorized as basic, share and collaborate, and advanced.
- Allows text formatting to make it colorful.

Tips

- None found.

Social Networking

- Not active on Facebook or twitter.
- Blog seemed to be dormant for months.

Primary Competitors

- MindMeister.
- Seavus Dropmind.
- MAPMYself.
- Mind42.

7.6.8   SWOT Analysis

Strengths

- Users can edit the maps remotely at the same time.
- Offline software can work when no Internet access exists.

- Export maps as pdfs, rtfs, *etc.* onto the computer.
- Chat among the collaborators.

Weaknesses

- Weak task management.
- Mapping only left to right. No vertical mapping.
- No personal note-taking.
- Retreating older version is complicated.

Opportunities

- Better task management system.
- Tips and advice for better brainstorming.
- Better version system.
- Better map management system and UI.

Threats

- No vertical mind mapping. Only left to right mapping.
- Not available for mobile phones and tablets.

7.6.8.1 MindMeister

Who Were They in 2012?

- MindMeister is published by MeisterLabs, Inc., and was first available to customers in late 2007.
- In March 2010, MindMeister debuted as one of the original partners selected by Google to take part in the launch of the Google Apps marketplace.

What Was It in 2012?

- Online collaborative mind-mapping and brainstorming tool with more than a million users.
- Available for Web, iPhone, iPad and Android.
- Used by some big companies like SAP, EA Games, Oracle, Philips, *etc*.

What Did the App Let Users Do in 2012?

- Create mind maps by forming topics and subtopics and making connections with branches.
- Collaborate, chat, and share maps with other users and friends.
- View timeline and revision history of changes made to the map.
- Customize the theme for the mind maps.
- Drag and drop additional files and images onto a topic on the map.
- Showcase the map remotely in form of a presentation using Presentation Mode.

Web Strategy

- Homepage has a small introductory video which shows how the service works.
- User can sign up after selecting the right plan for him (personal, pro, or business) using Gmail, Facebook, and Open Id single sign on.
- Allows user to make a free account easily, but the button to do that is not easily visible.
- The homepage changes to maps manage page on signing in.

Mobile Strategy

- Available for iOS and Android for free.
- User signs in with his account credentials and maps are synced.

User-friendliness

- Video tutorials and feature explanation available on Vimeo.
- No demo available before signing in or before starting.

Visual Analysis

- White background with blue base color.
- Maps are colorful and can be customized (Fig. 7.7).

Tips

- Available for iOS and Android for free.
- User signs in with his account credentials and maps are synced.

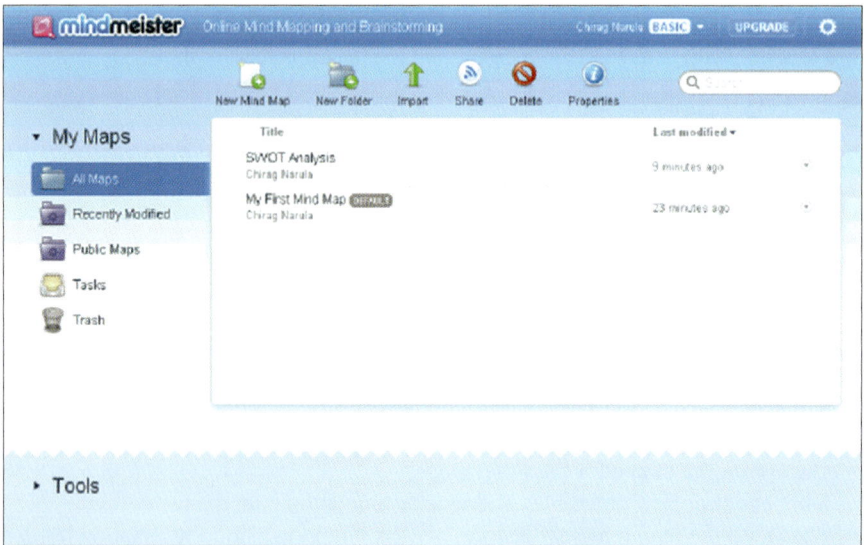

Fig. 7.7 Example screenshot of the MindMeister application screen in 2012 (Image: MeisterLabs, Inc.)

Social Networking

- Active on Facebook and Twitter.
- Active blog.

Primary Competitors

- Comapping.
- MAPMYself.
- Mind42.

7.6.9   SWOT Analysis

Strengths

- Clean user interface that allows horizontal mapping on both sides of a topic.
- Live collaboration where collaborators can work on the map concurrently.
- History revision showing changes made by each collaborator.
- Drag and drop attachments and images onto maps.
- Available for phones.
- Presentation Mode for presenting the maps.
- Maps management.
- Customization of the maps.

Weaknesses

- Trial period only after giving out credit card details. No demo available.
- No task management system.
- Downloading of map not possible.
- Data flow graph not possible (no vertical mapping).

Opportunities

- Feedback system implementation where viewer can comment on the map but not edit it.
- Vertical mapping.
- Task management system and project management system.

Threats

- No vertical mapping.
- No task management.
- No network management.

7.6.10   Project Management Applications

7.6.10.1   LiquidPlanner

Who Were They in 2012?

- LiquidPlanner was a project management software company based in Washington.
- The firm was founded in 2006 and launched its first beta release to the public in 2008.
- LiquidPlanner was founded by Charles Seybold and Jason Carlson, both former senior managers of Expedia.
- LiquidPlanner claims to be the first software as a service (SaaS)-based project management solution to allow users to express uncertainty in their task estimates using ranges.

What Was It in 2012?

- Mobile application available for iPads, iPhones, and Android.
- Project management app to help users keep up with their team, set timelines and goals, and track their progress.

What Did It Let Users Do in 2012?

- Manage projects and tasks.
- Collaborate and communicate with the team.
- See recent changes, task status, progress in a week, next tasks, and feed on the homepage.
- Create tasks, set deadlines and task status (at risk, active, warning), and view them in form of a calendar.
- Create weekly timesheet of the work every day.
- Make a weekly timesheet of the work every day for every task and submit to manager.
- Add new projects and project folders and organize tasks. LiquidPlanner schedules the tasks for the user by itself considering the priorities and work hours of each task and shows in form of a fuzzy logic graph.
- View all tasks and details of the tasks, comment on them, attach a file, and go through the workload (with the given tasks) and remaining work in form of a graph.
- Comment, post picture, attach file, and "watch" or follow a task.

Web Strategy

- Homepage gives a detailed tour of the service.
- Allows user to sign up for free for 30 days and start working after.
- User gets to the workspace after signing in, but homepage remains the same.

Mobile Strategy

- Available for iPhone, iPad, and Android.
- Apps could be downloaded for free and user can then sign in.

- iOS version requires iOS 3.2 or more.
- Customer rating was 2.5.

User-friendliness

- Demo projects on signing up to understand the system better.
- Help guide available to help through the steps.
- Training video available to understand the tools.
- Scheduling is complex and hard to use even after going through the tutorial.
- "Explain" links available in the service which then takes the user to the tutorials related.

Visual Analysis

- Web site uses conventional horizontal tab system (Home, Projects, Tasks, Timesheet, and Settings).
- Tab system (Home, Comments, Projects, Tasks, Recent) for mobile applications uses black and blue basic colors (Fig. 7.8).

Tips

- Available for iPhone, iPad, and Android.
- Apps can be downloaded for free and user can then sign in.

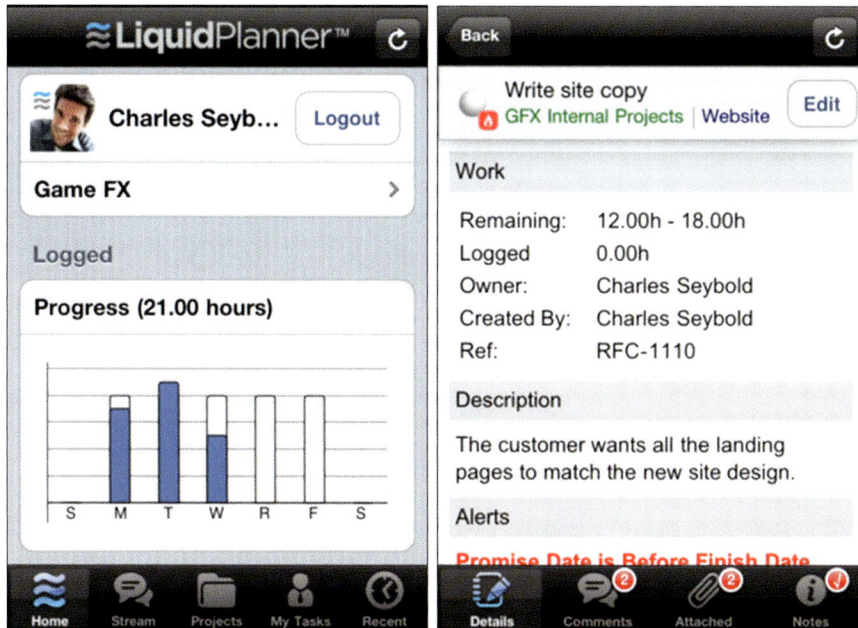

Fig. 7.8 Example screenshots LiquidPlanner application screen in 2012 (Images: LiquidPlanner, Inc.)

- iOS version requires iOS 3.2 or more.
- Customer rating was 2.5.

Social Networking

- 1000+ followers on twitter. Active.
- Active on Facebook with not many (450) followers.
- Active blog.

7.6.11   SWOT Analysis

Strengths

- Covers entire lifecycle of a project and a task.
- Available for iOS and web.
- Auto scheduling of tasks.
- Categorization of tasks.
- Stream communication among the members.
- 30 days trial available.

Weaknesses

- Web version is too complex and tough to understand initially.
- Assigning of tasks unclear.

Opportunities

- Better way to visualize the tasks and schedule.
- Remote team meeting or brainstorming sessions.
- Connections on the basis of projects; better networking.

Threats

- The software provides all the necessary tools for project management but shows no process.
- Feels complicated to use.
- No personal chat among the teammates.

7.6.12   Results of Competitive Analysis

From our investigations of roughly 10 brainstorming and idea management applications for mobile, Web, and desktop platforms, including those cited above, AM+A concluded that usable, useful, and appealing UX design must include incentives in order to prompt behavior change.

The proposed Innovation Machine needs to incorporate persuasion theory and provide better incentives. Persuasion theory is a discipline that has evolved over the years in relation to marketing and technology developments.

This powerful combination motivates users to change their behavior in the short and long term, thereby achieving more efficient and effective innovation processes. A good innovation management application for mobile devices such as smartphones and tablets should help users set goals, provide dynamic charts and illustrations, host competitions, and provide step-by-step instructions to motivate behavior change.

Extensive, up-to-date, searchable databases are another priority. These databases must be sufficiently user-friendly and flexible for users and their network of coworkers and professional acquaintances to add and update information easily. This adaptability is critical to maintaining and increasing usage and will inevitably provide a competitive advantage for the Innovation Machine.

The Innovation Machine must also encourage and strengthen team-oriented behavior change. Studies of persuasion theory show that social comparison is a superior incentive for behavior change in general. Cooperation and competition within and among teams can encourage greater restraint and behavior change. Virtual rewards provide strong motivation, and real financial rewards of $500 can prompt a significant change [*WSJ*, June 2010] in behavior.

Last but not least, the Innovation Machine should be fun to use. Gamification will serve as further incentive for users to learn about selecting wise investment/ expenditure combinations, controlling risk efficiently and effectively, among other fiscal techniques. The Innovation Machine should allow users to share these experiences with friends, family members, and the world, primarily through Facebook, Twitter, and blogs.

Based on these concepts and available research documents, we have proposed and are in ongoing development of conceptual designs for the multiple functions of the AM+A Innovation Machine. Subsequent evaluation will help us improve the metaphors, mental model, navigation, interaction, and appearance of all functions and data in the Innovation Machine's user interface. The resulting improved user experience will move the Innovation Machine closer to being a commercially viable product and service.

A well-designed Innovation Machine will be more usable, useful, and appealing to users in comparison with current products/services, which tend to be too narrowly focused, especially those that are entering the innovation space for the first time. Our objective is to provide a set of functions in an application that can reliably stimulate people towards being more effective and efficient innovators, with consequent benefits not only to their own company but also to themselves.

7.7 Persuasion Theory

As mentioned in Chap. 1, and in alignment with Fogg's persuasion theory (Fogg and Eckles 2007), we defined five key processes to create behavioral change via the Innovation Machine's functions and data:

- Increase frequency of using application.
- Motivate changing some living habits.
- Teach how to change living habits.
- Persuade users to plan short-term change.
- Persuade users to plan long-term change.

Each step has requirements for the application. Motivation is a need, want, interest, or desire that propels someone in a certain direction. Humans' sociobiological instinct is to compete with others for social status and success. We apply this theory in the Innovation Machine by prompting users to understand that every action in the process of innovation has the potential to have enormous consequence on their success.

We also drew from Maslow's *A Theory of Human Motivation* (Maslow 1943), which he based on his analysis of fundamental human needs. We adapted these to the Innovation Machine context:

- The *safety and security need* is met by the possibility to visualize the amount of progress made in innovation projects.
- The *belonging and love need* is expressed through friends, family, and social sharing and support.
- The *esteem need* is satisfied by social comparisons that display success in innovation activity, as well as by self-challenges that display goal accomplishment.
- The *self-actualization need* is fulfilled by the ability to visualize improvement and progress in innovation.

7.8   Information Architecture

7.8.1   Initial Concept Map

Based on all previous analysis, we laid out an initial concept map of the Innovation Machine. Its parts are discussed below, together with the impact on the information architecture and the persuasion process (Fig. 7.9).

7.8.2   Persuasion Process

7.8.2.1   Increase Use Frequency

Games and rewards are among the most common methods to increase use frequency. We developed several innovation game concepts for the Innovation Machine. In terms of rewards, the users might be awarded both virtual rewards (such as virtual badges and reputation titles) and real money rewards, funded by the companies that host innovation competitions to seek good ideas.

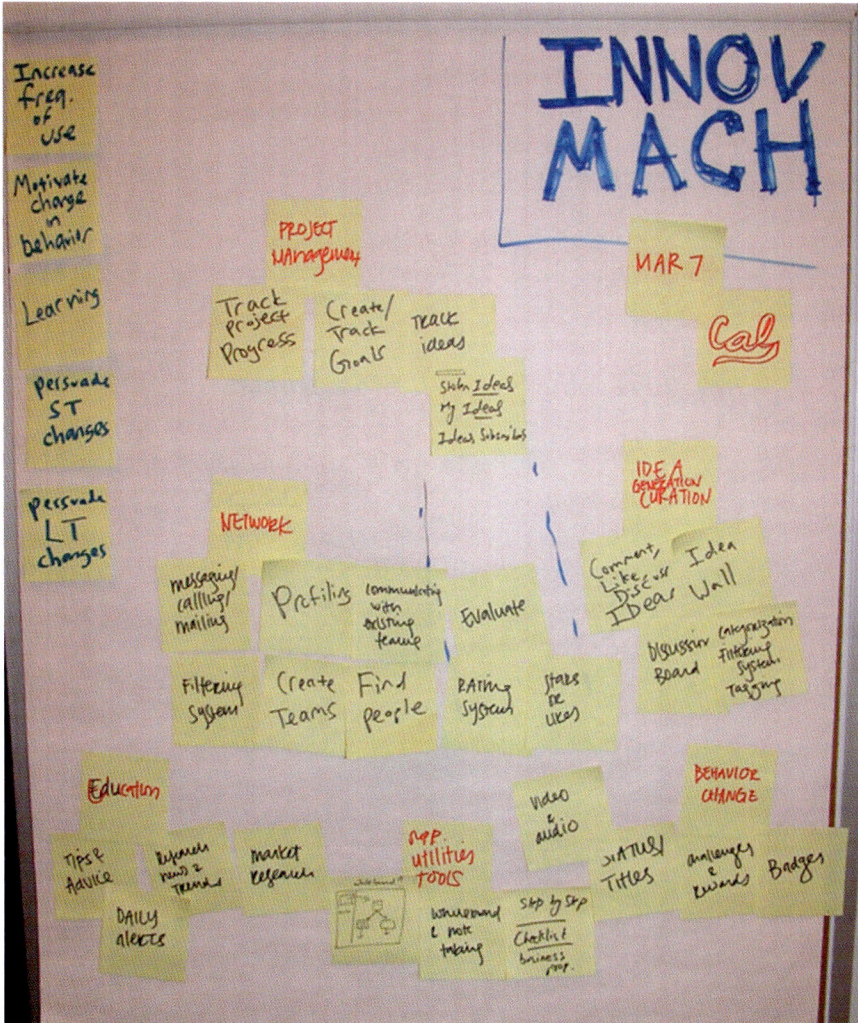

Fig. 7.9 Initial concept map based on previous analysis

7.8.2.2 Increase Motivation

In the Innovation Machine, we set the users' potential to develop new and innovative products and services as an important incentive for their behavior change. Being able to view their progress throughout a structured innovation process gives the user an understanding of important steps that are often overlooked in innovation, namely, those that involve the social and collaborative view of innovation. By providing an innovation timeline that users can follow, as well as tips and advice regarding particular steps, the Innovation Machine gives users a better sense of where they stand in the process of innovation.

In addition, Innovation Machine provides the user with tools that make these steps in the innovation process easier, such as a social network in which ideas can be discussed and teammates can be discovered. With our learning module, in which the user can find tips and advice regarding these steps, users receive suggested step-by-step plans of action in order to achieve each goal.

Gamification has also been found to be a strong motivator for behavior change. By providing small, virtual incentives for heightened activity or participation in competitions hosted by others within the Innovation Machine, users become more engaged and gain experience in innovation as a positive side effect.

7.8.3   Improve Learning

For the Innovation Machine, guiding the user through a structured process of innovation is crucial. To improve learning, the Innovation Machine integrates contextual tips on the following topics:

- Brainstorming for ideas and evaluating an idea.
- Developing an idea through discussion with others.
- Gathering resources necessary to move an idea into a project.
- Users can also choose to receive updates on latest research articles and news about innovation.

We seek to make the innovation process as explicit as possible to make it easy for users to develop their idea. Users may opt not to follow our structured process, but it is nonetheless available for users that seek the next step in developing their idea.

7.8.3.1   Information Architecture Modules

In designing the Information Architecture (IA) for the Innovation Machine, AM+A began by examining the IA for past Machines, including the Green Machine, the Health Machine, and the Money Machine (see earlier chapters). We discovered an overarching model for the IA that permeated throughout each of the past machines. In this model, there are five primary "modules" or branches of the IA, each of which is described below. While we altered slightly the details of each module to fit the needs and requirements of each respective machine, a generalized model was still evident. These modules are described below.

7.8.4   Dashboard

The Dashboard module is similar to a landing page for its respective Machine. It is an overview into the status of the users' behavior change. Here, users see a view of their goals and where they stand in achieving those goals.

7.8.4.1   Process

The Process View module is where the users get a more high-level view of the process and more details regarding each objective and goal. Users see the progress being made, as well as the next steps in achieving a particular goal.

7.8.4.2   Social Network

The Social Network module is an integral part of behavior change in modern software. Users engage in focused, subject matter-based connections with friends, family, and/or like-minded people that either share similar goals or wish to support others in achieving behavior-change objectives.

7.8.4.3   Tips/Advice

The Tips/Advice module provides focused knowledge about a given topic to give users insight into the habits they wish to either get rid of or adopt.

7.8.4.4   Incentives

The Incentives module presents users with a fun and engaging way to change their behavior. Gamification has proven to be a powerful tool in adopting users to try an application, even with virtual incentives; although in some cases, real incentives are provided. In addition, a leaderboard allows a user to compare his/her progress with others, tapping in the competitive nature of the human mind to create behavior change.

7.8.5   Innovation Machine Information Architecture

AM+A took the basic structure of the past Machines in creating the Innovation Machine information architecture. However, due to particular complexities that were unique to the Innovation Machine, we made the following primary changes:

- The Social Network module was split into two. A primary part of the Innovation Machine is the ability to discuss ideas and potential projects within a community or social network. However, equally important is the ability to find potential team members with whom to collaborate more intimately. Because the Innovation Machine added new depth to these features, we decided to split them into the "Discuss" and "Connect" modules.

Added Settings

- Provided a different menu navigation system in order to keep the screens uncluttered. The high-level basic modules are removed from most screen displays but may be quickly called back to enable users to navigate high-level modules.

7.8.6   Initial Diagram of the Information Architecture

The following is an initial diagram of the information architecture for the Innovation Machine (Fig. 7.10).

7.8.7   Information Architecture Components

The following describe components of the information architecture shown in Fig. 7.17 in greater detail.

7.8.8   Dashboard

My Profile

- Displays an overview of user's account information, including user's photograph, occupation, rankings and points, and social networking platforms.

Overview of Activity

- Displays user's recent innovations, including the user's innovation status (idea generation, discussion management, and project implementation). Also displays user's current challenges.

7.8.9   Innovate

Innovations

- Allows users to create private or public innovations.
- Allows users to take notes (private/collaborative, audio/visual/textual) and categorize those notes within the three steps to innovation: ideas (idea generation), discussions (discussion management), and projects (project implementation).
- Displays a list of all the user's innovations and their corresponding innovation status (ideas, discussions, projects).

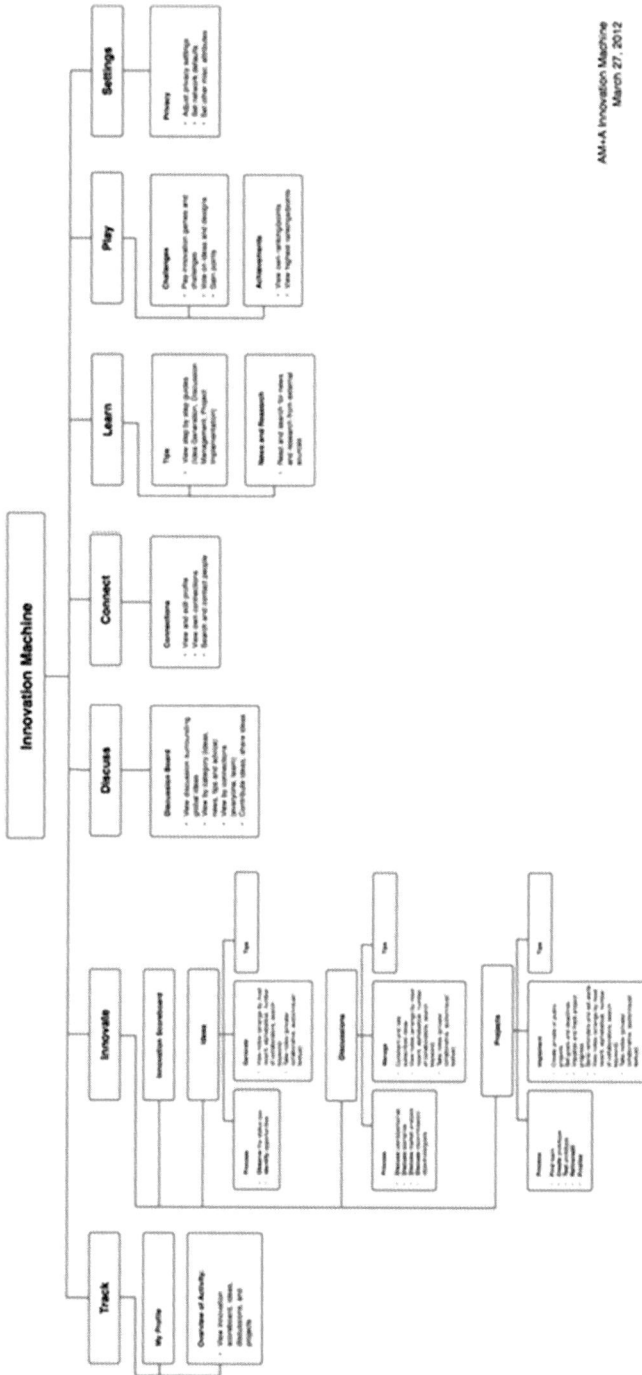

Fig. 7.10 Diagram of the Innovation Machine's information architecture

- By default, the innovations are listed from the most recently edited, the most recently updated, to the least recently updated (top to bottom). The innovations can also be sorted by keywords, format, and other filters.
- Each innovation displays a detailed description of the innovation, the innovation status bar, and the most recently updated ideas, discussions, and projects.

Innovation Status

- Displays a visual representation of user's innovation status according to the three steps to innovation: ideas (idea generation), discussions (discussion management), and projects (project implementation).
- Details the step-by-step process within each step. Includes links to tips and games.

7.8.10   Connect

Profile

- Displays user's photo, name, education, and relevant experience. Also displays user's innovator status and links to social networking platforms.

Network

- Displays list of contacts in user's network.
- Allows search according to contact's name or innovation group.

Activity

- Displays recent activity relevant to user including connection requests and messages from other users.

7.8.11   Discuss

Discussion Lists

- Displays lists of most popular discussions, most recent discussions, or discussion subscriptions. User can view number of comments and likes for each discussion. User can also subscribe to discussions.

Discussion

- User can create discussions that are private (specific networks) or public (anyone).
- Discussions include specific prompts or questions. They can also include details revealing the context of the prompt or question.
- Users can like, subscribe to, or comment on discussions.

- Users can also contact or collaborate with discussion owners.
- Discussions will aggregate discussion data and display comment highlights.

Learn

- Displays lists of technology, innovation, business, and design articles and sources. Lists are sorted according to user's subscriptions, top sources, and top articles.
- User can sort through sources and articles using keywords and other filters.
- User can subscribe to sources and save articles.
- User can read recent and trending articles, and user can share articles.

7.8.12   Play

Play Profile

- Displays user's photo, innovator status, and points.

Challenges

- Displays user's current challenges including relevant number of entries.
- Allows user to start new challenges. Challenges may be posted by other users or companies.

Leaderboard

- Displays top innovators, their photos, status, and points. User can also view the profiles of top innovators.

7.8.13   Settings

The following are specific adjustments to Settings:

- Allows user to adjust privacy settings. Specific settings may include visibility of user's profile and posts.
- Set network defaults, whether user posts are automatically global, private, or within a specific network.
- Enables user to set notification settings. Notifications are shown in email and on the landing page. They are less urgent than alerts, and the user checks them on their own time.
- Enables user to set alert settings. Alerts are important to know about right away and are received via text or through red exclamation points on top of the Innovation Machine application icon. Examples of alerts are upcoming deadlines, deadlines met, and contact requested.

7.9   Initial Designs

7.9.1   Initial Sketches

Based on the information architecture, AM+A prepared initial concept sketches of some key screens of the Innovation Machine. These "low-fi" "paper" prototypes were used to elicit feedback from potential users, prior to the development of mockup screens of higher fidelity (Fig. 7.11).

Fig. 7.11   Initial Innovation Machine screen-design mockups and conceptual designs

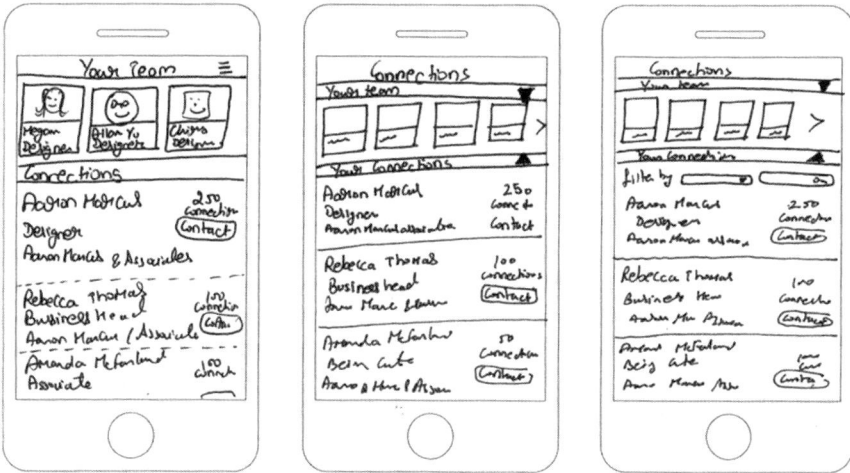

Fig. 7.11 (continued)

7.9.2   Dashboard v.1.0 Screen Design

In the Dashboard, users can view an overview of their profile as well as an overview of their most recent activity. Their activity is sorted into four tabs: All, Ideas, Discussions, and Projects. This set reinforces what AM+A has identified as the key components of the Innovation process: idea generation, discussion management, and project implementation. The user will be able to view two priority ventures within each category. These priorities are user-defined and can be adjusted in settings. Each user-created venture will also be accompanied by a status bar, revealing the user's innovation progress (Fig. 7.12).

7.9.3   Innovate v.1.0 Screen Designs

The Innovate module allows users to view more comprehensive information on each of their ventures. Whereas in Dashboard, users can only view their innovation progress within their current step (idea generation, discussion management, and project implementation), in Innovate, users can view their entire Innovation Status. In this section, users can view the tasks needed to complete each innovation step. Once they have completed each step, they can mark that it has been completed and their Innovation Status bar will be updated (Fig. 7.13).

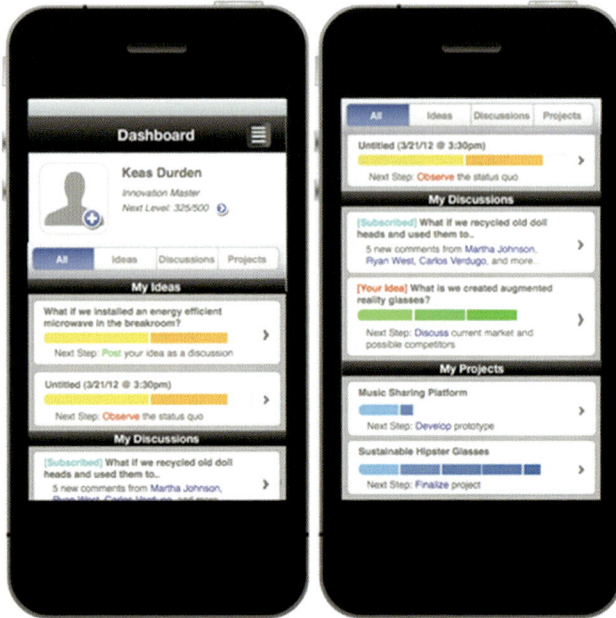

Fig. 7.12 Innovation Machine Dashboard v.1.0 screen designs

Fig. 7.13 Innovation Machine Innovate v.1 screen designs

7.9.4   Discuss v.1.0 Screen Design

In the Discuss module, users can connect with a larger innovation community through a discussion board. Here, users can post ideas for new discussions and comment on existing discussions within private, local, or global networks. Users can also subscribe to other users' discussions and follow these ideas within their own Dashboard (Fig. 7.14).

7.9.5   Connect v.1.0 Screen Designs

The Connect module allows users to manage their personal innovation network. Users can customize their personal profiles including education, experience, skills, and interests. Users can search for and contact other individuals of specific skills, experiences, and interests. Connections can be categorized into user-specified groups (Fig. 7.15).

7.9.6   Learn v.1.0 Screen Designs

In the Learn module, users can learn from recent activity in popular blogs related to innovation. The Innovation Machine aggregates all sources into this one module. Users can view top sources and top articles from these sources and can subscribe to

Fig. 7.14 Innovation
Machine Discuss v.1.0
screen design

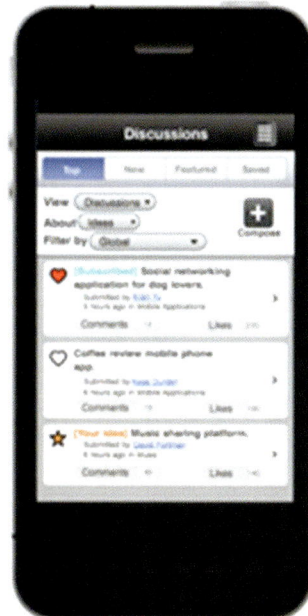

Fig. 7.15 Innovation Machine Connect v.1.0 screen designs

favorite sources. Icons represent each source. Articles can include tips or current events in the world of innovation. Within each source, users can view articles and authors and can save their favorite articles. Users can filter by recent, trending, and saved. They can connect with sources via Twitter, Facebook, and Google Plus (Fig. 7.16).

7.9.7   Play v.1.0 Screen Designs

In the Play module, users can track their point totals and innovation ranks. Here, they can view their most recently obtained points, their current challenges, and completed challenges. They can also search for new point opportunities and new challenges. Points are earned through methods that show the user is innovating (generating ideas, managing discussions, implementing projects) or contributing to others' innovations (liking ideas, contributing to discussions). Finally, they can compare their innovation rank to the innovation leaderboard (Fig. 7.17).

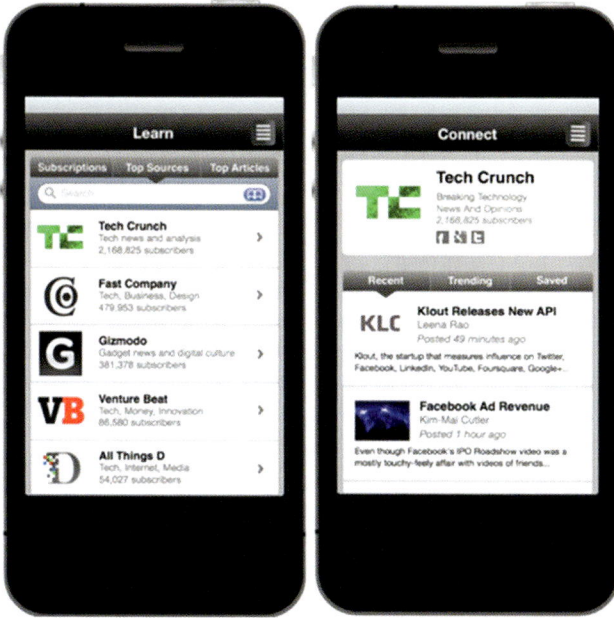

Fig. 7.16 Innovation Machine Learn v.1.0 screen designs

Fig. 7.17 Innovation Machine Play v.1.0 screen designs

7.9.8   Evaluation

AM+A evaluated the first round of designs by presenting a PowerPoint presentation with the concept, objectives, and initial screen designs with a small group of industry professionals. Although there was a consensus that the ideas and concepts behind the Innovation Machine were indeed novel, there was a general concern about the designs being too complicated, confusing, and lacking sufficient visual stimuli. With this feedback, AM+A sought to make the second round of mockups less textual and more visual by removing unnecessary words and adding appropriate icons. Overall, we agreed that there was a general lack of photos and icons and aimed to add more visual representations where possible. These objectives are represented in the screens below.

7.10   Revised Designs

With the comments drawn from the evaluation phase, AM+A aimed to make the Innovation Machine user experience more visual. In doing so, we found it necessary to add new screens to fill the gaps of certain flows and functions that were often missed in the one to two screens provided in our initial designs. In addition, we endeavored to ensure additional consistency in the design among the icons, symbols, labels, colors, and grid. With new design layouts and theme, as well as the additional mockups, we hoped to improve a better understanding of the interaction and general enhancement of the user experience of the Innovation Machine.

7.10.1   Dashboard v.2.0 Screen Design

Main Changes to the Dashboard 2.0 Module
- Added overview of user's profile with buttons to Facebook, Twitter, Google+, and messaging (Fig. 7.18).
- Main body shows two of the most recently updated innovations with a button with two arrows that lead the user to the full list of the user's innovations.
- Added section with "current challenges" that include the two most recently updated challenges, with a button below that leads user to full list of current challenges.
- Added icons for each innovations and challenges, as well as time-stamped them with when they were updated.
- Eliminated textual descriptions from each innovation.

Fig. 7.18 Innovation
Machine Dashboard v.2.0
screen design

7.10.2   Innovate v.2.0 Screen Designs (Fig. 7.19)

Main Changes to the Innovate 2.0 Module

- Added Innovate landing page, which displays list of all innovations. Each inno-vation has a corresponding icon, as well as a visible timeline that displays prog-ress for that innovation.
- Within an innovation screen, added an "overview" section at the top that also acts as a button to a more detailed overview.
- Aggregated content in main body to single list that is sortable by tab buttons: All, Ideas, Discussions, and Projects.
- Added screen for "Innovation Status." User can view status and upcoming steps in their innovation process. Can collapse or expand sections: Ideas, Discussions, and Project task checklists. Each task includes a checkbox, info (i) button, and gamification (puzzle) button. Info button leads user to tips and advice regarding that step in the checklist, while the puzzle button leads user to games and quizzes regarding that step (*i.e.,* quiz about team templates for the "find a team" step).

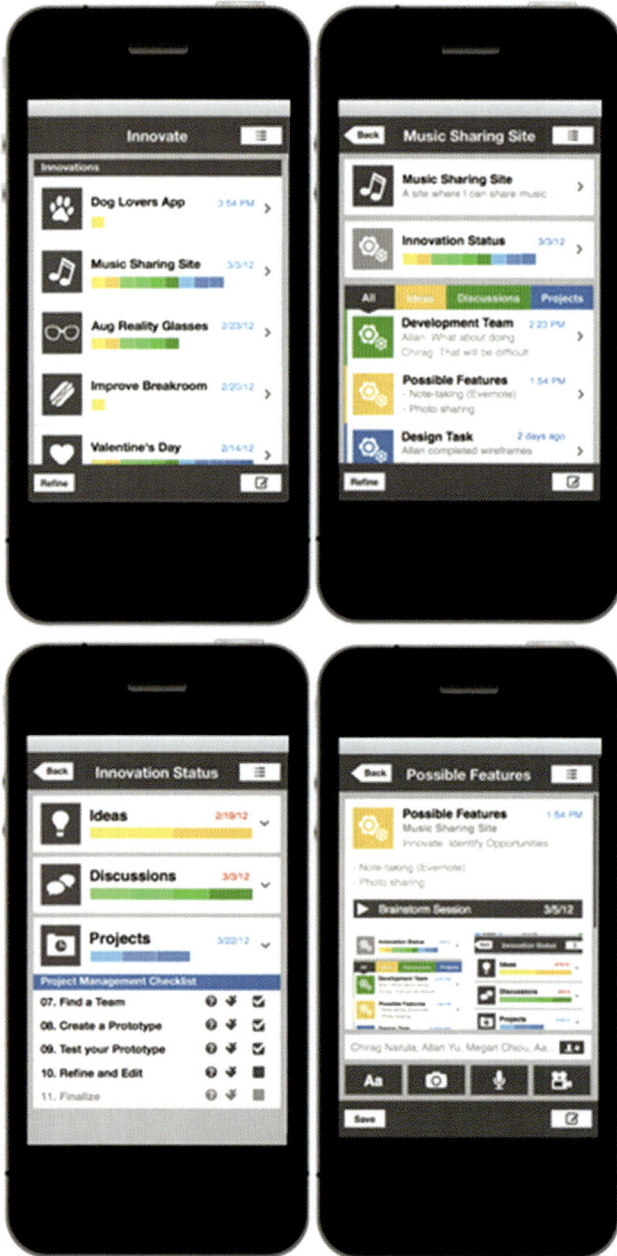

Fig. 7.19 Innovation Machine Innovate v.2.0 screen designs

7.10.3   Connect v.2.0 Screen Designs (Fig. 7.20)

Main Changes to Connect 2.0 Module

- Decreased size of photo to create an overarching grid, consistent across all interfaces.
- Added icons to items under "Education" and "Experience" sections.
- Added "Contact," "Add Connection," and "Innovation Reputation" buttons to profile section.
- Combined all alerts to appear in "Activity".
- Added a "Refine" button to list of contacts within user's network for filtering functionality.

7.10.4   Discuss v.2.0 Screen Designs (Fig. 7.21)

Main Changes to the Discussion 2.0 Module

- Changed tab button labels for filtering discussions from "Top," "New," "Featured," and "Saved" to "Most Popular," "Most Recent," and "Subscriptions."
- Decreased size of filtering system by removing drop-down menus and adding a search bar for quick filtering, as well as a "Refine" button.
- Eliminated "favorite" button, added icons for each discussion.

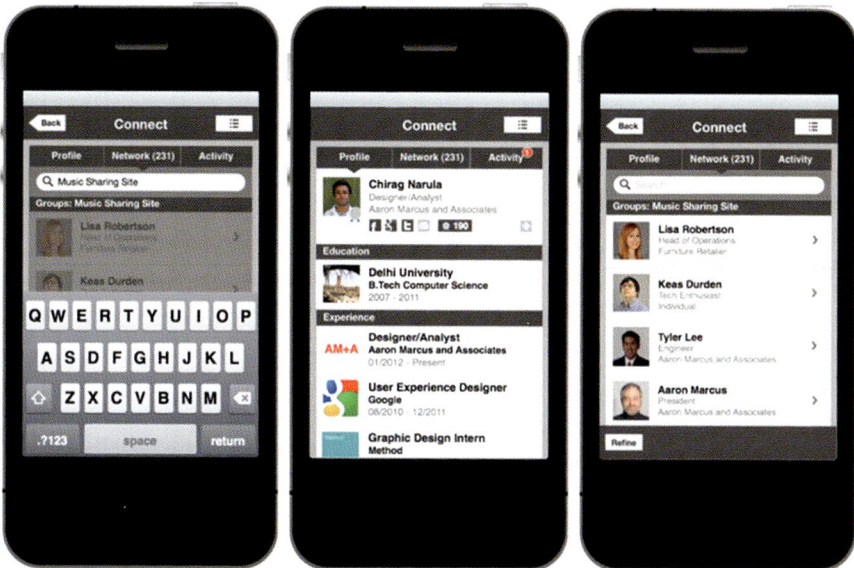

Fig. 7.20 Innovation Machine Connect v.2.0 screen designs

Fig. 7.21   Innovation Machine Discuss v.2.0 screen designs

- Words "Comments" and "Likes" were replaced with icons. Discussions that are liked or subscribed to by the user are marked by a depressed, white button, rather than gray, indicating that the discussion has either been liked or subscribed by the user.
- In a discussions page, the question is featured at the top, rather than the subject. An abridged version of the description is displayed, rather than the full version.
- Eliminated the collaboration opportunity message, integrated with the "Collaboration" button. When host of discussion is not looking for collaborators, the "Collaborate" button is grayed out.
- Aligned the "like," "message," "subscribe," and "collaborate" buttons.
- In a discussions page, added section for "Discussion Highlights" that displays most commonly used phrases/words among all comments. Button at the bottom of this section leads user to view all comments.
- Eliminated compose button on a discussion page. User must view all comments to access the compose button, so that user is somewhat accountable for reading some of the comments before posting an identical comment that someone else has already contributed.

7.10.5   *Learn v.2.0 Screen Designs (Fig. 7.22)*

Main Changes to Learn 2.0 Module

- Added refine button to both the "Learn" landing page and individual source page for filtering the list of article sources or articles.
- Added "Subscribe to source" button for the "Learn" landing pages.

7.10.6   *Play v.2.0 Screen Designs*

Main Changes to Play 2.0 Module

- Added a "Leaderboard" tab button, along with "My Play" and "New Play" (Fig. 7.23).
- Added medals to represent users' innovation reputation. (*i.e.,* Gold medal = Innovation Guru, silver medal = Innovation Master, bronze medal = Innovation novice.
- Moved "Recent Points" section to "Personal Points Overview".
- Displayed two current challenges, with button leading to "All Current Challenges".

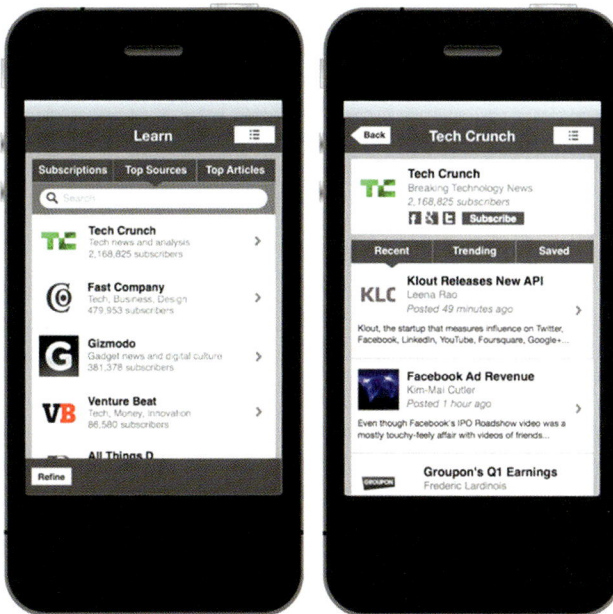

Fig. 7.22 Innovation Machine Learn v.2.0 screen designs

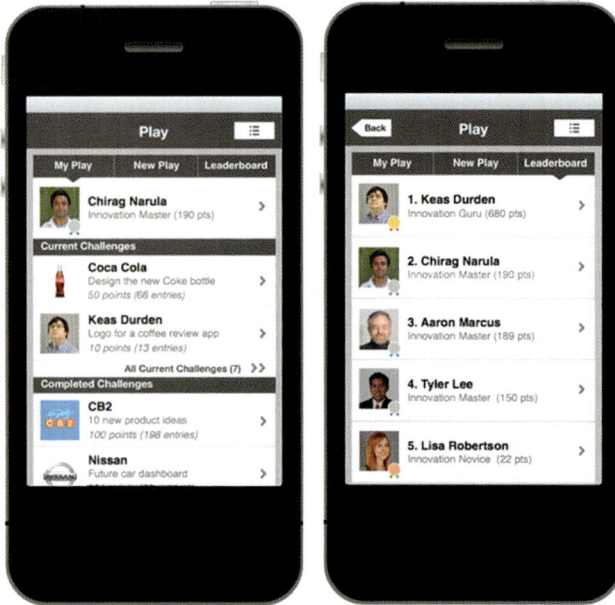

Fig. 7.23 Innovation Machine Play v.2.0 screen designs

- Displayed two completed challenges, with button leading to "All Completed Challenges".
- In leaderboard interface, added photos and medals to each leader, and corresponding "Innovation reputation" with the point values.

7.11 Conclusion

Using the user-centered design approach described above, AM+A plans to continue to improve the complete Innovation Machine development process, which would require significant time and funding from an outside source. Tasks include the following:

- Revising personas and use scenarios.
- Conducting user evaluations.
- Revising information architecture and look and feel.
- Building initial working prototype (*e.g.*, for iPhone or other tools and platforms).
- Redesigning, if appropriate, the Innovation Machine for different corporate and country cultures.
- Evaluating the Innovation Machine across different corporate and country cultures.

- Developing the Innovation Machine for enterprise use as well as personal use.
- Researching and developing improved information visualizations.

AM+A sought to incorporate information design and persuasion theory for behavior change into a mobile phone application. This self-funded work on the Innovation Machine project is current and ongoing and was undertaken to demonstrate the direction and process for such products and services. Though the design is incomplete, AM+A is willing to share the approach and lessons learned, in the interest of helping to solve the challenges of innovation.

At this stage, AM+A is seeking to persuade other design, education, and corporate groups to consider similar development objectives. We hope that our process and concept prototypes (the self-funding of which inevitably limited the amount of research, design, and evaluation) will inspire others and that they benefit from the materials provided thus far.

The process has already been demonstrated successfully with a previous project, the Green Machine (Marcus and Jean 2009), versions of which have been considered and used by SAP for enterprise software development (Marcus et al. 2011).

AM+A's long-term objective for the Innovation Machine is to create a functional working prototype to test whether the application can actually persuade people with innovation management challenges to improve the efficiency and effectiveness of innovation on the individual, team, and corporate level. If the theories are proven to be correct, this approach could create numerous new innovative products and services that will benefit many people in many different companies, organizations, countries, and cultures.

Acknowledgements The author thanks these AM+A Associates, all Designer/Analysts, for their significant assistance in planning, research, design, analysis, and documentation for this project.

- Ms. Megan Chiou.
- Mr. Chirag Narula.
- Mr. Allan Yu.

Further Reading

Allen E (2011) Designer founders: Stories by designers of tech startups. Retrieved from http://designerfounders.com

Boehret K (2011) A textbook case of iPad fun with studying. Wall Street J, February 8. Retrieved from http://online.wsj.com/article/SB10001424052970204369404577209142710109710.html

Cagan J, Vogel CM (2002) Creating breakthrough products: innovation from product planning to program approval. Prentice Hall PTR, Upper Saddle River

Carlson CR, Wilmot WW (2006) Innovation: the five disciplines for creating what customers want. Crown Business, New York

Corliss R. Learning linkedIn from the experts: how to build a powerful business presence on LinkedIn. Hubspot. Retrieved from http://www.hubspot.com/eBooks/learning-linkedin-from-the-experts/

Egger FN (2003) From interactions to transaction: designing the trust experience for business-to-consumer electronic commerce. Eindhoven University of Technology, Eindhoven

Gaffney G, Szuc D (2006) The usability kit. SitePoint, Collingwood (206)

Gasperini, Jim, & Marcus, Aaron (2011) *Social Media in The Enterprise.* Private unpublished AM+A publication

Gertner, Jon. (2012). True Innovation. The New York Times, February 25. Retrieved from http://www.nytimes.com/2012/02/26/opinion/sunday/innovation-and-the-bell-labs-miracle.html?pagewanted=all

Greenemeier L (2012) Innovation: professional seer. Sci Am 306 (5):80–83

Hartson R, Pyla PS (2012) The UX book: process and guidelines for ensuring a quality user experience. Elsevier, Waltham

Healey M (2012) Why employees don't like social apps. Information Week, January 30. Retrieved from http://informationweek.com/1322/social

Hof RD (2011) Building an idea factory. Business Week, October

Hotz RL (2012) When gaming is good for you. Wall Street J, March 5. Retrieved from http://online.wsj.com/article/SB10001424052970203458604577263273943183932.html

Human Factors International. Contextual innovation & design research services. Retrieved from http://www.humanfactors.com/downloads/documents/Contextual_Innovation.pdf

Jantsch J (2010) The referral engine: taching your business to market itself. Portfolio, New York

Jennings R (2012) Taiwanese tech start-ups open to new ideas. South China Morning Post, February 11

Jui S-L (2007) From "Made in China" to "Invented in China.". Publishing House of Electronics Industry, Beijing

Jui S-L (2010) Innovation in China: the Chinese software industry. Routledge, New York

Kolko J (2012) Wicked problems: problems worth solving. Austin Center for Design, Austin

Kuczmarski TD (1992) Managing new products: the power of innovation, 2nd edn. Book-Ends, Chicago

Lehrer J (2012) Groupthink: the brainstorming myth. The New Yorker, January 30. Retrieved from http://www.newyorker.com/reporting/2012/01/30/120130fa_fact_lehrer

Lehrer J (2012) How to be creative. Wall Street J, March 9. Retrieved from http://online.wsj.com/article/SB10001424052970203337060457726563220501586.html

Lin M (2012) Follow the flow: using mind-mapping to capture user feedback. UPA User Exp Mag 11:1

Lohr S (2012) The Yin and the Yang of the corporate innovation. The New York Times, January 26. Retrieved from http://www.nytimes.com/2012/01/27/technology/apple-and-google-as-creative-archetypes.html

Marcus A et al (2013) The travel machine. In: Marcus A (Ed.) (2013) Proceedings, design, user experience, and usabilty conference, Las Vegas, NV, 20–25 July 2013, pp 696–704. Springer, London

Marcus A, Jean J (2010) The green machine. Inf Des J 17(3, First Quarter):233–243

Martin C (2012a) Big man's $500 bet offers motivation for weight loss. Denver Post, 8 December 2012. Retrieved 8 December 2012. http://www.denverpost.com/styleheadlines/ci_20310588#ixzz2EUBYr2a6

Martin L (2012b) Motivating users: how to integrate persuasion into design. UPA User Exp Mag 11(1):22–25

Program in Open Innovation. Retrieved from http://openinnovation.berkeley.edu/reDesign. Retrieved from http://redesignresearch.com/

Quesenbery W, Brooks K (2010) Storytelling for user experience: crafting stories for better design. Rosenfeld Media, Brooklyn

Salvador T (2010) Session 3: Obstacles and opportunities along "The Way": Makiko Taniguchi, Curator: Heroic Complexity in Strategic Innovation. In: Ethnographic Praxis in industry conference proceedings, 2010:169–178

Shneiderman B (ed) (1993) Sparks of innovation in human-computer interaction. Ablex Publishing Corporation, Norwood

Silverman RE (2012) Doodling for dollars: firms try to get Gadget-Obsessed workers to look up – and Sketch ideas. Wall Street J, April 24. Retrieved from http://online.wsj.com/article/SB1000 1424052702303978104577362402264009714.html

Spender JC, Strong B (2010) Who has innovative ideas? Employees. Wall Street J, August 22. Retrieved from http://online.wsj.com/article/SB10001424052748704100604575146083310500518.html

References

Chesbrough H (2005) Open innovation: the new imperative for creating and profiting from technology. Harvard Business Review Press, Boston

Chesbrough H (2011) Open services innovation: rethinking your business to grow and compete in a new era. Josey-Bass, San Francisco

Dubberly Design Office (2007) [Concept map of innovation process]. A Model of Innovation. Retrieved from http://www.dubberly.com/concept-maps/innovation.html

Fogg BJ, Eckles D (2007) Mobile persuasion: 20 perspectives on the future of behavior change. Persuasive Technology Lab, Stanford University, Palo Alto

Jean J, Marcus A (2009) The green machine: going green at home. User Exp Mag 8(4), First Quarter 2009:20-22ff

Kelley T, Littman J (2005) The ten faces of innovation: IDEO's strategies for beating the Devil's advocate & driving creativity throughout your organization. Currency/Doubleday, New York

Marcus A (2010) Health machine. Inf Des J 19(1):69–89

Marcus A (2012a) The money machine: helping baby boomers to retire. User Exp Mag 11(2, Second Quarter):24–27

Marcus A (2012b) The story machine: a mobile app to change family story-sharing behavior. Workshop paper. In: Proceedings of CHI 2012, Austin, TX, May 2012, pp 1–4

Marcus A, Dumpert J, Wigham L (2011) User-experience for personal sustainability software: determining design philosophy and principles. In: Proceedings of design, user experience, and usability conference 2011, Orlando, FL, August 2011. Theory, methods, tools and practice. Lecture notes in computer science, vol 6769. Springer, New York, pp 172–177

Marcus A, Jean J (2009) Going green at home: the green machine. Inf Design J 17(3):233–243

Maslow A (1943) A theory of human motivation. Psychol Rev 50:370–396

Mobile/Website Competitor Applications URLs[1]

Bright Idea Innovation Suite: www.brightidea.com
CoMapping: www.comapping.com
Idea Scale: www.ideascale.com
Liquid Planner: www.liquidplanner.com
Mind Meister: www.mindmeister.com
Mindo: www.laterhorse.com/mindo/
Spigit: www.spigit.com

---

[1] For all these URLs, the retrieval date is 15 May 2012.

Chapter 8
The Driving Machine: Using Information Design and Persuasion Design to Change Driving Behavior and to Increase Safety and Sustainability

8.1 Introduction

A twenty-first-century global vehicle dashboard design challenge is to take advantage of technology to increase safety and to conserve energy. The context is this: advances in technology increase driving distractions, and global warming increases our desire to reduce our carbon footprint. In particular, the worldwide green movements have helped to increase people's awareness of sustainability issues and propelled development of innovative products to help decrease our ecological footprint.

The Driving Machine seeks to increase safe driving behavior and fuel-efficient driving by offering information, overviews, social networking, just-in-time knowledge, and incentives, including gamification, that can help to reduce, even prevent, vehicular accidents and promote more fuel-efficient driving. The question then shifts to how best to motivate, persuade, educate, and lead people to adopt safe-driving behavior and reduce their energy consumption. For our project, we researched and analyzed powerful ways to improve safe and green behavior by persuading and motivating people to become more alert drivers and to reduce their energy consumption through a vehicle dashboard application we call the "Driving Machine."

Dashboards and automotive-related applications are available to increase people's awareness of safety and the environment, but such technologies often do not focus on innovative data visualization, and they may lack persuasive effectiveness to encourage drivers to continue good driving behavior. Communicating one's carbon footprint, driving skills, and alertness helps build awareness and identity, but does not result automatically in behavioral changes. The question then becomes: How can we better motivate, persuade, educate, and lead people to become safer and more efficient drivers? Aaron Marcus and Associates, Inc. (AM+A) embarked on the conceptual design of a mobile phone-/tablet-based application, the Driving Machine, intended to address this situation.

© Springer-Verlag London 2015   429
A. Marcus, *Mobile Persuasion Design*, Human–Computer
Interaction Series, DOI 10.1007/978-1-4471-4324-6_8

The author's firm previously designed and tested similar concept prototypes that seek to change people's behavior, as described in previous chapters: The Green Machine *2009* (Jean and Marcus 2009; Marcus and Jean 2009) aimed to instill better sustainable practices, the Health Machine (Marcus, 2010) sought to encourage users to lead healthier lives, the Money Machine (Marcus 2012a) persuaded users to manage their wealth better, and the Story Machine (Marcus 2012b) enabled users to connect with their family history and community. The Driving Machine uses similar principles of combining information design/visualization with persuasion design.

The Driving Machine's objective is to combine information design and visualization with persuasion design to help users achieve their goals of driving more safely and efficiently by persuading users to adapt their driving behavior, for example, to follow traffic laws better and adopt carpooling behavior. The mechanism would be a tablet-like display either built into the car, like the Tesla, or simply provided by the user's tablet, which could be quickly and efficiently locked in place within the vehicle and later removed when departing.

AM+A applied user-centered design along with persuasive techniques to make the Driving Machine concept design application highly usable and to increase the likelihood of success in adopting new driving behavior. This chapter explains the development of the Driving Machine's user experience, user interface, information design, information visualization, and persuasion design.

8.2   Initial Discussion

As the amount of computing technology continues to increase in our cars and trucks, careful consideration must be given to dashboard design to ensure the safety and reliability of drivers, passengers, and vehicles. Increasingly states are passing laws that limit drivers' abilities to operate mobile phones or to read/send text messages while driving.

Recent research illustrates that even such laws may not go far enough, as cited by Paul Green (Green 2003), who describes how driving and using a mobile phone, regardless of having hands free or not, places drivers at greater risk of causing accidents than drivers who only talk to passengers inside their vehicles. The reason talking on the phone is a greater danger than talking to passengers is because passengers are more aware of current driving situations than people with whom one is communicating on the phone. One study by Redelmeier and Tibshirani, as cited by Green, notes that using a mobile phone increases the likelihood of a crash by up to 4.3 times versus those not using a mobile phone while driving. Estimates for distraction-related crashes in the United States typically come from a sample of about 5,000 police-reported crashes called the Crashworthiness Data System (CDS) (Green 2003). To overcome future problems that new technologies might have on driving, the National Highway Traffic Safety Administration (NHTSA) proposed a set of guidelines to test the impact of a specific task on driving performance and safety. If a task is deemed too distracting to a driver's focus based on the Visual-Manual NHTSA Driver

Distraction Guidelines for In-Vehicle Electronic Devices, NHTSA encourages auto-mobile manufacturers to prevent a driver from being able to perform the interfering task (National Highway Traffic Safety Administration 2012).

AM+A previously did research for BMW in a report titled "Future HMI Directions" (Marcus, et al 2002), in which AM+A thoroughly researched a driver-centered approach to HMI (human-machine interaction). Although the report is over 10 years old, the human factor issues are still highly relevant today as evidenced in Green's research and the NHTASA report.

In addition to safety, designing a system that encourages being environmentally conscious is an important attribute of our research. While fuel prices and the threat of global warming continue to rise, carpooling in the United States is at a very low 11 % (Johnson et al. 2010). Services such as Zimride seek to counter the low rate of carpooling by creating a social network where people can be drivers and passengers in carpools. Zimride also offers Lyft, which helps those who would normally travel alone in a taxi request on-demand ridesharing. Honda Motors developed its Ecological Drive Assist System to encourage efficient driving by supporting behavior change, in offering visual feedback via an ambient green or blue color, and also by gamifying driving behavior through the design of virtual leaves for more sustainable driving (Honda Motors 2008).

Our research shows that an innovative approach to vehicle dashboard design must account for the following: design for safety where a driver easily should be able to take a second glance at a display cluster and then refocus his/her attention on the road. Next, a display cluster must not increase the level of complexity that a driver encounters. For example, a focus on helpful rather than powerful features is important to ensure a reduction of complexity. A graphical user interface should not focus on visual complexity with an overabundance of graphics; rather, it must use graphics only if it enhances dynamic content that would otherwise be less visible. The user interface should not constrain the user to conform to a particular layout, but instead allow him/her to customize the available information present in the digi-tal dashboard. Lastly, the dashboard development must follow a user-centered design process.

8.3 User-Centered User-Experience Design

As noted in Chap. 1, the user-centered user-experience design (UCUXD) approach links the process of developing software, hardware, and user interface (UI) to the people who will use a product/service. UCUXD processes focus on users through-out the development of a product or service. The UCUXD process comprises these tasks, which sometimes occur iteratively:

Plan: Determine strategy, tactics, likely markets, stakeholders, platforms, tools, and processes.
Research: Gather and examine relevant documents, stakeholder statements.

Analyze: Identify the target market, typical users of the product, personas (characteristic users), use scenarios, competitive products.

Design: Determine general and specific design solutions, from simple concept maps, information architecture (conceptual structure or metaphors, mental models, and navigation), wireframes, look and feel (appearance and interaction details), screen sketches, and detailed screens and prototypes.

Implement: Script- or code-specific working prototypes or partial "alpha" prototypes of working versions.

Evaluate: Evaluate users, target markets, competition, the design solutions, conduct field surveys, and test the initial and later designs with the target markets.

Document: Draft white papers, user interface guidelines, specifications, and other summary documents, including marketing presentations.

AM+A carried out these tasks in the development of the Driving Machine concept design, except for implementing working versions.

To better understand the demographics and to focus on the creation of user-centered designs, AM+A incorporated the use of personas, which are described in the following section.

8.4   Personas

As noted in Chap. 1, personas are characterizations of primary user types and are intended to capture essentials of their demographics, contexts of use, behaviors, and motivations, and their impact on design solutions. Personas are also called user profiles. Typically, UI development teams define one to nine primary personas. For the Driving Machine personas, we identified three target markets: young drivers, early adopters, and elderly drivers. They were represented in the following personas.

8.4.1   Persona 1: Zoe Romero

- Age: 17.
- Race: Caucasian.
- Location: Salem, Massachusetts.
- Occupation: Student.
- Education: Some high school.
- Technology Usage: Smartphone, Windows laptop, and iPod Touch.
- Concern: Easily distracted with others in her car.
- Persona image credit: AM+A.

8.4.1.1   Textual Summary

Zoe is a rebellious teenager who often tries to push the limits/boundaries set by her parents. She recently was in a rear-end automobile accident where she was exiting a parking lot while leaving work and did not see a car speeding up behind her. The experience frightened her parents who implemented new safety rules for Zoe to follow. Her parents placed a tracking device inside her car to ensure that she neither speeds nor uses her mobile phone while driving. Her parents believe that, since she was trying to change the radio while she was driving, this activity could have been a plausible cause for the accident. Therefore, her parents desire a car that not only keeps her safe but also limits potential distractions that she might incur. Zoe works 3–4 days per week and often is tired the next day; consequently, she arrives late to school and often speeds to ensure that she arrives at class on time. Consequently, she is not as cautious as she should be and has been close to rear-ending cars on several occasions.

8.4.1.2 Design Implications Summary

Objectives

- Alert Zoe of potential collisions.
- Create progress reports on her driving.
- Earn badges for driving safely.

Context

- Young driver.
- Recently in a car accident.

Behavior

- Only drives to and from work and school.
- Occasionally brings a friend or two in her car.
- Changes the radio station whenever she hears a commercial.
- Forgets to check her surroundings when backing out.

Design Implications

- Allow her parents to limit the customization and controls she can utilize that are beyond those necessary to control an automobile.
- Notify user of potentially dangerous surroundings.

8.4.2 Persona 2: Jeff Park

- Age: 20.
- Race: Asian.
- Location: Los Angeles, California.
- Occupation: Student/pizza delivery driver.
- Education: Some college.
- Technology Usage: Uses his Android phone for all forms of communication (e.g., email, phone calls, messaging, social networks, etc.), Xbox 360, and a MacBook Air.
- Concern: Students jaywalk, talking/texting while driving.
- Persona image credit: AM+A.

8.4.2.1 Textual Summary

Jeff is funding his college tuition through loans and working as a pizza delivery driver. He must balance out his 25-h-per-week job while being a full-time student; consequently, Jeff has little time to try to understand new technologies. During his deliveries, Jeff's dispatcher often calls him about new deliveries, which can distract Jeff and prevent him from being aware of jaywalking pedestrians. Jeff therefore wishes to have a system that controls his ability to access such information while driving. Jeff's vision is color deficient; consequently, he wishes to have the option to customize the color scheme to make certain everything is legible and readable.

8.4.2.2 Design Implications Summary

Objectives

- Avoid vehicle collision with pedestrians.
- Allow his work to track his car's current position.
- Compare his fuel economy with fellow colleague's fuel economy.

Context

- On a tight budget.
- Drives around a college campus.
- Constantly in and out of his car.

Behavior

- Delivers pizza around campus.
- Needs to be alert of pedestrians.

Design Implications

- Significantly ease his interaction with a GPS to locate deliveries.
- Enhance his ability to see surrounding pedestrians.
- Encourage him to drive more efficiently.

8.4.3   Persona 3: Tina Romanski

- Age: 25.
- Race: Caucasian.
- Location: Chicago, Illinois.
- Occupation: Investment banker.
- Education: Bachelor of Science.
- Technology Usage: Multiscreen Bloomberg Terminal, iPhone (personal), Blackberry (work), iPad, and a Lenovo ThinkPad.
- Concern: Often exhausted on her drive home, because she often works 11+ hours per day.
- Persona image credit: AM+A.

8.4.3.1   Textual Summary

Tina is an investment banker in Chicago where she is usually the first to arrive and last to leave. In order to keep Tina's attention on the road, she needs a system that helps her avoid cognitive overload by limiting the actions that Tina can perform in responding and listening to new data. The Driving Machine incorporates persuasive techniques to change Tina's driving behavior to ensure that she only communicates with others when the circumstances allow. For example, the Driving Machine only allows Tina to receive notifications (e.g., phone calls, Facebook updates) when the traffic is at a safe level based on her past driving experiences.

8.4.3.2 Design Implications Summary

Objectives

- Find people with whom to carpool.
- Track her alertness and how often she is distracted.
- Learn more about her carbon footprint.

Context

- Tina lives in a suburb outside of Chicago where she works at the Chicago Board of Trade.
- She drives alone to her job 45 min each way.

Behavior

- Drives over 45 min each way to and from work.
- Often works long hours and is usually tired on the drive home.

Design Implications

- Ensure that Tina is awake and aware of where she is going.
- Limit the cognitive load on her given she is apt to be tired.
- Help Tina find people to carpool with her.

8.4.4 Persona 4: Chris Davis

- Age: 32.
- Race: White.
- Location: Austin, Texas.
- Occupation: Software developer.
- Education: Master of Science.
- Technology Usage: iPhone, iMac, iPad, and a Nintendo Wii.
- Concern: Tends to text and talk on his mobile phone while driving.
- Persona image credit: AM+A.

8.4.4.1   Textual Summary

Chris is an early adopter who constantly is reading the latest auto blogs to learn about new gadgets. Chris is a recent transplant to the Austin area, having lived the past 5 years in New York City. He is in the market for an affordable yet high-tech car that allows him automatically to convert his social life into his car. This conversion must occur in a safe manner that guarantees Chris's focus on the road. There should be as little cognitive distractions as possible. He wants to signify that this car is his and not just by the color of the car or its interior but by being able to personalize his dashboard. The Driving Machine provides a means to ensure that Chris can successfully operate his vehicle while at the same time have his devices present in the background. The Driving Machine is highly customizable because it allows Chris to create or pick from a selection of dashboard "skins."

8.4.4.2   Design Implications Summary

Objectives

- Customization of dashboard.
- Focus on driving by limiting distracting information.

Context

- Software developer who is new to the Austin area.
- Wants to meet new and interesting people.

Behavior

- Texts while driving.
- Calls/texts people while driving.
- Constantly changes through Pandora, XM Sirius Satellite Radio, and traditional radio.

Design Implications

- A way to limit device interaction depending on specific driving circumstances.
- Customization of dashboard design/layout.
- Allow Chris to meet people by carpooling.

8.4.5   Persona 5: Susan Li

- Age: 73.
- Race: Asian-American.
- Location: Seattle, Washington.
- Occupation: Retired.
- Education: Master of Arts.
- Technology Usage: Does not know how to use a computer and does not have a mobile phone.
- Concern: Drives under the speed limit and arthritis.
- Persona image credit: AM+A.

8.4.5.1   Textual Summary

Susan is a retired schoolteacher who often frets about trying and relying on new technologies. She continuously refuses to use a mobile phone or a computer, because she says she lived most of her life without either of them. She asks: Why, then, does she need them? Ms. Li is in the market for a new car, because her lease for her current one is ending. Because she is becoming older and her response time is slower than that of a younger person, she desires a car that includes sensors to ensure safe travel. The system should be able to automatically alert her of changes in speeds and make her aware of the surrounding traffic. The Driving Machine rewards Susan with badges whenever she avoids accidents, does not brake too hard, and drives safely.

8.4.5.2   Design Implications Summary

Objectives

- Drive more efficiently by being better aware of the speed limit and her surroundings.
- Hope to lower her insurance rates by allowing her insurer to track her driving behavior.

Context

- She often needs assistance when learning new technology.
- Her response time is slower because she is older.
- She does not have a mobile phone.

Behavior

Tends to drive under the speed limit.
Travels to her daughter's house every week day to watch her grandchildren.

Design Implications

- A user manual should not be necessary for ordinary use.
- Encouragement for safe driving with badges and lower insurance rates.

8.4.6   Persona 6: Wallace Jonick

- Age: 77.
- Race: African-American.
- Location: Columbus, Ohio.
- Occupation: Retired.
- Education: Medical School (Surgeon).
- Technology Usage: GPS and Samsung Galaxy 3.
- Concern: Cataracts and early dementia.
- Persona image credit: AM+A.

8.4.6.1 Textual Summary

Wallace is a recent retiree who suffers from early onset of dementia. Consequently, he needs a car that helps him to follow all traffic laws. The system must coach him to drive safely, but must not be too demanding, must not make Wallace feel uncomfortable, or make Wallace feel that the system controls him. On occasion, Wallace forgets where he is travelling from and where he lives. Because of his forgetfulness, the Driving Machine must be aware of Wallace's driving habits (e.g., where he traditionally drives to and from) to assist in reminding Wallace why he is travelling somewhere.

8.4.6.2 Design Implications Summary

Objectives

- Allow for him to feel independent while still being safely connected to others.
- Ensure Wallace is capable of driving by reminding him of the rules and regulations of driving (e.g., the speed limit).

Context

- Needs a visual system that allows for customization.
- Sends notifications to people that he arrives safely to places.
- Has early onset of dementia.

Behavior

- Only buys a new car once every 10 years.
- Often forgets why and where he is traveling.

Design Implications

- Provide a way to ensure he safely arrives to places.
- Offer a means to earn badges for alerting family members that he is driving to and from somewhere.
- Assist Wallace in case he is lost or forgets to remember why he is driving someplace.

8.5   Use Scenarios

As explained in Chap. 1, use scenarios are a user-centered UI development technique. In the development of an initial concept prototype, a use scenario helps determine what behavior to simulate. A scenario is essentially a sequence of task flows with actual content provided, such as the users' demographics and goals, the details of the information being exchanged, etc.

8.5.1   General Use Scenario

The following use scenario topics are drawn from the six preceding personas. However, some specific examples might be relevant only to a particular age group or driving experience. As noted in the Glossary of Chap. 1, an objective is a general sought-after target circumstance. A goal is more specific and is usually qualified by concrete, verifiable conditions of time, quantity, etc.

Carbon Footprint Monitoring

- Receive up-to-date articles, advice, and tips regarding monitoring current and past driving behavior.
- Set customizable alerts for driving, whether positive or negative.
- Receive unsafe alerts.
- Establish and maintain objectives (e.g., "I want to reduce my carbon footprint").
- Establish and maintain goals (e.g., specify the number of people you want to drive in a carpool). See the ramifications of this goal on current and past trends.
- Visualize and monitor the carbon footprint.

Carpooling

- Share current location with people nearby.
- Alert driver of any potential passengers.
- Visualize and monitor the number of and location of carpool passengers.

Social Media

- Post green/efficient driving achievements on the users' own walls and possibly their friends' walls, similar to a merit-badge system.
- Connect with insurance agent by automatically sending them status reports on your driving behavior.
- Share tips and strategies with specific friends or family.

- Import personal information from social media sites (e.g., race, sex, age). Users not connected to a social media site can add their information manually through the Driving Machine.
- Resolve any urgent ethical issues.

Gamification

- Set and use preexisting achievements to help reduce insurance premiums.
- Compare estimated fuel economy to actual fuel economy.
- Earn badges for being a driver in a carpool or on-demand carpool.
- Purchase carbon offsets by carpooling.
- Reward posting gas prices with gas reward cards.
- Develop an "economy of tipsterism," likes and dislikes, bribes and no bribes, objective vs. biased opinion, etc.

8.5.1.1   Persona Use Scenarios

The following specific use scenarios are for each of the previously described personas. For each one, expected behavior changes are described, also. Only three examples are provided as examples.

8.5.1.2   Zoe's Use Scenario

Zoe drives under difficult emotional circumstances of parental pressure, fatigue, and pressure from school. She likes using the Driving Machine because it consolidates her necessary information into one simple application. Zoe really enjoys the different view settings within the Driving Machine; she prefers the "Goth" or other backgrounds, because she is accustomed to them in her music and clothing styles. The Driving Machine helps Zoe keep track of her speeding and driving behavior. Last week she was alerted to a possible $300 fine for being 10 miles over the speed limit. The Driving Machine also helped Zoe with finding a better route to school. With the Driving Machine's navigation functions, she can drive more slowly but still get to school on time. Because her day is packed with work and school requirements, Zoe likes using the Driving Machine on the go. She received a replacement iPad-like dashboard as a gift recently and was able to use the Driving Machine's functions and data almost as soon as she began to use it.

8.5.1.3   Zoe's Behavior Changes

The Driving Machine gives Zoe's parents the peace of mind that they know she arrived at work and school safely.

The Driving Machine helps her keep track of how often she speeds. This function helps her save money on her car insurance by allowing them to track her driving behavior.

Zoe enjoys receiving badges that she can show off on her Facebook page to demonstrate her safe driving.

8.5.1.4   Chris's Use Scenario

Chris is new to the Austin metropolitan area and struggles to adapt to the less aggressive driving behavior of Austinites compared to the more aggressive driving behavior of New Yorkers. He subsequently uses the Driving Machine to coach himself regarding how to adopt a less aggressive driving style. He also uses the Driving Machine to find passengers to carpool. Chris enters his desired destination and how far he is willing to travel out of his way. Based on the information he provides, the Driving Machine determines not only how much money and fuel Chris would make and use by carpooling but also the estimated CO_2 emissions saved by carpooling. By carpooling, Chris also receives badges and fuel coupons from his employer for carpooling. When Chris receives a text/call during a dangerous situation, the Driving Machine automatically informs those who call/text him that he is driving and will respond back as soon as it is safe. The Driving Machine considers the situation's danger based on a variety of characteristics such as current speed, traffic, and previous driving behavior.

8.5.1.5   Chris's Behavior Changes

Chris meets new people and learns more about Austin's attractions.

He learns about the environmental impact of his driving.

The Driving Machine reduces Chris's likelihood of a traffic ticket by coaching him how to drive less aggressively.

Chris makes more informed decisions before driving alone because he knows the environmental impact of his driving.

8.5.1.6   Wallace's Use Scenario

Wallace was recently diagnosed with dementia. After informing his insurer, the insurer required Wallace to use the Driving Machine to guarantee continuation of his insurance policy. The Driving Machine not only helps Wallace to stay more alert and aware but also helps Wallace receive assistance to remind him where he is traveling. The Driving Machine visualizes Wallace's driving route and speed for his family, friends, and insurer to see. This is done to guarantee that Wallace is still cognitively capable to drive safely. The Driving Machine also analyzes Wallace's daily driving routine to look for patterns and anomalies in his driving behavior. His family particularly appreciates the peace of mind the Driving Machine's alerts provide. The alerts help the family members and friends know Wallace's speed and when he safely arrives at his destination.

8.5.1.7   Wallace's Behavior Changes

Wallace was able to keep his auto insurance.
Wallace is no longer lost or having trouble arriving to his destination.
The Driving Machine allows Wallace to be alert about his speed and traffic laws.
Wallace receives support from his friends and family when the Driving Machine informs them that he is not following his traditional driving route.

8.6   Market Research

Before undertaking conceptual and perceptual (visual) designs of the Driving Machine, AM+A studied six dashboard user interfaces. Through screen comparison-analysis and analysis of recent articles about trends in vehicle dashboard design, AM+A derived a synopsis of each dashboard's features, which contributed to improvements of initial ideas discussed internally among the development team for the Driving Machine's detailed functions, data, information architecture (metaphors, mental model, and navigation), and look-and-feel (appearance and interaction).

8.6.1   Audi A8 Dashboard

Audi's A8 dashboard incorporates a traditional instrument cluster with a large LCD between gauges. The system uses the Internet to grab fuel prices and find points of interest via Audi connect.

Positive

- Lane departure warnings.
- Limited cognitive distraction.
- Just-in-time knowledge for fuel prices and location data.
- Alerts driver of fatigue.

Negative

- No persuasion to drive efficiently (Fig. 8.1).

Fig. 8.1 Example of Audi 8 dashboard (Image: Audi AG.)

8.6.1.1   Cadillac User Experience

The Cadillac User Experience (CUX) incorporates an LCD display in lieu of a traditional analog instrument cluster to allow for greater user customization. The CUX unifies Cadillac's infotainment and telematics systems for a more uniform user experience.

Positive

- Adjustable user interface for the dashboard to fit personal preference.
- Integrated GPS in dash assists driving from having to reallocate his vision.
- Incorporating large buttons versus many smaller buttons allows faster user interaction.
- Only allows modifying the various dashboard layouts while car is in park.

Negative

- Capacitive controls, even with haptic feedback and physical ridges, afford little user feedback as to what the function actually performs.
- Requires a user to look at the controls to perform interaction.
- Allows driver to change data fields while driving, which can cause a driver to lose focus.
- Overabundance of steering wheel controls.

8.6.1.2   Ford SmartGauge with EcoGuide

The EcoGuide is a system that coaches a driver how to maximize his/her fuel economy by incentivizing driving behavior. A key component of coaching behavior change is Efficiency Leaves, which grows leaves by driving efficiently or shrinks leaves by driving less efficiently.

Positive

- Glancable fuel level and speed information.
- Incentivizes driving by watching leaves grow or decay.

Negative

- Potential for driver distraction in watching leaves grow.
- Inhibits a driver's ability to pay attention, with a myriad of functions, as discouraged in Miller's magic number 7 ± 2 or in other later studies promoting 4 ± 1.

8.6.1.3   Honda Ecological Drive Assist System

Honda's Ecological Drive Assist System incorporates three functions for greater fuel economy: an ambient color meter, a continuously variable transmission, and a scoring function. Together these three components seek to use persuasive techniques to encourage drivers to drive more environmentally friendly. The system's scoring function is visible via the ability to grow leaves depending on a driver's driving efficiency (Fig. 8.2).

Fig. 8.2 Honda Ecological Drive Assist System dashboard display (Image: Honda Motor Company, Ltd.)

Pros

- Gamification by trying to be a green driver by making the ambient meter green.
- Incentives for drivers to drive efficiently by watching leaves grow.

Cons

- No explanation behind what is green or blue driving.
- No means to compare driving styles with others.

8.6.1.4   Johnson Controls Multilayer Instrument Cluster

Johnson Controls' prototype dashboard user interface seeks to utilize spatial techniques to allow for prioritizing driving data such as speed and assistance information depending on the driving conditions.

Positive

- Limits cognitive overload by only showing relevant information depending on the driving conditions.
- Graphics to aid in distance to destination.

Negative

- Elaborated 3D graphics may cause sensory overload and be difficult to decipher information.

8.6.1.5   Nissan Leaf Dashboard

The Nissan Leaf Dashboard is a two-tier dash that separates driving diagnostics such as current speed and battery range. Nissan's system seeks to influence driving behavior by using persuasive techniques through which a driver is able to grow leaves depending on how efficient he or she drives (Fig. 8.3).

Positive

- Ensures driver safety with two-tier dashboard that allows for quick glances.

Negative

- Does not allow for customization of instrument panel, because it cannot show percentage of available battery capacity.
- Base 12 energy gauge instead of a base 10.

Fig. 8.3 Nissan Leaf dashboard display (Image: Nissan Motor Company, Ltd.)

8.7 Concept Design

From our investigations of dashboard and automotive-related applications, including those cited above, AM+A concluded that usable, useful, and appealing vehicle user interface (UI) design must include incentives to lead to behavior change. Safe and sustainable driving behavior is possible by providing incentives, such as games, and just-in-time systematic instructions to motivate people to change their behavior. The proposed Driving Machine needs to combine persuasion theory, provide better incentives, and motivate users to achieve short-term and long-term behavior change towards a Driving Machine everyday user.

Our Driving Machine concept assumes that the primary vehicle dashboard is one of approximately six screens that might be available in a vehicle:

- Driver's dashboard.
- Central screen often containing navigation controls and climate/media controls accessible to the driver the front-seat passenger.
- Front-seat passenger's screen, which might include a media-management system for the rear-seat screens.
- Backseat left, center, and right screens often on the back of the front seats and/or overhead in the center.

These screens are shown schematically in the accompanying Fig. 8.4.

The Driving Machine should be non-obtrusive, but encouraging to use. Well-designed games will serve as an additional appealing incentive to teach, i.e., to train the driver. Drivers should be able to receive badges for accomplishing certain tasks. The Driving Machine should allow users to share their experience with friends, family members, and the world, primarily through Facebook and Twitter. The Driving

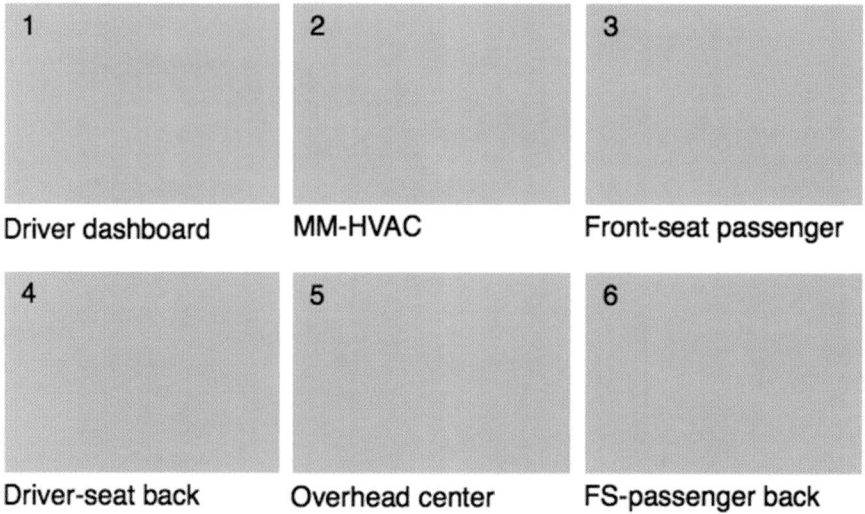

| 1 | 2 | 3 |
| Driver dashboard | MM-HVAC | Front-seat passenger |

| 4 | 5 | 6 |
| Driver-seat back | Overhead center | FS-passenger back |

Fig. 8.4 Diagrammatic depiction of the six screens likely to be available in a vehicle

Machine should also allow drivers to communicate their experiences with insurance companies to allow drivers to receive reduced rates and with family/friends.

Based on these concepts and available research documents, we have proposed and are developing conceptual designs of the multiple functions of the Driving Machine. Subsequent evaluation will provide feedback by which we can improve the metaphors, mental model, navigation, interaction, and appearance of all functions and data in the Driving Machine's user interface. The resultant improved user experience will move the Driving Machine closer to a commercially viable product/service.

In particular, we believe a well-designed Driving Machine will be more usable, useful, and appealing to memory-conscious users, especially those experiencing long- and short-term memory loss. Another objective is to provide a dashboard experience that can reliably persuade people to become safer and energy-conscious drivers.

8.8   Persuasion Theory

As mentioned in Chap. 1, and in alignment with Fogg's persuasion theory (Fogg and Eckles 2007), we defined five key processes to create behavioral change via the Driving Machine's functions and data:

- Increase frequency of using application.
- Motivate changing some living habits.
- Teach how to change living habits.
- Persuade users to plan short-term change.
- Persuade users to plan long-term change.

Each step has requirements for the application. Motivation is a need, want, interest, or desire that propels someone in a certain direction. Humans' sociobiological instinct is to compete with others for social status and success. We apply this theory in the Driving Machine by prompting users to understand that every action in the process of driving has the potential to have enormous consequence on their success.

We also drew from Maslow's *A Theory of Human Motivation* (Maslow 1943), which he based on his analysis of fundamental human needs. We adapted these to the Driving Machine context:

- The *safety and security need* is met by the possibility to visualize the amount of progress made in driving.
- The *belonging and love need* is expressed through friends, family, and social sharing and support.
- The *esteem need* is satisfied by social comparisons that display success in driving activity, as well as by self-challenges that display goal accomplishment.
- The *self-actualization need* is fulfilled by the ability to visualize improvement and progress in driving.

8.9   Impact on Information Architecture

All of the above has an impact on the information architecture of the Driving Machine, as described below.

Increase Use Frequency

- Games and rewards are the most common methods to increase use frequency. In terms of the rewards, the users will be awarded both virtual rewards (such as "star" nominations and new skins for blogs) and real rewards (such as insurance discounts or fuel discounts).
- In addition, we chose the social comparison as another incentive to increase use frequency. Users will form groups with families and friends and participate in competitions among different groups on diet control and exercise.

Increase Motivation

- In the Driving Machine, we set the users' future insurance premiums and environmental impact predication as an important incentive for their behavior change. Through viewing their current driving behavior and predicted future scenarios in 1–3 months, the users will have a stronger impression and awareness of their dangerous driving habits and carbon footprint.
- Because setting goals helps people to learn better and improves the relevance of feedback, the Driving Machine asks users how much carbon they are willing to reduce by carpooling. In accordance with the goal settings, the users will get suggested action plans to achieve each goal.

- Social interaction also has an important impact on behavior change. Another remarkable component of Driving Machine is to leverage social networkings and integrate features like those found in forums, Facebook, Twitter, or blogs. Users can send notes or messages to their social groups and share ideas with other people. The social ties will serve as an additional incentive to motivate behavior change.

Improve Learning

- For many drivers, learning how to drive responsibly and at the speed limit can be a challenge. To improve learning, the Driving Machine integrates contextual tips to explain how appropriate the distance is.

8.10   Information Architecture

The following is an initial diagram of the information architecture for the Driving Machine that we constructed. In this diagram, the information architecture is called a concept map and uses list structures rather than more typical boxes and links (Fig. 8.5).

The following are detailed descriptions of the components of the information architecture.

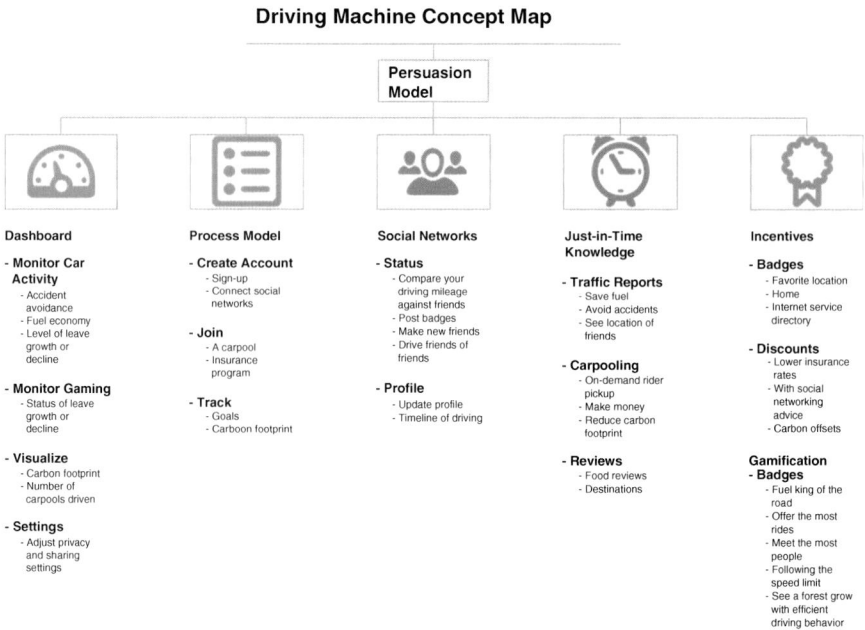

Fig. 8.5   Diagram of information architecture of the Driving Machine

8.10.1 Dashboard

Main Functions

- Displays user's driving behavior such as speed (e.g., going over and under the speed), carbon footprint in tables and trend charts.
- Records amount of money/time/fuel spent driving.
- Calculates and displays money/time/fuel used so far for the day, week, month, or year.
- Provides ability to see graphs comparing fuel usage and comparison to others with similar vehicles
- Ability to modify sharing and privacy settings.

Safety/Driving Record

- Displays user's speeding and traffic violations.
- Provides ability to see charts comparing driving record with those of similar experience/age/gender/location.
- Alerts family and friends of arrival to specific destinations.

Green/Sustainable Driving

- Displays overall carbon dioxide emissions for daily and historic driving.
- Provides ability to see more detailed expansion of contributing factors.
- Provides ability to see charts comparing environmental impact to other vehicles, driving situations (e.g., weather conditions, traffic conditions, number of passengers, etc.)

Driving Advisor

- Enables quick communication between user and driving coach.

Settings

- Enables user to set notification settings. Notifications are shown in email and on the landing page. They are less urgent than alerts and the user checks them on their own time.
- Enables user to set alert settings. Alerts are important to know about right away and are received via text or through red exclamation points on top of the Driving Machine application icon. Examples of alerts are speeding, breaking harshly, and fuel prices.

8.10.2 Process Model

Efficient Driving

- Displays collection of driving, travel, or other templates.
- Alerts others when traffic conditions or other circumstances cause the driver/ passenger to be late.

Goal Setting

- Records amount of money, time, and/or fuel allocated for specific goals.
- Calculates and displays money, time, and/or fuel put so far towards each goal compared to total amount needed.
- Provides ability to see charts of money/time/fuel for goals or a more detailed textual view.

Future Planning

- Displays collection of future planning calculators (e.g., lower carbon footprint, drive safer).
- Alerts when future plans have changed based on circumstances changing.
- Enables user to fill out a new future calculator or view/edit an existing, already filled out future calculator.

8.10.3   Social Networks: View and Make Social Media Achievement Postings

Announcements

- Displays announcements of posts from friends on the Driving Machine.
- Enables user to post own updates.
- Enables user to comment on posts or give them a "thumbs up".

Profile

- Displays user's profile.
- Includes user's picture, comments from friends, and updates from user.

Friends

- Displays list of friends on the Driving Machine.
- Displays privacy settings next to each friend.
- Allows search for a friend or clicking on a friend to go to their profile.

Driving Machine Wall

- Displays the main Driving Machine profile/wall.
- Includes updates and tips from the Driving Machine company.
- Allows comments from users of the app.

8.10.4   My Tips (Could Even Be from an Insurance Agent)

Traffic Updates

- Displays updates on current traffic conditions.

Tips

- Displays feed of tips from experts.
- Displays rating on tips and allows users to pick "thumbs up" or "thumbs down".
- Allows filtering based on popularity, type of person posting the tip, and time of posting.
- Allows search for tips on specific subjects.

Driving Machine Partnership

- Displays information about forming a partnership with the Driving Machine company.
- Displays updates on the partnership and opportunities to each more if the user is already in partnership.

8.10.4.1   (Incentives) Contests

Achievements Collection

- Displays pictures of achievements.
- Displays progress on each achievement through a progress bar.
- Enables clicking on picture to receive more information about a specific achievement.

Competitions

- Displays current competitions and progress of each user involved.
- Alerts about changes of user ahead in competition.
- Allows starting of new competitions.

Point Store

- Displays featured objects purchasable with points.
- Displays list of all objects and associated cost in points.
- Allows purchasing of objects with accumulated Driving Machine points.

Circles

- Displays list of circles.
- Displays user's joined circles.
- Allows looking at specific pages for already joined circles.

8.11   Designs

8.11.1   Initial Sketches

Based on the information architecture, AM+A has prepared initial concept sketches of some of the screen of the Driving Machine in several states of activity. AM+A sketched metaphor concepts, information architecture concepts, and specific screen design concepts before rendering detailed diagrams and screens. The

following are representative screen design examples with brief descriptions. In a
following Appendix, the screen designs are shown with specific logos for several
countries as examples of how the Driving Machine concept might be applied in
practice (Fig. 8.6).

This dashboard screen illustrates all permanent, pertinent information. The
outer rectangle showcases the automobile logo and the original equipment manu-
facturer (OEM) logo. The inner rectangle showcases (clockwise from top left)
external temperature (78 °ext), time (10:00 am), internal temperature (65 °int),
miles to empty (225 mte), compass (SW), and miles per gallon (17 mpg). In the
center, the current speed is shown in large numbers (70 mph) with the speed limit
in smaller numbers above (65 mph). All this information will always be present on
the dashboard, and this would be the default view in the case that nothing is wrong
with the car (Fig. 8.7).

This dashboard screen shows all possible indicators that could appear (e.g., low
gas, emergency brake engaged, maintenance required). These icons are industry-
standard and commonly used across all automobile companies (Fig. 8.8):

The dashboard with the left-turn signal (Fig. 8.9).
The dashboard with the left-turn collision signal. The red bar indicates that the car
   is close to colliding with an obstacle. The location of the red bar indicates the
   location of the obstacle. In this visual above, the obstacle would be to the left of
   the car (Fig. 8.10).
The dashboard with the collision signal. Again, the location of the red bar indicates
   the location of the obstacle (Fig. 8.11).
An alternative dashboard view, which emphasizes navigation capabilities. The
   dashboard showcases the same elements (clockwise from top left), external

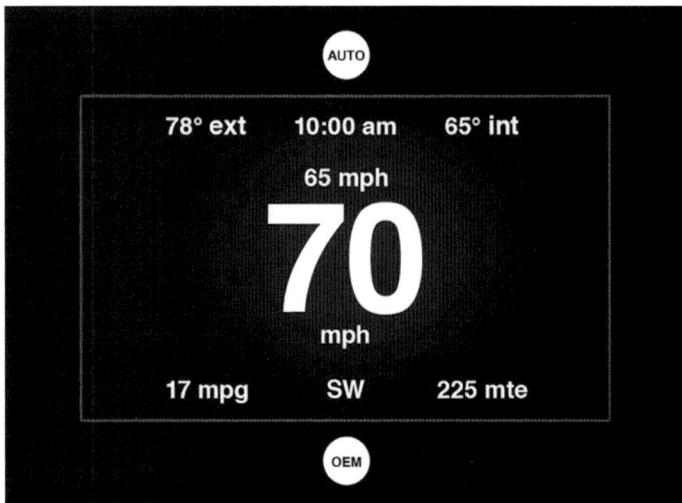

Fig. 8.6 Dashboard screen 1

Fig. 8.7 Dashboard screen 2

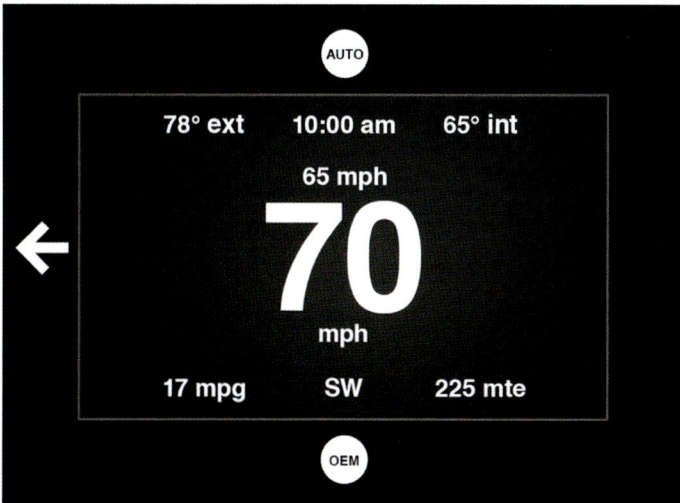

Fig. 8.8 Dashboard screen 3

temperature/internal temperature (78°/65°), time (10:00 am), compass (SE), miles to empty (225 mte), current speed (speed limit) (70 mph (65)), and miles per gallon (17 mpg) (Fig. 8.12).

A dashboard view with a themed background featuring the movie, "Little Shop of Horrors." Users would be able to shift and select different themed backgrounds or "skins" depending on their interests (Fig. 8.13).

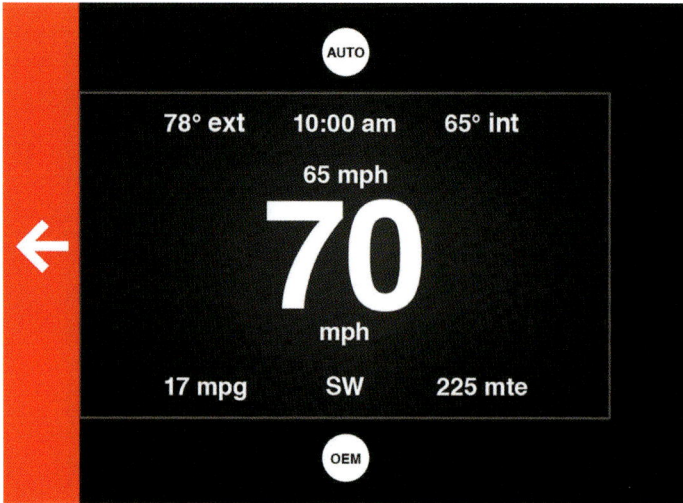

Fig. 8.9 Dashboard screen 4

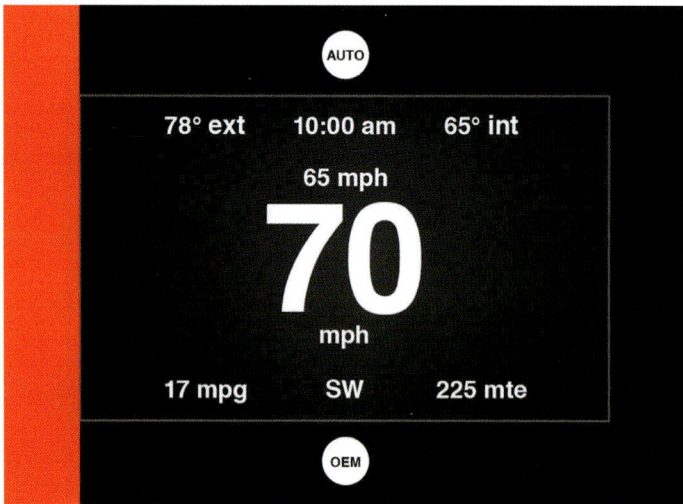

Fig. 8.10 Dashboard screen 5

A dashboard view with a "Pac man" themed background (Fig. 8.14).

A dashboard view with a themed background inspired by Swiss typographer, Wolfgang Weingart (Fig. 8.15).

In this version of the Weingart skin, the color of the sparkles indicates whether the driver is speeding. Here, the red sparkles indicate that the driver is above the speed limit (Fig. 8.16).

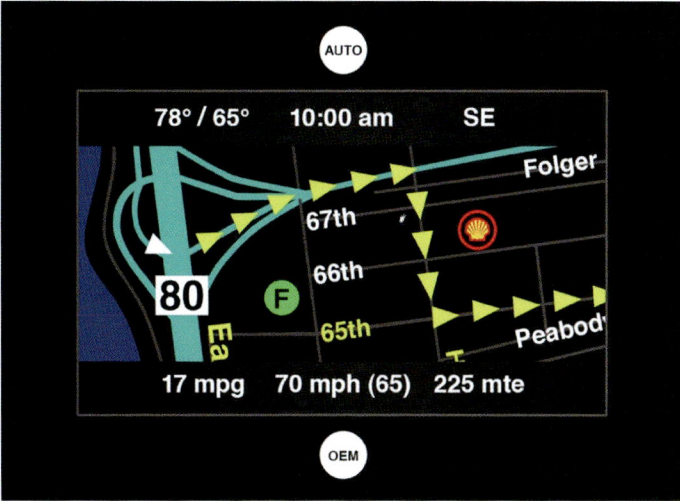

Fig. 8.11 Dashboard screen 5

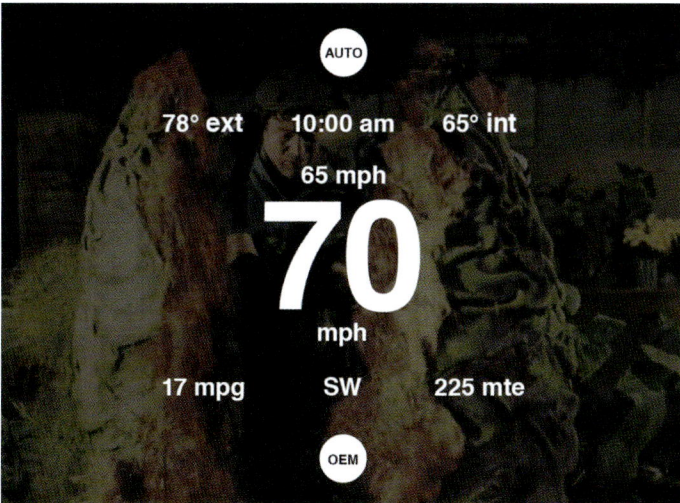

Fig. 8.12 Dashboard screen 6 (Image: public domain movie still)

A dashboard with a themed background that indicates the eco-friendliness of a driver's behavior (how much the driver brakes, how quickly the driver accelerates, etc.). Here, the more eco-friendly the driving, the more trees will appear (Fig. 8.17).

Another dashboard which indicates a driver's eco-friendliness. The outline around the inner rectangle indicates whether the driver is eco-friendly. Here, the green indicates that the driver is eco-friendly (Fig. 8.18).

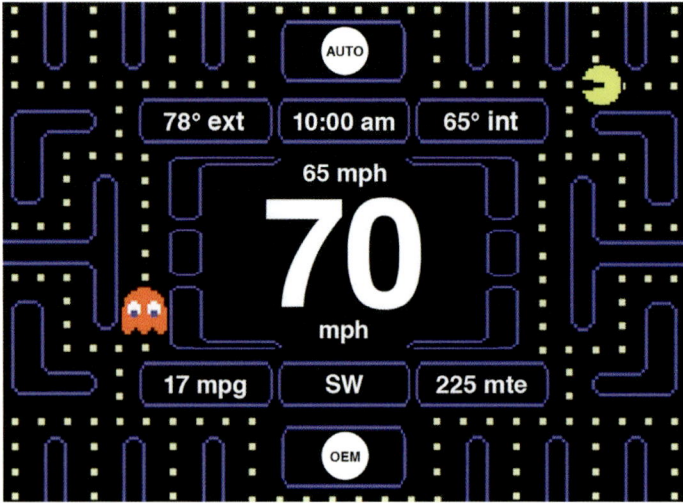

Fig. 8.13 Dashboard screen 7 (Image: AM+A design using Pac-Man-like image elements. Fair use of copyrighted materials claimed)

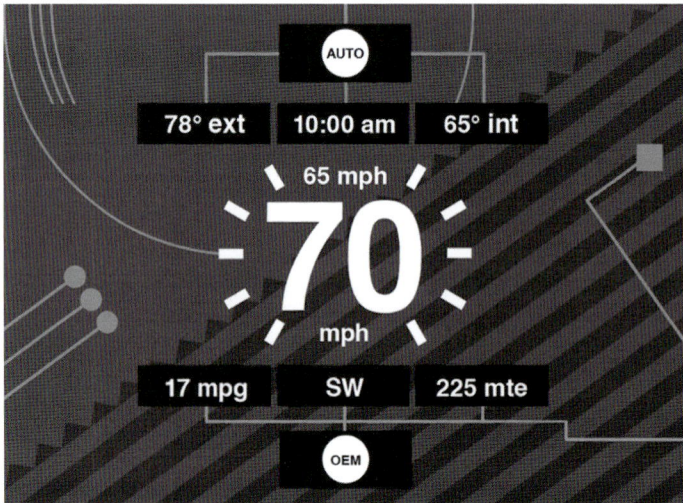

Fig. 8.14 Dashboard screen 8

The dashboard also indicates a driver's eco-friendliness. In this dashboard, the red outline indicates that the driver is not eco-friendly. The outline is also used to encourage the driver to change his/her behavior and drive in a more environmentally responsible manner (Fig. 8.19).

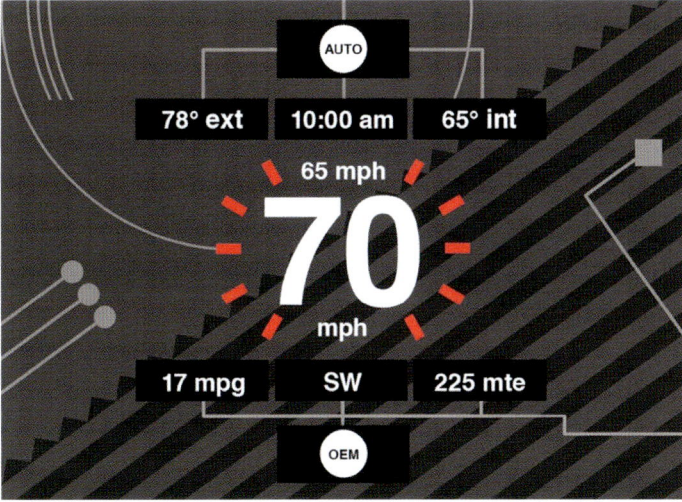

Fig. 8.15 Dashboard screen 9

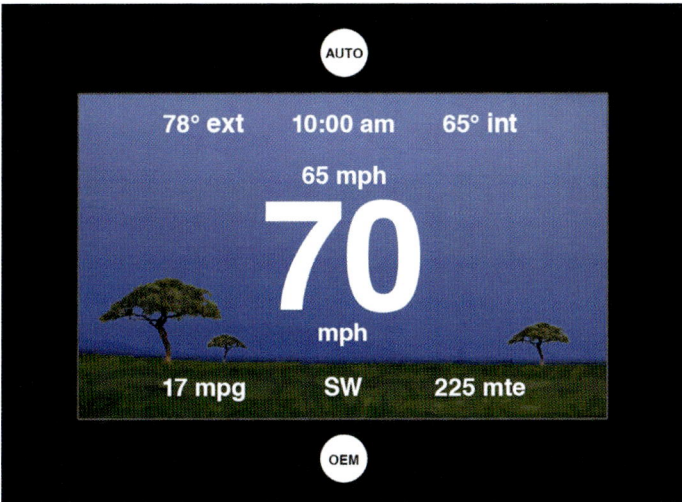

Fig. 8.16 Dashboard screen 10 (Image: Tree image by Megan Chiou)

The dashboard also indicates a driver's eco-friendliness. In this dashboard, the yellow outline indicates that the driver is making efforts to drive in an environmentally responsible manner, but he/she has room for improvement.

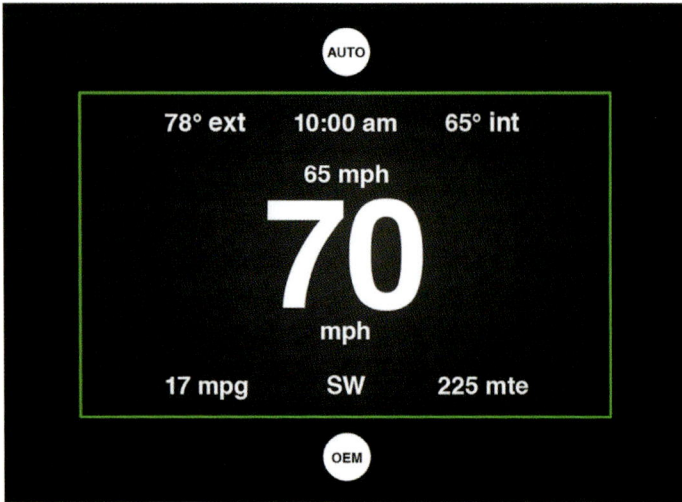

Fig. 8.17 Dashboard screen 11 (*Green*)

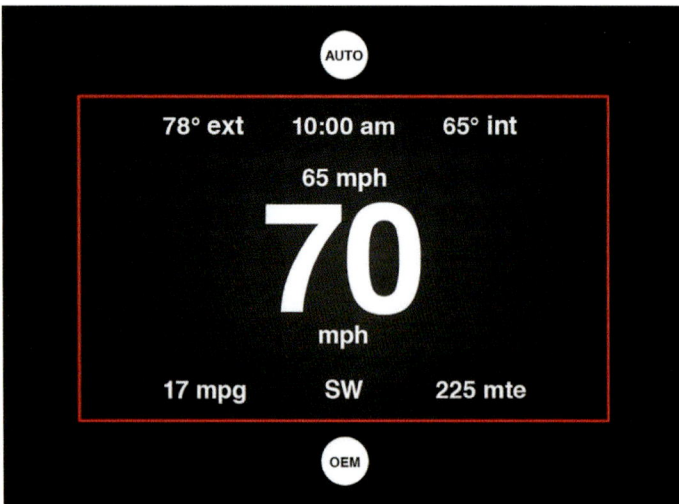

Fig. 8.18 Dashboard screen 12 (*Red*)

8.12   Next Steps and Conclusions

Because of time limitations, AM+A was not able to evaluate the initial designs and to revise them. Following the user-centered development process described above, AM+A plans to continue to improve the Driving Machine screen designs. The development is an ongoing process. Significant further effort would require additional time and funding from an outside source. Tasks include the following:

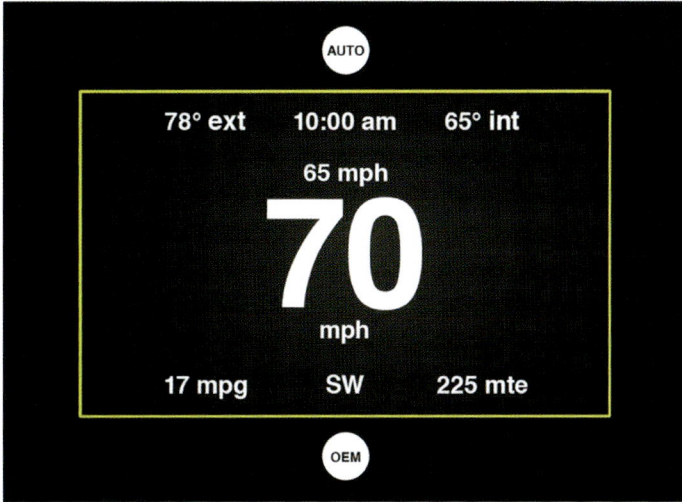

Fig. 8.19 Dashboard screen 13 (*Yellow*)

- Revise personas and use scenarios.
- Conduct user evaluations.
- Revise information architecture and look and feel.
- Build initial working prototype (e.g., for iPad, or other tools, platforms).
- Evaluate the Driving Machine across different demographics and cultures.
- Research and design improved information visualizations.

AM+A aimed to incorporate information design and persuasion theory for behavior change into a mobile tablet application that would constitute an advanced vehicle dashboard. This self-funded work on the Driving Machine project is current and ongoing and was undertaken to demonstrate the direction and process for such products and services. Though the design is incomplete, AM+A is willing to share the approach and lessons learned, in the interest of helping alleviate worldwide environmental problems and automobile safety challenges.

At this stage, AM+A is seeking to persuade other design, education, and automotive groups to consider similar development objectives. We hope that our process and concept prototypes (the self-funding of which inevitably limited the amount of research, design, and evaluation) will inspire others and that they benefit from the materials provided thus far.

The process has already been demonstrated successfully with a previous project, the Green Machine (Marcus and Jean 2009), versions of which have been considered and used by SAP for enterprise software development (Marcus et al. 2011).

AM+A's long-term objective for the Driving Machine is to create a functional working prototype to test whether the application can actually persuade people who experience driving challenges to exercise greater vehicle control, increase safety, and reduce their carbon footprint. If the theories are proven to be correct, this approach could have significant implications that will benefit millions of people in the United States and abroad by reducing the number of automobile accidents and decreasing greenhouse gas emissions.

Acknowledgements The author thanks these AM+A Associates, all designer/analysts, for their significant assistance to plan, research, design, analyze, and document the content of this project:

- Mr. Scott Abromowitz.
- Ms. Megan Chiou.

Further Reading

Audi of America (2012) "Audi A8". http://models.audiusa.com/a8. Retrieved 10 Aug 2012
Brand Values and Strategic Goals for a Future HMI (presented to Aaron Marcus and Associates, Inc. in 2001

Bibliography of Relevant References

Arsenault R, Ware C (2000) Eye-hand co-ordination with force feedback. CHI Lett 2(1):408–414
Ashley S (2001) Driving the info highway. Sci Am 285(4):52–58
Baddeley AD (1986) Working memory. Oxford University Press, New York, Specific page unknown
Bennett J (1984) Managing to meet usability requirements. In: Bennett J, Case D, Sandelin J, Smith M (eds) Visual display terminals: usability issues and health concerns. Prentice-Hall, Englewood Cliffs
Bertone CM (1982) Human factors considerations in the development of a voice warning system for helicopters. In: Behavioral objectives in aviation automated systems symposium. Society of Automotive Engineers, Warrendale
Boff KR, Kaufman L, Thomas JP (eds) (1986) Handbook of perception and human performance, volume II: cognitive processes and performance. Wiley-Interscience, New York
Brewster SA, Wright PC, Edwards ADN (2001) Guidelines for the creation of earcons. Glasgow Interactive Systems Group, Glasgow, Scotland
Briziarelli G, Robert WA (1989) The effect of a head-up speedometer on speeding behavior. Percept Mot Skills 69:1171–1176
Campbell and Hershberger. (1988). Automobile head-up display simulation study: effects of image location and display density on driving performance. Unpublished manuscript, Hughes Aircraft Company
Campbell JL, Carney C, Kantowitz BH (1997) Human factors design guidelines for Advanced Traveler Information Systems (ATIS) and Commercial Vehicle Operations (CVO) (Technical report FHWA-RD-98- 057). U.S. Department of Transportation, Federal Highway Administration, Washington DC
Campbell JL, Carney C, Kantowitz BH (1998) Human factors design guidelines for Advanced Traveler Information Systems (ATIS) and Commercial Vehicle Operations (CVO). Retrieved 9 November 2001, from http://www.fhwa.dot.gov/tfhrc/safety/pubs/atis/index.html
Carroll JM, Mack RL (1984) Learning to use a word processor: by doing, by thinking, and by knowing. In: Thomas JC, Shneider ML (eds) Human factors in computer systems. Ablex, Norwood
Carswell CM, Wickens CD (1985) Lateral task segregation and the task-hemispheric integrity effect. Hum Factors 27:695–700
Cialdini RB (2001) The science of persuasion. Sci Am 284:76–81

Commission of the European Communities (1999) Commission recommendation of 21 December 1999 on safe and efficient in-vehicle information and communication systems: A European statement of principles on human machine interface. European Union, Brussels

Constantine LL, Lockwood LAD (1999) Software for use: a practical guide to the models and methods of usage-centered design. ACM Press, New York

Cooper A (1995) About face: the essentials of user interface design. IDG Books Worldwide, Foster City

Cooper A, Saffo P (1999) The inmates are running the Asylum: why high tech products drive us crazy and how to restore the sanity. Sams Publishing, Indianapolis

Curbow D (2001) User benefits of connecting automobiles to the internet. In: CHI 2001 Proceedings

De Waard D (1996) The measurement of drivers' mental workload. The Netherlands: University of Groningen, Traffic Research Centre, Haren

Denmerlein JT, Martin DB, Basser C (2000) Force-feedback improves performance for steering and combined steering-tru·geting tasks. In CHI 2000 conference proceedings: conference on human factors in computing systems, ACM Press, pp 423–429

Dingus TA, Hulse MC (1993) Some human factors design issues and recommendations for automobile navigation information systems. Transp Res Part C: Emerg Technol 1(2):119–131

Dingus T, Hulse M, Jahns S, Alves-Foss J, Confer S, Rice A et al (1996) Development of human factors guidelines for Advanced Traveler Information Systems (ATIS) and Commercial Vehicle Operations (CVO): Literature Review, Technical report FHWA-RD-95- 153. U.S. Department of Transportation, Federal Highway Administration, Washington, DC

Doll TJ (1986) Synthesis of auditory localization cues for cockpit applications. In: Human factors society 30th annual meeting. Human Factors Society, Santa Monica

Draper SW (1993) The notion of task in HCI. ACM Press, New York

Ewing J, Mehrabanzad S, Sheck S, Ostroff D, Shneiderman B (1986) An experimental comparison of a mouse and arrow-jump keys for an interactive encyclopedia. Int J Man–Mach Stud 24(1):29–45

Fleming J (1998) Web navigation: designing the user experience. O'Reilly and Associates, Cambridge, MA

Gaver W (1997) Auditory interfaces. Hum Comput Interact

Gill J (2000) Which button. Royal National Institute for the Blind, London

Green P (1998) Visual and task demands of driver information systems (Technical report UMTRI-98-16). The University of Michigan Transportation Research Institute, Ann Arbor

Green P (2000) Crashes induced by driver information systems and what can be done to reduce them. In: Convergence 2000 conference proceedings (SAE Publication P-360). Society of Automotive Engineers, Warrendale

Green P (2001) Safeguards for on-board wireless communications. Paper presented at the second annual plastics in automotive safety conference. Troy, MI

Green P, Williams M (1992) Perspective in orientation I navigation displays: a human factors test. In: Vehicle navigation and information systems confuence proceedings. Society of Automotive Engineers, Warrendale, pp 221–226

Green P, Levison W, Paelke G, Serafin C (1993) Suggested human factors design guidelines for driver information systems (Technical report FHWA-RD-94-087). The University of Michigan Transportation Research Institute, Ann Arbor

Greenland AR, Groves DJ (1991) Head-up display concepts for commercial trucks. In: Future transportation technology conference and exposition (SAE Technical Paper Series Number 911681). Society of Automotive Engineers, Warrendale, PA

Greenstein J, Arnaut L (1987) Human factors aspects of manual computer input devices. In: Salvendy G (ed) Handbook of human factors

Helander MG (1987) Handbook of human factors. Wiley-Interscience, New York

Held R, Estanthiou A, Green M (1966) Adaptation to displaced and delayed visual feedback from the hand. J Exp Psychol 72:887–891

Helsen WF, Elliott D, Starkes JL, Ricker KL (1998) Temporal and spatial coupling of point of gaze and hand movement in aiming. J Mot Behav 30(3):249–259

Hewett TT (1998) Cognitive factors in design: Basic phenomena in human memory and problem solving. CHI, April

Hoffman P (1999) Accommodating color blindness. Usabil Interf 6(2). Retrieved 17 November 2001, from http://www.stcsig.org/usability/newsletter/9910-color-blindness.html

Hooey BL, Gore BF (1998) Advanced Traveler Information Systems (ATIS) and Commercial Vehicle Operations (CVO): components of the intelligent transportation systems: head-up displays and driver attention for driver navigation information (Technical report FHWA-RD-96-153). US Department of Transportation Federal Highway Administration, Washington, DC

Horton W (1994) The icon book: visual symbols for computer systems and documentation. Wiley, New York

Human Factors International. Retrieved in November 2001 from: http://www.humanfactors.com/wording/acronyms.asp

Human centered transportation system of the future. (1997, June) Presentation of the national science and technology council's human centered transportation safety team at ITS America 7th annual meeting. Washington, DC

Hutchins EL, Hollan JD, Norman DA (1986) Direct manipulation interfaces. In: Norman DA, Draper SW (eds) User centered system design: new perspectives on human-computer interaction. Erlbaum, Hillsdale

IBM. Out of box experience guidelines. Retrieved November 4, 2001 from http://www-3.ibm.com/ibm/easy/eou_ext.nsf/Publish/77.

Jacob RJK (1996) Human-computer interaction: input devices. ACM Comput Surv 28(1):177–179

John BE, Kieras DE (1996) The GOMS family of user interface analysis techniques: comparison and contrast. ACM Trans Comput Hum Inter 3(4):320–351

Johnson J (2000) GUI bloopers – don'ts and do's for software developers and web designers. Morgan Kaufmann, Burlington

Kahn P, Lenk K (1998) Principles of typography for user interface design. Interact Mag, November/December

Kamada T, Kawai S (1991) A general framework for visualizing. University of Tokyo

Kantowitz BH, Sorkin RD (1983) Workspace design. Human factors: understanding people-system relationships. Wiley, New York

Kline DW, Scialfa CP (1996) Visual and auditory aging. In: Birren JE, Schaie KW (eds) Handbook of psychology of aging, 4th edn. Academic Press, New York

Kowlaski LA, Ph.D., McMurtrey K, Wickham DP (2001) Improving the out of box experience: A case study. Society of Technical Communication.

Laurienti PJ, Burdette JH, Wallace MT, Yen YF, Field AS, Stein BE (2002) Deactivation of sensory-specific cortex by cross-modal stimuli. J Cogn Neurosci 14(3):420–429

LeCompte DC (1996) Irrelevant speech, serial rehearsal, and temporal distinctiveness: A new approach to the irrelevant speech effect. J Exp Psychol (22)

LeCompte DC (1999) Seven, plus or minus two, is too much to bear: three (or fewer) is the real magic number. In: Proceedings of the human factors and ergonomics society 43rd annual meeting. HFES, Santa Monica, pp 289–292

Little C (1997) The intelligent vehicle initiative: advancing "human-centered" smart vehicles. Public Roads 61(2):18–25

Luce PA, Feustel TC, Pisoni DB (1983) Capacity demands in short-term memory for synthetic and natural speech. Hum Factors 25:17–32

Lund A (2001) Measuring usability with the USE questionnaire. Usability interface: newsletter of the society for technical communication usability SIG

Maclean KE, Snibbe SS, Levin G (2000) Tagged handles: merging discrete and continuous manual control. CHI Lett 2(1):225–232

Mandel T (1997) The elements of user interface design. Wiley, New York

Manes P, Green P (1997) Evaluation of a driver interface: effects of control type (knob versus buttons) and menu structure (depth versus breadth) (Technical report UMTRI-97-42).

Marcus A (2000) User interface design for a vehicle navigation system. In: Bergman E (ed) Information appliances and beyond, interaction design for consumer products. Morgan Kaufmann, San Francisco

Marcus A (2012a) The money machine: helping baby boomers to retire. User Exp Mag 11:(2, Second Quarter):24–27

Marcus A (2012b) The story machine: a mobile app to change family story-sharing behavior. Workshop paper. In: Proceedings of CHI 2012, Austin, TX, May 2012, pp 1–4

Marcus A et al (2013a) The innovation machine. In: Marcus A (Ed.) (2013) Proceedings, design, user experience, and usability conference, Las Vegas, NV, 20–25 July 2013, pp 67–76. Springer, London

Marcus A et al (2013b) The travel machine. In: Marcus A (Ed.) (2013) Proceedings, design, user experience, and usability conference, Las Vegas, NV, 20–25 July 2013, pp 696–704. Springer, London

McAteer S (1998) Internet everywhere. Jupiter strategic planning services, Web technology strategies

McGrenere J (2000) Bloat: the objective and subjective dimensions (Student Poster). In: CHI 2000 proceedings

Meier BJ (1988) Ace: a color expert system for user interface design. In: Proceedings of the ACMSIGGRAPH symposium on user interface software. ACM Press, New York

Meister D, Sullivan DJ (1969) Guide to human engineering design for visual displays (Technical report N0014-68-C-0278). Office of Naval Research, Washington, DC

Michael SG, Casali JG (1995) Auditory prompts: effects on visual acquisition time and accuracy in a dashboard-mounted navigational display task. In: Proceedings of the 3rd annual mid-Atlantic human factors conference

Microsoft Corporation (1999) Microsoft windows user experience: official guidelines for user interface developers and designers. Microsoft Press, Redmond

Miller GA (1956) The magical number seven, plus or minus two: some limits on our capacity for processing information. Psychol Rev (63)

Mills C, Weldon L (1987) Reading text from computer screens: human-computer interaction laboratory. University of Maryland, College Park

Miner N, Gillespie B, Caudell T. Examining the influence of audio and visual stimuli on a haptic interface

Mullet K (1995) Designing visual interfaces: communication oriented techniques. Prentice-Hall, New York

Muratore D (1987) Human performance aspects of cursor control devices (6321). Mitre, Houston

National Science and Technology Council's (NSTC) Human centered transportation safety team presentation at ITS America 7th Annual Meeting in Washington, DC. (June 1997). The Human Centered Transportation System of the Future

Nielsen J (1994) Heuristic evaluation. In: Nielsen J, Mack RL (eds) Usability inspection methods. Wiley, New York

Nielsen J (1999) Designing web usability: the practice of simplicity. New Riders Publishing, Indianapolis

Norman D (1988) The psychology of everyday things. Basic Books

Nowakowski C, Utsui Y, Green P (2000) Navigation system evaluation: the effects of driver workload and input devices on destination entry time and driving performance and their implications to the SAE recommended practice (Technical Report UMTRI-2000-20). The University of Michigan Transportation Research Institute, Ann Arbor

Paap KR, Cooke NJ (1997) Chapter 24: "Design of Menus". Handbook Hum Comput Interact

Patterson RD, Milroy R (1979) Existing and recommended levels for auditory warnings on civil aircraft. Medical Research Council Applied Psychology Unit, Cambridge, UK

Popp MM, Farber B (1991) Advanced display technologies, route guidance systems, and the position of displays in cars. In: Gale AG (ed) Vision in vehicles—III. Elsevier Science, North-Holland

Preece J, Rogers Y, Sharp H, Benyon D (1994) Human-computer interaction. Addison-Wesley, Reading

Rober P, Hofmeister, J. (2001). An analysis of human computer interaction in vehicles. In: Systems, social and internationalization design aspects of human-computer interaction: Vol. 2 of the Proceedings of HCI International 2001

Rosenfeld L, Morville P (1998) Information architecture for the world wide web
SAE (2000) Navigation and route guidance function accessibility while driving (SAE Recommended Practice J2364). Society of Automotive Engineers, Warrendale
SAE (2001) Calculation of the time to complete in-vehicle navigation and route guidance tasks (Technical report SAE J2365). Society of Automotive Engineers, Warrendale
Salvendy G (1987) Handbook of human factors. Wiley Interscience, New York
Sanders MS, McCormick EJ (1993) Human factors in engineering and design, 7th edn. McGraw-Hill, New York
Segen J, Kumar S (1999) Dextrous interaction with computer animations using vision-based gesture interface. In: Bullinger H (ed) Human-computer interaction: ergonomics and user interfaces
Sheridan TB (1997) Trains, planes and automobiles. The Reflector, May
Shneiderman B (1983) Direct manipulation: a step beyond programming languages. IEEE Comput 16(8):57–69
Shneiderman B (1992) Designing the user interface: strategies for human-computer interaction. Addison-Wesley, Reading
Shukla M, Denzil J, Jampani A (2001) Measuring awareness and distraction caused by change in information density. Department of Computer Science, Virginia Tech, Blacksburg
Simpson CA, Marchionda, Frost K (1984) Synthesized speech rate and pitch effects on intelligibility of warning messages for pilots. Hum Fact 27
Simpson CA, McCauley ME, Ronald EF, Ruth JC, Williges BH (1987) Speech controls and displays. In: Salvendy G (ed) Handbook of human factors. Wiley, New York
Smith BA, Ho J, Ark W, Zhai S (2000) Hand eye coordination patterns in target selection. IBM Almaden Research Center, Almaden
Snowberry K, Parkinson SR, Sisson N (1983) Computer display menus. Ergonomics 26(7):699–712. Specific page unknown
Society of Automotive Engineers (2000) Navigation and route guidance function accessibility while driving, SAE recommended practice J2364. Society of Automotive Engineers, Warrendale
Solso RL (1988) Cognitive psychology. Allyn and Bacon, Boston
Sorkin RD (1987) Design of auditory and tactical displays. In: Salvendy G (ed) Handbook of human factors. Wiley, New York
Standing L (1973) Learning 10,000 pictures. Q J Exp Psychol 1973(25):207–222
Stokes A, Wickens C, Kite K (1990) Display technology – human factors concepts. Society of Automotive Engineers, Warrendale
Stricker A, Shea, B (1999) Cognitive technologies (Information Report CITL IR-002) Cognition and Instructional Technologies Laboratories Texas AandM. Retrieved 18 November 2001 from http://citl.tamu.edu/cognitive-tech.htm
Tarrière C, Hartemann F, Sfez E, Chaput D, Petit-Poilvert C (1988) Some ergonomic features of the driver-vehicle-environment interface (Technical Report SAE 885051)
Tufte ER (1983) The visual display of quantitative information. Graphics Press, Chesire
Tufte ER (1990) Envisioning information. Graphics Press, Chesire
United States Department of Defense (1989) Human engineering design criteria for military systems, equipment and facilities, Military Standard MIL-STD-1472D. Naval Forms and Publications Center, Philadelphia
Usher DM (1982) A touch-sensitive VDU compared with a computer aided keypad for controlling power generated man-machine systems (IEE conference publication no. 212)
Walker N, Fain WB, Fisk AD, McGuire CL (1997) Human Factors 39(3)
Watson WE (1981) Human factors design handbook. McGraw-Hill, New York
Weintraub DJ (1987) HUDs, HMDs, and common sense: polishing virtual images. Hum Fact Bull 30(10)
Weintraub DJ, Haines RF, Randle RJ (1984) The utility of head-up displays: eye-focus vs. decision times. In: Proceedings of the human factors and ergonomics society 28th annual meeting. Human Factors and Ergonomics Society, Santa Monica

Weintraub DJ, Haines RF, Randle RJ (1985) Head-up display (HUD) utility. II. runway to HUD transition monitoring eye focus and decision times. In: Proceedings of the human factors and ergonomics society 29th annual meeting. Human Factors and Ergonomics Society, Santa Monica

White J (2011) The car dashboard that wants to be an iPad. Wall Street J, December 14. Retrieved from http://online.wsj.com/article/SB10001424052970204903804577082203288412734.html#

Wickens CD (1990) Engineering psychology and human performance. Scott Foresman, Glenview

Woodhead M (1957) Effects of bursts of loud noise on a continuous visual task (RNP No. 57/891). Royal Navy

Woodruff A, Landay J, Stonebraker M (1998) Constant information density in zoomable interfaces. Department of Electrical Engineering and Computer Sciences, University of California, Berkeley

Woodson WE (1981) Human factors design handbook. Joint Army- Navy-Air Force Steering Committee. McGraw-Hill, New York

*Yoo H, Tsimhoni O, Watanabe H, Green P, Shah R (1999) Display of HUD warnings to drivers: determining an optimal location (Technical report UMTRI-99-9). The University of Michigan Transportation Research Institute, Ann Arbor

Zerweck P (1999) Multidimensional orientation systems in virtual space on the basis of finder. In: Bullinger H (ed) Human-computer interaction: ergonomics and user interfaces

Zetie C (1995) Practical user interface design – making GUIs work. McGraw-Hill, London

References

Fogg BJ, Eckles D (2007) Mobile persuasion: 20 perspectives on the future of behavior change. Persuasive Technology Lab, Stanford University, Palo Alto

Green P (2003) The human-computer interaction handbook. In: Jacko JA, Sears A (eds) L. Erlbaum Associates Inc., Hillsdale, pp. 844–860. Retrieved from http://dl.acm.org/citation. cfm?id=772072.772126

Honda Motors (2008) Honda develops ecological drive assist system for enhanced real world fuel economy: implementation on all-new insight dedicated hybrid in Spring 2009 [Press Release]. Retrieved from http://world.honda.com/news/2008/4081120Ecological-Drive-Assist-System/

Jean J, Marcus A (2009) The green machine: going green at home. User Exp Mag 8(4), First Quarter 2009:20-22ff

Johnson T, Jones S, Silverman A (2010) Programs hope to reverse skid in car pooling – USATODAY. com, August 5. Retrieved 9 August 2012, from http://www.usatoday.com/news/nation/2010-08-04-carpooling-down_N.htm

Marcus A (2010) Health machine. Inf Des J 19(1):69–89

Marcus A, Jean J (2009) Going green at home: the green machine. Inf Design J 17(3):233–243

Marcus A, Chen E, Brown K, Ball L (2002) BMW: future HMI directions. Aaron Marcus and Associates, Inc

Marcus A, Dumpert J, Wigham L (2011) User-experience for personal sustainability software: determining design philosophy and principles. In: Marcus A (ed) Design, user experience, and usability. Theory, methods, tools and practice, lecture notes in computer science, vol 6769, pp 172–177. Springer, Berlin/Heidelberg. Retrieved from http://www.springerlink.com/content/f8lm37795r3743v7/abstract/

Maslow A (1943) A theory of human motivation. Psychol Rev 50:370–396

National Highway Traffic Safety Administration (2012) Visual-manual NHTSA driver distraction guidelines for in-vehicle electronic devices. Department of Transportation National Highway Traffic Safety Administration Docket No. NHTSA-2010-0053 Visual-Manual NHTSA Driver Distraction Guidelines for In-Vehicle Electronic Devices

Bibliography of Vehicle User Interfaces URLs[1]

Audi A8 > Audi of America. (2012). Retrieved 10 August 2012, from http://models.audiusa.com/a8

Berman, B. (2010). EV Expert Says Nissan LEAF's Dashboard Lacks Most Important Number, PluginCars.com, December 21. Retrieved 10 August 2012, from http://www.plugincars.com/ev-expert-says-nissan-leaf-dashboard-lacks-most-important-number-106590.html

Ford's Smartgauge With Ecoguide Coaches Drivers to Maximize Fuel Efficiency on New Fusion Hybrid | Ford Motor Company Newsroom, October 28. (2008). Retrieved August 10, 2012, from http://media.ford.com/article_display.cfm?article_id=29300

Multilayer Instrument Cluster | Johnson Controls Inc (n.d.) Retrieved 10 August 2012, from http://www.johnsoncontrols.com/content/us/en/about/our_company/featured_stories/multilayer_instrument.html

Ziegler C (2012). Cadillac CUE: driving is safer (and more dangerous) than ever, The Verge, August 8. Retrieved 9 August 2012, from http://www.theverge.com/2012/8/6/3220366/cadillac-cue-safety

---

[1] Retrieval date: 9 August 2012 for all URLs.

Chapter 9
The Learning Machine: Combining Information Design/Visualization with Persuasion Design to Change People's Learning Behavior

9.1 Introduction

"Think about how different the world is today in terms of the media, in terms of medicine, in terms of the way people really experience their lives, and education is stuck in a 19th-century model" (Mr. Joel Klein, as cited in "A High-Tech Fix for Broken Schools" (Williams 2012)). As Mr. Klein says, most schools around the world are undoubtedly out-educated and in crisis nowadays. Especially for the U.S., current education systems revealed the terrible cost of losing young minds to failing schools (US college debt, nearly $1 trillion, is bigger than housing or credit-card debt.). Dropout rates are particularly high among minority children in urban schools. Approximately 7,000 students drop out of U.S. high schools each school day, as cited in "High School Dropouts in America" (Alliance for Excellent Education 2010). However, even parents in the best suburban schools are alarmed by the fact that the US now ranks 31th world-wide in math, 23rd in science, and 17th in literacy among 65 participating nations in the latest survey (Williams 2012).

Society has significant challenges with schools, and need new ideas about how to fix them. Deep changes are needed in the attitude toward teaching and learning. Digital learning gives young minds a shot at educational excellence. The modestly funded schools in Mooresville, N.C., which is best known as "Race City, USA," are drawing national attention because they give every student from third grade through high school a laptop computer and have achieved a major success in recent years. With digital learning being used by schools, "our teachers are better informed, our parents are better informed, and our students are understanding what they're doing and why they're doing it." Mr. Edwards notes as cited in "A High-Tech Fix for Broken Schools" (Williams 2012). Moreover, the *Wall Street Journal* also indicates the competition between newly increasing online courses and name-brand schools that "revolutionaries outside the ivy walls are hammering their way not onto campus but straight into class" (Williams 2012). The author points out the thrilling collegiate coup by recalling this fact: all of the 210 students, who got a perfect overall in the

© Springer-Verlag London 2015
A. Marcus, *Mobile Persuasion Design*, Human–Computer Interaction Series, DOI 10.1007/978-1-4471-4324-6_9

written assignments and exams, came from the online group. Obviously, it is time to build a new education/learning model with a new attitude and new technology now.

Besides the new attitude toward learning, it is also essential to apply scientific methods of learning to the new model. According to learning theories, learning is not compulsory, it is contextual. Learning does not happen all at once, but builds upon and is shaped by what people already know. To that end, learning may be viewed as a process, rather than a collection of factual and procedural knowledge. Thus, understanding how to harness and improve the learning process is vital to improve a student's behavior and to achieve successful results.

Therefore, AM+A created the Learning Machine, which aims to lead the nation, even the world in a different way: by using digital technology combining with learning theories, user-experience design, information-design, and persuasion design, to improve public education. As a tablet concept-prototype design, the Learning Machine is intended to guide users through the learning process more efficiently and effectively, with greater usability, usefulness, and appeal. The Learning Machine is a suite of functions and data, tools and content, best practices and templates, that are tailored to help users obtain necessary resources, manage knowledge and track progress, connect to peers and experts, find tips and advice regarding the learning process, and to seek out and enjoy appropriate incentives to behavior change.

Within this virtual space, users can easily search for necessary books, articles, resources, or topics via Internet, a Bookstore and a Discussion area. Within this virtual space, learning communities develop within and among universities, teachers, and students, in which people can discuss, share, and exchange knowledge more frequently than ever before.

Current applications and services tend to be specialized and seldom address all aspects of the learning process. Moreover, after reviewing existing mobile applications, the authors found opportunities for further adaptations and improvements to serve users' needs better. Finally, existing products seem not engaging enough and do not provide an overall "persuasion path" to change users' short-term and long-term behavior. Such a path is essential for leading users to improved learning habits.

The Learning Machine is intended for use by people from any university/college, country, and culture. It is intended to assist in the complete learning process, from joining in/participating in learning to evaluating the results of learning. The Learning Machine will be a platform that facilitates open discussion about knowledge among learners and teachers all over the world. The Learning Machine's primary objectives are the following:

- Combine information design/visualization with persuasion design.
- Persuade users to adapt their behavior and lifestyles to include better understanding of engagement, exploration, explanation, extension, and evaluation in the learning process.
- Apply user-centered design along with persuasive techniques to make the Learning Machine highly usable, useful, and appealing, thereby increasing the efficiency and effectiveness of users' efforts of knowledge-acquisition and retention behaviors.

9.2   User-Centered Design

As noted in Chap. 1, the user-centered design (UCD) of user-experiences approach links the process of developing software, hardware, and user-interface (UI) to the people who will use a product/service. UCD processes focus on users throughout the development of a product or service. The UCD process comprises the following tasks, which sometimes occur iteratively. The entire approach mirrors or is similar to innovation processes previously described in AM+A's Innovation Machine project in a previous chapter:

- *Plan*: Determine strategy, tactics, likely markets, stakeholders, platforms, tools, and processes.
- *Research*: Gather and examine relevant documents, stakeholder statements.
- *Analyze*: Identify the target market, typical users of the product, personas (characteristic users), use scenarios, competitive products.
- *Design*: Determine general and specific design solutions, from simple concept maps, information architecture (conceptual structure or metaphors, mental models, and navigation), wireframes, look and feel (appearance and interaction details), screen sketches, and detailed screens and prototypes.
- *Implement*: Script or code specific working prototypes or partial "alpha" prototypes of working versions.
- *Evaluate*: Evaluate users, target markets, competition, the design solutions, conduct field surveys, and test the initial and later designs with the target markets.
- *Document*: Write white papers, user-interface guidelines, pattern descriptions, specifications, and other summary documents, including marketing presentations.

AM+A carried out most of these tasks in the development of the Learning Machine concept.

9.3   Learning Theory

What is learning? Is it a change in behavior or understanding? Or is it a process? Learning has not been officially defined by scholars. Generally, it refers to "acquiring new, or modifying existing knowledge, behaviors, skills, values, or preferences and may involve synthesizing different types of information" (Wikipedia: "Learning" 2012). To be more specific, human learning may occur as part of education, personal development, schooling, or training. Learning may be goal-oriented and may be aided by motivation; it may occur consciously or without conscious awareness.

To describe how information is absorbed, processed, and retained during learning, AM+A studied conceptual frameworks, *i.e.,* learning theories, which mainly include three domains: behaviorism, cognitivism, and constructivism.

According to those theories, majority of scholars believe that learning is shaped by what people already know and should be viewed as a process (Kolb 1984). Popular belief states that learning is influenced by not only internal factors but also

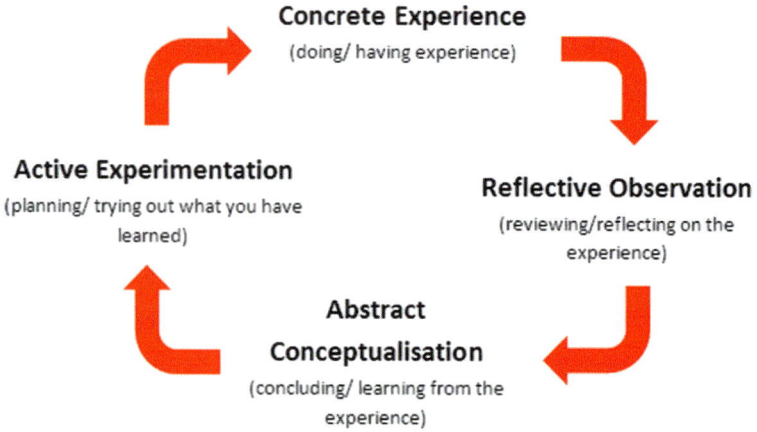

Concrete Experience
(doing/ having experience)

Active Experimentation
(planning/ trying out what you have learned)

Reflective Observation
(reviewing/reflecting on the experience)

Abstract Conceptualisation
(concluding/ learning from the experience)

Fig. 9.1 Kolb's learning cycle

by external factors (Sundberg 2014). The Learning Machine planned to infuse these philosophies in a tablet application to help students, teachers, and administrators learn more successfully and understand learning better by using a structured learning process combining internal and external resources, in order to fix the education system, which some critics consider "broken" (Fig. 9.1).

9.3.1 Learning Models

AM+A explored two popular models of learning, Neil Fleming's "VARK model" (Finn 2012) and "Learning Pyramid" (Fleming 2012).

   The first learning model is now one of the most common and widely used categorizations of the types of learning styles. This model, which expanded upon earlier neuro-linguistic programming models, was developed by Neil Fleming, who theorized that combinations of different communication characteristics determine five learning styles: Visual, Aural/Auditory, Read/Write, Kinesthetic, and Multimodal (VARK) (Finn 2012).

   In the VARK model, Neil Fleming (2012) proposes that individuals learn in different ways with a predilection for certain methods. The theory states that there are five distinct learning styles and self-knowledge of one's preferred communication characteristics is an effective way to improves learning. This model focuses on how people learn and indicates the importance of internal influence.

   Another model, the Learning Pyramid, which was developed by National Training Laboratories (Fleming (2012), shows an important learning principle that is supported by extensive research: people learn best when they are actively involved in the learning process. In the accompanying Fig. 9.2, which describes the Learning Pyramid, the lower down the cone people go, the more people learn and retain.

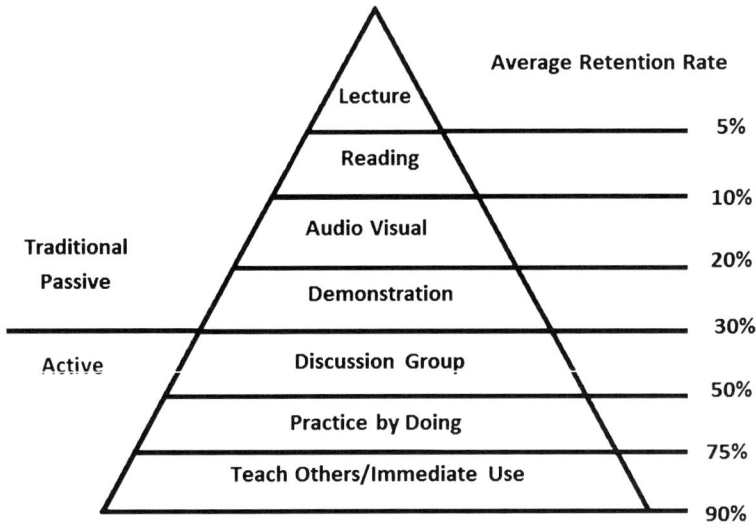

Fig. 9.2 Learning pyramid

In this model, the greatest methods of learning retention are at the base of the pyramid. This learning retention is achieved through *discussion groups, practice-by-doing, sharing ideas, and teaching others*, all of which contribute to significant effectiveness in achieving a deep understanding and transformational learning. In addition, based on the results of marketing surveys, students' primary concern about on-line education is the interaction with teacher and students. Therefore, the Learning Pyramid illustrates the fact that interaction is crucial in one's learning process as one of the external influences.

The VARK model and the Learning Pyramid model build together a philosophy that the learning process is influenced by both internal and external factors. Both effective learning methods and active collaboration between people are imperative for learning. The Learning Machine planned to utilize these ideas by creating a collaborative learning community with rich resources. As a result, people of different learning styles can study individually in an efficient way, and can share, validate, and discuss knowledge more frequently than ever before.

9.3.2 E-Learning Theory

From the didactical point of view, there are numerous approaches to learning, such as learning by observation, learning by enquiry and investigation, learning by doing, learning individually, learning face-to-face with individuals and in groups, learning by experiment learning by evaluation, and learning by reflection. E-learning includes all forms of electronically supported learning and teaching. The information and communication systems, whether networked learning or not, serve as specific media to implement the learning process.

E-learning environments that exploit interactive multimedia are of special interest. According to Barbara Buckley et al. (2004), interactivity fosters active learning; the sensory-rich nature of technology facilitates the engagement of additional powerful cognitive processes and integration of assessment tools into the environment can provide students with feedback, encouragement, etc. However, society still lacks information and information communications technology (ICT) use in schools throughout the world, despite the possible educational benefits and social learning opportunities they promote. Though several factors are important in the lack of e-learning, for example, technical constraints, integration into the teaching process, innovation of teaching methods, general acceptance, *etc.*, one of the most important reasons is the attitude toward e-learning. With regard to e-learning, people are still afraid of lacking "social presence," which is defined as "the degree of awareness of another person in an interaction and the consequent appreciation of an interpersonal relationship." (Walther 1992)

Social presence in a computer-mediated communication environment refers to the user's degree of feeling, perception, or reaction being connected to another personal and intellectual entity (McIsaac 2002), which involves a subjective quality of the communication medium related to the concepts of intimacy and immediacy. Earlier research showed that effective management of the social presence in user-interface design can improve user engagement and motivation. Enhancing social presence in an e-learning environment seems to instill the learner with an impression of a quality learning experience. One benefit is to induce and sustain the learners' motivation. The enhancement of social presence can create a successful learning experience in situations involving learners and instructors in online environments. Besides, computers could also be perceived as a social actor to improve involvement and motivation while a single learner participates in computer learning activity with no instructor involved [Text edited and adapted from "Designing Social Presence in e-Learning Environments: Testing the effect of interactivity on children" (Tunga and Deng 2006)].

In conclusion, learning is a social activity that requires a close connection to achieve better quality. The social connection is important in an online environment due to the more likely isolated nature of the instructional settings. Thus, attention must be paid to the social demand of the users in the design of computerized learning. Creating a social interface for an e-learning environment can help counterbalance the negative effects that the lifeless computer environment may have on users.

9.3.3 Learning Process

AM+A's research of various models of learning show that learning is related to both internal and external factors. Alternately, both people and environment are capable of contributing to learning process. We wanted to present a learning process that was clear and comprehensive, suitable for users of varying backgrounds and expertise. One compelling approach was "5E's Learning Cycle Model," an instructional model for constructivism (educational constructivism, not art historical

Fig. 9.3 5E's model
(Image: http://
designingepd.weebly.com/
framework.html)

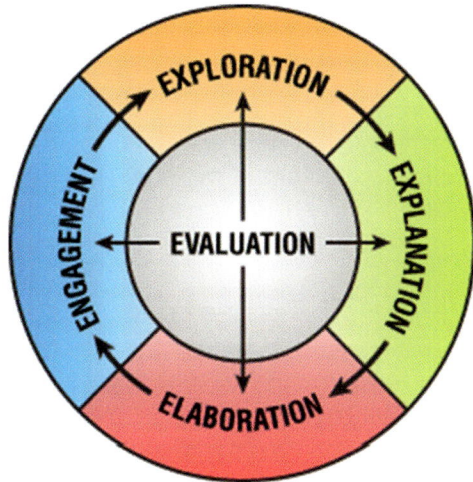

constructivism) developed by The Biological Science Curriculum Study (BSCS), a team led by Principal Investigator Roger Bybee (see, for example Bybee 1989).

As mentioned before, learning is not compulsory, it is contextual. Learning does not happen all at once, but builds upon and is shaped by what people already know. To that end, learning may be viewed as a process, rather than a collection of factual and procedural knowledge.

This learning cycle rests on learning or educational "constructivism," which is "a dynamic and interactive model of how humans learn" (Bybee 1989). Numerous studies have shown that the learning cycle as a model of instruction is far superior to transmission models in which students are passive receivers of knowledge from their teacher (Bybee 1989). In this model, learning is divided into five stages: Engagement, Exploration, Explanation, Elaboration (Extension), and Evaluation (see Fig. 9.3).

In studying this cycle model, AM+A utilized different approaches to create a process that contains these five aspects in learning. The phrases in parentheses in the titles below and some of the terms in the text refer to components of the Learning Machine's information architecture, which will be explained later. Here they are for future reference.

9.3.4 Part 1: Engagement (Course Content)

The first step is to engage in the learning materials. Engagement is a time when the teachers are on center stage. The teacher poses the problem, pre-assesses the students, helps students make connections, and informs students about where they are heading, so that a students' interests are captured or the topic is established.

9.3.4.1   Identify Learning Styles

According the so-called design schools or d-school's modes of design thinking, an approach to product/service innovation-process, empathy is the foundation of a human-centered design process. Thus, in order to truly understand users and their needs, one must first observe the status quo. By examining the learning styles and communication characters with VARK models, both students and teachers can better understand the context in which learning might occur.

9.3.4.2   Capture Interest

After identify the learning style, teachers have a better understanding of students' needs and how those needs are currently being addressed. Through this understanding, teachers can better establish topics to capture students' interest.

9.3.5   Part 2: Exploration (Course Content)

Once students' interests or topics have been identified, in the case of a new topic, students are enabled to construct knowledge in the topic through facilitated observation and questioning. Students listen to the lectures; watch the videos, read books and articles, practice in the lab, etc. This exploration is imperative for students in the learning process.

9.3.6   Part 3: Explanation (Social Media)

There are a number of essential questions that must be discussed to generate further and deeper thoughts. To answer these questions, one must engage and explore in a number of discussions. After the exchange of ideas and thoughts, students are asked to explain what they have discovered, and the instructor leads a discussion of the topic to refine the students' understanding through the *Discussion Board* and *Group Management*. Also, through discussing with a larger community, such as *Forum* and *Blog*, one can check assumptions and gain feedback.

9.3.7   Part 4: Extension (Research)

Students extend by themselves to gain more in the learning process. They can search for more information, or apply what they have learned in different but similar situations, and the instructor guides the students toward the next discussion topic.

9.3.8   Part 5: Evaluation (Course Overview)

The instructor observes each student's knowledge and understanding, and leads students to assess whether what they have learned is true. Evaluation should take place throughout the cycle, not within its own set phase. This part has a different user interface for different users.

9.3.8.1   Track

Students can track their performance in the course, including the progress of tasks and projects, and the result of tests. By checking the progress, students are clearly informed of the evaluation of their learning behavior.

9.3.8.2   Monitor

Teachers can monitor the task progress of students, as well as the questions and discussions from students. More important, teachers get to know students' situation exactly using the database in the Learning Machine. For example, where most students stop or go back in the lecture video, which questions most students answered wrong in the test, what key words are most common in students' posted questions and searching process, etc.

9.4   Personas

The UCD process emphasizes determining personas and use scenarios for future product/service development of successful user experiences. Personas are characterizations of primary user types and are intended to capture essentials of their demographics, contexts of use, behaviors, motivations, and the impact of all the preceding on design solutions. Personas are also called user profiles. Typically, UX/UI development teams define one to nine primary personas.

For the Learning Machine, the following personas were determined by analyzing available data.

9.4.1   Persona 1: Paul Moore

- Age: 19.
- Race/Gender: Caucasian, male.
- Occupation: Undergraduate.
- Location: New York, New York.
- Persona image credit: AM+A.

9.4.1.1   Textual Summary

Paul is a college student studying animation in New York University. Paul has multiple interests, and he is a smart person who is always curious about unknown things and is fond of learning. For that reason, learning is the process of discovering the wonderful world for Paul. Paul prefers to discuss with others, no matter teachers or friends, in order to share and exchange the information. Thus, he doesn't like an online video lecture without interaction.

Recently Paul would like to learn computer programming, which might help him better perform in his major. Because he is busy with college life, he would like a flexible schedule of learning at night. It is not necessary for him to get a certification or degree; just to learn something useful. In addition, he prefers more access to the teacher or experts, so that he can understand better by discussion.

9.4.1.2 Design Implications Summary

Objectives

- Obtain more learning resources in a quick and convenient way.
- Gain access to teachers or experts in various fields.
- Find means to brainstorm and discuss with a broader network.
- Arrange a flexible schedule for learning as a full-time student on campus.
- Keep up with the world and the latest news in different industries.

Context

- College student majoring in animation.
- Has multiple interests and pure love of learning.
- Discovers unknown things in spare time.

Behavior

- Auditory learner.
- Accepts online courses, only when they include interaction with people.
- Uses social media, *e.g.,* YouTube, Facebook, and Twitter.
- Owns an iPad and an iPhone.

Design Implications

- Access to learning resources quickly and easily.
- Communicate instantly with a broader network with different background.
- Arrange a flexible schedule for learning at night.
- Stay updated with the world's latest news.

9.4.2   Persona 2: Shuang Li

- Age: 22.
- Race/Gender: Asian (Chinese), female.
- Occupation: Graduate.
- Location: Chicago, Illinois.
- Persona image credit: AM+A.

9.4.2.1   Textual Summary

Shuang is an introverted young woman who is not good at interacting with people. Yet, her dream is to become a teacher in the future. Although Shuang has gained a lot of teaching skills, she still needs a platform to learn better communication and teaching skills in practice.

Over the past year, Shuang has been studying in China, but recently she came to the U.S. for a master's degree in education. It is difficult for a shy young woman to adapt to the open discussion and active participation in American classrooms in a different education environment. However, to become a teacher, it is imperative for her to learn to be more open and to communicate better with others. Now Shuang is looking for an appropriate way for herself to communicate with others and adjust herself to a different learning culture.

9.4.2.2   Design Implications Summary

Objectives

- Adapt to a different learning environment.
- Find means to learn better communicate with others.
- Find ways to practice teaching skills.
- Keep up with the latest news in the field.

Context

- Foreign graduate majoring in education.
- Introverted and not good at interacting with others.
- Improve communication and teaching skills in practice.

Behavior

- Kinesthetic learner.
- Likes online course.
- Uses social media, likes to communicate by email.
- Owns an iPad.

Design Implications

- Adapt to a different learning environment.
- Communicate better with others.
- Learn how to teach others in practice.
- Stay updated with the world's latest news.

9.4.3   Persona 3: Siddharth Shankar

- Age: 28.
- Ethnic Group/Gender: Indian-American, male.
- Occupation: Sales Representative.
- Location: San Francisco, California.
- Persona image credit: AM+A.

9.4.3.1   Textual Summary

Siddharth is a sales representative of Inspira, a 4-year-old San Francisco Bay area-based marketing company that produces and sells skin-care products. He has been working in the marketing company in San Francisco for several years since he graduated from the university. He is now one of the most experienced sales representatives in his team. Yet, Siddharth expects an opportunity of promotion to become a team leader. Therefore, he decides to get a secondary degree of MBA at a business school in Los Angeles.

However, it takes him hours' drive there every weekend, which makes him exhausted. Another problem is that he cannot gain access to teachers in a fast and convenient way when he has problem. Sometimes busy teachers even seem not to notice his email until several days later. Therefore, Siddharth is looking for some more efficient and effective method of learning.

9.4.3.2   Design Implications Summary

Objectives

- Obtain more learning resources in a quick and convenient way.
- Save traffic time and make best use of his weekend time.
- Gain convenient access to teachers anytime outside the classroom.
- Keep up with the latest news in the field.

Context

- Sales representative in a marketing company.
- Work for years and want to be promoted.
- Study to get a secondary degree at nights and on the weekend.

Behavior

- Read and write learner.
- Interested in online courses.
- Uses email to communicate.
- Owns an iPad and a smartphone.

9.4.3.3   Design Implications

- Access to learning resources quickly and easily.
- Save traffic time.
- Communicate instantly with teachers and peers.
- Stay updated with the world's latest news.

9.4.4   Persona 4: Luis and Gabriela

- Age: 20 and 26.
- Race/Gender: Hispanic, male and female.
- Occupation: House wife and worker.
- Location: El Paso, Texas.
- Persona image credit: AM+A.

9.4.4.1   Textual Summary

Gabriela and Luis are a young Hispanic couple. Gabriela married Luis early at the age of 18, just after she graduated from high school. Now they have a two-year-old daughter. Gabriela is busy taking care of her baby and family. Yet, she would like to go back to gain university education because she is still young and did not complete her schooling.

The husband, Luis, is a 26 year-old entry-level customer-service employee at a small company of ethnic handicraft. Luis's job is learned from his grandparents, so that he had to start immediately after he got married. Now he wants to continue education to get a better job in order to earn more and better support his family.

Both Gabriela and Luis have little time; however, they still want to save some weekend time for the family. Besides, they cannot leave their little daughter alone at home without parents. The best way for them is to take online course so that they spend time with their family and meanwhile become students in a university.

9.4.4.2 Design Implications Summary

Objectives

- Obtain more learning resources in a quick and convenient way.
- Save traffic time in weekend for the family.
- Arrange a flexible schedule for learning.
- Control the expense on education.
- Find means to better communicate with teachers and peers.
- Keep up with the latest news in the field.

Context

- Hispanic young couple with a daughter.
- Not well educated and plan to go back to university.
- Take care of family and study simultaneously.

Behavior

- Multimodal learner.
- Accepts online course.
- Uses email to communicate.
- Owns an iPad.

Design Implications

- Access to learning resources quickly and easily.
- Save traffic time for family.
- Arrange a flexible schedule for learning.
- Control the expense on education.
- Communicate instantly with teachers and peers.

9.4.5   *Persona 5: Aidan Robertson*

- Age: 18.
- Race/Gender: African American, male.
- Occupation: Undergraduate.
- Location: Detroit, Michigan.
- Persona image credit: AM+A.

9.4.5.1   Textual Summary

Born and raised in Detroit, Michigan, Aidan has been drawing and designing since high school. Now Aidan has graduated and plans to continue his education in college. Aidan is interested in high technology and expects to design applications by himself. However, the 18-year-old young man has concerns about the high cost in the university, because his parents were divorced 2 years ago, and it is difficult for Aidan's parents and for himself to afford such an expense in university education.

Furthermore, Aidan is not able to settle down in the near future. In the current circumstances, he follows his father to change work from city to city. Therefore, the best choice for Aidan might be online education, which is less expensive and appropriate for his changing physical location.

9.4.5.2 Design Implications Summary

Objectives

- Obtain learning resources without access to university.
- Learn from a distance, not limited to a certain city or university.
- Arrange a flexible schedule for learning.
- Find means to better communicate with teachers and peers.
- Keep up with the latest news in the field.

Context

- Graduate from high school.
- Become well educated without going to university.
- Concern about the high cost in university.

Behavior

- Visual learner.
- Be fond of online courses.
- Uses various social media.
- Owns an iPad and a tablet.

Design Implications

- Access to learning resources outside the university.
- Learn anywhere in the world, not limited to the location.
- Arrange a flexible schedule for learning.
- Communicate instantly with teachers and peers.
- Stay updated with the world's latest news.

9.4.6   Persona 6: Marie Laudati

- Age: 32.
- Race/Gender: Caucasian, female.
- Occupation: Teacher.
- Location: Seattle, WA.
- Persona image credit: AM+A.

9.4.6.1   Textual Summary

Marie is a professor who has been teaching in Seattle University for 5 years. She is fond of technology and attempts to find a way of combining traditional teaching and modern technology.

According to her experience in the past 5 years, she feels that there are many problems in the existing education system, *e.g.*, the high percentage of dropouts arising from the high cost of education. Marie is worried about the situation and trying to come up with a solution in online education.

9.4.6.2   Design Implications Summary

Objectives

- Combine traditional teaching with modern technology.
- Find solutions to fix the problems in education system.

- Improve teaching techniques according to students' situation and feedback.
- Keep up with the latest news in the field.

Context

- Caucasian young teacher with rich experience.
- Want to improve the effect of teaching and communication.
- Adjust new technology into the traditional teaching.

Behavior

- Experienced teacher.
- Likes new technology and online course.
- Uses email to communicate.
- Owns various technical tools, such as iPad.

Design Implications

- Communicate instantly and remotely with students.
- Track students' performance in accuracy with database.
- Improve teaching techniques with database.
- Keep up with the latest news in the field.

9.5   Use Scenarios

As described in Chap. 1, user-centered design (UCD) techniques prescribe the use of use scenarios in designing a successful user experience. Use scenarios are a user-centered UI development technique used during the development of an initial concept prototype. A use scenario helps determine what behavior to simulate. A scenario is essentially a "mini-story" or sequence of task flows with actual content provided, such as the user's demographics and goals, the details of the information being exchanged, etc. Use scenarios are "real world" and "every-day language" descriptions.

9.5.1   General Use Scenario

The following use scenario topics are drawn from the five preceding personas. However, some specific examples might be relevant only to a particular age group, education level, gender, or culture. Note the general usage of the terms "objective" and "goal." An objectiv e is a general sought-after "qualitative" target circumstance. A goal is more specific and is usually quantified by concrete, verifiable conditions of time, achievements, deliverables, etc.

The following use scenarios implement the 5E's model: "Engagement, Exploration, Explanation, Elaboration, Evaluation" into the Learning Machine. The terms in bold in parentheses refer to items in the Learning Machine's information architecture, which will be described in a following section of this document.

9.5.2   Learning Management (Engagement, Exploration)

- Take a short pre-test to tell the student's learning style (based on Neil Fleming's VAK/VARK model). Student and his/her teachers will be informed of the test result. **(Pre-Test).**
- Receive notification of announcement and updates of subscribed articles, blog posts, and resources via push notifications. *(Announcement).*
- Keep up with the latest news, activities, events, and receive daily tips and advice for learning more efficiently and effectively. *(Announcement).*
- Push notifications can be customized according to students' different learning styles. *(Announcement).*
- Remind user of the of course schedule, including basic information of time, location, professor, etc. *(Calendar).*
- Review information about the enrolled university, courses, and lectures.
- View an introduction to the professor. *(Course Information).*
- View information about enrolled courses: general objectives, specific intentions, course content, evaluation policies, bibliography, calendar of activities, etc. *(Course Information).*
- Manage documents of course, which include collection of course slides, examples, projects, etc. *(Course Document).*
- Watch online lectures and while taking notes and adding comments. *(Lectures).*
- Use message box to post questions and get feedback from teacher anytime.*(Lectures).*

9.5.3   Social Network (Explanation)

- View public profile with contact information of teacher and classmates. *(Profile).*
- Exhibit subjects, majors, university, or interested topics, activities, projects in public profile, and view others' information as well. *(Profile).*
- Connect to teachers and peers enrolled in the same course via Board, while access to others in different major, university, or countries via Forum and Blog. *(Board, Forum, Blog).*
- Post questions or topics for discussion, collaborate with others, and hold brainstorming sessions remotely using a virtual tool. (Board).
- Discuss and share ideas, news, knowledge, and information based on different topics and fields. *(Forum).*
- Interact with peers and learn from each other with peer-to-peer tutoring. *(Forum).*
- Broadcast thoughts on blog, as well as viewing others, comments, and sharing with others' articles and other information on the blog. *(Blog).*
- View and track the discussion user has posted, commented on, or topics and blogs to which user is subscribed to. *(My Track).*

Library Resource (Extension)

- Access online library with rich resources of books, articles, magazines, newspapers, journals, etc. *(Library).*

- Purchase books or articles with students' discount. Subscribe to interesting magazine, newspapers, journals, and keep updated with the latest news. *(Library)*.
- Find and filter through related discussion topics, projects, and blog articles using tags in the Learning Machine. *(Search Internal)*.
- Make use of search engine based on Google and combine with resources from Apple and Amazon. *(Search External)*.

Gamification (Evaluation)

- Students review the information of tasks (assignment, project, and activities, etc.) and receive deadline alerts. *(Task)*.
- Students report the progress of tasks and complete tasks given by teacher before deadline. *(Task)*.
- Students keep track of scores, grades, and results (mistakes and errors) with teachers' comments of all the assignment, test, and project during the course. *(Grade)*.
- Students gain digital achievements (*e.g.*, badges, awards, discounts for purchases, etc.) for making deadlines, completing goals, etc. *(Achievement)*.
- Students can check the status (not scores) and achievement of their friends. **(Rank).**
- Students are rewarded for particularly helpful or insightful posts and articles with points that are redeemable for reputation status and titles. *(Forum, Blog)*.
- Teachers give students tasks and assignment, set goals and deadlines for project. For individuals, teachers can modify the format of task to better adjust to the learning style of a student. *(Task)*.
- Teachers monitor and keep track of the visualized progress of ongoing tasks. *(Task)*.
- Teachers have access to a database of students' mistakes and errors, and can keep track of students' learning questions and problem. Thus, teachers are able to change and arrange the lectures and exercises to make them better. *(Grade)*.
- Teachers, who are evaluated by students, can also gain digital achievements (e.g., badges, awards, discount coupons, etc.) for a high evaluation score. *(Achievement)*.
- Teachers can check the status of students enrolled in the class and achievement of friends. *(Rank)*.
- Teachers are rewarded for particularly helpful answers to students' questions, or insightful articles on blog with points, which are redeemable for reputation status and titles. *(Forum, Blog)*.

9.5.4   Persona Use Scenarios and Behavior Changes

9.5.4.1   Paul's Use Scenario

Paul has multiple interests, and he is avid learner who spends most of his time on different topics, from programming to history, besides his own major animation. Paul gains access to post questions to teachers or professors in various fields (*e.g.,* computer science), and get to see what they think of these topics. Paul makes use of the Learning Machine to get a sense of experts' sentiments.

Paul also uses the Learning Machine to broadcast topics of what he learns recently to a relevant network in Forum and to publicize his newest thoughts on his learning blog. From all of his reviews, Paul has accumulated a number of ideas and comments from people with different backgrounds, which help him broaden his horizon and interests.

Paul carries the portable iPad and studies with the Learning Machine anywhere anytime he wants to. No matter if he is in the library or café, no matter if it's the noon break or night-time at home, Paul can decide the location and schedule for learning.

Paul is interested in discovering the world, and he likes to keep himself updated with what's new in the world. With the Learning Machine, Paul is able to quickly overview the news in different industries. Paul also relies on the Learning Machine to keep him updated with the latest animation trends and technology news.

Paul's Behavior Changes

- Paul had always watched online videos on YouTube to learn, but they didn't work well, because there were no experts to answer questions he encountered. Paul makes use of Chat World (messenger) to seek for professors' help and interact with teachers. He is satisfied with the access to teachers in various fields and the results: he always gets answered quickly. Because Paul is passionate about learning and discovering, the Learning Machine acts as an outlet for his passion.
- In the past, Paul was limited to communicate with classmates and peers in the same background and with similar values. In the circumstance, he sometimes feels ideas are blocked. Now, he uses the social network in Learning Machine, which leads him to a broader network of people in various fields and backgrounds. The Learning Machine also enables him to develop more interests and learn easier by talking with different people than ever before.
- As a full-time college student, Paul had his own college life on campus. Therefore, to take a classroom-based course often conflicted with his course schedule in the university. Besides, he didn't want to be officially enrolled in a course, because sometimes he just audited to learn for interest. Now, Paul can study whenever he is free, at any place he wants to sit. It's no longer necessary for him to arrange a specific time for a course.
- Because Paul was interested in many things, he used to catch up with the news on different topics, subjects, and industries by looking through numerous newspaper, magazines, and blogs, which turns out to be a massive challenge. Now, the Learning Machine allows Paul to subscribe to as many topics as he wants. Moreover, he can easily sort the information by different subjects or courses, which helps Paul organizes his information related to different fields.

9.5.4.2 Shuang's Use Scenario

Shuang relies on the Learning Machine to adapt to a new learning environment, which is totally different from her own country. Shuang finds many articles about how to get though the rocky transition as a new foreign student on the Forum and

Blog. Besides, many native students and foreign students in a similar situation provide useful advice and suggestions, which help her make a successful transition into school life.

Shuang is able to use the Learning Machine, not only as an information- gathering platform, but she is also able to use it to brainstorm privately as well as collaboratively with others. Shuang and her team members in one project use the Group Management functions in the Learning Machine to track project progress. They keep track of their discussion automatically with the Learning Machine, and also post the topic on the Forum and Blog to gain others' opinions. In her final report to her professor, Shuang's team submits the report along with the brainstorm transcript and project timeline, which wins a good score for them on the course.

Shuang is active in the Forum and Blog, where she answers others' questions and post reviews or comments on the blog articles. So many people find her reviews helpful that Shuang has obtained Guru status in the Learning Machine. Besides, she also uses the Learning Machine to receive tips on teaching method and skills. Shuang benefits a lot in the process as she practices both the way of communication with people and the skills of teaching others through the forum in the Learning Apps.

Then Shuang teaches, guides, and tutors her peers in the Peer-to-Peer Tutoring section using Learning Machine to practice her teaching skills. In the peer-to-peer tutoring section, Shuang needs to act like a real teacher, who gives lectures and posts and answers questions by students, sometimes even needs to give assignment and tasks.

Shuang's Behavior Changes

- As a new student with a different cultural background, Shuang used to be uncertain of the environment and to be afraid to speak up within her school. However, she is now conquering the difficulties and getting comfortable with the different learning culture with the help of advice and suggestions from the Learning Machine, which provides a platform for people to exchange all information related to learning.
- Shuang is not good at expressing herself, because of her shy personality and use of a foreign language. However, now she feels able to share her comments freely, especially in online chat. She never feels nervous and participates in the discussion actively in front of the screen. Shuang has found an appropriate way for her to discuss and communicate with others online.
- Shuang used to feel lack of confidence sometimes; especially when she was having problems explaining something to others. Yet, she enjoys posting on the Forum and Blog and comments on others' topics and questions. Because English is not her first language and she's shy by nature, Shuang likes this platform much better than the face-to-face meetings she typically attended earlier.
- No matter how much Shuang learned from textbooks, she didn't really know how to teach until she really did teaching. Because Shuang can sign in on the Learning Machine as a volunteer to be the tutor in the Peer-to-Peer Tutoring, she begins to learn how to be a teacher and how to teach or tutor others in practice instead of relying just on "book knowledge" or theory. Shuang even becomes more confident because she gains others' respect and praise during the process.

9.5.4.3   Siddharth's Use Scenario

Having success with sales, Siddharth works in a fast-paced environment. To become a team leader, he will be involved heavily in all aspects of his company in the marketing field. Therefore, Siddharth customizes his news feed to aggregate all relevant materials, which is helpful for him to efficiently catch up with the latest news in field and uses the vast number of resources within the Learning Machine to learn more about marketing trends as well as management practices.

Siddharth just stays at home and uses the Learning Machine on an iPad to take the online courses. It works for him to listen to the lectures, take notes, ask questions, and submit assignments as a normal student on campus. He also appreciates saving much time for learning instead of being stuck in traffic jams. As a result, he is able to have a good rest on weekends.

Taking online courses with Learning Machine, which is easier for Siddharth. Siddharth gains convenient access to teachers anytime outside the classroom. Communication with teachers is available from Monday to Friday, not limited to the weekend classes. Besides, he can contact with teacher for a longer time online instead of end the talk in several minutes in a hurry after class.

As a sales representative, Siddharth spends most of his workday with customers and clients instead of his co-workers in the marketing field. After using the Learning Machine, Siddharth gets along well with his classmates and connects to a broader network around the world. He even get to know some partners in nearby neighborhoods and cities.

Siddharth's Behavior Changes

- Siddharth used to read newspapers and journals every day to catch up with the latest news. He spent a lot of time on this activity, but it is somewhat inefficient. Having the Learning Machine, Siddharth now subscribes to all of his interested news and receives notification of the updates daily easily with one app. He gains rich resources in a quick and convenient way and easily keeps up-to-date with the recent news within his field.
- As a part-time student of a University in Los Angeles, Siddharth was constantly driving between company and university across the city, often spending one hour minimum on each trip. Siddharth felt exhausted, because he had to drive four hours to Los Angeles to take courses every weekend. Siddharth now turns to take online courses, which saves him traffic time and make the best use of his weekend time. He even makes more progress on his study because of having a good rest.
- Siddharth was limited to the time and location of the courses in classroom-based courses. He cannot access to the teacher easily when he had questions on weekdays. He could not discuss with other classmates in a convenient way because they were not in the same city. Now he just goes nowhere and easily learns anytime he wants to and discusses with anyone he wants. He is no longer limited to the time and location.

- Besides the client and customers, Siddharth would like to extend his social network, for example to his classmates in the courses. Given his schedule, Siddharth had difficulty keeping contact with his classmates during a short time in classroom. Now he is heartened that he is not alone in his sentiments. He contacted other classmates through the contact information posted in profiles. He also has opportunities of making friends outside his workspace through the social network in the Learning Machine.

9.5.5   Luis and Gabriela's Use Scenario

Luis and Gabriela use the vast number of resources within the Learning Machine to learn more about the world. Luis subscribes to news and topics related to handicraft, while Gabriela prefers fashion magazines and career news for housewives. Through the Learning Machine, Gabriela also visits the individual profiles of people within her classes in which she enrolled, and learns from her classmates.

Using the Learning Machine, Luis and Gabriela arrange a flexible schedule for learning. Gabriela learns when she is on the bus, or waits at the bus stops, while Luis can learn at home after work every night. Moreover, both of them save traffic time on weekends for more time with their family.

The cost of online education is much less than enrolling in an on-campus university. Using the Learning Machine, Luis and Gabriela figure out the cost is less than half of the money they would pay for the same education in a classroom. They perfectly control the budget for education expense and will get certification as well as becoming enrolled in the university with the help of the Learning Machine.

The young couple, especially Luis, finds it is a better way to connect to others and resources. Through the Discussion Board, Luis has also learned about an interesting, new handicraft startup. Luis realized that one of the start-up founders was his classmate from this handicraft course. Using the messaging functions on the Learning Machine, Luis contacted his classmate. It turned out that the start-up team was looking for someone with technical experience, so after a few discussion sessions the start-up offered Luis a spot on their core team. After accepting the position, Luis quit his boring, albeit cushy, corporate job and is loving the excitement and challenges that come with start-up life.

Luis and Gabriela's Behavior Changes

- In the past, Luis and Gabriela did not really know where to get reliable information, so they would only refer to one or two publications. Now, they can quickly scan headlines and news with the Learning Machine's comprehensive database. Luis and Gabriela also share interesting articles and useful information with each other in the Learning Machine.
- Luis and Gabriela planned to go back to university several times in the past, but they could not both stay in the university without anyone to take care of their little daughter who is only two years old. Especially for Gabriela, she cared

about her family and wanted to spend more time with them. Now they find the solution with the Learning Machine. They don't have to leave the house and sit in the classroom; instead, they just stay at home and study online.

• Though Luis and Gabriela decided to go back to university, they do not have a large sum of money for college to be deducted from their income to support their family. The Learning Machine helps them to save the expense on education, which decreases the amount to less than half of their original budget for college. It is possible for the young couple to be educated while supporting the family at the same time.

• Gabriela, as a housewife, spends most of her time taking care of her little daughter in the past. Luis, as a customer-service operator, is busy all day with his headset and a voice on the other end of the line. They used to feel disconnected from the world outside. Now, both of them are better connected to others and to the world. With the Learning Machine, they keep updated with the news via push notification. They also share their feelings and thoughts after dinner. Both of them gain much benefit during the learning process.

9.5.6 Aidan's Use Scenario

A priority concern for Aidan is the expense to attend a university. The lower cost of online courses via the Learning Machine helps Aidan to spend much less to enroll as a full-time student. Aidan saves a large amount of the cost of education fees and the cost of books through the discounts and coupons provided in the Learning Machine.

Aidan enrolled in courses about design in the Learning Machine as a full-time student. He finds it even more convenient and valuable compared to on-campus college education. Aidan obtains efficiently the desired learning resources, which come from libraries and universities all over the world, instead of limited ones at a specific university.

Aidan can learn easily anytime, anywhere he wishes. He uses the Learning Machine to take lectures from-a-distance during his trip, not limited to a certain city or country. His schedule for learning is also flexible, he can learn when he is waiting at the bus stop or sitting on a train, he can learn during a break or on the weekend.

Aidan uses his newfound fame through the Learning Machine's social networking to publicize his own ideas about design and an app-review blog. Despite not having the tech background, Aidan's impressive Learning Machine status and "achievement-badge" collection validates him as someone passionate about technology. Aidan uses the Learning Machine to connect to people with similar interests in design and technology.

Aidan's Behavior Changes

• Many of Aidan's friends had to drop out after they graduated from high school because of the high cost of education. Therefore, Aidan's father was worried about the money for Aidan's education in a university. However, Aidan can afford it by himself after he decided to enroll in the online courses available in the Learning Machine, which provides an opportunity for continuing his education.

- Aidan used to spend much time and feel the low efficiency of doing research in the library in school. Even with the Internet, he still had limited access to resources. Now, Aidan enjoys the powerful search engine within the Learning Machine, which connects him to libraries and universities around the world. Moreover, instead of resources from only one or two publishers in the past, Aidan can read about materials from a variety of notable resources. He prefers to learn using the Learning Machine's technology, too.

- Aidan was restricted to his high school near his neighborhood, which lacks education resources. Now, he can select the course(s) and university not limited to the location. Aidan is applying for his dream school in France, which he had never considered before.

- Aidan was looking for people with promising ideas to collaborate with, but he did not have a platform he could use to look for such opportunities outside his current location. Now, Aidan connects with like-minded design-passionate folks using the community within the Learning Machine. He can access those people, not only his classmates, but also a broader group across the world via the Forum and Blog in the Learning Machine.

9.5.7 Marie Laudati's Use Scenario

With the push notification, Marie can reply to students' questions immediately during her working time. She can also post the topics and discuss with a group of students at the same time in the Discussion Board. In addition, Marie and her students share interesting articles with each other through the Forum and Blog. Not only students, but also Marie as the professor, benefit much in the process of teaching and communication.

Marie is informed of the learning style of her students at the beginning of the course. Sometimes she might adjust the assignment with advice about the course based on students' different preference/skills. In addition, she can track in the Learning Machine's database how students watch her video lectures, or the questions for which most students make mistakes in a test. As a result, she can explain more effectively in the Discussion Board or post her own blogs.

Marie has been a professor at Seattle University for a few years. While she enjoys her job, she recognizes that it is a large organization that has a tendency to move more slowly. As such, she feels that it is difficult to stay up-to-date with all the activities within the school and the field. With the Learning Machine, she is able to subscribe to the latest news and communicate with other professors around the world on exactly the topics and issues that are of highest interest to her.

Through the Learning Machine, Marie is able to see a concise visual summary of all of her students' current tasks. Through this visual summary, she can see how different components within those projects are developing within an overarching timeline. Once students update their work on the Learning Machine, Marie receives timely updates about the progress of individual project members. Besides, Marie also uses this platform to challenge her students to brainstorm ways to streamline their reporting processes. Marie can set specific deadlines for the students. As deadlines grow closer, she is notified by alerts and can choose to send reminders to students.

9.6   Market Research

In order to have a clearer vision of the target market for the Learning Machine, AM+A conducted market research with potential customers, in collaboration with marketing students of the University of California at Berkeley (UC/B), International Diploma Program (IDP). One main objective of the research was to find out more about students' willingness to use online learning applications as a way of studying, graduating, and obtaining certification, instead of real-time, face-to-face education. The second objective was to get a better understanding of their learning behavior and their attitude toward the online education.

The research process included an exploratory part and an active part. First secondary research was carried out in order to find greater understanding about the topic and summarizing existing knowledge, theory and practical examples. The secondary research also included an exploratory interview of typical users, which, together with the findings of the data research, led to a guideline for the qualitative part of the research project. As qualitative research, the marketing research team conducted six in-depth interviews, including techniques such as probing, storytelling, word association, and photo association. The findings of the in-depth interviews led to the construction of a questionnaire, which was accessed online via SurveyMonkey. The questionnaire included 10 questions and was posted in a Facebook group of 130 IDP students of UC/Berkeley. The survey achieved 43 respondents; the main results of the research can be summarized as follows:

9.6.1   Qualitative Research

The in-depth interviews were carried out with 6 participants, 3 male and 3 female, who were aged between 18 and 32. The interviews particularly focused on students' learning behavior, expectations of the Learning Machine, and the attitude toward online education. Some primary findings from the interviews are the following:

Learning Behavior

- Interviewees explained they have similar learning habits, such as studying from books, highlighting most important phrases, listening to teachers, and interacting with other students.

Attitude Toward Online Education

- Most people did not have experience with online learning.
- In general, the research showed that people can imagine taking an online course; their answers and attitude toward online education was more positive than negative.
- Social interaction in classroom with students/teachers is highly valued.
- Interviewees had a fear that body language and enthusiasm a good professor can deliver would be missing in an online education.
- More respondents disagreed that online education has the same quality than face-to-face education.

Expectation of Online Education

- Downloadable content and live lectures were mentioned to be extremely necessary for a possible online learning tool.
- People would prefer using a desktop or portable PC rather than a tablet for an online learning tool.
- Most people think the price of an online course should be less than half of a face-to-face education.

The marketing research showed that students emphasized interaction with other students and teachers (body language), and the social interaction is highly appreciated in the classroom. Thus, to develop the Learning Machine, we need to design active social interaction for users to communicate with others.

9.6.2   Quantitative Research

In order to enrich the market research with quantitative data, AM+A prepared a questionnaire that was distributed among 43 UC/Berkeley Extension IDP students, most of them are undergraduates or graduates aged between 18 and 34 years. The questionnaire can be found in the appendices of the present document.

More than half of the respondents (29 out of 43) claimed that they have a tablet or want to buy one. Only a small number of the respondents (12 out of 43) disagreed with the statement that they can imagine attending an online course using a tablet (Fig. 9.4).

Quite interestingly nearly half of the respondents who own a tablet were neutral on this question. This result may be due to the lack of good learning-related applications in the current market (Fig. 9.5).

According to the survey results, one can determine that the main concern about online education is the course content. However, most of the learning applications

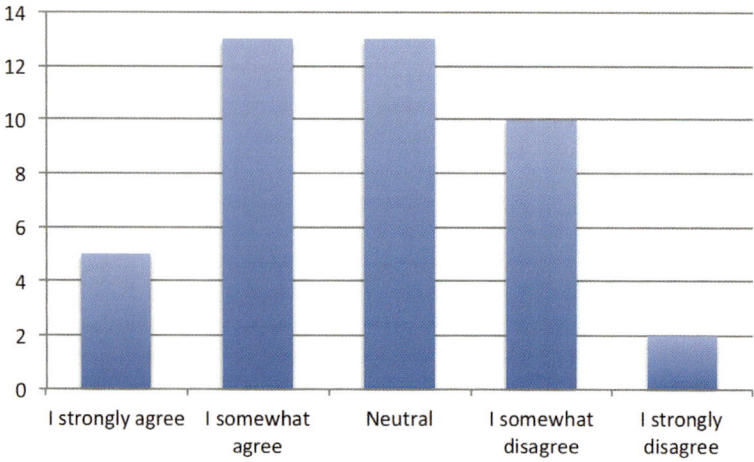

Fig. 9.4 Q2. "I can imagine attending an online course using a tablet."

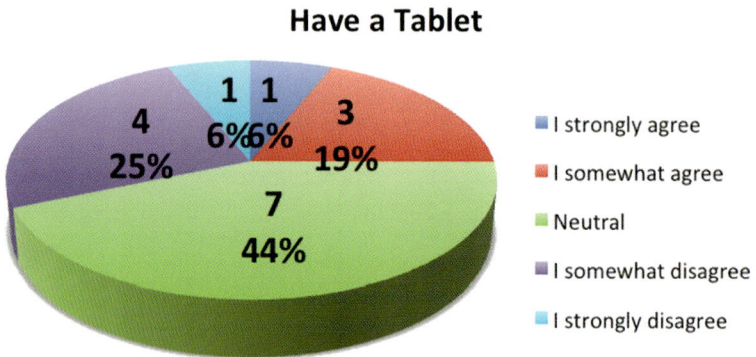

Fig. 9.5 Cross analysis of the quantitative survey results

that already exist in the market lack resources or seldom update their content. Regarding the students' primary need in education, these products seem likely not to be able to satisfy users. That situation might also explain why nearly half of the respondents (26/43) have not taken an online course before (Fig. 9.6).

The marketing research that AM+A conducted via UC/B Extension, IDP, offered interesting findings about students' attitudes toward traditional, face-to-face education and new forms of online education. The findings of the research were useful to AM+A in developing the Learning Machine and led to better development of the application suite.

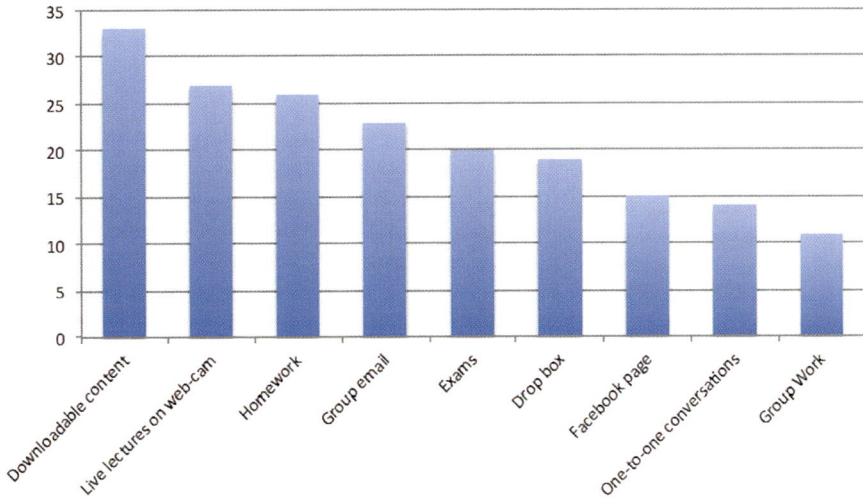

Fig. 9.6 Q5. "Which features would you expect this application to have?"

9.7   Competitive Analysis

9.7.1   Comparison Studies

Before undertaking conceptual and perceptual (visual) design of the Learning Machine prototype screens, AM+A studied approximately seven online-education Websites and learning-related tablet applications. Through screen comparison analysis and customer-review analysis, AM+A determined these applications' major benefits and drawbacks. This in-depth analysis helped further to develop initial ideas about the Learning Machine's detailed functions, data, information architecture (metaphors, mental model, and navigation), and look and feel (appearance and interaction)

9.7.2   Learning Applications

9.7.2.1   BlackBoard

Who Were They in 2012?

- Blackboard was founded by CEO Michael Chasen and chairman Matthew Pittinsky in 1997 and became a public company in 2004.
- The company provides education, mobile, communication, and commerce software and related services to clients including education providers, corporations,

and government organizations. The company has also been a leader in the development of Internet-based education software.

- As of December 2010, Blackboard software and services were used by over 9,300 institutions in more than 60 countries.

What Was It in 2012?

- Blackboard's seven platforms are: Learn, Transact, Engage, Connect, Mobile, Collaborate, and Analytics.

Blackboard Learn is a learning management system that comprises four modules:

- A learning system that provides online course delivery and management for institutions.
- A community and portal system for use in creating online campus communities.
- A content management system for centralized control over course content.
- A system to record and analyze student assessment results.

Blackboard Collaborate launched in July 2010, providing users with Web-based video and audio conferencing.

Blackboard Mobile offered two applications, which provide students with access to teaching and learning content and campus information through native mobile applications.

What Did the App Let Users Do in 2012?

- *Blackboard Learn:* There are many functions for students and teachers in Blackboard Learn:
- Students can access course materials, view announcements, check grades, participate in discussions, save and upload document by link Dropbox account, etc.
- Student can receive notification for new announcement, new grades items, and new tests being posted.
- Students can take Mobile Test (multiple choice, hot spot, fill-in-the-blank).
- Students can read blog posts and interact with each other by posting comments and uploading media as an attachment.
- Student can reflect on their course journals through Blackboard Mobile Learn, as well as comment on peers.
- With the class roster, students can quickly view their entire class list, making and organizing study groups.
- Instructors can use the tool "Journal" to comment on student journals.
- Instructor can use "Task" to help students track and manage the progress of various tasks, from turning in homework.
- *Blackboard Collaborate* is used by K-12 schools and higher education institutions for professional development, and distance learning through virtual classroom settings in which courses are held online and students are able to interact with each other and the course teacher in real time. The platform is also used by businesses for distance learning and for conferencing.

Web Strategy

- Website looks user friendly.
- Simple and easy screen flow.
- Connected with Dropbox account.

Mobile Strategy

- Free service to all users.
- Students and instructors can access documents in multiple formats.
- Create threaded discussion posts; upload media as attachments to discussion boards and blogs.
- Mobile device is interconnected with university platform.
- Users can get the activity feed on the mobile application.

User-Friendliness

- Platform is easy to use and understandable.
- Easy reach of content.

Visual Analysis

- Screen flow seems intuitive.
- Data visualization seems busy and often redirects to internet browser. There is difficulty in the screen flow, especially sometimes the application will suggest going to the browser because of a problem with visualization.
- Users cannot change the language directly from the application.
- There are too many folders and it seems hard for users to find content.

Tips

- None.

SWOT Analysis

Strengths

- Strong functions and features that are helpful.

Weaknesses

- User interface is too busy.
- Application is not user-friendly enough.

Opportunities

- Cooperate with university around the world, so there is a large user group for Blackboard.

Threats

- Lack of functionality with social networking and creating connections, groups, and teams.

9.7.2.2   Khan Academy

Who Were They in 2012?

- Academy is a non-profit educational organization created in 2006 by Bengali-American educator Salman Khan, a graduate of MIT and Harvard Business School.
- With the stated mission of "providing a high quality education to anyone, anywhere," the Website supplies a free online collection of more than 3,400 μ-lectures via video tutorials stored on YouTube teaching mathematics, history, healthcare and medicine, finance, physics, chemistry, biology, astronomy, economics, cosmology, organic chemistry, American civics, art history, macroeconomics, microeconomics, and computer science.

What Was It in 2012?

Khan Academy is an organization on a mission. It is a not-for-profit with the goal of changing education for the better by providing a free world-class education for anyone anywhere. People view Khan Academy as:

- A complete custom self-paced learning tool.
- A dynamic system for getting help.
- A custom profile, points, and badges to measure progress.
- A real-time class report for all students.
- A better intelligence for doing targeted interventions.

What Did the App Let Users Do?

- Students can make use of an extensive video library, interactive challenges, and assessments from any computer with access to the Web.
- Users can download videos to take individual videos or entire playlists to watch offline at their own pace.
- Users can easily watch videos with features: follow along, skip ahead, or go back by navigating through subtitles.
- Users can log in to track their progress, including credits they received for watching videos, and achievements they won.
- Coaches, parents, and teachers have visibility into what their students are learning and doing on the Khan Academy.

Mobile Strategy

- The materials and resources are available to users completely free of charge.
- Rich resource with various subjects available in video library. Each video is a digestible chunk, approximately 10 min long, and especially purposed for viewing on the computer.

User-Friendliness

- Good user interaction and feedback.
- Easy to use.

- Similar to YouTube.
- User friendly, *e.g.*, subtitles incorporated in the videos, a 30-s rewind button in the video.

Visual Analysis

- Presents simple frame on three levels.
- Seems understandable, with clear flow in screen layout.
- User friendly; gathers different Internet tools.
- Seems to lack social networking functions (Fig. 9.7).

Tips

- Improves e-learning system distance education.

SWOT Analysis

Strengths

- Free application.
- Simple and understandable interface.
- Easy to use with step-by-step method.

Weaknesses

- Only English speaker.
- Lack of courses (working in progress).

Opportunities

- The opportunity to learn from any place with an internet connection.
- Any person from any socioeconomic status can learn.
- Application and Website are still building in progress; it might be more helpful after it's completed.
- Provide exercises in the next version.

Threats

- Free and downloadable resources might lead to followers or competitors in the near future.

9.7.2.3   MIT Open Course Ware (OCW)

Who Were They in 2012?

- The concept of MIT Open Course Ware grew out of the MIT Council on Education Technology, which was charged by MIT provost Robert Brown in 1999 with determining how MIT should position itself in the distance learning/e-learning environment.

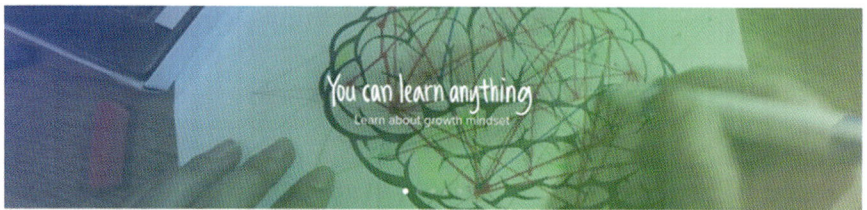

Fig. 9.7 Example screenshots of the Khan Academy app as it looks now in 2015 (Images: Khan Academy, updated images used at companies request)

- In September 2002, the MIT Open Course Ware proof-of-concept pilot site opened to the public, offering 32 courses. By September 2004, 900 MIT courses were available online. The project was announced in October 2002 and uses Creative Commons Attribution-Noncommercial-Share Alike license.
- MIT OCW is supported by MIT, corporate underwriting, major gifts, and donations from site visitors.

What Was It in 2012?

- MIT OCW is a large-scale, Web-based publication of MIT course materials.
- MIT OCW is an initiative of the Massachusetts Institute of Technology (MIT) to put all of the educational materials from its undergraduate and graduate level courses online, partly free and openly available to anyone, anywhere.
- MIT OCW is to provide a new model for the dissemination of knowledge and collaboration among scholars around the world and contributes to the "shared intellectual commons" in academia, which fosters collaboration across MIT and among other scholars.
- Please note that at the time of going to press this app is now defunct.

What Did the App Let Users Do in 2012?

- Watch free video course online.
- Use MIT mobile as a essential institute services for life on campus.
- Campus news and push notification of campus emergency information.
- Live shuttle tracking with push notification for predicated stop times.

  - Searchable campus map and self-guided campus tour.
  - Calendar or events, exhibits, holidays, and academic calendar.
  - Stellar course guide with push notifications for class announcements.
  - Contact list and searchable directory of faculty staff, and students at MIT.
  - MIT libraries account management and book search.
  - QR code and barcode scanner.

Mobile Strategy

- Free mobile application.
- Partly free and openly resource available to anyone, anywhere.

User-Friendliness

- Take time to register.

Visual Analysis

- Seemed simple and clear.
- Seemed well organized (Fig. 9.8).

Tips

- None.

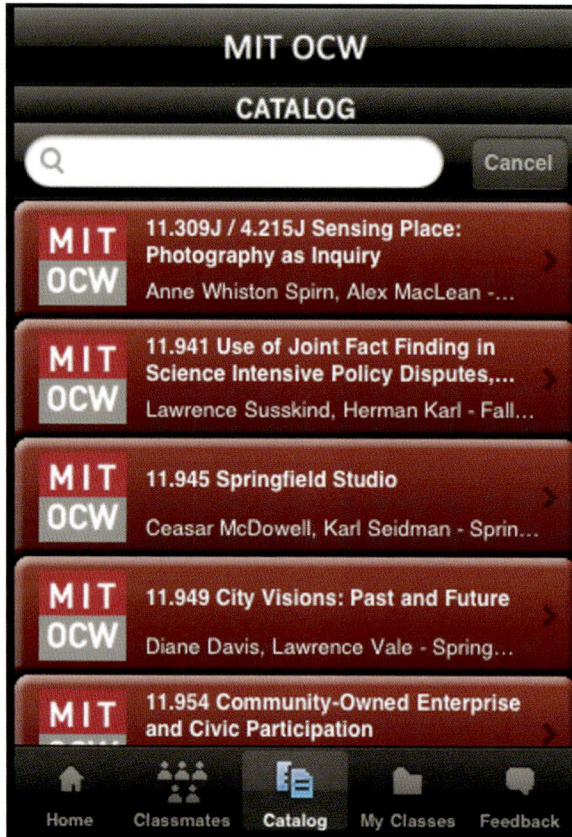

Fig. 9.8 Example screen of MIT OCW application (Image: Massachusetts Institute of Technology)

SWOT Analysis

Strengths

- A majority provided homework problems and exams (often with solutions) and lecture notes.
- Some courses also included interactive Web demonstrations in Java, complete textbooks written by MIT professors, and streaming video lectures.
- The videos are available in streaming mode, but may also be downloaded for viewing offline. Many video and audio files are also available from iTunes.

Weaknesses

- Course resource online is limited and updated slowly.
- A few of these online courses are limited to chronological reading lists and discussion topics.

Opportunities

- Initiative to launch such online system.
- Supported by MIT.

Threats

- Lack of resource.
- Competition in this market.

9.7.2.4 TED

Who Were They in 2012?

- TED is a nonprofit devoted to Ideas Worth Spreading. It started out (in 1984) as a conference bringing together people from three worlds: Technology, Entertainment, Design.
- Since 1984 TED's scope has become ever broader. Along with two annual conferences, the TED Conference in Long Beach and Palm Springs each spring, and the TED Global conference in Edinburgh UK each summer, TED includes the award-winning TEDTalks video site, the Open Translation Project and TED Conversations, the inspiring TED Fellows and TEDx programs, and the annual TED Prize.

What Was It in 2012?

- TED is an application presenting talks from some of the world's most fascinating people: education radicals, tech geniuses, medical mavericks, business gurus, and music legends.
- TED is best thought of as a global community, welcoming people from every discipline and culture who seek a deeper understanding of the world.

What Did the App Let Users Do?

On iPad:

- The entire library is at users' fingertips.
- Watch TEDTalks in high- or low-resolution formats based on users' network connectivity.
- Play videos on users' devices or send to their home entertainment system via AirPlay.
- Curate users' own playlist. Watch the videos later, even when they can't be online.
- Listen to an on-demand playlist of TEDTalks audio.
- Sort views by recency or popularity.
- Find something by tags, themes, or related talks.
- Share favorites with users' friends.
- Tell the application how much time users have so that the applications can guide users to a delightful playlist.

On iPhone and iPod touch:

- The entire library is available for browsing and searching in both online and offline modes.

Mobile Strategy

- Riveting talks by remarkable people, free to the world.
- Free mobile application with free resource (videos) that are open to anyone around the world.
- Update fast and frequent: there are more than 1,100 TED talk videos (with more added each week) on the official TED app – now for both iPad and iPhone.

User-Friendliness

- Simple and clear. Easy to use.
- Appealing to users with screen captures of the videos.

Visual Analysis

- Seemed simple and clear.
- Seemed well organized.

Tips

- Once users tell the application how much time they have, the applications can guide users to a pleasing playlist and recommended videos selection.

SWOT Analysis

Strengths

- Rich resource and increasing updates every week.
- User-friendly interface.

Weaknesses

- Limited to video resource only.

Opportunities

- Well known and popular among users, which would enable company to offer users more.

Threats

- Free for download, which makes a low entry-barrier.

9.7.2.5   Together Learn

Who Were They in 2012?

- Developed by Net Power and Light Inc. and designed uniquely for the iPad.
- TogetherLearn is the way to experience Together Learning with top universities and educators around the globe.

What Was It in 2012?

- TogetherLearn is an application to make learning incredibly fun by bringing users an appealing, vivid experience of inspired things with other people, and amplifying them all to the very highest level.

What Did the App Let Users Do in 2012?

- Enables users to bring people, ideas, and conversations together live so they can learn effectively, become inspired together, and feeling the fun of learning.
- Enables users to make their own master class, full of admired and engaging minds. Users can pause the class to engage in a lively discussion, backup if users don't understand something, or fast forward to the best parts.
- Enables users to invite friends from Facebook or personal address book and enjoy a vivid and real-time learning together experience. With ChalkTalk™ users can highlight interesting ideas, make notes on the screen for everyone to see, give a shout out, or point out the obvious. Users can pick their own colors or change colors on the fly. If users are very appreciative, they can give "standing ovations," with Standing O™, just as if users were present.

Mobile Strategy

- Free and available to anyone, anywhere.
- Spread by users' invitation to friends to learn together.

User-Friendliness

- Users need to learn how to use the user interface and features with very little guidance.
- Once users get used to the user interface, it seemed powerful and convenient.
- Interactive user interface for users to communicate.

Visual Analysis

- Seemed pleasant and appealing (Fig. 9.9).

Tips

- None.

SWOT Analysis

Strengths

- Easy to invite friends from Facebook or personal address book.
- Present users with vivid interaction and communication.

Weaknesses

- Simple way of learning (only video), lack of diversity.

Opportunities

- Nice and appealing interface and interaction.

Fig. 9.9 Example screen of TogetherLearn application in 2012 (Image: Net Power and Light, Inc.)

Threats

- Lack of systematic resource of learning.

9.7.2.6   Shufflr

Who Were They?

- Shufflr is a cool social video app where videos find users. The new Shufflr Daily Fix picks the best videos from users' world and brings them to users, every day.

What Is It?

- Shufflr overall seemed like a "Pinterest for videos," it's a video aggregator, with social networking functions well integrated. It's the best way to find, watch, and share videos on iPhone and iPad.

What Do They Let Users Do?

- Enable users to see what videos their friends are watching, celebs are sharing, to discover other interesting people with similar tastes, to know trending videos, breaking news and more.

- Share what users are watching on Facebook and Twitter on the tap of a button.
- Queue videos from anywhere with the "ShufflrQ" bookmarklet and watch it on the go.
- Personalized Daily Video Fix presented in a refreshing and multi-touch UI concept.
- Display "Daily Fix Timeline" – a window into your world of videos that went past.
- Autoplay controls for a relaxed, "hands-free" video viewing.

Mobile Strategy

- Free application to anyone, anywhere.
- Aggregate everything users wanted from social video at one place, which enable users spend less time searching and more time watching.

User-Friendliness

- The interface seems good, clear, appealing.
- The video-find features are user friendly and simple to use.

Visual Analysis

- The general appearance is very professional.
- The interface appeals to younger people who get their view of the world (for better or worse) from peer videos and occasionally from knowledgeable-expert videos (Fig. 9.10).

Tips

- None.

SWOT

Strengths

- Interactive way of searching and providing resources.

Weaknesses

- Only for entertainment instead of education industry.

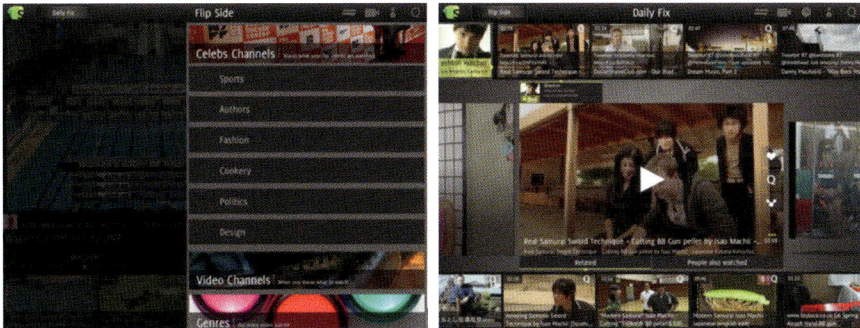

Fig. 9.10 Example screenshots of the Shufflr application in 2012 (Images: Althea Systems)

Opportunities

• Expand to various industry based on users' expectation and needs.

Threats

• Limited market. Since the application might most appeal to younger users, who might be interested in videos.

9.7.3   Online Education Websites

9.7.3.1   School Loop

Who Were They in 2012?

• School Loop is created by Mark Gross in 2004, together with his co-founders Tom Burns and Dede Tisone, to help educators keep students in school and on track.
• School Loop now serves over 3600 schools in 30 states and includes among its customers the districts of San Francisco, Long Beach, Los Angeles, Kansas City, and Albuquerque. [cited in URL: http://www.schoolloop.com/company/mission/]

What Was It in 2012?

• School Loop is created by an online application for elementary school, middle school, and high school students to view their grades online and communicate with teachers.

What Did the App Let Users Do in 2012?

• Teachers post homework for students, school announcements, and downloads. Class-specific information can be provided in a teacher's personal School Loop webpage.
• Students can see their homework assignments, school announcements, and group discussions on School Loop, and they can also discuss assignments with other students and teachers through assignment discussions. Students are allowed to store files in a digital locker for later use. Groups can be formed by students with the permission of a teacher or a staff member.
• Both teachers and students can e-mail teachers through a site email system known as LoopMail. The system provides a calendar system for assignment due-dates.
• Staffs are allowed to upload a student's grades onto the Internet for immediate viewing.
• Parents can sign up for their own School Loop account to keep track of their child's progress and check homework, assignments, and grades. They may also receive weekly emails about their child's grade and may contact the teachers.

Web Strategy

- Free website strategy.
- A large amount of participating schools.

User-Friendliness

- Easy to register.
- Video demo for instruction.

Visual Analysis

- Simple and clear.
- Well organized (Fig. 9.11).

Tips

- Parents can sign up to receive weekly emails about their child's grade and may contact the teachers.

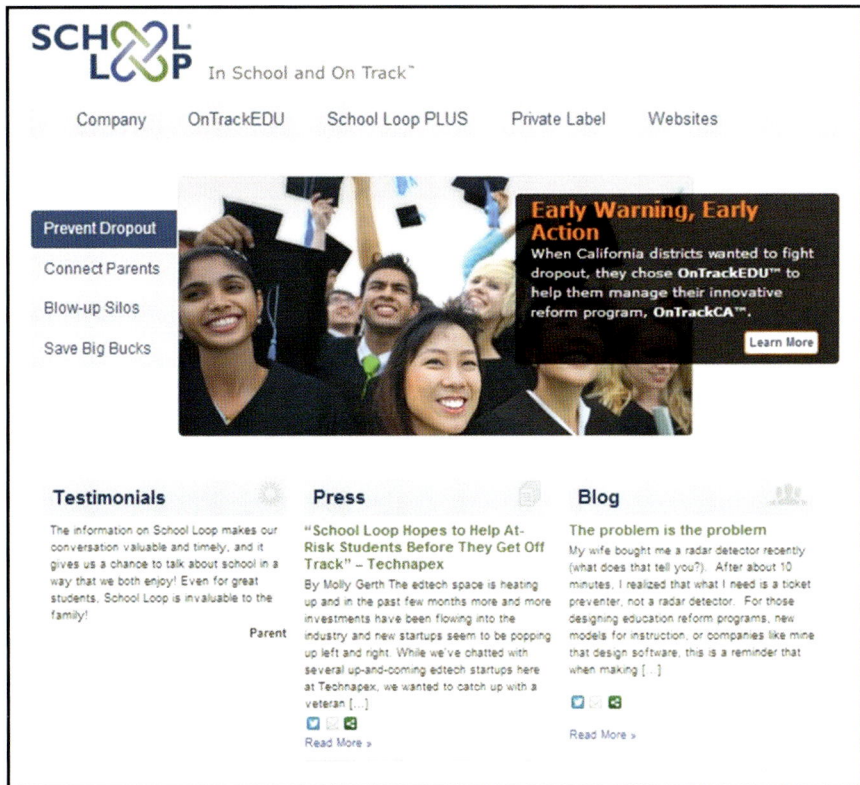

Fig. 9.11 Example screen of School Loop Website in 2012(Image: School Loop, Inc.)

SWOT Analysis

Strengths

- New methods of communication in education, which improve the work efficiency of teachers and stuffs and guarantee the transparent level of students' performance.

Weaknesses

- Complicated ways to execute the process which combines online and offline.

Opportunities

- Schoolloop provides a new clue for traditional education. It's might be a possibility for the future school.

Threats

- The website seems more like a subsidiary method of traditional education instead of an independent online education system.

9.7.3.2   Coursera

Who Were They in 2012?

- Coursera is an educational technology company founded by computer science professors Andrew Ng and Daphne Koller from Stanford University. As of November 2012, more than 1,900,241 students from 196 countries have enrolled in at least one course.

What Was It in 2012?

- Coursera is a social entrepreneurship company that partners with the top universities in the world to offer courses online for anyone to take, partly for free.
- Coursera wants to empower people with education that will improve their lives, the lives of their families, and the communities they live in.

What Did the App Users Do in 2012?

- Use technology to enables the best professors to teach tens or hundreds of thousands of students.
- Give everyone access to the world-class education that has so far been available only to a select few.
- Help users to master the material and new concepts quickly and effectively. Key ideas include mastery learning, to make sure that users have multiple attempts to demonstrate their new knowledge; using interactivity, to ensure student engagement and to assist long-term retention; and providing frequent feedback, so that users can monitor their own progress, and know when they've really mastered the material.

Web Strategy

- Coursera partners with various universities and makes a few of their courses available online free for a large audience.
- Coursera offers courses in a wide range of topics, spanning the Humanities, Medicine, Biology, Social Sciences, Mathematics, Business, Computer Science, and many others to cater users' various interests.

User-Friendliness

- Easy to understand and use.
- Clear categories with detailed sub-titles.

Visual Analysis

- Simple, clear, professional (Fig. 9.12).

Tips

- None.

SWOT Analysis

Strengths

- The efficacy of online learning.
- The importance of retrieval and testing for learning.

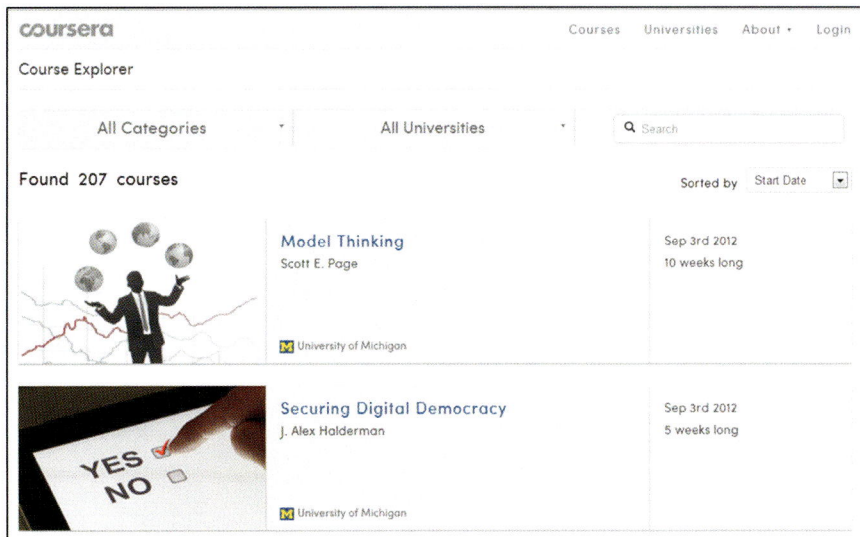

Fig. 9.12 Example screenshot of Coursera Website in 2012 (Image: Coursera, Inc. Used with permission from University of Michigan)

Weaknesses

- Lack of active interaction in the online system, which is a disadvantage compared to many other online learning application now.

Opportunities

- Many of their partner institutions are planning to use the capabilities of their platform to provide their on-campus students with a significantly improved learning experience. It's an opportunity for online education and on-campus courses to combine.

Threats

- Competitive from increasing online education in current market.

9.8 Persuasion Theory

In alignment with Fogg's persuasion theory (Fogg and Eckles 2007), we have defined five key processes to create behavioral change via the Learning Machine's functions and data:

- Increase frequency of using application.
- Motivate changing some living habits.
- Teach how to change living habits.
- Persuade users to plan short-term change.
- Persuade users to plan long-term change.

Each step has requirements for the application. Motivation is a need, want, interest, or desire that propels someone in a certain direction. Humans' sociobiological instinct is to compete with others for social status and success. We apply this theory in the Learning Machine by prompting users to understand that every action in the process of innovation has the potential to have enormous consequence on their success.

We also drew from Maslow's A Theory of Human Motivation (Maslow 1943), which he based on his analysis of fundamental human needs. We adapted these to the Learning Machine context:

- The **safety and security need** is met by the possibility to visualize the amount of progress made in tasks, assignments, and projects in courses.
- The **belonging and love need** is expressed through classmates, teachers, friends from Forum and Blog, and social sharing and support.
- The **esteem need** is satisfied by social comparisons that display success in learning activity, as well as by self-challenges that display goal accomplishment.
- The **self-actualization need** is fulfilled by the ability to visualize improvement and progress in learning.

9.9   Impact of Persuasion Theory on Information Architecture

Considering persuasion theory as described above, it has the following impacts on the information architecture of the Learning Machine.

9.9.1   Increase Use Frequency

Reward is among the most common methods to increase use frequency. In terms of rewards, the users might be awarded both virtual rewards (such as virtual badges and reputation titles) as well as real money (scholarships), provided by related schools and organizations that aim to encourage hard working students.

9.9.2   Increase Motivation

In the Learning Machine, we set the users' potential to make progress and improve themselves as an important incentive for their behavior change. Being able to view their progress of all the tasks, assignments, and projects, users are clearly informed the status and encouraged to make progress. By providing a learning timeline that users can follow, as well as tips and advice regarding particular parts, the Learning Machine gives users a better sense of what they can improve standing in the step of learning process.

In addition, Learning Machine provides the user with tools that make these steps in the learning process easier, such as a social network in which questions can be discussed and topics can be established. With our Research module, in which the user can find tips and advice regarding these steps, users receive suggested step-by-step plans of action in order to achieve each goal.

Gamification has also been found to be a strong motivator for behavior change. The Leader Board gives the user an understanding of important steps that are often overlooked in learning, namely those that involve the social and collaborative view of learning. Besides, by providing small, virtual incentives for heightened activity or participation in discussion with others within the Learning Machine, users become more engaged and gain experience in learning as a positive side-effect.

9.9.3   Improve Learning

For the Learning Machine, guiding the user to participate in each part of the learning process is crucial. To improve learning, the Learning Machine integrates contextual tips on the following topics:

- Learning styles and strategies for different styles.
- Learning method and techniques.
- Course and project information related to users' field.

Users can also choose to subscribe to interesting topics and receive updates on latest news about learning.

9.10   Information Architecture

In designing the Information Architecture (IA) for the Learning Machine, AM+A began by examining the IA for past Machines, including the Green Machine, the Health Machine, and the Money Machine. We discovered an overarching model for the IA that permeated throughout each of the past Machines. In this model, there are five primary "modules," or branches of the IA, each of which is described below. While we altered slightly the details of each module to fit the needs and requirements of each respective Machine, a generalized model was still evident. These modules are described below.

9.10.1   Dashboard

The Dashboard module is similar to a landing page for its respective Machine. The Dashboard is an overview into the status of the user's behavior change. Here, the user gets a view of his/her goals and where s/he stands in achieving those goals.

9.10.2   Process View or Road Map

The Process View module is an overview or roadmap in which the user sees a more high-level view of the process and more details regarding each objective and goal. The user sees the progress being made, as well as the next steps in achieving a particular goal.

9.10.3   Social Network

The Social Network module is an integral part of behavior change in modern software. Users engage in focused, subject-matter-based connections with friends, family, and/or like-minded people that either share similar goals or wish to support others in achieving behavior-change objectives.

9.10.4 Tips/Advice

The Tips/Advice module provides focused just-in-time knowledge about a given topic to give users insight into the habits they wish to either get rid of or adopt.

9.10.5 Incentives

The Incentives module presents users with enjoyable and engaging ways to change their behavior, including competitions, awards, rewards, mementos, trophies, leader boards, and specific games. Gamification has proven to be a powerful tool in encouraging users to try an application, even with virtual incentives, although in some cases, real incentives may be provided. A leaderboard allows a user to compare his/her progress with others, tapping in the competitive nature of the human mind to create behavior change.

These components are organized into the diagram depicted in Fig. 9.13. They are described in further detail below and appear in the screens depicted later. The following lists the content and purpose of the components of the Learning Machine's

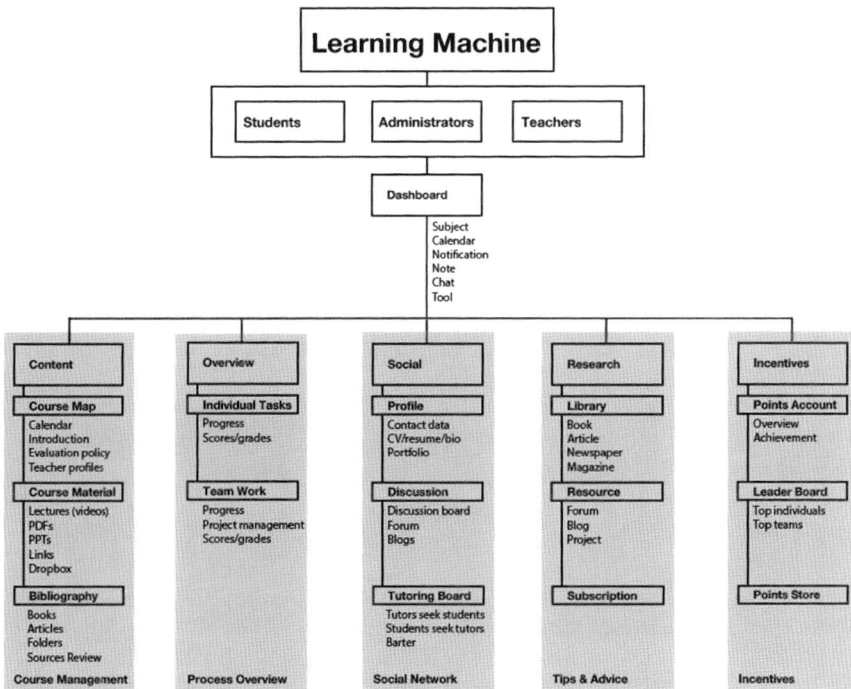

Fig. 9.13 Information architecture of the learning machine

information architecture. The objective of this application is to provide a suitable tool between teachers and students, one which enables them to download, share, comment on, and visualize data.

9.10.6   Dashboard Component

This is the entry page of the Learning Machine for users to keep updated with all the latest news, messages, and notification.

9.10.7   Tool Button (At Top)

Tool collection for users to customize their own selections and use essential tools anytime during the learning process, like such as calculator, dictionary, etc.

9.10.8   Search Button (At Top)

Quick search engine for users to search for resource within the Learning Machine.

9.10.9   Course List (At Top)

Name list of users' curriculum allows users to customize the displayed or hidden courses and to select a course for study.

Course Pane

- *Subject:* Display all the information of subjects in which users have enrolled.
- *Calendar:* Display users' course schedule and task deadlines. Users can add or delete the content in the calendar.
- *Notification:* Rolling notification of updates of users' subscription and alterations of courses in which users have enrolled, e.g. messages or announcements from teachers.

9.10.10   Note Slide (Right Side)

Note slide is always at the right side. Users can take notes after clicking the slide. Users also can review and filter the notes based on date, subject, tag, and location.

9.10.11   Chats Button (Bottom)

Users can send a message/email or chat with friends, classmates, group members, professors by clicking the name in the contact list which is divided into different groups. Users can customize the group and will get a notification whenever they get a message.

9.10.12   Urgent Notification (Bottom)

Users will be informed immediately when they have urgent messages, for example, the approaching deadline of a project, or the announcement of schedule changes.

9.10.13   Content: Specific Course Component

Once users select a course for study, they enter the Specific Course Page of the Learning Machine. In this Specific Course module, there is a menu including five functions: Content, Overview, Social, Research, Incentives.

9.10.14   Content

Users can check all the course information and course material in the Content.

9.10.15   Course Map

Provides all the related course information, introduction, and teachers' profiles:

- *Calendar:* Course schedule, which includes the course time, location, etc.
- *Introduction:* Course description, e.g. course content, objectives, assignments, activities, *etc.*
- *Evaluation policy:* Illustration of course grade, and teachers' standards.
- *Teacher profiles:* Introduction of teachers.

9.10.16   Course Material

Maintain all the course materials, including lectures, slides, files, and links for users to review and download.

Lectures (Videos):
This session is interactive:

- Students watch video lectures, take notes, and share comments with classmates in the same time. Teachers post questions and track students' feedback during the lectures. This section works like a live lesson in classroom. During the videos, teachers can upload quiz or questionnaire to keep up the attention of the student and improve on their concentration about the lecture.
- Students can be asked to give their opinion about a topic or about their knowledge and background experience, share this information with teachers and students to create a debate table, and gather all discussions as a data source. Students have to answer questions to go on with the lectures, which aims to avoid students' lack of attention or passive attitude. Teachers can make sure that students are following and understanding the lesson by monitoring students' status of answering questions.
- Students are able to stop and play the video again, go forward or backward on their pace to follow teachers' lectures. Teachers can track the moments at which students seem to have problems of understanding. For example, they have stopped the video several times in the same point of the lecture, and they went back to review the explanation.
- The videos are divided in chapter paragraph and highlights, reachable every moment during the reproduction by a carousel gallery placed on the bottom of the video that shows just when the user placed a finger on that exactly area close to the timeline of the video (YouTube is an example).

- Students can visualize and play the video (lecture). The lessons start the video and reproduce the images and the speech or dialogue, which is placed at the bottom or at the top of the video image.

Files (PDF):
Students can download files in PDF format (and possibly others).

Slides (PowerPoint):
Course presentations are updated and stored for students to review and download.

Links:
Links, which can be articles, videos, lectures, topics, etc., are put in this part for students to study further.

Dropbox:
Linked with Dropbox account to make it easy to pass along files.

Bibliography:
The course bibliography recommended by professors. This text may be sorted by author, topic, or some other scheme. Annotated bibliographies give descriptions about how each source is useful to an author in constructing a paper or an argument. These descriptions, usually a few sentences long, provide a summary of the source and describe its relevance. Reference management software may be used to keep track of references and generate bibliographies as required.

Books:

Book list provided by teachers. Students can directly click on the book name to access to the detailed information. Also, students can choose to purchase a digital one using the bottom "buy" on the right side.

Articles:

Article list provided by teachers. Students can download the files or access to the article links on the Internet

Folders:

Folders to store the downloaded files or purchased books for users to review.

Sources Review:

- Students can rate the reading materials with five stars, can rank them with "Worth reading" or "Not worth reading," and make comments on them. Students can also recommend reading material for the course or other classmates by write reviews with links in the Sources Review.
- Teachers should adjust and update the reading materials according to students' feedback. Teachers should also check students' Sources Review to figure out new and interesting material for course in the future.

9.10.17 Overview Component

Users can keep track the progress of the individual assignment and team project in the Overview.

9.10.18 Individual Tasks

- *Progress:* Track progress of individual assignments or experiment. Once students finish the tasks, they can directly submit/upload the files to complete the progress.
- *Scores/Grades:* Review scores or grades of assignment, quiz, exams, etc. Students can also check the mistakes in exams and comments on assignment.

9.10.19 Team Work

- *Progress:* Track progress of team projects.
- *Project management:* Manage multi-tasks in different teams. Students can also easily exchange or share their work between team members.
- *Scores/Grades:* Review scores or grades of projects, etc. Students can also review the comments from teachers and other classmates on their project presentation.

9.10.20   Social Component

Users can interact with peers and teachers here through Discussion board and Forum. Also, they post and share blog articles with others. Moreover, they can be a tutor for peers or try to seek for tutor in the Board.

Profile

- *Contact data:* Contains all useful contact data for peers and community.
- *CV/Resume/Bio:* Contains useful background information.
- *Portfolio:* Contains useful examples of work.

9.11   Discussion

- *Discussion Board:* Students who enrolled in a specific course can use the Discussion Board to discuss topics related to the course. Teachers should review the discussion, answer the questions, establish the topics, and monitor the class status.
- *Forum:* Students can communicate in this digital community with people in different background. Students can choose different subjects, topic, or key words, etc., to filter the content and people.
- *Blogs:* Students can post their own blogs and comments or subscribe to others' blog.

9.11.1   Tutoring Board

This section offers barter for peer-to-peer tutoring. A student can register as a tutor and post his information or find other students to tutor him.

- *Tutors seek students:* Enables teachers/mentors to find available students.
- *Students seek tutors:* Enables students to find teachers/mentors who can assist them.
- *Barter:* Enables community exchange of knowledge, products, services.

9.11.2   Research (Tips and Advice) Component

Users can do internal and external research with the Learning Machine.

9.11.2.1   Library

Access to library and resources in universities around the world.

- Books.
- Articles.
- Newspapers.

- Magazines.
- Websites.

9.11.2.2 Resource

Access to resources in the Learning Machine.

- *Forum:* Use key words to search for related topics in Forum.
- *Blog:* Use key words to search for related topics in Blog.
- *Project:* Use key words to search for related projects which are open to public. This includes projects proposal, projects in progress, and completed projects reports.

9.11.2.3 Subscription

Manage the search alerts and push notifications.

9.11.2.4 Incentives Component

Users can do internal and external research with the Learning Machine.

9.11.2.5 Points Account

- *Overview:* Overview of points earned in good test, good project, reading, discussion, sharing, and so on.
- *Achievement:* Track digital blade and achievement unlocked.

9.11.2.6 Points Store

Users can exchange gift or coupons with earned points in the store.

9.11.2.7 Leader Board

The board to display the top 100 users who wins most points or top 50 teams who do the best projects with the highest rating

- Individual.
- Team.

9.12   Designs

Based on the information architecture, AM+A prepared initial concept sketches of some key screens of the Learning Machine. These "low-fi" concept prototypes were used to elicit feedback from potential users, prior to the development of mockup screens of higher fidelity (Fig. 9.14).

9.12.1   Evaluation

AM+A presented early versions of screen sketches to an internal group within AM+A at a point just after the first round of conceptual screens had been completed. Although there was a consensus that the concepts behind the Learning Machine were indeed novel, there was a general concern about the screen designs being too complicated, confusing, and lacking sufficient visual stimuli. With this feedback, AM+A sought to make a second round of mock-ups less textual and more visual by removing unnecessary words and adding appropriate icons and images. Overall, we agreed that there was a general lack of photos and icons and aimed to add more visual representations where possible. These objectives are represented in the revised screens presented below.

9.12.2   Revised Concept/Visual Designs

With the comments drawn from the evaluation phase, AM+A aimed to make the Learning Machine user experience more visual. In doing so, we found it necessary to add new screens to fill the gaps of certain flows and functions that were often missed in the one to two screens provided in our initial designs. In addition, we endeavored to ensure additional consistency in the design among the icons, symbols, labels, colors, and grid. With new design layouts and theme, as well as the additional mockups, we hoped to improve a better understanding of the interaction and general enhancement of the user experience of the Learning Machine. The revised screens appear below with comments (Fig. 9.15).

9.13   Next Steps and Conclusions

Using the user-centered design approach described above, AM+A plans to continue to improve the complete Learning Machine development process, which would require significant time and funding from an outside source. Tasks include the following:

Fig. 9.14 Initial sketches of screens for the learning machine

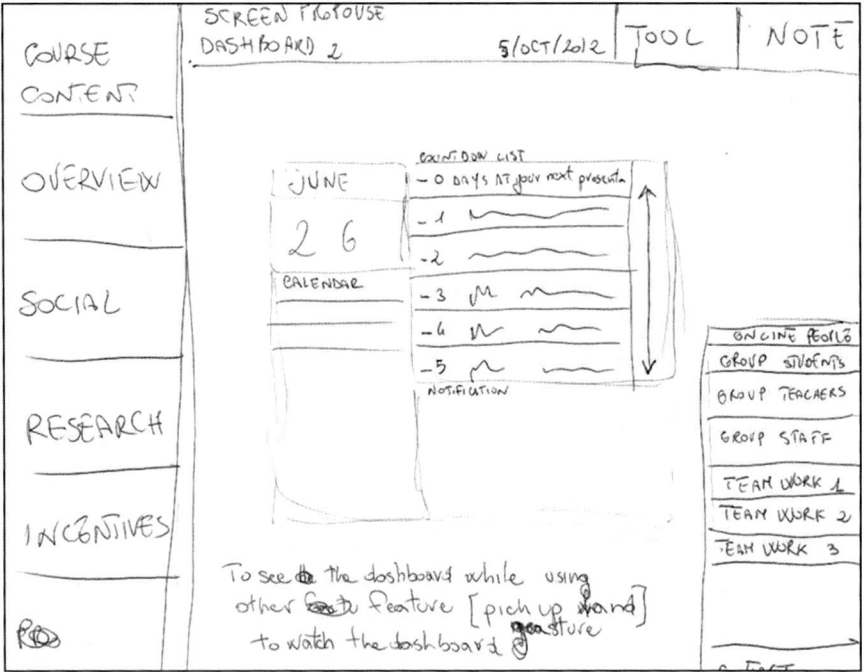

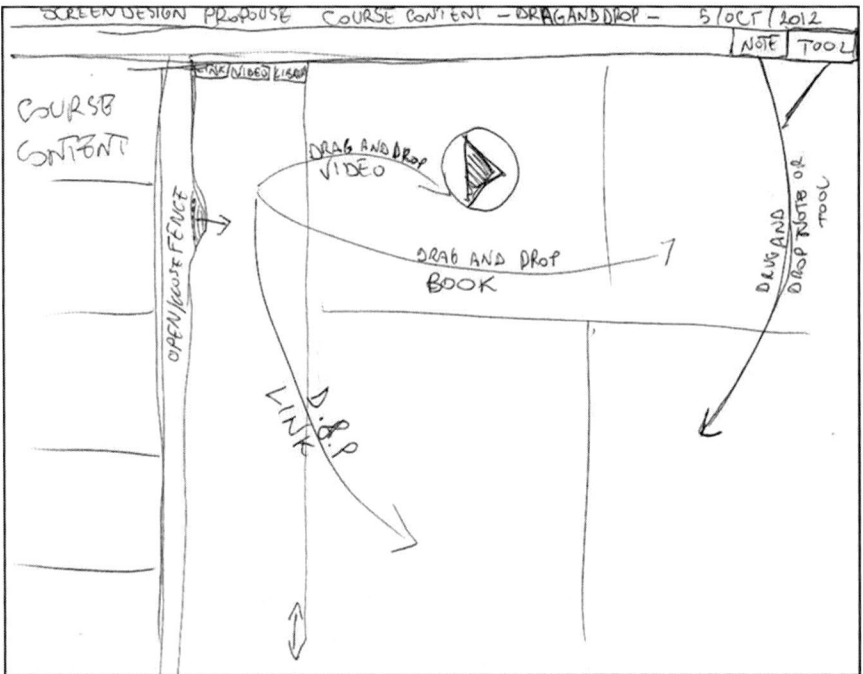

Fig. 9.14 (continued)

Fig. 9.14 (continued)

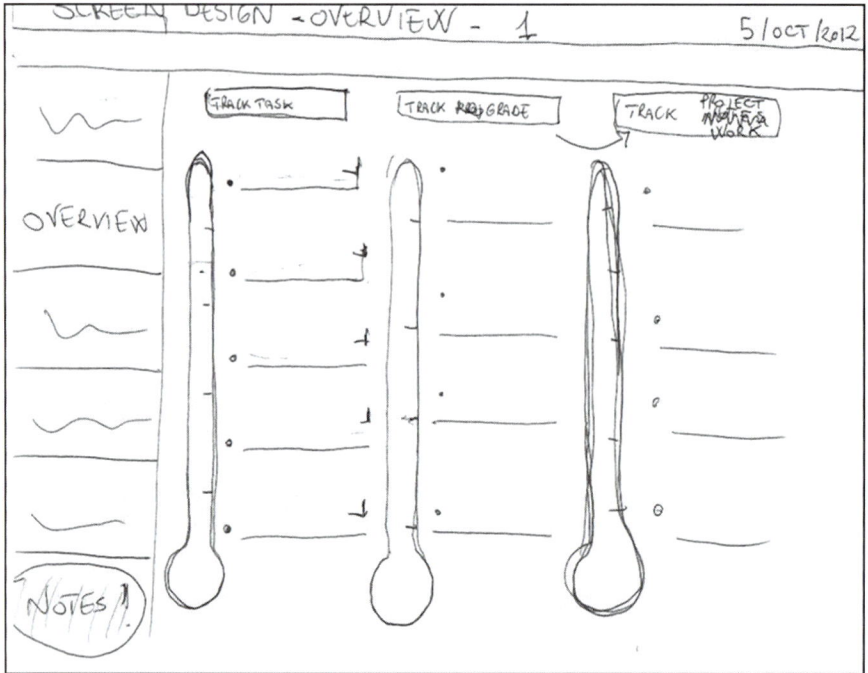

Fig. 9.14 (continued)

Fig. 9.15 Revised screens show typical key moments in navigation. The first screen shows the main "dashboard" of the Learning Machine

- Revising personas and use scenarios.
- Conducting user evaluations.
- Revising/completing the information architecture and look-and-feel.
- Building initial working prototype (e.g., for iPad, or other tools, platforms).

- Re-designing, if appropriate, the Learning Machine for different corporate and country cultures.
- Evaluating the Learning Machine across different corporate and country cultures.
- Developing the Learning Machine for enterprise use as well as personal use.
- Researching and developing improved information visualizations.

AM+A sought to incorporate information design and persuasion theory for behavior change into a mobile tablet application. This self-funded work on the Learning Machine project is current and ongoing and was undertaken to demonstrate the direction and process for such products and services. Though the design is incomplete, AM+A is willing to share the approach and lessons learned, in the interest of helping to solve the challenges of innovation.

At this stage, AM+A is seeking to persuade other design, education, and corporate groups to consider similar development objectives. We hope that our process and concept prototypes (the self-funding of which inevitably limited the amount of research, design, and evaluation) will inspire others, and that they benefit from the materials provided thus far.

The process has already been demonstrated successfully with a previous project, the Green Machine (Marcus and Jean 2009), versions of which have been considered and used by SAP for enterprise software development (Marcus et al. 2011).

AM+A's long-term objective for the Learning Machine is to create a functional working prototype to test whether the application can actually persuade people with learning management challenges to improve the efficiency and effectiveness of learning on the individual, team, and on the corporate level. If the theories are proven to be correct, this approach could create numerous new innovative products and services that will benefit many people in many different companies, organizations, countries, and cultures.

Acknowledgments The author thanks his AM+A Associates, all Designers/Analysts, for their significant assistance in planning, research, design, analysis, and documentation.

- Ms. Yuan Peng.
- Mr. Nicola Lecca.
- Mr. Peter Rinzler.

The author also thanks Prof. Bob Steiner, Marketing Faculty, University of California/Berkeley Extension, International Diploma Program, Marketing, Advisor, and his marketing students:

- Ms. Doris Masser.
- Mr. Bru Perez.
- Mr. Marcelo Rocha.
- Ms. Tor-Einar Sandvik.

The author acknowledges the publication of a subsequent paper (Marcus 2013d) based on the original white paper on which this chapter is based.

Further Reading

[No author] (2012) Teachers pay teachers. Bloomberg Businessweek, pp 56–57, 29 October 2012
[No author] (2012) The big three, at a glance. Retrieved November 2, 2012, from NYTimes.com: http://www.nytimes.com/2012/11/04/education/edlife/the-big-three-mooc-providers.html

Agile Mind, Inc., Charles A. Dana Center (2002, March 22). The role of advanced placement courses in enabling high achievement in every high school: a research paper. The University of Texas and the AVID Center

Berman DK (2012, October) In the future, who will need teachers? Wall Street Journal, pB1

Buckley (1999) Model-based teaching and learning with BioLogica™: what do they learn? How do they learn? How do we know? (Citations: 32)

Canton N (2012) Cell phone culture: How cultural differences affect mobile use. CNN. Retrieved from http://edition.cnn.com/2012/09/27/tech/mobile-culture-usage/index.html, 28 September 2012

Duneier M (2012) The world is his classroom. Princeton Alumni Weekly, p 13, 10 October 2012

Efrati A (2012) Start-up expands free course offerings online. Wall Street Journal, 12 April 2012

Fleming N (2012) VARK: a guide to learning style. Retrieved from http://business.vark-learn.com/about-varkl introduction/, 3 October 2012

Gelernter D (2012) The friendly, neighborhood internet school. Wall Street Journal, 9 August 2012

Greene JP (2012) Jay Greene: the imaginary teacher shortage. Wall Street Journal, pA15, 8 October 2012.

Hollander S (2012) Online Holdouts no more. Wall Street Journal, pA4, 26 September 2012

Holzinger A (2004) Rapid prototyping for a virtual medical campus interface. IEEE Computer Society. http://business.vark-learn.com/about-vark/introduction/

Hua-Min Chang, Chia-Cheng Liu,Fang-Wu Tung, Yen-Chuan Lai, Chun-Yu Chen, Ko-Hsun Huang, Yi-Shin Deng. Apply learning probes on developing domestic digital learning environment. National Chiao-Tung University, Institute of Applied Arts

Koller D, Ng A (2012) Log on and learn: the promise of access in online education. Forbes, 19 September 2012. Retrieved from http://www.forbes.com/sites/coursera/2012/09/19/log-on-and-learn-the-promise-of-access-in-online-education/

Marcus A (2010) Health machine. Inf Des J 19(1):69–89

Marcus A (2012a) The money machine: helping Baby Boomers to Retire. User Experience Magazine, 11:2, Second Quarter 2012, pp 24–27

Marcus A (2012b) The story machine: a mobile app to change family story-sharing behavior. Workshop Paper. Proc. CHI 2012, Austin, TX, May 2012, pp 1–4

Marcus A et al (2013a) The innovation machine. In: Marcus A (Ed.) (2013) Proceedings, design, user experience, and usability conference, Las Vegas, NV, 20–25 July 2013, pp. 67–76. Springer, London

Marcus A et al (2013b) The travel machine. In: Marcus A (Ed.) (2013) Proceedings, design, user experience, and usability conference, Las Vegas, NV, 20–25 July 2013, pp. 696–704. Springer, London

Marcus A et al (2013c) The driving machine. In: Marcus A (Ed.) (2013) Proceedings, design, user experience, and usability conference, Las Vegas, NV, 20–25 July 2013, pp. 140–149. Springer, London

Marcus A, Jean J (2010) The green machine. Inf Des J 17:3, First Quarter, 2010, pp. 233–243

Mossberg W, Mossberg S (2012) A dragon that takes dictation and controls a Mac by Voice. Wall Street Journal, 9 October 2012, pA13A

Pappano L (2012) The year of the MOOC. Retrieved November 2, 2012, from NYTimes.com: http://www.nytimes.com/2012/11/04/education/edlife/massive-open-online-courses-are-multiplying-at-a-rapid-pace.html?pagewanted=all

Pivec M, Baumann K (2004) The role of adaptation and personalisation in classroom-based learning and in e-learning. J Univ Comput Sci 10(1):73–89

Polovina S (2011) The learning pyramid. Retrieved October 3, 2012, from http://homepages.gold.ac.uk/polovina/learnpyramid/about.htm, 16 November 2011

Reif R (2012) What campuses can learn from online teaching. Wall Street Journal, pA17, 2 October 2012

Tunga F-W, Deng Y-S (2007) Increasing social presence of social actors in e-learning environments: effects of dynamic and static emoticons on children. Displays 28(4–5):174–180

Wikipidia (2012) E-learning. Retrieved October 3, 2012, from http://en.wikipedia.org/wiki/E-learning

References

Buckley B, Gobert JD, Kindfield ACH, Horwitz P, Tinker RF, Gerlits B, Wilensky U, Dede C, Willett J (2004) Model-based teaching and learning with BioLogica™: what do they learn? How do they learn? How do we know? J Sci Educ Technol 13(1):23–41

Bybee RW et al (1989) The 5 E learning cycle model., Retrieved October 3, 2012, from http://faculty.mwsu.edu/west/maryann.coe/coe/inquire/inquiry.htm

Finn H (2012) Watching the ivory tower topple. Wall Street Journal, p A13A, 25 March 2012

Fogg BJ, Eckles D (2007) Mobile persuasion: 20 perspectives on the future of behavior change. Persuasive Technology Lab, Stanford University, Palo Alto

Kolb D (1984) Experiential learning: experience as the source of learning and development. Prentice Hall, Englewood Cliffs. See, also: "Kolb's Learning Styles and Experiential Learning Cycle" http://nwlink.com/~donclark/hrd/styles/kolb.html. Retrieved November 20, 2014

Marcus A (2013d) The learning machine. Proceedings, design, user experience, and usability conference, Las Vegas, NV, 20–25 July 2013, pp. 247–256. Springer, London

Marcus A, Dumpert J, Wigham L (2011) User-experience for personal sustainability software: determining design philosophy and principles. In: Marcus A (ed) Design, user experience, and usability. Theory, methods, tools and practice, lecture notes in computer science, vol 6769, pp 172–177. Springer, Berlin/Heidelberg. Retrieved from http://www.springerlink.com/content/f8lm37795r3743v7/abstract/

Marcus A, Jean J (2009) Going green at home: the green machine. Inf Design J 17(3):233–243

Maslow AH (1943) A theory of human motivation. Psychol Rev 50:370–396

McIsaac MS (2002) Online learning from an international perspective. Educ Media Int 39(1):17–22

Sundberg P (2014) Concept map of learning theories (Paul Sundberg). http://courses.education.illinois.edu/EdPsy317/sp03/learning-maps/sundberg-learning-theories.gif. Retrieved November 20, 2014

Tunga F-W, Deng Y-S (2006) Designing social presence in e-learning environments: testing the effect of interactivity on children. Interac Lear Environ 14(3):251–264

Walther JB (1992) Interpersonal effects in computer-mediated interaction: a relational perspective. Comm Res 19(1):52–90

Wikipidia (2012) Learning. Retrieved October 3, 2012, from http://en.wikipedia.org/wiki/Learning

Williams J (2012) A high-tech fix for broken schools. Wall Street Journal, pA11.14 August 2012

Mobile/Website Competitor Applications URLs[1]

Coursera: https://www.coursera.org/about
Khan Academy: http://itunes.apple.com/us/app/khan-academy/id469863705?mt=8; http://www.khanacademy.org/downloads
MIT Mobile: http://mitx.mit.edu
Schoolloop: http://www.schoolloop.com/
Together Learn: https://itunes.apple.com/us/app/togetherlearn/id548411521?mt=8

---

[1] (Retrieval date for URLs except Khan Academy: 26 October 2012, Khan Academy retrieval date 29th May 2015)

Chapter 10
The Happiness Machine: Combining Information Design/Visualization with Persuasion Design to Change Behavior

10.1 Introduction

The happiness industries are a significant component of most economies. These industries include business sectors such as tourism and travel, banking and financial institutions, consumer electronics and games, entertainment, and computer technology. In addition, most of the world's religions, most politicians, most psychologists and healthcare providers, even many educational institutions, seek to make people happier. Sometimes the businesses and organizations call their target markets customers, members, or constituents.

The primary objective of the Happiness Machine is to change the above-described people's behavior, to ease, enhance, and enrich the experiences of a person enabling them to be happier in a deep way, not a superficial way.

By combining information design and persuasion and motivation theory, with a particular focus on the works of Maslow's theory about basic human needs (Maslow 1943) and on Fogg's captology theory (Fogg 2003; Fogg et al. 2007), the use of the Happiness Machine should prompt people to change their behavior and to make out of their lives a deeper and more personally enriching experience. Above all, this dimension of personal learning that will distinguish this application from the majority of mobile apps that already exist in the field of happiness.

The success and effectiveness of the described approach, i.e., especially, the combination of information design with persuasion design in order to promote behavioral change of mobile application users, has already been studied and realized in several previous projects of the author's company: the Green Machine (Jean and Marcus 2009; Marcus and Jean 2010), the Health Machine (Marcus 2011), the Money Machine (Marcus 2012a, 2013b), the Story Machine (Marcus 2012b), the Innovation Machine (Marcus 2013a), the Driving Machine (Marcus 2013c), and the Learning Machine (Marcus 2013d). All of these past Machine projects and the current Happiness Machine rely on the conceptual design of an application through a user-centered user-experience development process, as defined and explained below.

© Springer-Verlag London 2015   539
A. Marcus, *Mobile Persuasion Design*, Human–Computer
Interaction Series, DOI 10.1007/978-1-4471-4324-6_10

This chapter describes two versions of the designs, Happiness Machine 1.0 and 2.0, conducted primarily with the assistance of Mr. Hsu-Shin (Jonathan) Liu and Ms. Min Lee during three-month internships as Designer/Analysts. As of this writing, additional groups have designed further versions during one-week short courses, Happiness Machines 3.0, 4.0, and 5.0: graduate design students at the Institute of Design, Illinois Institute of Technology, Chicago, IL, USA and at the School of Design, Hong Kong Polytechnic University, Hong Kong, China; and participants in the Mobile User-Experience Design workshop at the De Tao Academy, Shanghai, China. Altogether, about 60 participants in 14 teams have researched and designed additional versions, which may be published at a later date.

In the context of this chapter describing the Happiness Machine, metaphors may be termed the "concepts" of the Machine, and the mental model may be termed the "information architecture." The discussion below about the initial and revised screen designs will describe, also, the interaction and appearance, especially as the designs move from conceptual designs (so-called wire-frame versions) to perceptual designs (so-called look-and-feel versions). One unique approach of the Machines is to combine information-design and information-visualization (tables, forms, charts, maps, and diagrams) with persuasion design.

As described in Chap. 1, AM+A carried out user-centered user-experience design involving most of the traditional development steps (planning, analysis, research, design, evaluation, documentation, and training) and including all of the standard user-interface components (metaphors, mental models, navigation, interaction, and appearance). AM+A also incorporated aspects of persuasion theory, as outlined in Chap. 1.

10.2   Impact of Persuasion Design on Information Architecture

Applying the techniques of persuasion design that were introduced in Chap. 1 and demonstrated in previous chapters focusing on different Machine projects, AM+A observed the following impacts of persuasion design on the Happiness Machine information architecture.

10.2.1   Increase Use Frequency

Incentives such as games, awards, rewards, recognition, and nostalgic objects (souvenirs or treasures) are the most common methods to increase use frequency. In the Happiness Machine, we have developed many tricks, random good deeds, funny challenges, and other happiness-increase-related knowledge. Moreover, we developed another game that is related to increasing smiling and to having more fun with people. In terms of rewards, users will be provided virtual rewards (such as "expert" statuses and new skins for blogs), titles (such as "X is now a four-star Mr. Happy"), and smiling faces that they can give to others, which they can communicate, for example via Facebook, LinkedIn, or Twitter. In addition, we chose social comparison as another incentive to increase use frequency. Users will be able to consult others' diary entries (with permission), and they will compete with them on the degree of expert status and on the number of accomplished challenges.

10.2.2   Increase Motivation

Because setting goals helps people to learn better and improves the relevance of feedback, one of the main features of the Happiness Machine is the happiness advisor, Dr. Happy, which assists the traveler by providing information, by giving suggestions and ideas, by posing challenges, and by offering regular feedback. The advisor motivates, it pushes the user towards the experience, the experimentation, and observation of personal idiosyncrasies at a destination, by suggesting a range of different activities in the form of challenges or goals. The incentive of self-accomplishment will encourage users to achieve these challenges, which, in turn, will change and improve their understanding of issues blocking happiness, and techniques that enrich their experiences with their own attitudes, behavior, and relation to people, objects, contexts, obstacles, fear, and anger. This process of learning and self-discovery is designed to change the use's approach towards in the long term.

In addition, the creation of a personal happiness (or gratitude) diary will increase users' awareness of their steady progress and will motivate them to add further material in the form of visual or textual records of documents (photos, texts, voice, music, videos, etc.), experiences, encounters, and lessons learned.

Social interaction has an important impact on behavior change. Therefore, another important component of the Happiness Machine is to leverage social networks and integrate features like those found in forums, Facebook, Twitter, or blogs. Users can send notes or messages to their friends or others, and they can share their personal diaries with trusted individuals. Furthermore, they can consult others' diaries and can ask them for information and tips. These social ties will serve as an additional incentive to motivate behavior change.

10.2.3   Improve Learning

The central objective of the Happiness Machine is to trigger a learning process that is both informative and reflective, and that combines textual, visual, and sonic elements. Thus, the advisor provides a range of practical (techniques, tricks, games, etc.), current (news, events, etc.), general (history, religion, politics, etc.) and more culture-specific (values, symbols, behavior, etc., relevant to a culture) information, but it also proposes a range of activities that are supposed to make the user discover happiness particularities and characteristics related to specific goals. A primary concept is to support, to assist, and stimulate users during their daily life, in order to make them observe certain things, to make them reflect actively about their experiences, and to perform incremental behavior changes that deliver positive change.

We also seek to make the education process both informative and entertaining. Therefore, we have proposed games that refer to specific vocabulary knowledge of users and their contexts, to destination-related general and more culture-specific information, as well as to challenges posed by Dr. Happy, the happiness advisor. Through playing games featuring educational information, users will be able to increase their knowledge and understanding of their own motivations, behavior, and results, without getting bored.

10.3 Happiness Theory: Theoretical Background and Research

10.3.1 Happiness Definition

What behavior should people change that will give their lives deeper and more person-ally enriching experiences? Even before that, what behavior makes people happy in the first place? This is a very broad topic, and people have different answers to this question. Therefore, we plan to design the Happiness Machine by, first, understanding Happiness Theories and happiness in general. We shall combine our version of happiness theory with information design and persuasion design (Fig. 10.1).

10.3.2 Exploratory Survey

To gain basic insights from people regarding what happiness is for them, we conducted an exploratory survey via social networking services (SNSs). We asked people, "What makes you happy?" (Fig. 10.2).

Over a period of two weeks, 10+ professionals/graduate students participated, with ages ranging from 25 to 45 and with multicultural backgrounds. They gave diverse answers, ranging from eating food, spending time with friends, shopping, love, and, for particularly for young males, having sex with women.

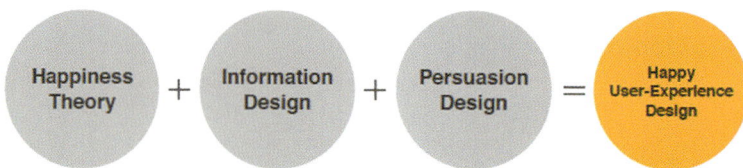

Fig. 10.1 Diagram of how happiness theory, information design, and persuasion design combine to enable a happier user-experience design

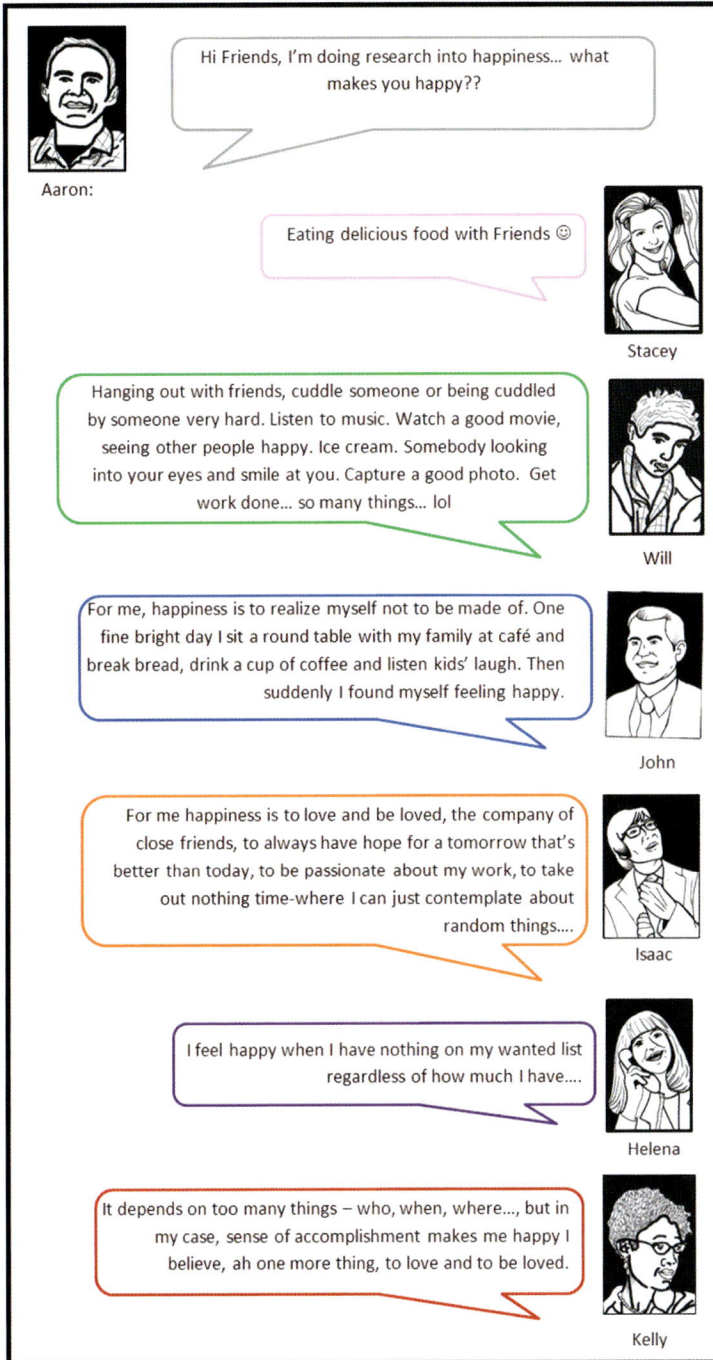

Fig. 10.2 Sample survey results of happiness inquiry (Reproduced due to copyright)

As we can assume from the survey, happiness can be defined as follows:

- Happiness is relative, and happiness is achieved differently for different people. [URL: Greater Good Science Center].
- Happiness is a person-dependent (or subjective) emotion. This term means that what could make someone happy may not be of any importance to someone else. [URL: Achor].

10.3.3   Secondary Research

Happiness is highly subjective and, therefore, requires further study in the respective fields that deal with such subject matter, such as positive psychology, happiness theories, philosophy of happiness, emotional design, and endorphins.

We examined writings of and about the following people:

- Shawn Achor: Happiness Advantage.
- Stuart Brown: Benefits of Play.
- Bernie deKoven (Dr. Fun): Deep Fun.
- Pieter Desmet: Design for Happiness.
- Viktor Frankl: Holocaust Survivor, Philosophy on Happiness.
- Carl Jung: Happiness in Human Mind, Personal Archetypes [see URL: Jung].
- Jane McGonigal: Digital game to real-world collaboration.
- Martin Selgiman: Positive psychology.

According to a study by Pohlmeyer at the Institute of Positive Design at Delft, there are two kinds of happiness: Deep Happiness and Shallow Happiness. [URL: Desmet] She defines Shallow Happiness as a mere satisfaction of immediate desire, such as physical, cognitive, emotional, or spiritual needs (i.e., feeling good). Deep Happiness, however, allows people to pursue intrinsic values and objectives/goals in life guiding them to live well and sustain happiness in a meaningful life.

Sonja Lyubomirsky [Lyubomirsky] studied sources of happiness in people. According to her theory (see Fig. 10.3), happiness is 50 % a result of genetic

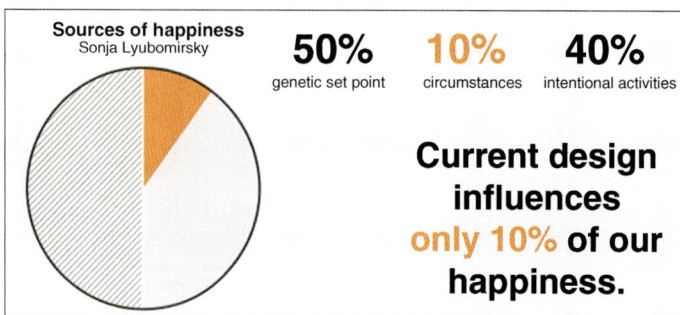

Fig. 10.3 Chart of Sources of Happiness, per Sonia Lyubomirksy, in document cited in the text (Image courtesy of Sonja Lyubomirksy)

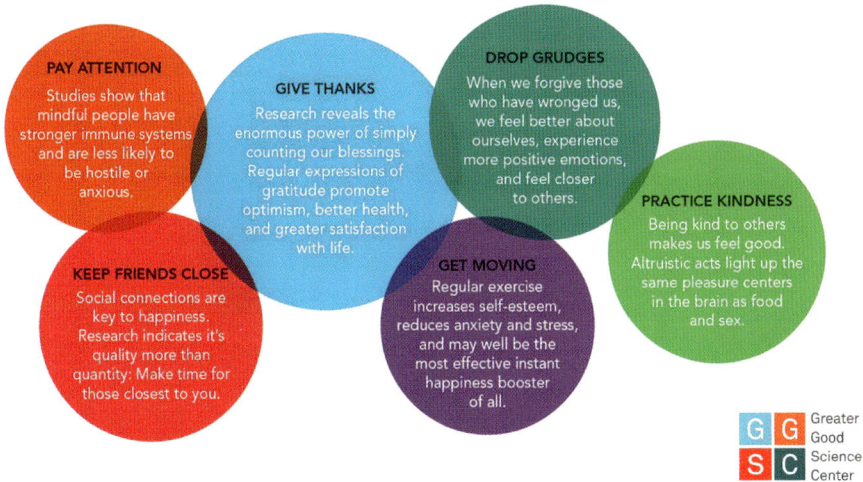

Fig. 10.4 Habits of happiness according to University of California at Berkeley's Greater Good Science Center (Image: University of California at Berkeley)

make-up, only 10% due to circumstances (i.e., environment), and 40% a result of guided, intentional activities. The result opens up a significant opportunity (40 %) to achieve greater happiness: people can change their behavior to make life happier.

The University of California at Berkeley's Greater Good Science Center [greatergood.berkeley.edu/] presents six habits of happiness worth cultivating (see Fig. 10.4):

- *Pay attention*: Studies show that mindful people have stronger immune systems and are less likely to be hostile or anxious.
- *Keep friends close*: Social connections are key to happiness. Research indicates it's quality more than quantity. Make time for those closest to you.
- *Give thanks*: Research reveals the enormous power of simple counting our blessings. Regular expressions of gratitude prompt optimism, better health, and greater satisfaction with life.
- *Drop grudges*: When we forgive those who have wronged us, we feel better about ourselves, experience more positive emotions, and feel closer to others.
- *Get moving*: Regular exercise increase self-esteem, reduces anxiety and stress, and may well be the most effective instant happiness booster of all.
- *Practice kindness*: Being kind to others makes us feel good. Altruistic acts light up the same pleasure centers in the brain as food and sex.

Shawn Achor, a "happiness guru" from Harvard University, also talks about ways to be happy and be successful in life (Achor 2010). *The Happiness Advantage: The Seven Principles of Positive Psychology that Fuel Success and Performance at Work*. New York: Crown Publishing presents his theory. Some of the activities that he mentions are these:

Table 10.1 Shawn Achor's key activities to increase happiness, per source cited in the text (Achor 2010)

Small Changes Ripple Outward
▫ Creating lasting positive change
▫ 3 Gratitudes (Emmons & McCulough, 2003)
▫ Journaling (Slatcher & Pennebaker, 2006)
▫ Exercise (Babyak et al., 2000)
▫ Meditation (Dweck, 2007)
▫ Random Acts of Kindness (Lyubomirsky, 2005)

- Meditate.
- Find something to look forward to (the most enjoyable part is the anticipation).
- Commit conscious acts of kindness.
- Infuse positivity into your surroundings (turn off TV's).
- Exercise.
- Spend money on positive experiences or learning (not on things).

His argument is that by capitalizing on positivity to improve productivity and performance-return, one can train one's brain to be more positive (Table 10.1).

Pieter Desmet from the Institute of Positive Design at Delft has challenged designers with the comment that we may surround ourselves with design that makes us feel good, but may not make us happy [Desmet]. He argues that, as designers, we should focus on positive design that unlocks the potential that is inside of us. One example of his designs is Tiny Tasks [Tiny Tasks] a set of key chain tags that have printed on them small tasks for users to complete, which will bring to users positive experience in doing the activities. He points out that there are four ingredients for such meaningful activities: Does the activity use talents and skills that the person possesses? Does it contribute to someone else/something bigger? Does it bring pleasure in doing the activity itself? What value can it bring?

We found numerous blogs, and online articles with tips about how to get happy or become happier. One blog on happiness and health listed several ways to release the "happy hormones," or endorphins [URL Happy Hormones]: smile, do vigorous exercise, eat chocolate, consume spicy food, expose oneself to sunshine, be optimistic and happy, laugh and cry, and be silly. Another, similar blog by Beth Cooper was entitled "10 Simple Things you Can Do Today that Will Make you Happier, Backed by Science" [Cooper].

10.3.4   Primary Research

Inspired by one of the Shawn Achor's habits, journaling, we looked into the relationship between happiness and journaling. Our research objectives were as follows:

- To understand what makes people happy and what they do to achieve it.

- To understand if keeping a diary or journal for self-reflection helped users become happier, and how.
- To understand why people find it hard to keep up with happiness-inducing activities and thus do not become happier.

Participant selection criteria for the study were people who have experience in keeping a diary/journal previously; professionals and students who own smartphones and use them daily; and people who have an interest in living a happier life.

Key questions for interviews were the following:

- Tell me about a time when you became happy/positive after doing or changing something in your life.
- Tell me about journals you keep or kept in the past.
- Do you think about how happy you are and how to improve your overall quality of life?
- Are there any other things that you did to become positive and happier?
- What made it difficult for you to keep up with such activities?

Selected quotes from people included the following:

- "[When journaling,] I know that there's certain things I do pretty much every day, but I haven't set the goals to do it." –Meg, 34.
- "List will be useless when I get too many things on the list. So I want to keep it as brief as possible." –HS, 25.
- "When I get stressed, I play sports. But usually, sports like tennis or soccer require other people who can play with me." –Jack, 30.
- "If I'm outdoors, I would play sports. Other times, I watch TV at home because I'm indoors." –Jack, 30.

Based on the surveys, we analyzed transcripts and organized our findings (see Figs. 10.5 and 10.6).

10.3.5 Process for Analysis

After conducting research, we identified insights and design principles, which are indicated below. Our analysis process includes transcribing interviews and gathering observations. We then organized research findings using these different methods and frameworks:

- The 5E's: Entice, Enter, Engage, Exit, Extend.
- The 4A's: Activity, Attitude, Ambition, Anxiety.
- User empathy: Thinking, Feeling, Saying, Doing.
- Affinity clustering: Compare and contrast.

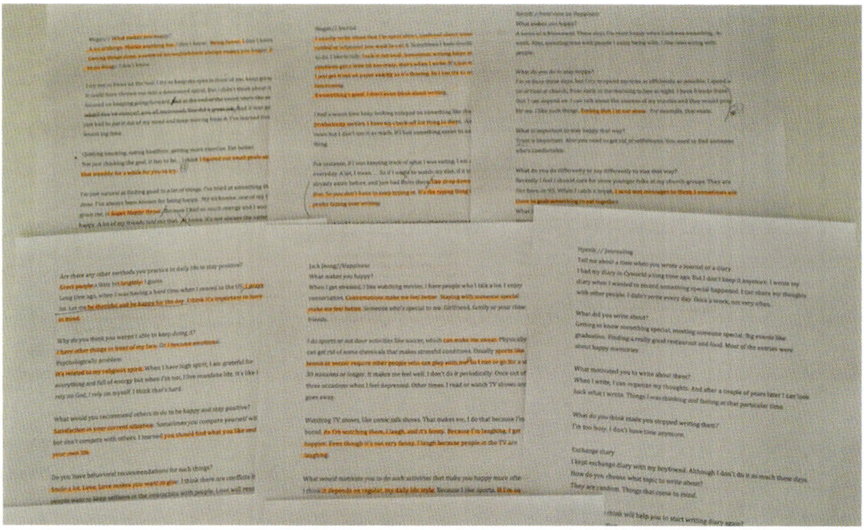

Fig. 10.5 Insights generation: transcripts

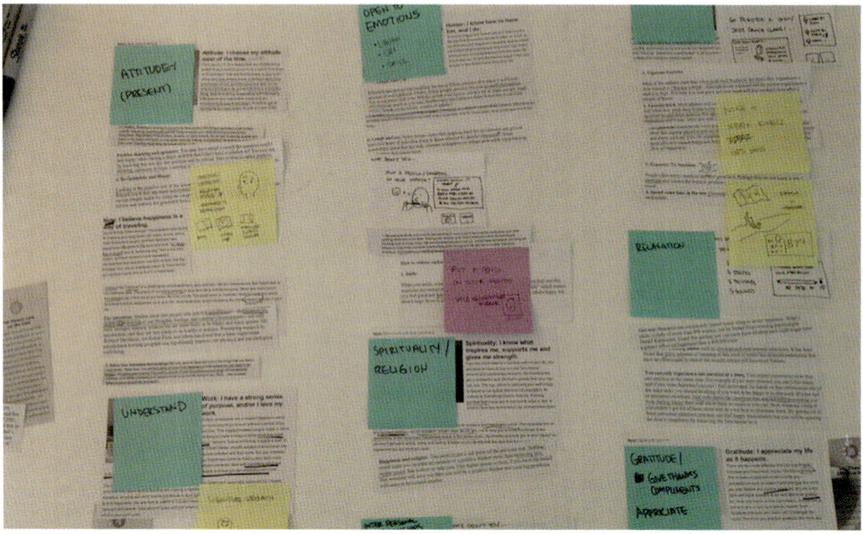

Fig. 10.6 Analysis: clustering research findings

Our organized research findings became the basis for our structure or model of the user-interface (UI) design or user-experience (UX) design of the Happiness Machine.

10.4   Insights and Design Principles

10.4.1   Insight 1. Happiness Is an Objective

Design principles: The Happiness Machine must be able to assist users to achieve objectives and goals and be personalized to the individual "journey-through-happiness" in perceived steps. Directions should be "chunkable" so that people can easily remember what to do.

10.4.2   Insight 2. Happiness Is Discipline; a Constant Reminder to Change Attitude

Design principles: The Happiness Machine must function as a reminder to be mindful about happiness. The Machine should expedite change by helping to reflect.

10.4.3   Insight 3. Happiness Is in a Constant State of Flux

Design principles: The Happiness Machine must be adjusted for or customized to different users in different life stages. Directions should be suitable/adaptable to each user.

10.4.4   Insight 4. Balancing Engagement and Persuasion Is Key

Design principles: The Happiness Machine should persuade users to do certain activities with the least friction possible.

10.4.5   Mental Model for Deep Happiness

We developed the following model for deep happiness based on our earlier research (Fig. 10.7).

10.5   Practices for Deep Happiness

The five practices mentioned in the model above are the following (Fig. 10.8).

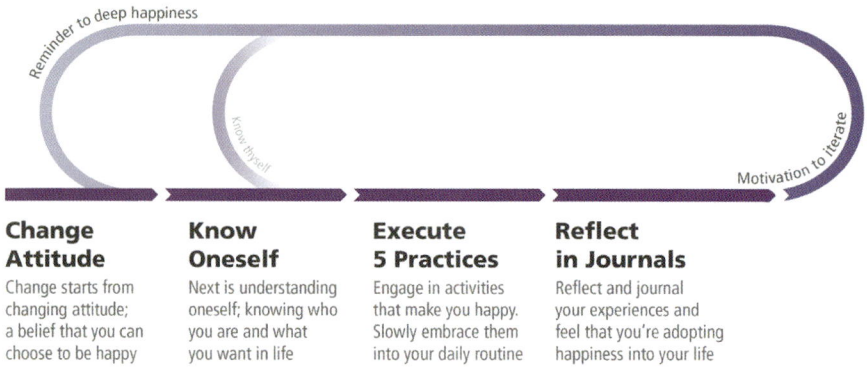

Change Attitude

Change starts from changing attitude; a belief that you can choose to be happy

Know Oneself

Next is understanding oneself; knowing who you are and what you want in life

Execute 5 Practices

Engage in activities that make you happy. Slowly embrace them into your daily routine

Reflect in Journals

Reflect and journal your experiences and feel that you're adopting happiness into your life

Fig. 10.7   Model of deep happiness

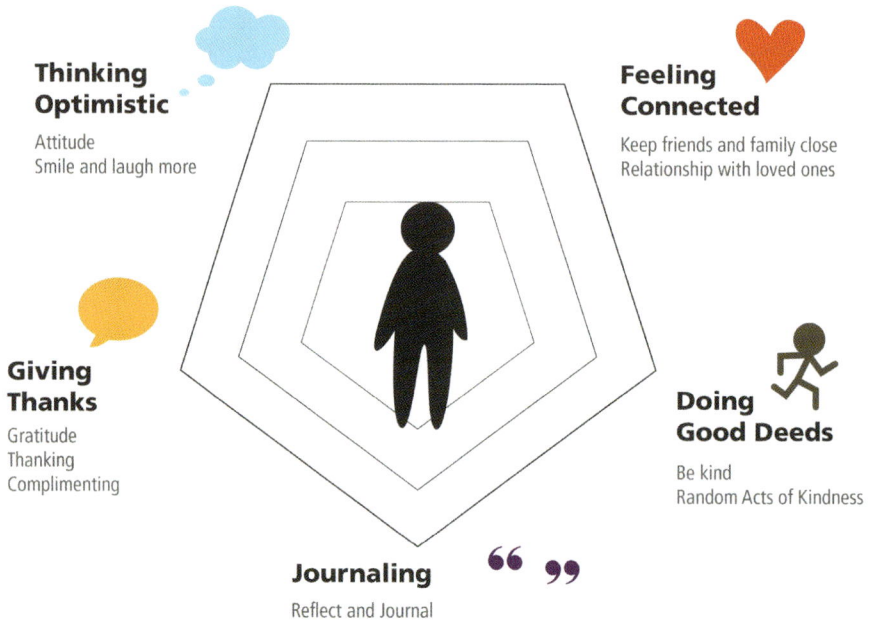

Thinking Optimistic

Attitude
Smile and laugh more

Feeling Connected

Keep friends and family close
Relationship with loved ones

Giving Thanks

Gratitude
Thanking
Complimenting

Doing Good Deeds

Be kind
Random Acts of Kindness

Journaling

Reflect and Journal

Fig. 10.8   Five practices based on the model of deep happiness

10.5.1   Tactics for Happiness Machine

Based on our research, we developed the following measures that could be considered relevant and practical for the Happiness Machine:

- Involvement.
- Personalization.
- Goal/habit-centered.

- Entertainment.
- Connectedness.
- Accomplishment.

10.6   Market Research

In order to have a clearer vision of the target market for the Happiness Machine, AM+A conducted two comprehensive market research activities with potential customers, through collaboration with graduate marketing students of the University of California at Berkeley Extension's, International Diploma Program (Aboelwafa et al. 2013). One main objective of the research was to find out more about people's general usage of smartphones and mobile applications. A second objective was to get a better understanding of their "happy behaviors," especially when using smartphones and mobile applications.

Our market research has gone through the following steps (Fig. 10.9).

10.6.1   Secondary Data

In a first step, secondary data was collected in order to get a general overview of people's mobile usage and happy behaviors. As is shown below, 80% of the world's population now has a mobile phone, which accounts for about 5 billion people. Currently, around 1.08 billion of these mobile devices are smartphones, i.e., approximately 20%. Current tendencies seem to indicate, however, that the number of smartphones will rise considerably during upcoming years. Solely in the first quarter of the fiscal year 2012, Apple, for example, sold a total of 37.04 million devices, which corresponds to about 377,900 mobile phones per day (Fig. 10.10).

Interestingly, not only teenagers and young people make use of smartphones. In fact, most users are between 18 and 44 years old. They use their phones mainly to write text messages (92%), to go online (84%), to write emails (76%), to play games (64%) and to use downloaded apps (69%) (Fig. 10.11).

In order to gain further information, qualitative research was carried out in the form of a focus group discussion and in-depth interviews.

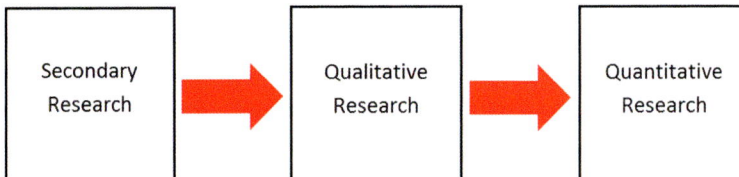

Fig. 10.9 Marketing research steps

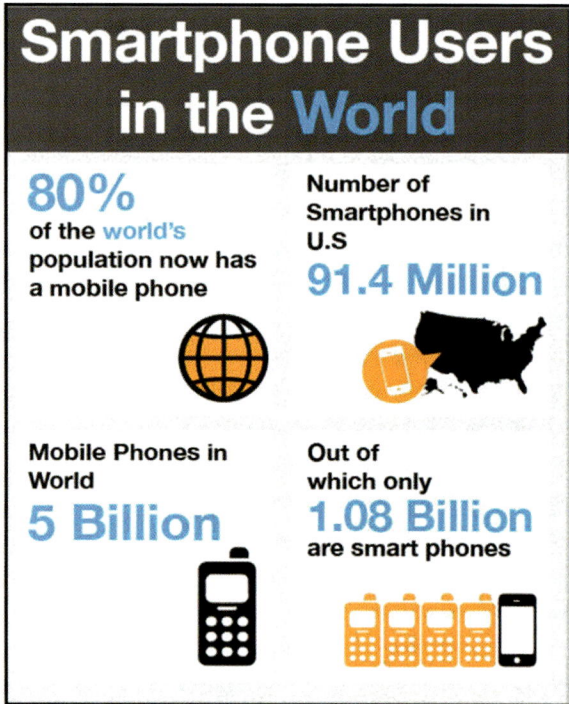

Fig. 10.10 Smartphone users in the world (Aboelwafa et al. 2013)

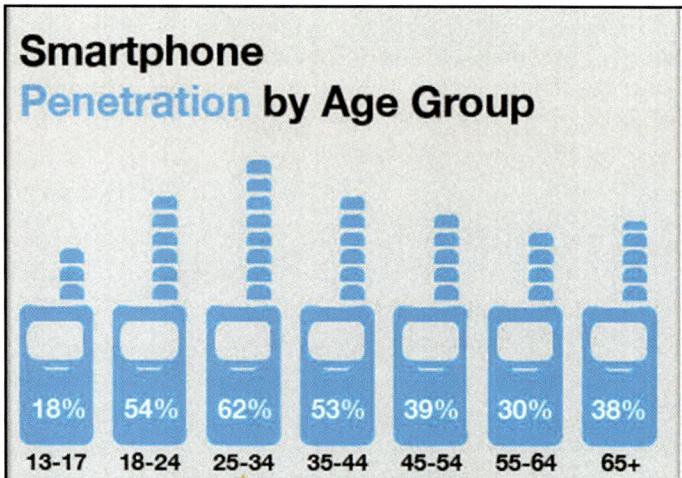

Fig. 10.11 Smartphone penetration by age group (Aboelwafa et al. 2013)

10.6.2   Qualitative Research Techniques

We used the following in conducting qualitative research:

- In-depth interviews.
- Deeper view to the subject.
- Probing done.
- Focus groups.
- Group dynamics.
- Thought leaders can affect others.
- Thought-cloud (mental map) diagrams.
- Making respondents think deeper without help from others.
- Word association.

10.6.3   Quantitative Research

We asked the following sample questions of participants:

- Please tell us what things make you happiest.
- On a scale of 1 to 9, how happy do you feel? 1 is the least, 9 is the highest.
- How often do you think of how happy you are or how to improve my overall quality of life?
- How do you look for new applications you want to download? You can choose multiple options.
- Assume you have downloaded an application that helps you increase your happiness, what content would you prefer?
- If the application tracked your state of happiness how often would go to the application to look for tips and track progress.
- Every day, Once a week, Once in 15 days, Once a month.
- Will you be interested in an application that helps you be happier?
- Would you like to share your happiness progress/tips on social networks (Facebook, Twitter etc.).
- How much would you be willing to pay (in USD) for the application?
- Age?

10.6.4   Results and Conclusions

What makes people happy? The following "word-cloud" figure represents the results of the questions posed to the participants (Fig. 10.12).

Almost 77% of the respondents said they think about how happy they are quite frequently (Fig. 10.13).

Fig. 10.12 Word cloud representing participants views about what made them happy (Aboelwafa et al. 2013)

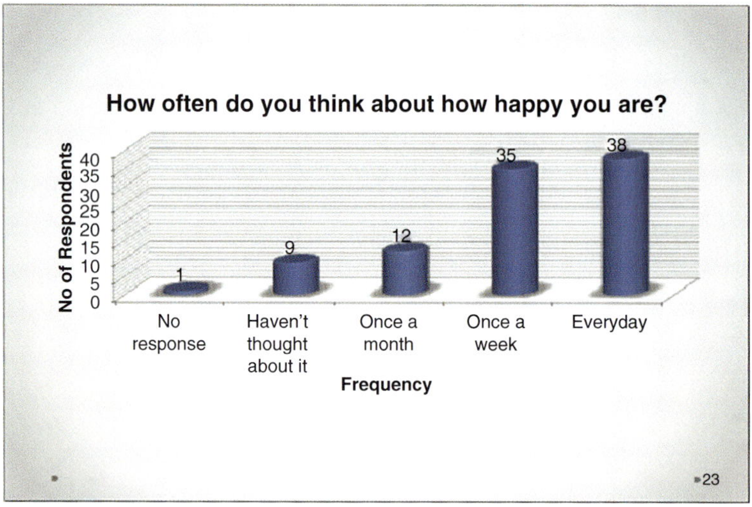

Fig. 10.13 Frequency of thinking about happiness (Image courtesy of Aboelwafa et al. 2013)

More than 50% of the respondents expressed that the most common way to choose new apps is relying on recommendations from relatives/friends (Fig. 10.14). 65% of all the respondents want the application to be free (Fig. 10.15).

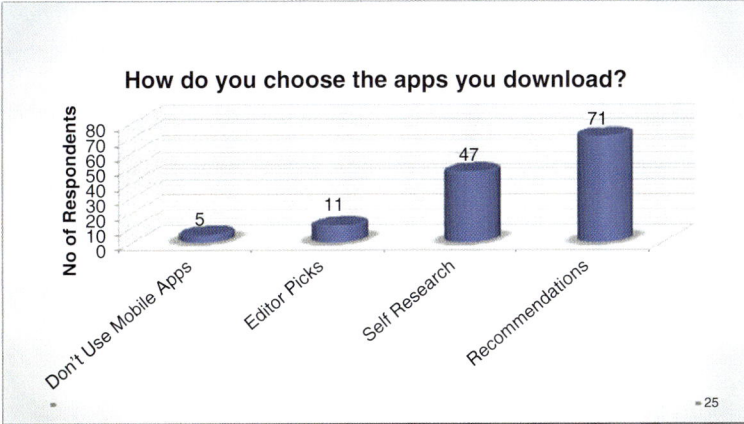

Fig. 10.14 Reasons for choosing applications to download (Aboelwafa et al. 2013)

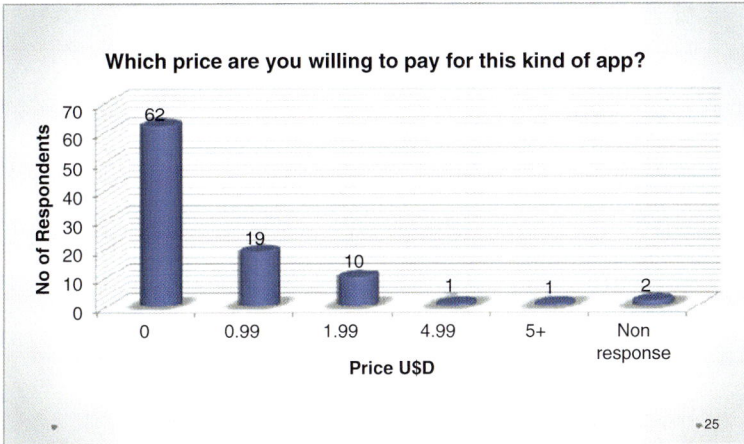

Fig. 10.15 The price participants were willing to pay for such an application (Aboelwafa et al. 2013)

The preference for a free application doesn't change among age groups (Fig. 10.16).
85% of the interested respondents will not be willing to pay more than one dollar for the application (Fig. 10.17).
51% of all the respondents were interested in the application.
29% of them will actually download it (Fig. 10.18).

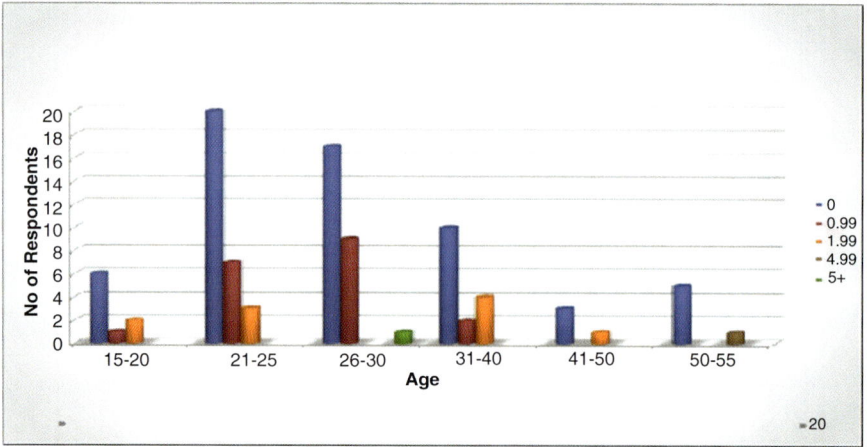

Fig. 10.16 Age of the respondent vs. price (Aboelwafa et al. 2013)

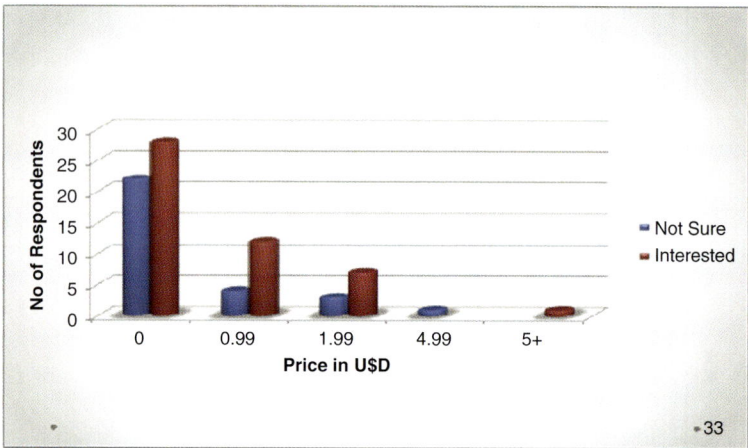

Fig. 10.17 Interested people vs. price for such an application (Aboelwafa et al. 2013)

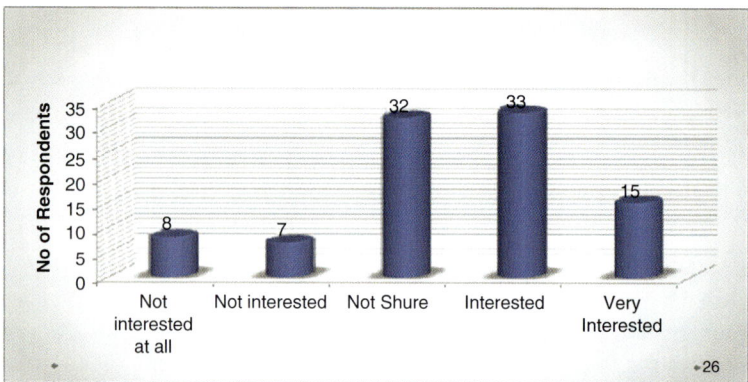

Fig. 10.18 Level of interest in downloading this kind of application (Aboelwafa et al. 2013)

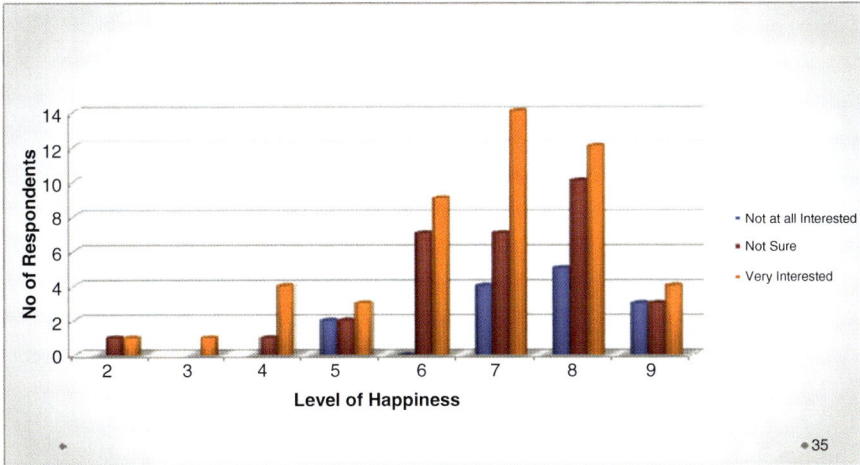

Fig. 10.19   Interest in downloading vs. level of happiness [Aboelwafa et al. 2013)

Contrary to our initial forecasts, the people who showed more interest in down-loading the app were those who were said to be more happy (Fig. 10.19).

10.6.5   Content of the Application

The following trends emerged for the content of the application based on the market research:

- Interested respondents want to see data refreshed everyday or every week.
- About 60% of the interested respondents did not want to share information on social media.
- The majority of the interested respondents wanted to determine what makes them happy, and wanted tips/info based on these preferences.

Interested respondents want to see data refreshed everyday or every week (Fig. 10.20).

About 60% of the interested respondents didn't want to share information on social media (Fig. 10.21).

The majority of the interested respondents wanted to determine what makes them happy and wanted tips/info based on these preferences (Fig. 10.22).

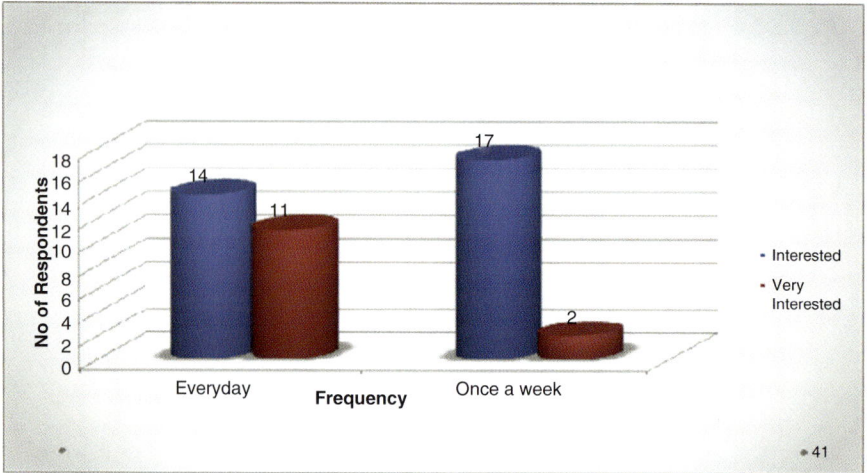

Fig. 10.20 Interested people vs. frequency of the tracking of the level of happiness (Aboelwafa et al. 2013)

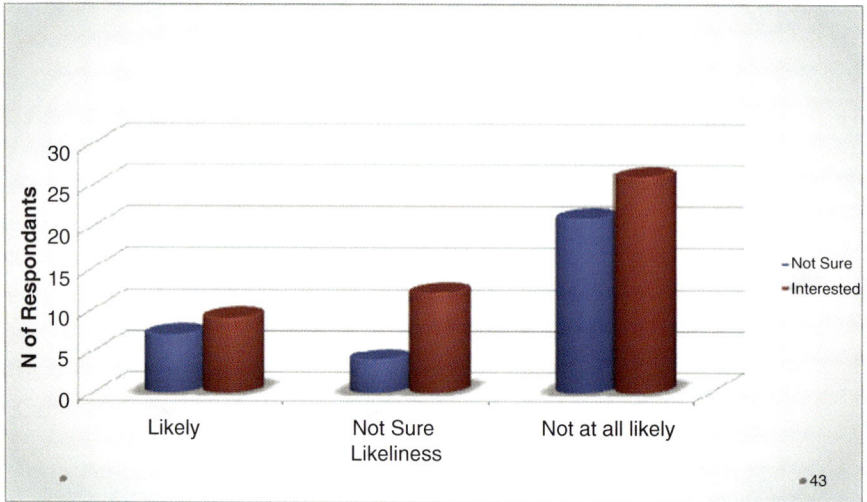

Fig. 10.21 Interested people vs. likeliness to post achievements on any social media (Aboelwafa et al. 2013)

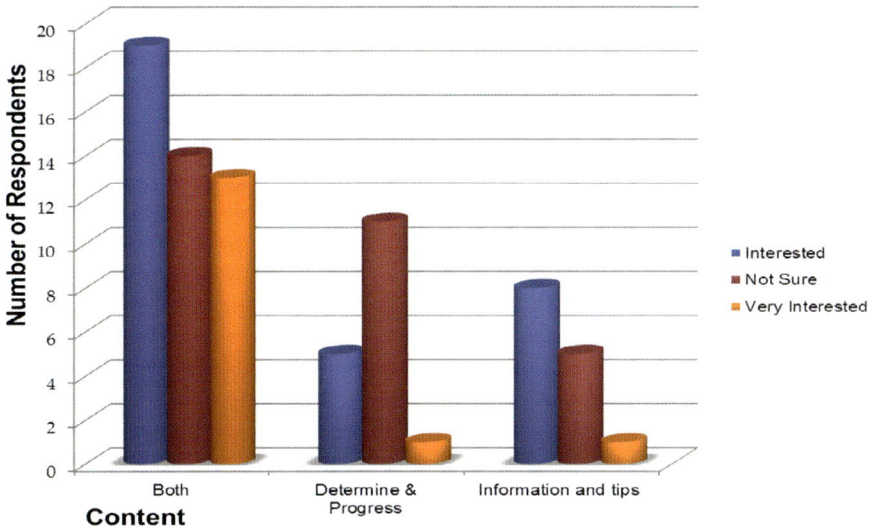

Fig. 10.22 Interested people vs. desired content of the application (Aboelwafa et al. 2013)

10.7 Competitive Analysis

Before undertaking conceptual and perceptual (visual) designs of the Happiness Machine prototype, we carried out comparison studies. AM+A studied in detail 11 smartphone applications. Through screen comparison and customer review analyses, AM+A derived these applications' major benefits and drawbacks. This in-depth analysis helped us to develop further initial ideas for the Happiness Machine's detailed functions, data, information architecture (metaphors, mental model, and navigation), as well as look and feel (appearance and interaction). The following summarize our findings on these products as of mid- to late-2013.

10.7.1 Facebook

Facebook is the most popular online social network (Fig. 10.23).

Pros

- Share update, photo and video and see your friend's, family's, or verified celebrity's update, photo and video.
- Discover and learn about great place nearby.
- User can create event that connect people who have the same goal or interest together.

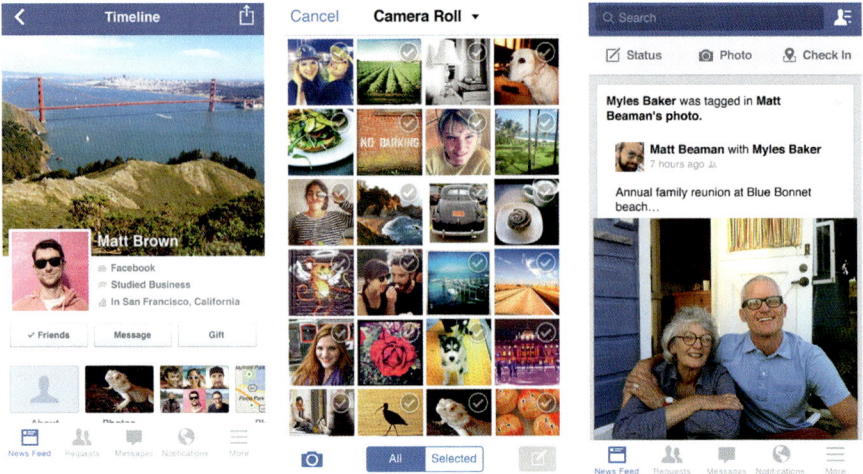

Fig. 10.23 Example screenshots of Facebook on a mobile device (Images courtesy of Facebook https://www.facebookbrand.com/)

Cons

• User shares both good and bad things or mood on Facebook.

10.7.1.1   Happier

Happier is an application with which people can share their happiness moments by uploading photos into different categories of content (Fig. 10.24).

Pros

• Users can see happiness moments from all over the world.

Cons

• There seems little difference from Facebook.
• Seeing a stranger's happiness moments is not necessarily an effective way to make many people happier.

10.7.1.2   Foursquare

Foursquare enables people to discover and learn about great places nearby, search for what they've been craving, and get deals and tips along the way (Fig. 10.25).

Pros

• Discounts and freebies nearby.

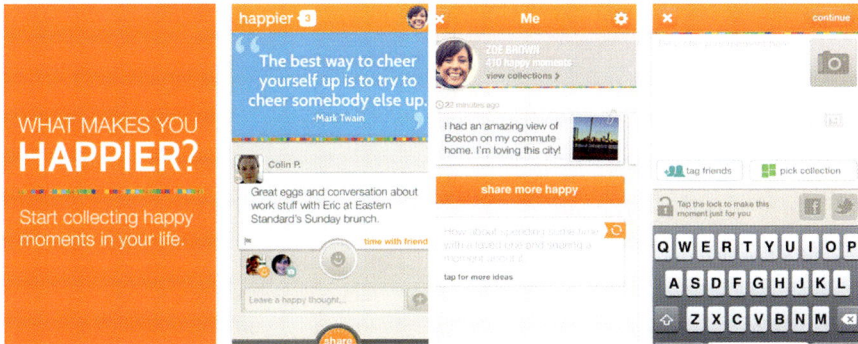

Fig. 10.24 Happier screen images in 2013 (Images: Happier, Inc.)

Fig. 10.25 Foursquare screen images (Images: Foursquare.)

- Platform that assists users in gathering local information such as food, shopping, sights, arts, trending, and recently opened with trusted advice from people all over the world, and in posting reviews or other travel-related content.
- Platform is easy to use and understand.

Cons

- Users cannot narrow down research.
- No information about local services, events, and transportation.

10.7.1.3 Path

Path is a personal social network designed to bring people closer with family and friends.

Pros

- Users can choose different stickers to convey their emotions.

• Path provides users with fast, fun, and private messaging.

Cons

• The main function is very similar to Facebook.

10.7.1.4   Snapchat

Snapchat is a real time picture chatting application.

Pros

• The app offers a fun way to interact with friends through real-time photo.
• Users feel surprised and enjoy happiness moments for a limited time.

Cons

• Users take advantage of this "magical function" to do some wacky things such as embarrassment moments.

10.7.2   *Conclusions*

A large number of mobile applications already exist in the field of happiness, most of which emphasize the social connection aspect of achieving greater happiness. However, few if any of the applications take a holistic approach to

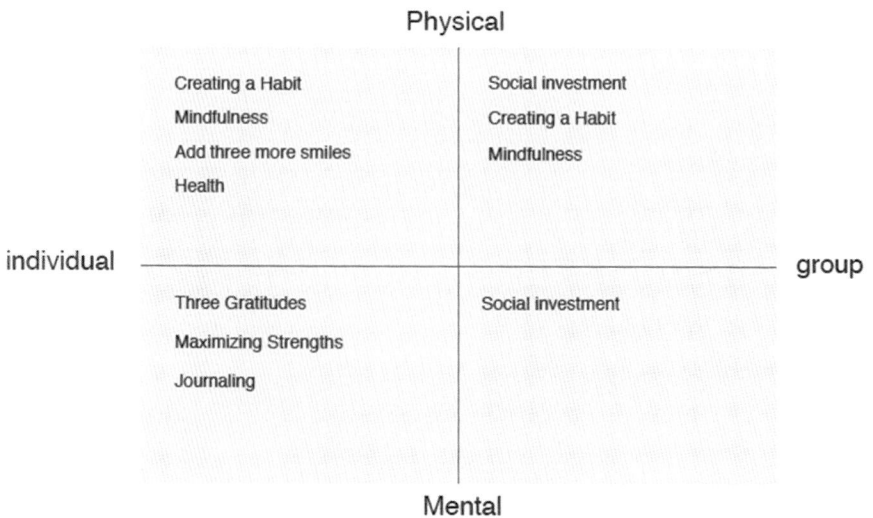

Fig. 10.26 Diagram of opportunity space for the Happiness Machine based on considerations of research

help users realize and achieve deep happiness. The Happiness Machine may be able to find an opportunity in its personal, intercultural approach that the app can provide, distinguishing itself from existing applications that are more about passing time (Fig. 10.26).

10.8   Personas

As mentioned in Chap. 1, personas are characterizations of primary user types, and are intended to capture essentials of their demographics, contexts of use, behaviors, motivations, and subsequent impacts on design solutions. Personas are also called user profiles. Typically, UI development teams define one to nine primary personas. For the Happiness Machine 1.0, we determined the following persona by analyzing available data.

10.8.1   Persona 1: Joey Chen

10.8.1.1   Textual Summary

Joey is a 25-year-old Chinese girl, pursuing graduate school at University of Southern California, school of communication. She loves food, art, museums, and tourist attractions. She would like to share this information with her friends through social networks. She also enjoys finding interesting/weird stories, images, and video on the Internet from all around the world. She uses her smartphone often during a long commute from her home to workplace. Also, she loves to get instant information from apps and check for the latest updates when she's waiting for a long line, bus, or even walking on the street.

She is currently doing an internship as communication coordinator at Yahoo and lives in San Francisco. She uses some mobile applications, such as Foursquare and Yelp, a lot to find some interesting things to do and share her happiest moments on Facebook or Instagram when she is stunned by awesome food, things, people and place. Nevertheless, she usually finds some difficulty in finding the real interesting things to do, because the information on the apps is so abundant and overwhelming, so she has a hard time deciding what to do; she has not found the right approach so far. Just reading some comment online easily bores her. Instead, she is interested in feeling, living, experiencing the happiness of people, place, and things so that she can understand better these interesting stories, images, and videos. If she finds truly happy moments, she wants to share them by creating her own stories.

10.9   Use Scenario

As mentioned in Chap. 1, use scenarios are a UI development technique whereby, in the development of a prototype to simulate the major characteristics of a software product, a use scenario is written to determine what behavior will be simulated. A scenario is essentially a sequence of task flows with actual content provided, such as the subject user's demographics and goals, the details of the information being worked with, etc. The following use scenario is drawn from the preceding persona.

10.9.1   Joey Chen's Use Scenario and Behavior Change: World-Wide Happiness + Happy Queue

As usual, Joey goes to a Starbucks Sunday morning next to her apartment on Mission Street. She takes a look on her smartphone while waiting in a long line. Suddenly, she notices that there are some photos of her friends' happiness moments being displayed on her screen. She sees the photo of herself and one of friends eating dinner together yesterday. Joey sends a Happy Note: "Thanks for introducing great restaurant." Then she sees one of her best friends riding her bike along the edge of lake Michigan in Chicago. She realizes she hasn't herself ridden the bike for such a long time. So she opens the Happiness Machine and clicks the Search box, which then pops up ten categories of happy things to do (Food, Cool Spots, Place, Little things, etc.) that she can do based on her location. She clicks Cool Spots because she wants to find some outdoor activities to do. There are some happy moments shared by people world-wide who have been to San Francisco. These moments are displayed in a beautiful photo. Joey became obsessed with one of the photos of a Cool Spots, Golden Gate Bridge. People were riding their bikes on the Golden Gate Bridge. She decides to have a good time there. She orders her coffee and waits for her order. At the same time, her friend Peter shared a video of himself performing magic trick via C2P (Card to Phone). She is very interested in how this trick surprised people. So, she downloads the C2P app and learns this fast magic trick. She shares this happy moment on the Happiness Machine and calls it: "Learning my first Magic trick at Starbucks."

10.10   Information Architecture

The following text describes the information architecture of the first version of the Happiness Machine, 1.0.

10.10.1   Machine Information Architecture

In designing the Information Architecture (IA) for the Happiness Machine, AM+A began by examining the IA for past Machines, including the Green Machine, the Health Machine, the Money Machine, the Driving Machine, and the Innovation Machine (see descriptions in previous chapters and Marcus citations for past Machines in the Bibliography). We discovered an overarching model for the IA that permeated each of the past machines. In this model, there are five primary modules, or branches, of the IA, each of which is described below. While we altered slightly the details of each module to fit the needs and requirements for the Happiness Machine, we maintained the same general approach as in all the Machines. These modules are described below.

10.10.2   Dashboard

The Dashboard module is similar to a landing page for a Machine. It provides an overview into the status of the user's behavior change. Here, the user gets a view of his/her goals and where she/he stands in achieving those goals.

10.10.3   Process

The Process View module is where the user gets more high-level views, or a "road map" of the process and more details regarding each objective and goal. The user sees the progress being made, as well as the next steps in achieving a particular goal.

10.10.4   Social Network

The Social Network module is an integral part of behavior change in modern software. Users engage in focused, subject-matter-based connections with friends, family, and/or like-minded people that either share similar goals or wish to support others in achieving behavior-change objectives.

10.10.5   Tips/Advice

The Tips/Advice module provides focused knowledge about a given topic to give users insight into the habits they wish either to get rid of or adopt.

10.10.6   Incentives

The Incentives module presents users with a fun and engaging ways to change their behavior. Gamification has proven to be a powerful tool in adopting users to try an application, even with virtual incentives, although in some cases, real incentives are provided. In addition, a leaderboard allows a user to compare his/her progress with others, tapping in the competitive nature of the human mind to create behavior change.

10.10.7   Information-Architecture Diagram

AM+A adapted the basic structure of the past Machines in creating the Happiness Machine 1.0 information architecture. However, due to particular complexities that were unique to the Happiness Machine, we made the following primary changes:

- The respective tabs were attributed different labels: Dashboard, Diary (process), Advisor (tips/advice), Fellows (social network), and Challenge (incentives).
- A different menu navigation system was provided, in order to keep the screens uncluttered. The high-level basic modules are removed from most screen displays, but may be quickly called back to enable users to navigate high-level modules.

The accompanying Fig. 10.27 is a diagram of the information architecture for the Happiness Machine 1.0.

10.10.8   Happiness Machine Objectives

Combining happiness theories, information design, and persuasion design, we defined the objectives of the Happiness Machine as follows:

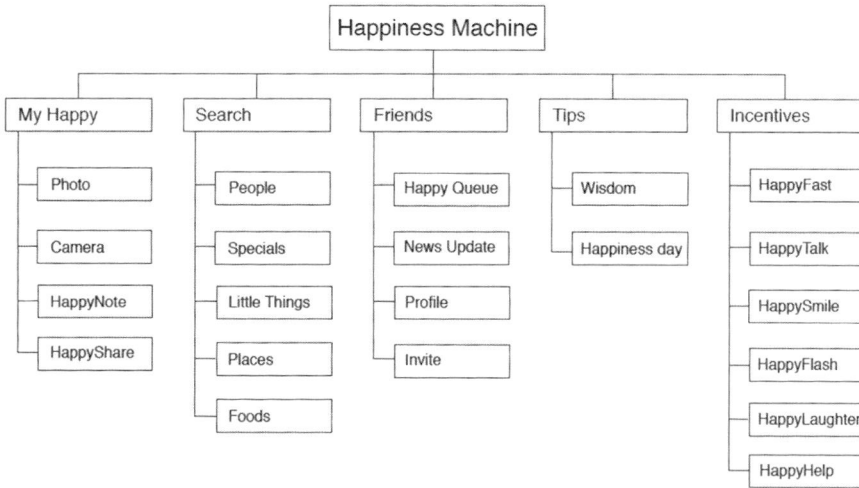

Fig. 10.27   Information architecture diagram of the Happiness Machine 1.0

- Shift users from boredom to having an interest in something.
- Help users to enjoy people and connect with people when users feel lonely.
- Enable users to do good deeds when they are without purpose or goal.

10.10.9   Happiness Machine 1.0 Use Scenarios

10.10.9.1   Definition

As explained in Chap. 1, use scenarios are a user-interface/user-experience development technique whereby, in the development of a prototype to simulate the major
characteristics of a software product, one determines typical behaviors to be
simulated and writes brief descriptions of them. A use scenario is essentially a
sequence of task flows with actual content provided, such as the subject user's
demographics and goals, the details of the information being worked with, etc.

10.10.9.2   General Use Scenario

Use scenario topics are drawn from the previously defined personas. Some specific
examples might be relevant only to users of a particular age group, education level,
gender, or culture. Note the conventions for usage of the terms "objective" and "goal."
An objective is a general sought-after target circumstance. A goal is more specific and
is usually qualified by concrete, verifiable conditions of time, quantity, etc.

10.10.9.3   Specific Use Scenario Example

The following persona-based use scenario leading to behavior change is based on one specific use scenario and features specific suggestions for future functionality: World-Wide Happiness and Happy Queue.

As usual, Joe goes to a Starbucks Sunday morning next to his apartment on Mission Street. He takes a look on his smartphone while waiting in a long line. Suddenly, he notices that there are some photos of his friends' happiness moments being displayed on the screen. He sees the photo of himself and one of his friends eating dinner together yesterday. Joe sends a Happy Note: "Thanks for introducing me to a great restaurant." Then, he sees one of his best friends riding his bike along the edge of Lake Michigan in Chicago. He realizes he hasn't himself ridden his bike for a very long time. So, he opens the Happiness Machine and clicks the Search box, which then pops up ten categories of happy things to do (Food, Cool Spots, Place, Little things, etc.) that he can do based on his location. He clicks Cool Spots because he wants to find some outdoor activities to do. There are some happy moments shared by people world-wide who have been to San Francisco. These moments are displayed in beautiful photos. Joe becomes obsessed with one of the photos of a particular Cool Spots, Golden Gate Bridge. People are riding their bikes on the Golden Gate Bridge. He decides to have a good time there. He orders his coffee and waits for his order. At the same time, his friend Peter shares a video of himself performing magic trick via C2P (Card to Phone). He is very interested in how this trick surprised people. So, he downloads the C2P app and learns this fast magic trick. He shares this happy moment on the Happiness Machine and calls it: "Learning my first Magic trick at Starbucks."

10.10.10   Metaphors and Mental Model

Based on the UCD UX approach, on the results of the market research, especially that carried out by UC Berkeley Extension Marketing graduate students under the direction of Prof. Bob Steiner (Aboelwaffa et al. 2013; Baruggia et al. 2013), and product, comparison studies, as well as on the persuasion and motivation theories of Fogg et al. (2007) and Maslow (1943), AM+A conceived of an innovative mobile application, the Happiness Machine, which combines information design and visualization with persuasion design.

Our main UX design objectives for the Happiness Machine are to ease, enhance, and enrich the experiences of people in daily life, to foster a deeper process of self-development, of happiness understanding and learning, and to thereby contribute to short- and long-term changes of typical behavior via a shift from fear and anger, to acceptance and fun; from passive to active behavior; from isolated to socially immersive behavior.

Therefore, the focal point of the Happiness Machine, as opposed to the majority of existing mobile applications oriented to simple fun, happiness, joy, mirth, psychology, life-path, mindfulness, religion, and other disciplines, is supposed to be an emphasis on the user, and not on the destination itself, which, of course, will nevertheless play an important role in the concrete implementation and conception of the software.

Although able to serve people of all ages, genders, and other demographics, this Happiness Machine is targeted especially for young US males approximately 20–30 years of age, from higher to average economic and educational demographics, who have a general interest in and openness towards fun, change, cultures, and people.

To put it simply, the Happiness Machine aims to answer the following two critical questions:

- How can information design/visualization present persuasive information to promote a personally enriching, profound, as well as educative short- and long-term happiness-oriented behavior change?
- How can mobile technology assist in presenting persuasive information and promote behavior change of medium to young, primarily male, somewhat high income, and educated people?

The Happiness Machine will contain several key components. The many detailed functions are grouped into five sets or "tabs": a dashboard (1), an individual diary (2), a mobile travel advisor (3), a tab related to social comparisons and social interactions, (4) and a section dedicated to incentives (5). The sections below explain each tab.

10.10.10.1   Tab 1: Dashboard/Profile

The Happiness Machine will contain a dashboard, which can be considered as the "homepage" of this mobile application. The dashboard will contain the most important information users may need at any time during their daily life. Therefore, potential elements of the dashboard will be the following:

- Current time and date.
- Current weather and weather forecasts.
- High quality camera and video: This function helps users capture and share their happiness moment via beautiful photo and video on Happiness Machine.
- HappyNote: Users can see a visible and valuable compliment from friend, family, colleague, boss, or future co-corker.
- HappyBank: (See (Estonia's Happiness Bank)). We apply the idea of "Bank of happiness, as an online market for good deeds". The concept of the bank of happiness started in Estonia is that people do nice thing for each other, just because no payments or products are involved. Thus, the HappyBank helps users make lists of strengths and needs. Users in HappyBank use their strengths to help each other.

All of these elements will be optional for Happiness Machine users, which means that users can personalize and customize the dashboard according to their needs and interests, by deciding which of the above-mentioned features to show and which ones to hide.

10.10.10.2   Tab 2: Search

The Happiness Machine provides access to world-wide happiness, a collection of happiness stories and wisdom. The concept of the HappySearch is that users are standing somewhere, e.g., the lobby of a building or a street corner, and they are bored – how can they find some happiness things to do?

- People: Users can find an artist, a celebrity, or a famous designer's joyful experiences that they have created based on the users' location. Thus, users can see something, learn about something, or became engaged in some interesting, playful, and/or enjoyable experience based on where they are now. Then, they can create their own interesting story and leave it for others to enjoy.
- Specials, little things, and place.
- Food: HappySearch is also an "augmented reality happy telling." When users walk on the street, they can use their HappySearch to browse surrounding restaurants to see some awesome food people have shared from their previous happy experiences.

10.10.10.3   Tab 3: Friends

The above-described features and functions of the Happiness Machine could be significantly enhanced in their persuasive and motivational power by incorporating a social dimension into the application, i.e., by providing possibilities for social interaction, comparison, and competition. All functions dedicated to these aspects appear under a tab named "Friends".

It is an acknowledged and well-proven fact in the field of persuasion theory (Fogg 2003) that this social component can become a superior incentive for behavior change of individuals. By leveraging the features of successful social networks such as Facebook, by creating specific fora, or a specific web portal for the Happiness Machine, several ways of social interaction and exchange can be considered:

- Users can share their happiness moment with friends or with other users at the same, or also at other places. Taking up again the idea of cultural and digital storytelling, it becomes clear that one of the main motivations to tell a story is also to have someone listening to it, so that users will have an interest in sharing their happy user experiences, in receiving feedback and comments from their followers or "subscribers", and as well in giving recommendations to others.
- At the same time, a social platform for exchanges and interactions could give Happiness Machine users the possibility to look at others' Happy user experience, to follow or "subscribe" to them. Thus, they can see how they create happiness

moment, which curious and interesting things they discovered, which photographs they uploaded, and so on and so forth.

- The attribution of points and virtual rewards in the below-described games section (see tab 5) can then trigger social competition among application users, and can further stimulate their motivation to discover the happiest moment, and to proceed in their learning process. The more points and rewards a user accumulates, the higher he or she will rank in the list of happy experts, and the more likely it is that other user will ask him or her for experience, and for hints. Happiness Machine users will be even more motivated if they can communicate their current expert status on their Facebook profile, or via Twitter.

- The possibility of social interaction can also serve as a way to animate discussions among users. These exchanges would represent another form of reflective activity, and they would thus enhance the individual sharing process. In this case, reflection would occur within a community, and through interaction with others.

10.10.10.4   Tab 4: Tips

Happiness Machine users will be assisted and accompanied during their life by a mobile happiness advisor. It is important to stress that this advisor will be essentially different from the typical push services, which are based on both the current location of travelers, and on the preferences they have expressed in advance. Such services belong to the field of so-called personalized hypermedia: Based on users' characteristics, needs, and preferences, on usage behaviors, and on the respective usage environment, they provide personalized and tailor-made offers and recommendations.

10.10.10.5   Tab 5: Incentives

The users' happiness and sharing experiences will be enhanced and enriched in the form of games that will appear under the tab "Incentives". This form of interactive game with friends will serve as a major incentive: It will have a stimulating effect on users, and will increase their motivation to create and share happiness moment. Especially the attribution of virtual rewards, in the form of titles (e.g., "X is now a four star happy expert!") or medals (e.g., bronze, silver and gold), represents a strong incentive and increases use frequency. Moreover, real rewards are given to successful Happiness Machine users. The Happiness Machine will offer points as incentives, like airport-parking or airline-mileage clubs. Having achieved a certain number of happiness moments about a HappyNote or HappyLaff, users could be rewarded with an eBook, with free subscription to an e-magazine, or with a tangible present, such as a coffee mug. Companies and corporations involved as sponsors or partners in the Happiness Machine may even offer coupons or price reductions for future purchases.

The integration of games will, at the same time, add a component of fun and play to the application, and will thus make the life experience not only informative and reflective, but also entertaining.

Such incentives, as well as the prospect of rewards and of positive feedback, will motivate application users to become active, wiser and happier. Incentives include the following:

- HappyFast: Learning new thing will make people happier, because people can share the experience of learning and achievement with their friends. HappyFast teaches you fast and easy magic when you have free time.
- HappyTalk: [see for example, Mouthoff, https://itunes.apple.com/us/app/mouthoff/id306588353?mt=8] We apply the Mouthoff to the Happiness Machine. The concept of Mouthoff is to hold the phone in front of the mouth and talk, shout, laugh, scream, or sing to get the selected to animate in time with the sounds the speaker makes. Thus, in HappyTalk users can hold the Happiness Machine in front of their mouths and talk to get the mouth to animate in time with the sounds they make. Moreover HappyTalk will capture your voice and interactive-animation between the users and the Happiness Machine. Users can share this fun interactive-experience on the Happiness Machine.
- HappySmile: The happiness machine encourages users to smile more and help users to practice smile. The target user of HappySmile will be the waiter in the restaurant. The waiter's behavior will influence customer's mood. The HappySmile enables waiters to check their smile before leaving the kitchen. The waiter with a natural smile has a good performance. The good performance has a great influence on customer's emotion. So the customers are willing to give more tips and come to this restaurant again.
- HappyLaff: HappyLaff is a collection of people's laughter. Users record at least one-minute video about happiness moment. HappyLaff helps them to select the best amazing 30 s that is focused on laughter. Through the selected 30-s video, users can capture their own and others' laughter, and of their happy moments. The video is able to convey the awesome feelings of happiness moment more than just the photo.
- HappyHelp: HappyHelp, based on Johnson and Johnson's *Donate a Photo* (http://www.donateaphoto.com/), generates different kinds of good deeds based on the users' location. Consider users' photos, for example. If users visit a park and take beautiful photos, HappyHelp will give users a hint that there is a good deed users can do easily, like "One photo helps restore a public park." When users share a photo on HappyHelp, they will make a difference by raising money and awareness for causes they care about. Furthermore, they can share with their friends and they will help their causes meet their goals faster. So with HappyHelp, users can realize and recognize what they can do based on their location. Thus, it's an effective and efficient way for users to do something meaningful and then have deeper happiness.
- The Happiness Machine 1.0 and 2.0 versions were focused on illustrating three key features: HappyNote, HappyLaff, andHappySearch.

10.10.10.6   Happy Note

Problem Statement

People love compliments. Compliments from the heart make people happier. However, people sometime hesitate to give a compliment and/or do not know what is the appropriate way of expressing compliments. Users' friends and family are often curious about the users' values, contributions, experiences, and joyful moments. Strangers may be affected strongly and in a positive way by "random acts of kindness."

Ideation and Synthesis

How might one make verbal compliments visible and valuable? What if users can see their friends' contributions and valuable experiences? How might we transform happy interaction experiences into a digital space? How might users start sending a HappyNote to someone who is their close friend and make a compliment or even start sending HappyNote to people whom they don't know well?

Use Scenario

Strong Ties: for people the user knows
    Users can see their compliments given to their friends (from the people who make a compliment to their friends). The users can know more about the happy experiences of their friends. Also, users can share their own compliments received with friends. Those who receive the HappyNotes can interact with their friends in a larger ecosystem, like gaining points on a leader board.
    Weak Ties: people users don't know well
    Users who have received other people's favors can send them a HappyNote. Consider the Wayz app for example [see https://www.waze.com]: Users can get the best route every day, with real-time help from other drivers who can send them a HappyNote to encourage their good deeds and make Happiness moments count.

10.10.10.7   HappyLaff

Problem Statement

Happiness moments are hard to capture and share with others. Only if there is an effective, efficient, satisfying, and appealing tool will users take advantage of the Happiness Machine frequently.

Ideation and Synthesis

How might we use people's laughter (users and their friends) as a good start? What if we can have a collection of people's laughter to encourage them to capture, relive, and enjoy happy moments? What if we can capture happiest moments of our laughter and joyful experiences? Laughter is a powerful catalyst that will help us to remember and capture a moment of happiness rapidly. People can't stop laughing while listening to their friends' laughter.

Use Scenario

Users record at least one-minute video about a happiness moment. HappyLaff helps them to select the best amazing 30 s that is focused on laughter. Through the selected 30-s video, users can capture their own and others' laughter, and of their happy moments. Also the "cover image" or "PR still-frame" of the video will be featured as the image that reminds them and that they can see and select to view the video. Thus, they can tap a smile of their friends to enjoy the laughter and see the happy memory again via video.

10.10.10.8   HappySearch

Problem Statement

People come to a spot, have some fun, and then leave. There's nothing valuable for future visitor. How can we leave a "Happiness legacy" for others to enjoy?

Ideation and Synthesis

You are standing somewhere, e.g., the lobby of a building or a street corner, and you are bored. How can you find some happiness at this moment and location?

Use Scenario

Users can find an artist, a celebrity or a famous designer's joyful experiences that they have created based on the users' location. Thus, users can see something, learn about something or became engaged in some interesting, playful, and/or enjoyable experience based on where they are now. Then, they can create their own interesting story and leave it for others to enjoy.

10.10.10.9   Further Possible Implementations

By further enhancing the above described features linked to HappyNote and social interaction, the Happiness Machine and its main functions could significantly enrich the overall happy user-experience especially in enterprise usage that makes employees happier.

Another possible application of the HappyNote could be applied to promote self-strengths. Why do some users get a compliment from other people? The reason is that users achieve some accomplishments like doing good research. Then the skill "research" would become one of the users' strengths (Happiness Machine could make a list of users' strengths).

10.11   Designs

The following subsections discuss initial and revised screen designs for Happiness Machine 1.0.

10.11.1   Initial Designs

Based on the information architecture, AM+A prepared initial designs of some key screens of the Happiness Machine. The following describes the representative initial screen designs of the Happiness Machine.

10.11.2   Landing Screen

The Landing Screen typifies the main menu and app instruction throughout the Happiness Machine. This page displays five major functions with descriptions below (Fig. 10.28).

10.11.3   HappyNote

The development of HappyNote applies Cialdini's, Fogg's, and Maslow's theories cited earlier to innovate thoughts in happy user experience design. Sending/Receiving a joke via e-mail will make people feel more enjoyable, interested, and surprised. However it is not a deep happiness. The HappyNote you received is from

Fig. 10.28 Initial Landing Screens for Happiness Machine 1.0

people who care about you or love you. This is what Maslow's hierarchy of needs, emphasizes: including the level of love/belonging, esteem, and self-actualization. HappyNote allows users to make a compliment, review their valuable experience, and, more importantly, make people more deeply happy. The following shows how it works.

10.11.4   Creating HappyNote

Provide users with HappyNote, tools to write, check-in function, and different shapes of Notes (Fig. 10.29).

10.11.4.1   Select People and Make a Compliment

Click the bubble "Make someone happy" to choose whom you want to give a compliment (Fig. 10.30).

Fig. 10.29 Initial Design
for HappyNote for the
Happiness Machine 1.0

Fig. 10.30 Initial
Compliment Screen of
Happiness Machine 1.0

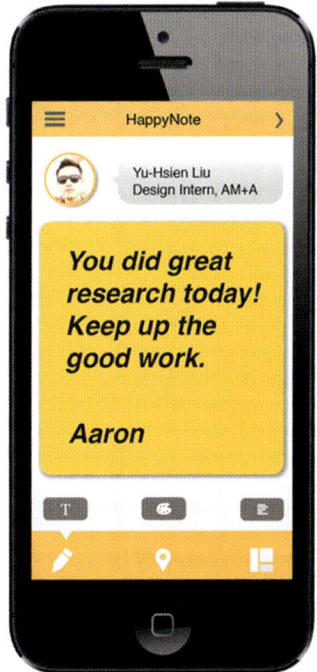

10.11.4.2   Post It

The HappyNote will become an icon appearing on the user's camera. Thus, the user can choose any background or context desired to post this HappyNote (Fig. 10.31).

10.11.4.3   Add Your Feeling

Putting different kinds of stickers on your HappyNote makes it possible for the user to express feelings naturally (Fig. 10.32).

10.11.4.4   Receiving HappyNote

A user who received the HappyNote can collect the happiness moment by means of words and photos (Fig. 10.33).

Fig. 10.31   Initial "Post it" screens of Happiness Machine 1.0

Fig. 10.32 Initial Sticker
screen of Happiness
Machine 1.0

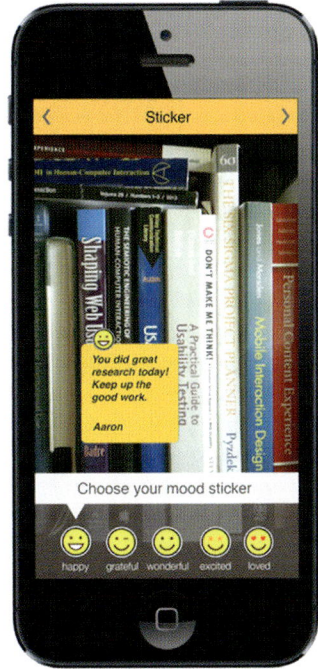

Fig. 10.33 Initial screens
for Receiving a HappyNote
for the Happiness
Machine 1.0

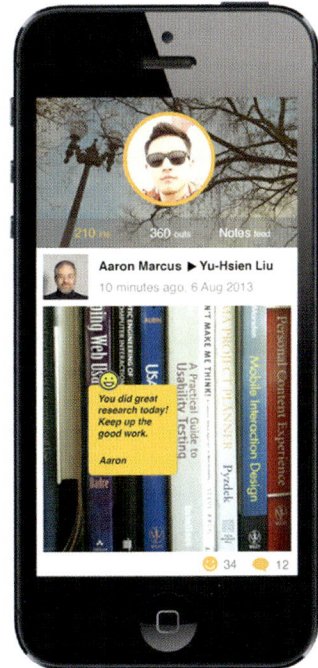

10.11.4.5   Browsing HappyNote

The social network enables the user receiving the HappyNote to have more happy interaction experiences from friends and family (Fig. 10.34).

10.11.4.6   Full-size photo and HappyNote

Users can view larger versions of the HappyNotes (Fig. 10.35).

10.11.4.7   HappyLaff

HappyLaff helps users to capture their happiness moments when laughing.

10.11.4.8   Video recording

Provides users with video recordings to capture happiness moments. HappyLaff helps users to select the best 30 s video that captures their own laughter (Fig. 10.36).

Fig. 10.34 Initial Browse screen of Happiness Machine 1.0

Fig. 10.35   Initial screens for photos with the HappyNote function for the Happiness Machine 1.0

Fig. 10.36   Initial
HappyLaff screens of
Happiness Machine 1.0

10.11.4.9   Choose Smile

Provides users with three different smiles that are from the selected 30 s set (Fig. 10.37).

10.11.4.10   Choose Filter

The filter allows users to make their smiles even more beautiful and powerful (Fig. 10.38).

10.11.4.11   30-s laughter

Users' friends can enjoy users' videos by just clicking on the users' smiles. They can also easily enjoy other people's laughter by just clicking other smiles (Fig. 10.39).

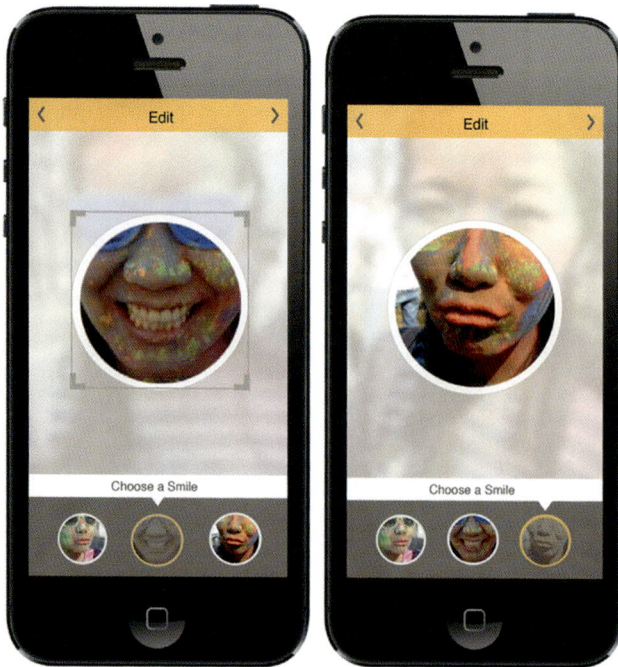

Fig. 10.37 Initial screens for editing an HappyLaff image for Happiness Machine 1.0

Fig. 10.38 Filters for
HappyLaff images

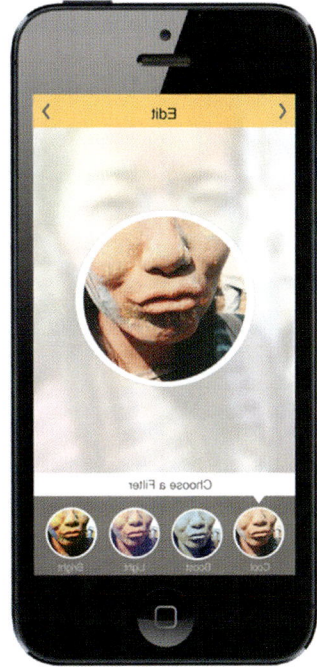

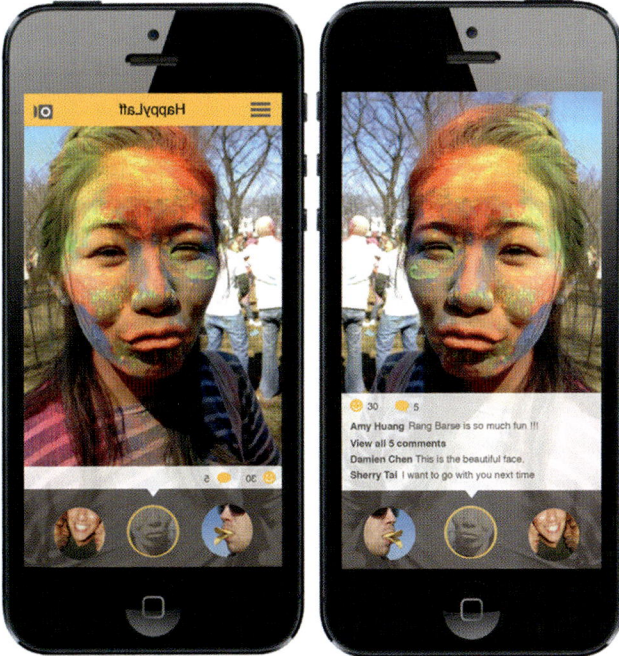

Fig. 10.39 Initial screens of HappLaff conversation function for the Happiness Machine 1.0

10.11.5   Future Extension Laughter Display

Instead of just seeing who likes your video, users can listen to their friends' selected laughter (Fig. 10.40).

10.11.5.1   HappySearch

Users can discover happiness moment wherever they go. So they can learn about something, create their happiness moment, and leave it for others to enjoy.

10.11.5.2   Search option

HappySearch provides user with five options to discover (Fig. 10.41).

10.11.5.3   Happy people

Users can find an artist, a celebrity, or a famous designer's joyful experiences that they have created based on the users' location (Fig. 10.42).

Fig. 10.40 Initial screens for Laughter Play function for the Happiness Machine 1.0

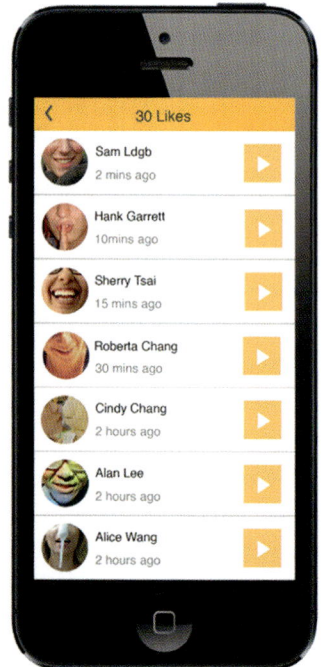

Fig. 10.41 Initial screens
of Happiness Search
function for the Happiness
Machine 1.0

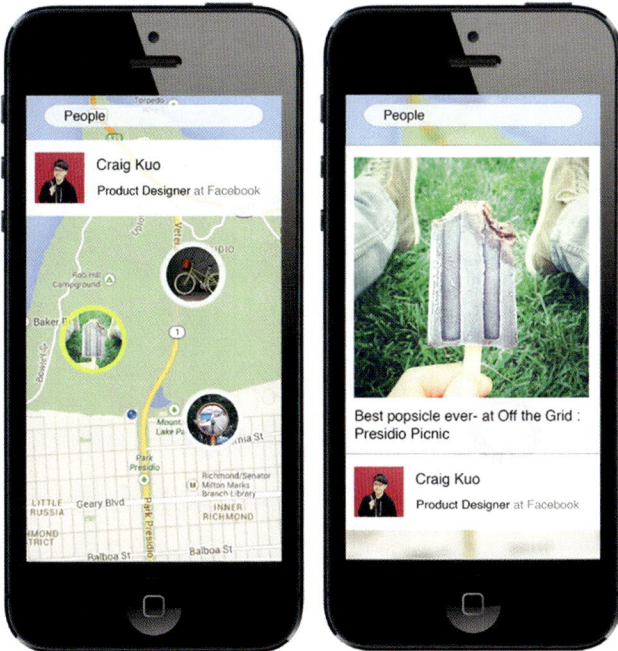

Fig. 10.42 Initial screens of Happy People function for the Happiness Machine 1.0

10.12   Happiness Machine 2.0

10.12.1   Personas

Note: Personas were developed for additional versions of the Happiness Machine. Upon reflection, AM+A felt that the first version could be explored further (usually it takes about three times to get it "about right"). The subsequent use scenarios, information architect, conceptual designs, and perceptual designs differed from the earlier version.

10.12.1.1   Definition

Because people experience happiness in many different ways, we wished to demonstrate different use scenarios of relatively different functions of Happiness with different personas. To create user profiles, we considered different demographic groups, behavioral segmentation, as well as three of Carl Jung's 12 archetypes to create three personas. (Jung 2013) for which we found the following:

"In Jung's psychological framework, archetypes are innate, universal prototypes for ideas, and may be used to interpret observations. Jung treated the archetypes as psychological organs, analogous to physical ones in that both are morphological constructs that arose through evolution." (Fig. 10.43)

For the Happiness Machine 2.0, we emphasized the Jungian archetypes and devised the following three personas.

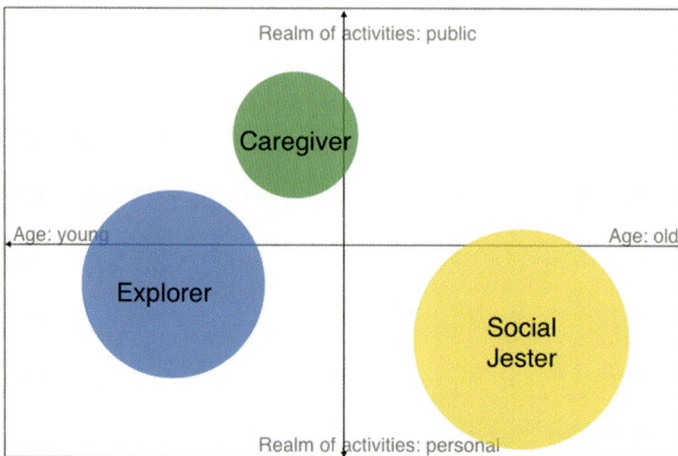

Fig. 10.43 Diagram of personas

10.12.1.2   Persona 1: Joe (Explorer)

Persona Descriptions

Explorer/Seeker/Wanderer.
Ambitious, autonomous, curious individual who is always looking for challenges.

Joe's Story in Brief

Engineer working in San Francisco.
When not working, loves to go out and try things.

Feature Recommendations

Ideas of adventurous activities/Mini games.
Bite-size tips and advice to live a happier life.
Mini games.

10.12.1.3   Persona 2: Carol (Care Giver)

Persona Descriptions

Enjoys helping others.

Carol's Story in Brief

Artist, craft maker, and cook in Austin, Texas.
Loves to use her skills, and give back to community.

Feature Recommendations

Volunteer connector.
Online platform to do good deeds to random people.
Ideas for doing random acts of kindness.

10.12.1.4   Persona 3: Margaret (Jester)

Persona Descriptions

Values play/fun, emotional connection with people.

Margaret's Story in Brief

Laid-back old lady who isn't much good with technology.
Misses good time with her 5year-old grandchild, Simon, since they moved to another state.

Feature Recommendations

Easy ways to connect with loved ones.
Makes you smile/laugh by sharing funny faces.

Following are tables we prepared for each persona: explorer, caregiver, and social jester. Each table has five tactics/guidelines to consider for the Happiness Machine on top of the bar. We filled out each box on both the persona's conceptual (strategic) and tactical/UX solutions (Table 10.2).

Table 10.2 Tables of conceptual solutions and tactical/UX solutions for the three personas

Joe: Explorer	Involvement	Habits	Entertainment	Accomplish-ment	Connectedness
Conceptual Solutions	Active, challenges, games, competition, escape, learn, new	Planning, managing time, seeking new and exotic adventures	Discover the new, learning the self, winning	Proof of self worth, self value, mastery, communicating accomplishments	Lighter connection, connecting with new people, showing dominance over others
Tactical/UX Solutions	Puzzles, games, unique accessories, challenge them	Adventure planning tool, pictures of cool ideas, journaling	Delivering insight, data, self knowledge, background information	Trophies, passport stamps, collectables	Comparing to others, bring them in touch with like-minded poeple, new people

Carol: Caregiver	Involvement	Habits	Entertainment	Accomplish-ment	Connectedness
Conceptual Solutions	Taking care of others, generosity	Donate, altruism, volunteer, active community involvement, caring, protecting, sacrificing	Seeing others take care of one another, gratitude	Helping others and making a difference, helping world become a better place live	Doing things for others, nurturing others, helping people in need
Tactical/UX Solutions	Remiders, organization, opportunity to help others, gratitude journal	Taking pictures of good deeds, sharing ideas for Random Acts of Kindness, pictures and videos of philanthropy	Thank you message, sharing food recipes	Digital happy cash for donation, Charitable activities, Good Deeds Journal	Giving gifts to a friend, thanking others, community stats/facts

Margaret: Jester	Involvement	Habits	Entertainment	Accomplish-ment	Connectedness
Conceptual Solutions	Jokes, clever tips, ways to connect, make others laugh, advice for others	Planning for parties, capture memeories, remembering good times	Live in the moment, quick/cheap fun, laid-back, enjoy life	Entertaining others, social skills	Reputation, how they are seen in the eyes of others
Tactical/UX Solutions	Tips to make people laugh, prank-generator, happy alarm clock	Family/friends get together planner, video and picture editing tool, collection of picture by events	New jokes, making funny faces, making funny sounds, mimicking funny gestures, drawing funny pictures	Journal for self-reflection, how many times they let themselves get sad or down	Sharable jokes and contents, including friends and family in life

10.13   Happiness Machine 2.0: Use Scenarios

As noted in Chap. 1 and in earlier sections, use scenarios are a UI development technique whereby, in the development of a prototype to simulate the major characteristics of a software product, a use scenario is written to determine what behavior will be simulated. A scenario is essentially a sequence of task flows with actual content provided, such as the subject user's demographics and goals, the details of the information being worked with, etc.

10.13.1   Use Scenario

Introduced by his/her friends, users download the application on their mobile devices. Users learn Happiness Machine's objectives and how it works. In the application tutorial, a slogan informs users: "It's your choice to be happy." This slogan gives users an idea what Happiness Machine is about. The intent is to help them understand the Happiness Machine helps to achieve deep, sustainable happiness (eudaemonia), not shallow, short-term happiness. Happiness Machine will suggest tasks to do to make them feel happy. Push notifications will show up on the phone and prompt them to do certain activities as a part of their daily activities. Users will remember tasks to do throughout the day for them to practice. Before going to sleep, users will be encouraged/reminded to write/update their journals. Journaling helps them to reflect upon their day, which will help them to adopt happiness-promoting behaviors in their daily lives over time.

10.13.2   Further Possible Implementations

By further enhancing the above described features linked to Happiness Machine and social interaction, the Happiness Machine and its main functions could significantly enrich the overall happy user-experience especially in commercial enterprises, organizations, education centers, and governments that want to help their employees, members, and/or participants to be happier.

Another possible application of the Happiness Machine could be applied to promote their own strengths. Why do users receive a compliment from other people? The reason is that users achieve some accomplishments, like doing a good research. Then the skill "research" would become one of the users' strengths (Happiness Machine could make a list of users' strengths). This approach is similar to LinkedIn and other social media's trends in recent years.

10.13.3   Happiness Machine 2.0: Information Architecture

Following the completion of the first version of the Happiness Machine, AM+A carried out further analysis and design for a second version, Happiness Machine 2.0, taking into account all previous work. The sections below provide additional information about Happiness Machine 2.0.

In designing the Information Architecture (IA) for the Happiness Machine 2.0, AM+A began by examining the IA for past Machines, including the Green Machine, the Health Machine, the Money Machine, the Driving Machine, and the Innovation Machine. We discovered an overarching model for the IA that permeated throughout each of the past machines. In this model, there are five primary 'modules', or branches of the IA, each of which is described below. While we altered slightly the details of each module to fit the needs and requirements of each respective machine, a generalized model was still evident. These modules are described below (Fig. 10.44).

10.13.4   Set Up

Users will go through initial set up page, which will require them to answer questions about their personal preferences for happiness (e.g., the application might displays 10 images and ask users to select ones that best represent their state of happiness).

10.13.5   Dashboard Concepts

10.13.5.1   Happy Score

Happy Score quantifies users' happiness to give overview of happiness.

10.13.5.2   Happy Points

Happy Points are earned after users complete personalized tasks. These points can be used as virtual "digital cash" by users to purchase games and other items. Points are also used for virtual donations.

10.13.5.3   Happy Tracker

Happy Tracker visualizes users' overall happiness in five categories. Happy Tracker is a dashboard for Happy Score. Happy Tracker gives an overview into the status of users' behavior change. Here, users can see a view of their objectives and goals and where they stand in achieving those goals.

Happiness Machine 2.0

Set Up

Happy Tracker	Happy Journal	Happy Tasks	Happy Tips	Settings
Visualizes user's overall happiness in 5 categories. Dashboard for Happiness Score.	Prompts users to write journal, a process-flow of past, present, and future events about happiness	Motivates users to be happier by suggesting personalized tasks to do in the form of Achievements and Challenges	Instructs how to use Happiness Machine and provides focused knowledge about happiness	Edit your personalized preference, your profile, and manage your friends, social network, etc.

▪ Relating to others (Contacting others) ▪ Doing things (Physical and good deeds) ▪ Being optimistic (Smile and laugh) ▪ Showing gratitude (Thanking) ▪ Journaling	▪ Question-based journal entry ▪ Sort by: Dates, people, topics, geolocation, mood	▪ Mini Game Good Deeds Call Monitor ▪ Happy Points (Digital currency for Happiness Machine users)	▪ Happier life tips Relating to others Doing Things Being Optimistic Showing Gratitude Journaling ▪ App tutorial What is Happiness Machine? What are Happy Scores? What is Happy Wall? What is Happy Journal?	▪ Edit My Profile Profile Settings ▪ Manage Friend's List Friend's updates Invite friends

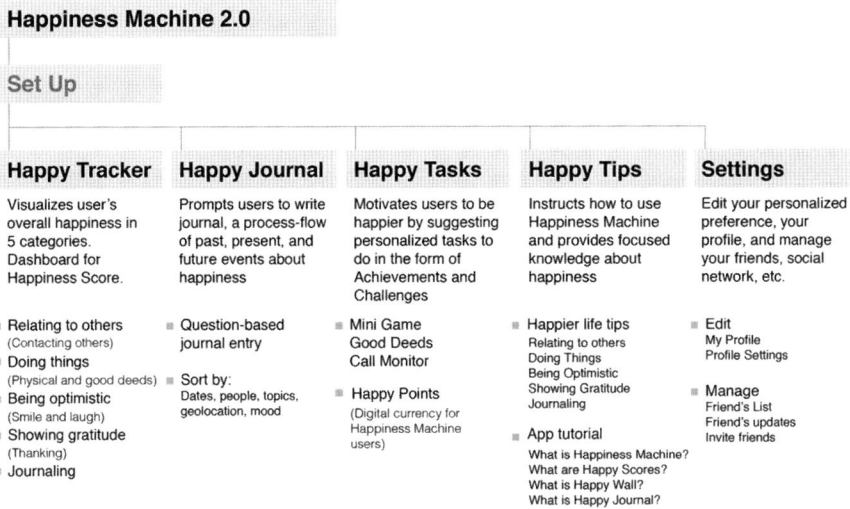

Fig. 10.44 Information Architecture of the Happiness Machine 2.0

This is a process viewing module, in which users can access more high-level views of the process and more details regarding each objective and goal. Users see the progress being made, as well as the next steps in achieving a particular objective or goal.

10.13.6   Journey Map Concepts

10.13.6.1   Happy Journal

Unlike normal journals or diaries, Happiness Journal has questions on top to guide users what to write about. Joe chooses Relating to Others to write about people he met during his day.

10.13.6.2   Happy Tasks

Lists of activities to do for users. Tasks vary depending on user's preferences and interests.

10.13.7 Tips/Advice Concepts

10.13.7.1 Happy Tips

The Tips/Advice module provides focused knowledge about a given topic to give users insight into the habits they wish to either get rid of or adopt. Just-in-time knowledge. Crowd-sourced, user-generated tips constantly updated.

10.13.8 Social Networking Concepts

10.13.8.1 Social Network

The Social Network module is an integral part of behavior change in modern software. Users engage in focused, subject-matter-based connections with friends, family, and/or like-minded people that either share similar goals or wish to support others in achieving behavior-change objectives.

10.13.9 Incentives Concepts

10.13.9.1 Incentives

The Incentives module presents users with a fun and engaging way to change their behavior. Gamification has proven to be a powerful tool in adopting users to try an application, even with virtual incentives, although in some cases, real incentives are provided. In addition, a leaderboard allows a user to compare his/her progress with others, tapping in the competitive nature of the human mind to create behavior change.

10.14 Happiness Machine 2.0: Initial Screen Designs (Look-and-Feel)

The following subsections discuss initial screen designs. Based on the information architecture, AM+A prepared initial designs of some key screens of the Happiness Machine 2.0. The following describes the representative initial screen designs of the Happiness Machine 2.0.

10.14.1 Landing Screen

The Landing Screen typifies the main menu and app instruction throughout the Happiness Machine. This page displays five major functions with descriptions below (Fig. 10.45).

Fig. 10.45 Initial visual designs of Happiness 2.0 screens

- Relating to Others: How much users use his/her phone to call and connect with others?
- Doing things: How much users spend time to go out and do new activities?
- Smile-o-Meter: How much users smile/laugh? (captured by camera)
- Gratitude: How much users write thanks to others in emails, notes, etc.?
- Journaling: How much users write journals?

10.14.1.1 Happy Tasks

Sending/Receiving a joke via e-mail will make people feel a little happier, interested, and surprised. However this emotional change is not deep happiness. The Happy Note users receive is from people who care about the users or love the users. These people affect the users according to Maslow's hierarchy of Needs, including the level of love/belonging, esteem, and self-actualization. HappyNote allows users to make a compliment, review their valuable experiences, and, more importantly, make other people feel deeply happier. The following show how the process works (Fig. 10.46).

10.14.1.2 HappyTips

Provide users with HappyNote, tools to write, check in function, and different shape of Note (Fig. 10.47).

Fig. 10.46 Initial visual designs of Happiness 2.0 screens

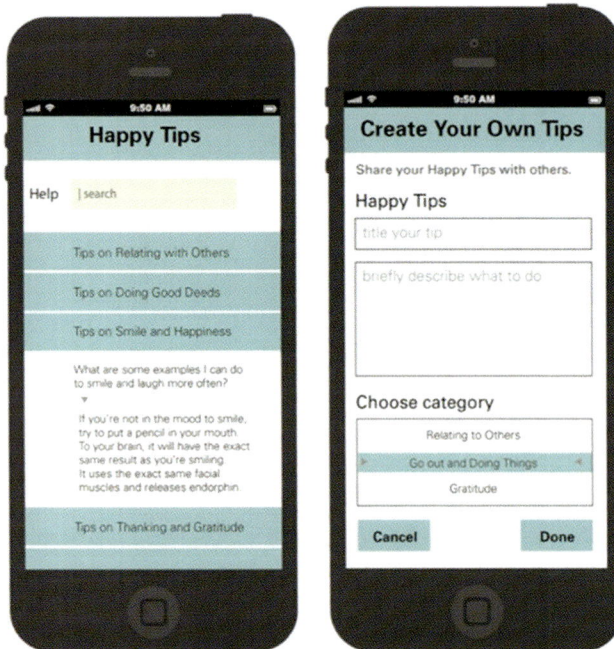

Fig. 10.47 Initial visual designs of Happiness 2.0 screens

10.14.1.3 HappyJournal

Unlike normal journal or diary, HappinessJournal has questions on top to guide users what to write about. Joe chooses Relating to Others to write about people he met during his day (Fig. 10.48).

Fig. 10.48 Initial visual
designs of Happiness 2.0
screens

Fig. 10.49 Initial visual
designs of Happiness 2.0
screen

10.14.1.4   MiniGames

Playful game that Joe can swipe to make different faces for Lego person. Joe saves
and shares his mood with friends (Fig. 10.49).

10.14.1.5   Revised Screen Designs

Based on previous screen designs and internal discussion, AM+A revised the screens to be more in keeping with Apple iOS screen-design guidelines and to integrate color, typography, sign design, and layout amongst all of the screens. In addition, AM+A attempted to style the screens with the layout paradigm of "Swiss-German" graphic design of the 1960s, which has been adapted recently by Microsoft Windows 8 layout paradigm. The first screen shows the revised screen-design guidelines covering spatial grid, type sizes, colors, and alert-colors (Figs. 10.50 and 10.51).

10.15   Evaluation, Usability Tests, and User Feedback

Lack of available time prevented us from doing extensive user evaluation and testing of the Happiness Machine. The study of happiness theory and incorporation of the theory into the information architecture, conceptual designs, and perceptual designs took more time than expected. AM+A's approach recommends such testing,

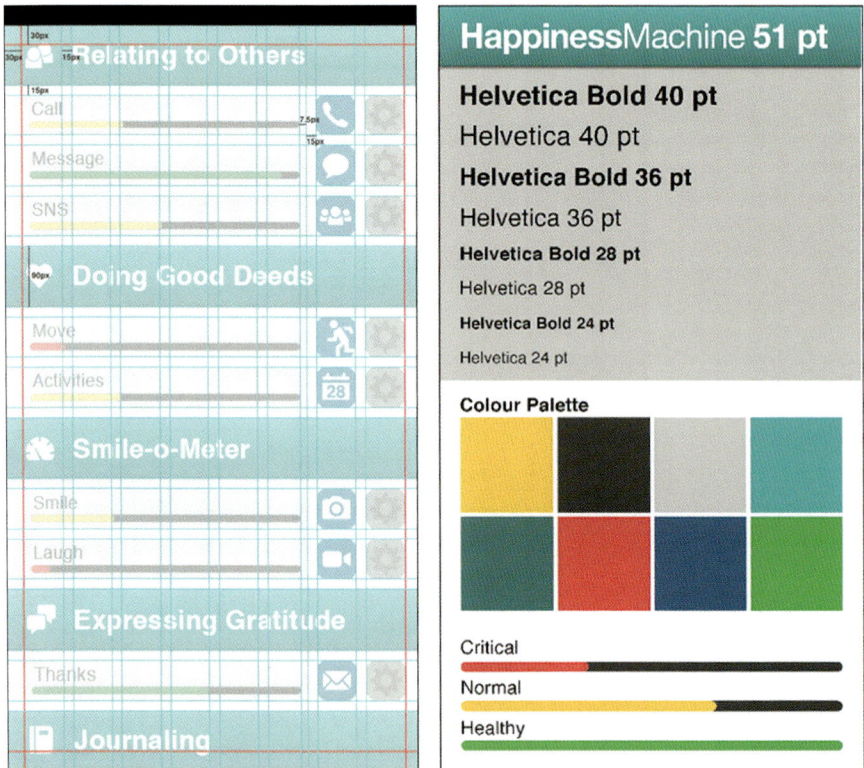

Fig. 10.50   Revised-screen design-guidelines for spatial grid, type fonts/sizes, colors, and alert colors

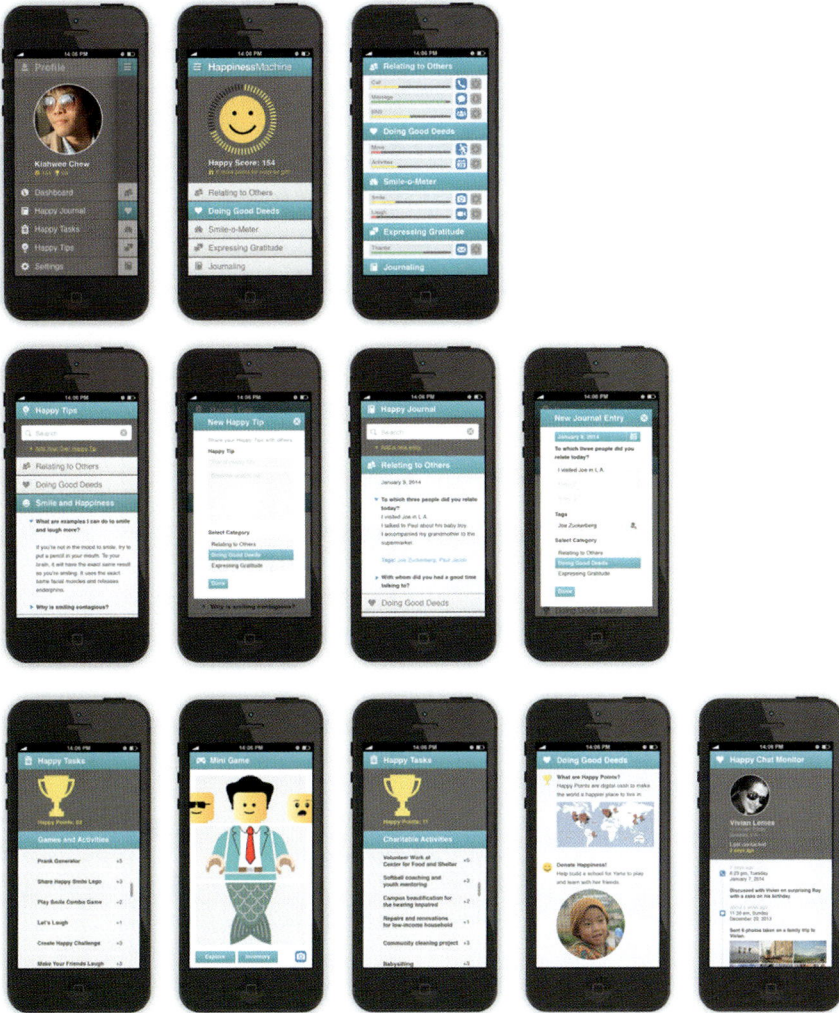

Fig. 10.51 Revised HM 2.0 screen designs

especially with different cultural groups who may have different key components of happiness, different personas, and use scenarios, and hence different information architectures and conceptual/perceptual screen designs.

Regarding user feedback: Users could collect their happiness moments/experiences by receiving and sending many of HappyNotes. The Happiness Machine could collect information about individuals' experiences in real time in their natural environments (see, e.g., Bruno Frey's Experience Sampling Method (ESM) (Frey 2010a, b). In the method mentioned users are asked to indicate the intensity of their feelings. One could then calculate happiness through the aggregation of instantaneous statements of affect.

10.16   Conclusions

The Happiness Machine is conceived to intensify users' happy experiences, and to motivate them to think positively during daily life. The application incorporates persuasive and motivational elements in order to stimulate users to reflect on and change their behavior; it arouses their interest in something when they felt bored, enjoy connecting people when they feel lonely, and find good deeds to do when they seem without purpose.

The Happiness Machine is intended to be entertaining and fun to use, even though at the same time it provides information, and educates users, leaving space for reflective activities and personal discovery of other happy experiences with people, places, activities, and objects (such as food). Especially thanks to the HappyNote, HappyLaff, and HappySearch and the possibility of social interaction and comparison, users' behavior could thus be changed not only in the short term, but also in the long term.

AM+A plans to continue development of the Happiness Machine. In fact, versions 1.0 and 2.0 were followed by two teams of graduate design students at the Institute of Design, Illinois Institute of Technology (IIT), Chicago, Illinois, USA, developing two further versions, Happiness Machine 3.0, in October 2013 in a one-week workshop about mobile user-experience design.

Following that success, AM+A led four teams of graduate students to develop four further versions, Happiness Machine 4.0, at the Polytechnic University (HKPU), Hong Kong, China, in a similar one-week workshop about mobile user-experience design during November 2013. Those designs were targeted specifically for China (three teams) and India (one team).

Subsequently, AM+A presented its one-week mobile user-experience design workshop to about 40 professionals, faculty, and graduate students from across China at the De Tao Academy in Shanghai during December 2013. The eight teams developed specific solutions for China. As occasions permit, this work will be published or made available.

Acknowledgments   For the Happiness Machine 1.0 and 2.0, the author acknowledges the primary assistance of Mr. Yu-Hsien (Jonathan) Liu and Ms. Min Lee, AM+A Designer Analysts, for their assistance in the preparation of the text and images (then a student and a graduate of the Institute of Design, IIT, Chicago); of Ms. Megan McQuade, AM+A Business Development Assistant, for her assistance in editing the text; and Mr. Kia Hwee Chew, AM+A Designer/Analyst for his re-design of visual screens for Happiness Machine 2.0.

The author also thanks Prof. Bob Steiner, Marketing Course leader, University of California at Berkeley Extension, International Diploma Program, and his graduate marketing students, who assisted on this project (Khaled Aboelwafa, Fernanda Aboelwafai, Alicia Cheng, Benitua Hung, Meghana Panth, Augustin Roggio, Cindy Suryadinata, and Basak Yolgecti. Their assistance included carrying out market research, including surveys and interviews for the Happiness Machine 1.0 and 2.0, and preparing both reports and presentations about their efforts, which they made available to AM+A.

The author also acknowledges these publications Marcus 2014a, b) that have appeared subsequent to the AM+A white paper on which this chapter is based:

Marcus, Aaron (2014a). "The Happiness Machine: Mobile Behavior Change." *Proc.*, Design, User Experience, and Usability Conference, Design, User Experience, and Usability 2014, in Marcus, Aaron, Editor, User Experience Design for Diverse Interaction Platforms and Environments, Lecture Notes in Computer Science, Volume 8518, 2014, pp. 258–268. Springer, London.

Marcus, Aaron (2014b). "The Happiness Machine." Proc., User-Experience Professionals Association/China and User Friendly Conference 2014, Wuxi, China, 13–16 November 2014. Unnumbered pages available on memory stick.

Further Reading

Achor S (no date) Five ways to train your brain. Retrieved from https://wiki-health-fabulous.wiki-spaces.com/file/detail/5+ways+to+train+your+brain+wksht+Achor.pdf. Accessed 30 Aug 2013

Aristotle (1962) Nicomachean Ethics. Translated, with Introduction and Notes by Martin Ostwald. The Bobbs-Merrill Company, Indianopolis

Barbugiai F, Cheng A, Hung B, Suryadinata C (2013) The happiness machine: marketing research. University of California/Berkeley Extension, International Diplomoa Program, Marketing Research Class Project Report, 21 October 2013

Bernstein E (2013) How to be a better conversationalist. The Wall Street Journal, 12 August 2013. Retrieved from http://online.wsj.com/. Accessed 21 Aug 2013

Cialdini RB (2001a) The science of persuasion. Sci Am 284:76–81

Cialdini RB (ed) (2001b) Influence: science and practice, 4th edn. Allyn and Bacon, Boston

de Botton A (2006) The architecture of happiness. Pantheon Books, New York

Delft Institute of Positive Design (2012). Retrieved from http://studiolab.ide.tudelft.nl/diopd/. Accessed on 30 Aug 2013

Eco U (1970) Semiologie des messages visuels. Communication 16:11–52

Eco U (1976) A theory of semiotics. Indiana University Press, Bloomington

Eco U (1985) The semantics of metaphor. In R Innis (1985), Semiotics: an introductory anthology. Indiana University Press, Bloomington, p 262

Facebook (2012) Sutro media. Retrieved from http://www.facebook.com/. Accessed on 21 Aug 2013

Frey BS (2010a) Experience sampling method (ESM)..http://www.bsfrey.ch/articles/C_516_Happiness_and_Public_Choice_LV.pdf.

Frey BS (2010b) Happiness: a revolution in economics (Munich lectures in economics). The MIT Press, Cambridge, MA

Hartson R, Pyla PS (2012) The UX book. Process and guidelines for ensuring a quality user experience. Elsevier, Waltham

Hipmunk (2012) About. Retrieved from http://www.hipmunk.com/. Accessed on 21 Aug 2013

Innis RE (1985) Semiotics: an introductory anthology. Indiana University Press, Bloomington

Kim R (2012) Kleiner-backed Noom coaches you to better health. GigaOm, 28 June 2012. Retrieved from http://gigaom.com/. Accessed on 21 Aug 2013

Lakoff G, Johnson M (1980) Metaphors we live by. Chicago University Press, Chicago

Law R, Fuchs M, Ricci F (2011) Information and communication technologies in tourism 2011. Springer, Vienna

Levi-Strauss C (2000) In: Jacobson C, Schoepf B (eds) Structural anthropology. Basic Books, New York

Marcus A (2002) Information visualization for advanced vehicle displays. Inf Visual 1(2):95–102

Marcus A (2013) The money machine. Inf Design J 20(3):228–46

Maslow AH (1943) A theory of human motivation. Psychol Rev 50:370–396

Maslow AH (2006) Motivazione e personalità. (Ed. 11). Armando Editore, Roma

Mossberg WS (2013) Creating and sharing videos that are not too long and not too short: two services let you gather photos and video clips and upload them for automated video editing. The Wall Street Journal. Retrieved from http://online.wsj.com/. Accessed on 21 Aug 2013

Naik G (2012) Next cameras come into view. The Wall Street Journal, 21 June 2012. Retrieved from http://www.online.wsj.com/. Accessed on 21 Aug 2013

Peirce CS (1933a) Existential graphs. In: Hartshorne C, Weiss P (eds) Collected papers of Charles Sanders Peirce, vol 4. Harvard University Press, Cambridge, MA, pp 293–470

Peirce CS (1933b) The simplest mathematics. In: C Hartshorne and P Weiss (eds) Collected papers of Charles Sanders Peirce, vol. 4. Harvard University Press, Cambridge, MA. Accessed on 21 Aug 2013

Tristram C (2001) The next computer interface. Technology Review 53–59

Wagner M, Armstrong N (2003) Field guide to gestures: how to identify and interpret virtually every gesture known to man. Quirk Books, Philedelphia

Wang D, Xiang Z (2012) The new landscape of travel: A comprehensive analysis of smartphone apps. In: Fuchs M, Ricci F, Cantoni L (eds) Information and communication technologies in tourism 2012. Springer, New York, pp 308–319

Wikipedia (2012d) Wikitude. Retrieved from http://en.wikipedia.org/. Accessed on 21 Aug 2013

References

Aboelwafa K, Panth M, Roggio A, Yolgecti B (2013) The happiness machine: marketing research. University of California/Berkeley Extension, International Diplomoa Program, Marketing Research Class Project Report, 26 June 2013

Achor S (2010) The happiness advantage: the seven principles of positive psychology that fuel success and performance at work. Crown Publishing Group, New York

Estonia's Bank of Happiness, Kindness is the currency: http://www.npr.org/blogs/parallels/2013/07/18/200869850/At-Estonias-Bank-Of-Happiness-Kindness-Is-The-Currency. Accessed on 21 Aug 2013, and http://www.onnepank.ee Accessed on Sept 2013

Fogg BJ (2003) Persuasive technology: using computers to change what we think and do. Morgan Kaufmann, Amsterdam

Fogg BJ, Eckles D, Bogost I, Consolvo S, Holmen E, Spasojevic M,…White S (2007) Mobile persuasion: 20 perspectives on the future of behavior change. Stanford University Press, Palo Alto

Jean J, Marcus A (2009) The green machine: going green at home. User Exp 8(4):20–29

Jung C (2013) Carl Jung's Archetypes, http://en.wikipedia.org/wiki/Archetype. Accessed on 28 Oct 2013

Marcus A (1998) Metaphors in user-interface design. ACM SIGDOC 22(2):43–57

Marcus A (2011) Health machine. Inf Des J 19(1):69–89

Marcus A (2012a) The money machine: helping Baby Boomers to Retire. User Experience Magazine, 11:2, Second Quarter 2012, pp 24–27

Marcus A (2012b) The story machine: a mobile app to change family story-sharing behavior. Workshop Paper, presented at CHI 2012 Conference on Human Factors in Computing Systems. 5–10 May 2012, Paris, France, pp 1–4

Marcus A et al (2013a) The innovation machine. In: Marcus A (Ed.) (2013) Proceedings, design, user experience, and usabilty conference, Las Vegas, NV, 20–25 July 2013, pp. 67–76. Springer, London

Marcus A et al (2013b) The travel machine. In: Marcus A (Ed.) (2013) Proceedings, design, user experience, and usabilty conference, Las Vegas, NV, 20–25 July 2013, pp. 696–704. Springer, London

Marcus A et al (2013c) The driving machine. In: Marcus A (Ed.) (2013) Proceedings, design, user experience, and usabilty conference, Las Vegas, NV, 20–25 July 2013, pp. 140–149. Springer, London

Marcus A et al (2013d) The learning machine. In: Marcus A (Ed.) (2013) Proceedings, design, user experience, and usabilty conference, Las Vegas, NV, 20–25 July 2013, pp. 247–256. Springer, London

Marcus A (2014a) The happiness machine: mobile behavior change. In: Proceedings Design, User Experience, and Usabiity Conference, Design, User Experience, and Usability 2014, in Marcus, Aaron, Editor, User Experience Design for Diverse Interaction Platforms and Environments, Lecture Notes in Computer Science, Volume 8518, 2014, pp. 258–268. Springer, London

Marcus A (2014b) The happiness machine. In: Proceedings of User-Experience Professionals Association/China and User Friendly Conference 2014, Wuxi, China, 13–16 November 2014. Unnumbered pages available on memory stick

Marcus A, Jean J (2010) The green machine. Information Design Journal 17:3, First Quarter, 2010, pp 233–243

List of Website URLs

4 Design Secrets Behind Dots, the Insanely Addicting iPhone Game: http://www.wired.com/design/2013/07/can-quit-playing-dots-heres-why/?cid=9730814. Accessed on 21 Aug 2013

Achor S: www.ted.com/talks/shawn_achor_the_happy_secret_to_better_work.html. Accessed on Sept 2013

Animoto: http://animoto.com/. Accessed on 21 Aug 2013

Arianna Talks Third Metric on BBC's Newsnight: http://www.huffingtonpost.com/2013/08/06/arianna-third-metric_n_3713446.html. Accessed on 23 Aug 2013

Coca-Cola Happiness Machine: http://www.youtube.com/watch?v=lqT_dPApj9U. Accessed on 21 Aug 2013

Cooper B. 10 Simple Things you Can Do Today that Will Make you Happier, Backed by Science, [http://blog.bufferapp.com/10-scientifically-proven-ways-to-make-yourself-happier. Checked on 6 Aug 2013

Desmet P. Design for Happiness, http://www.youtube.com/watch?v=jTzXSjQd8So. Accessed on Sept 2013

Donate a Photo. Johnson and Johnson: http://www.donateaphoto.com/

Design and Emotion Society, The: http://www.designandemotion.org/en/home/. Accessed on 21 Aug 2013

Exclusive Preview on iOS 5 – iPad Zauberer Simon Pierro: http://www.youtube.com/watch?v=LAhP-yLJJ9s. Accessed on 21 Aug 2013

Facebook for iPhone: https://itunes.apple.com/us/app/facebook/id284882215?mt=8. Accessed on 21 Aug 2013

Foursquare for iphone: https://itunes.apple.com/us/app/foursquare-find-restaurants/id306934924?mt=8. Accessed on 21 Aug 2013

Good Deeds Day: http://gdd.goodnet.org/doing_good. Accessed on 21 Aug 2013

Greater Good Science Center, UC Berkeley, http://greatergood.berkeley.edu/pdfs/happycircle-ggsc.pdf. Accessed on Sept 2013

Happier for iPhone: https://itunes.apple.com/us/app/happier/id499033500?mt=8. Accessed on 21 Aug 2013

Happiness Institute, The: http://www.thehappinessinstitute.org/alive/. Accessed on 21 Aug 2013, and http://www.scribd.com/doc/88359632/Happiness-Pieter-Desmet. Accessed on 21 Aug 2013

Happiness Advantage, The: The seven principles of positive psychology that fuel success and performance at work: http://100habits.com/?p=514. Accessed on 21 Aug 2013, http://learnby-blogging.com/?p=2535. Accessed on 21 Aug 2013

Happiness Advantage with Shawn Achor, The: http://www.tpt.org/?a=kits&id=36. Accessed on 21 Aug 2013

Happiness: The Pursuit of Happiness Project: https://www.facebook.com/ThePursuitofHappiness Project?ref=stream&hc_location=stream. Accessed on 23 Aug 2013

Happy Hormones "Endorphins", How to Release: http://healthymind786.blogspot.com/2013/04/how-to-release-happy-hormones-endorphins.html. Accessed on Sept 2013

Happy News: http://www.happynews.com/. Accessed on 23 Aug 2013

Holden, Dr. Robert, Oprah.com, Take Action! 10 Ways to Increase your Happiness, http://www.oprah.com/spirit/10-Ways-to-Increase-Your-Happiness. Accessed on Sept 2013

How to disrupt mindless accumulation and increase happiness: http://news.efinancialcareers.com/145468/how-to-disrupt-mindless-accumulation-and-increase-happiness/. Accessed on 21 Aug 2013

Jung's Archetypes: http://en.wikipedia.org/wiki/Archetype. Accessed on Oct 2013

Laugh and sounds: http://www.psy.vanderbilt.edu/faculty/bachorowski/laugh.htm. Accessed on 21 Aug 2013

Lyubomirsky, Sonja, The How of Happiness: http://thehowofhappiness.com. Accessed on 25 Nov 2014

Magic iPad – Best Christmas Gift Ever (by iSimon): http://www.youtube.com/watch?v=–FAN7nFk6Xo. Accessed on 21 Aug 2013

Magisto: http://www.magisto.com/. Accessed on 21 Aug 2013

Make a wish foundation of America: http://wish.org/content/global-search?keyword=help%20dying%20children. Accessed on 21 Aug 2013

Measure Happiness, How To: http://mitpress.typepad.com/mitpresslog/2012/08/how-to-measure-happiness.html. Accessed on 21 Aug 2013

Mouthoff - app for iPhone: https://itunes.apple.com/us/app/mouthoff/id306588353?mt=8. Accessed on 21 Aug 2013

Newsnight Happiness Mpeg 4: http://www.youtube.com/watch?v=u73pEZInshI

Path for iPhone: https://itunes.apple.com/us/app/path/id403639508?mt=8. Accessed on 21 Aug 2013

People, not apps: Facebook's Home event in under 5 min: http://www.youtube.com/watch?v=NHuRNsh8Sbc. Accessed on 21 Aug 2013

Snapchat for iPhone: https://itunes.apple.com/us/app/snapchat/id447188370?mt=8. Accessed on 21 Aug 2013

Tiny Tasks, http://designinghappiness.wordpress.com. Accessed on 21 Aug 2013

UC Berkeley Institute for Greater Good: http://greatergood.berkeley.edu. Accessed on 21 Aug 2013

Wayz app: https://www.waze.com. Accessed on 25 Nov 2014

What the Magisto?: http://www.youtube.com/watch?v=2YL41fGxzvY. Accessed on 21 Aug 2013

World, please meet happier: http://blog.happier.com/2013/02/07/hello-world/. Accessed on 21 Aug 2013

Chapter 11
The Marriage Machine: Combining Information Design/Visualization with Persuasion Design to Change Behavior and to Improve Couples' Relationships

11.1 Introduction

Marriage is a significant component of all cultures. In the United States, 37.5 % of the US population between 15 years and older are married. More importantly, 11.7 % of population are separated or divorced (American Community Survey/American FactFinder 2012). These figures are somewhat debatable as shown by Pew Research figures: 51 % of the US adult population are married (Cohn 2013) and 14 % of married adults are separated or divorced (Cohn 2013). The state of marriage has declined since World War 2, and some consider the institution significantly challenged. Whatever the statistics, those who are married and their family relations and friends are naturally concerned with strengthening the bonds between members of committed couples, whatever their gender and whatever their legal status. AM+A decided to investigate how mobile technology could assist them to strengthen their bonding.

By combining information design and persuasion/motivation theory, with a particular focus on the works of Maslow's theory about basic human needs (Maslow 1943) and Fogg's theory of persuasion (Fogg 2003; Fogg et al. 2007), the use of the Marriage Machine should prompt couples to change their behavior, to make their relationship with each other more enjoyable and personally enriching. Above all, it is this dimension of personal learning that distinguishes the Marriage Machine from the majority of mobile apps that already exist in the field of marriage and/or relationship maintenance.

The success and effectiveness of the described approach, i.e., the combination of information design with persuasion design in order to promote behavioral change of mobile application users, has already been studied and realized in several previous projects of the author's company: the Green Machine (Marcus and Jean 2009), the Health Machine (Marcus 2011), the Money Machine (Marcus 2012a), the Story Machine (Marcus 2012b), the Innovation Machine (Marcus et al. 2013a), the Driving

© Springer-Verlag London 2015

A. Marcus, *Mobile Persuasion Design*, Human–Computer
Interaction Series, DOI 10.1007/978-1-4471-4324-6_11

Machine (Marcus et al. 2013b), the Learning Machine (Marcus et al. 2013c), and the Happiness Machine of 2014 (Marcus 2014a, b). All of these past Machine projects and the current Marriage Machine rely on the conceptual design of an application through a user-centered user-experience development process, as defined and explained below.

11.2   User-Centered User-Experience Development

As mentioned in Chap. 1, user experience (UX) can be defined as the "totality of the […] effects felt by a user as a result of interaction with, and the usage context of, a system, device, or product, including the influence of usability, usefulness, and emotional impact during interaction, and savoring the memory after interaction" (Hartson and Pyla 2012). That definition means the UX goes well beyond usability issues, entailing also social and cultural interaction, value-sensitive design, emotional impact, fun, and aesthetics.

Also as mentioned in Chap. 1, the user-centered design (UCD) approach links the process of developing software, hardware, and user interface (UI) to the people who will use a product/service. UCD processes focus on users throughout the development of a product or service. The UCD process comprises these tasks, which sometimes occur iteratively: plan, research, analyze, design, implement, evaluate, document, and train.

AM+A carried out many of these in the development of the Marriage Machine concept design (with the specific exception of implementation).

Over the past two and one-half decades in the user-interface (UI) design community, designers, analysts, educators, and theorists have identified and defined a somewhat stable, agreed-upon set of user-interface components, i.e., the essential entities and attributes of all user interfaces, no matter what the platform of hardware and software (including operating systems and networks), user groups, and contents (including vertical markets for products and services). As explained in Chap. 1, these components comprise the following: metaphors, mental models, navigation, interaction, and appearance.

For the Marriage Machine, AM+A considered all of these components. In the context of the Marriage Machine, metaphors may be termed the "concepts" of the Machine, and the mental model may be termed the "information architecture." The discussion below about the initial and revised screen designs will describe, also, the interaction and appearance, especially as the designs move from conceptual designs (so-called "wire-frame" versions) to perceptual designs (so-called "look-and-feel" versions). One unique approach of the Machines is to combine information design and information visualization (tables, forms, charts, maps, and diagrams) with persuasion design.

11.3   Persuasion Theory

As explained in Chap. 1, persuasion design involves multiple aspects to attract and motivate users to do something. Cialdini (2001a, b distinguishes six basic phenomena in human behavior, which are supposed to favor positive reactions to persuasive messages of others: reciprocation, consistency, social validation, liking, authority, and scarcity. These tendencies in the social influence process are characteristic of human nature and are thus valid across national boundaries; nevertheless, cultural norms, traditions, and experiences can have an impact on the relative weight of each of the six mentioned factors.

In addition, Fogg has made significant contributions to persuasion-related research (Fogg 2003; Fogg et al. 2007). In alignment with Fogg's persuasion theory, we defined six key processes to achieve behavioral change via the Marriage Machine's functions and data:

- Attract user via business marketing.
- Increase frequency of using application.
- Motivate changing some habits: interaction with and openness towards others, experience and observation of differences, ways of journaling, or diary documentation.
- Teach how to change habits.
- Persuade users to change habits (short-term change).
- Persuade users to change general approach to objectives, people, objects, contexts, obstacles, fear, sadness, and anger (long-term or lifestyle change).

Each step has requirements for the application.

Motivation is a need, want, interest, or desire that propels someone in a certain direction. From the sociobiological perspective, people in general tend to maximize reproductive success and ensure the future of descendants. We apply this theory in the Marriage Machine by making people understand that a determinate behavior can fundamentally enrich their daily experiences, that it can increase their knowledge and understanding, and that it can ultimately trigger a process of relationship building to improve marriage success.

We also drew on Maslow's Theory of Human Motivation (Maslow 1943), which he based on his analysis of fundamental human needs. We adapted these needs to the Marriage Machine context:

- *Safety and security*: Met by the assistance of an advisor and by the provision of obstacles, fun, human contact, and other related information and tips.
- *Belonging and love*: Expressed through friend, family and social sharing, and support.
- *Esteem*: Satisfied by social comparisons that display progress and destination expertise, as well as by self-challenges that are suggested by the advisor, and that display goal accomplishment processes.
- *Self-actualization*: Fulfilled by being able to follow and retrace continuous progress and advancement in personal diary.

As noted with caution earlier in Chap. 1: Persuasion design must have a strong underlying ethical basis; otherwise, it can be misused, e.g., in propaganda directed to inhuman treatment of others. Another aspect to consider is that much practical persuasion theory has been used in addition programs. There is a danger that some users will simply substitute the Machine for independent cognition, speech, and behavior, exchanging one addiction for another, namely, the benefits, answers, or rewards that the Machine can provide. These are somewhat extreme examples, but one should remain vigilant against misuse of the techniques by developers as well as by users.

11.3.1   Increase Use Frequency

Incentives such as games, awards, rewards, recognition, and nostalgic objects are the most common methods to increase use frequency. In the Marriage Machine, we have developed many tricks, random good deeds, funny challenges, and other relationship-building-related knowledge. In terms of rewards, users will be provided virtual rewards (such as "points" they can use to purchase a gift for their partners).

11.3.2   Increase Motivation

Because setting goals (as opposed to objectives, which are without time lines or measurable results) helps people to learn better and improves the relevance of feedback, one of the main features of the Marriage Machine is Marriage Services, which assists the user by providing information, by giving suggestions and ideas, and by offering regular feedback. These tips and advice motivate users towards a better understanding of issues blocking the improvement of their marriages and teach them techniques that enrich their experiences with their own attitudes, behavior, and relation to people, objects, contexts, obstacles, fears, and angers. This process of learning and self-discovery is designed to change the users' approach in the long term.

In addition, the creation of a "marriage health" meter will increase users' awareness of their steady progress they are making towards their marriage and will motivate them to add further material in the form of visual or textual records of documents (photos, texts, voice, music, videos, etc.), experiences, and encounters with their partners.

11.3.3 Improve Learning

The central objective of the Marriage Machine is to trigger a learning process that is both informative and reflective and that combines textual, visual, and sonic elements. Thus, the advisor provides a range of practical (techniques, tricks, games, etc.), current (news, events, etc.), general (history, religion, politics, etc.), and more culture-specific (values, symbols, behavior, etc.) information, but it also proposes a range of activities that are supposed to make the user discover marriage particularities and characteristics related to specific goals. A primary concept is to support, assist, and stimulate users during their daily lives, in order to make them observe certain things, to make them reflect actively about their experiences, and to perform incremental behavior changes that deliver positive change.

We also seek to make the education process both informative and entertaining. Therefore, we have proposed games that refer to specific knowledge of users' partners and their contexts. Through playing games featuring educational information, users will be able to increase their knowledge and understanding of their partners and their marriages without getting bored.

11.4 Theoretical Background and Research

11.4.1 Marriage Definition

As a social phenomenon, marriage has been present throughout history and in all cultures. Over time, the concept of marriage has undergone a series of transformations. The traditional style refers to a solemn act between two people, acknowledged by and empowered with religious and/or civil legitimacy. However, the term has also come to mean couples who opt for a serious relationship without legal and/or religious accreditation. We adopt the latter definition of marriage, intending our application to be for committed couples who may or may not be "legally" married per the rules of specific religions, states, or national laws.

11.4.2 Marriage UX Design

What behavior should couples change that will make their relationships more enjoyable and personally enriching? This topic is very broad, and popular wisdom and professional knowledge have different views about this question. Therefore, we plan to design the Marriage Machine by understanding marriage theories (Fig. 11.1).

Marriage is highly subjective and, therefore, requires an interdisciplinary approach to fields that deal with such subject matter, such as positive psychology,

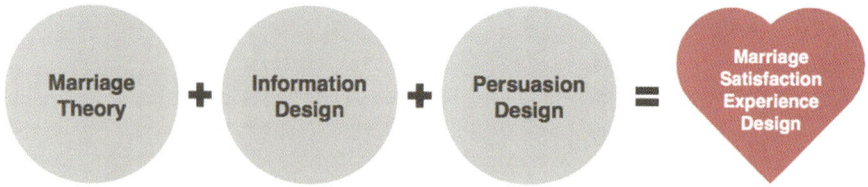

Fig. 11.1 Conceptual diagram of the Marriage Machine

marriage theories, philosophy of marriage, and emotional design. As such, we examined the following articles in our research (also cited in the Bibliography):

- Córdova, J. V., Gee, C. G., and Warren, L. Z. (2005). Emotional skillfulness in marriage: Intimacy as a mediator of the relationship between emotional skillfulness and marital satisfaction. Journal of Social and Clinical Psychology, 24, 218–235.
- Gottman, J. and Gottman, J. (2011, Winter). How to keep love going strong: 7 principles on the road to happily ever after. YES! Magazine, (56), 38–39.
- Karney, B. (2010, February). Keeping marriages healthy, and why it's so difficult. Retrieved from http://www.apa.org/science/about/psa/2010/02/sci-brief.aspx
- Tsapelas, I., Aron, A., and Orbuch, T. (2009). "Marital boredom now predicts less satisfaction 9 years later." Psychological Science, 20 (5): 543–545.
- Luo, Shanhong, and Klohnen, Eva C (2005). "Assortative Mating and Marital Quality in Newlyweds: A Couple-Centered Approach," Shanhong Luo and Eva C. Klohnen, University of Iowa; *Journal of Personality and Social Psychology*, Vol. 88, No. 2, pp. 304–326.

From our analysis of the aforementioned literature, we have been able to distill ingredients of a healthy marriage into the following principles:

Enhance Your Love Map

- Use periodic marriage checkups.
- Know each other's objectives, goals, hopes, and worries.

Nurture Fondness and Admiration

- Remind users of partner's positive qualities and express fondness and admiration.
- Enhance users' and partner's physiological intimacy.
- Enhance users' and partner's emotional connection.

11.4.3   Turn Towards Each Other

- Make "bids" for partner's attention, affection, humor, or support.

Let Your Partner Influence You

- Share power and decision making with partner.

Resolve Conflicts

- Avoid conflicts.
- Solve solvable problems.

Create Shared Meaning/Value

- Marriages that have shared sense of meaning result in less conflicts and decrease the likelihood that problems will to lead to gridlock.

Cultivate Excitement

- Try new and exciting activities together to rekindle feelings.

11.5   Insights and Design Principles

11.5.1   Insights

From our reading, we developed these insights:

Insight 1. Marriage improvement is an objective.

Design principles: The Marriage Machine must be able to assist users to achieve specific goals in a "journey-through-marriage" viewed as perceived steps. Directions should be "chunked," so that people can easily remember where they are and sense what to do.

Insight 2. Marriage is discipline, a constant reminder to change attitude.

Design principles: The Marriage Machine must function as a reminder to be mindful about marriage. The Machine should expedite change by helping users to reflect.

Insight 3. Marriage is in a constant state of flux.

Design principles: The Marriage Machine must be adjusted for or customized to different users in different life stages. Directions should be suitable/adaptable to each user at that point in time of use.

Insight 4. Balancing engagement and persuasion is key.

Design principles: The Marriage Machine should persuade users to do certain activities with the least friction possible.

11.5.2   Tactics for Marriage Machine

Based on our research, we developed the following measures that could be considered relevant and practical for the Marriage Machine and which are explored below:

- Involvement.
- Personalization.
- Goal/habit-centered.
- Entertainment.
- Connectedness.
- Accomplishment.

11.6   Market Research

In order to have a clearer vision of the target market for the Marriage Machine, AM+A conducted comprehensive market research with potential customers, through collaboration with graduate marketing students of the University of California at Berkeley's Extension Program for International Students (Arora et al. 2014). One main objective of the research was to find out more about people's general usage of smartphones and mobile applications. A second objective was to get a better understanding of their "relationship behaviors," especially when using smartphones and mobile applications.

Market research has gone through the following steps (Fig. 11.2).

11.6.1   Secondary Research

In a first step, secondary data was collected in order to get a general overview of people's mobile usage and relationship behaviors. As is shown below, 80 % of the world's population now has a mobile phone, which accounts for about 5 billion people. Currently, around 1.08 billion of these mobile devices are smartphones, i.e., approximately 20 %. Current tendencies seem to indicate that the number of smartphones will rise considerably during the upcoming years. Around 80 % of time spent on smartphones is inside mobile applications, and 2/3 of the US consumers have spent money on mobile applications at least once (ABI Research) (Fig. 11.3).

Interestingly, not only teenagers and young people make use of smartphones. In fact, most users are between 18 and 44 years old. They use their phones mainly to write text messages (92 %), to go online (84 %), to write emails (76 %), to play games (64 %), and to use downloaded apps (69 %) (Fig. 11.4).

Fig. 11.2 Marketing research steps

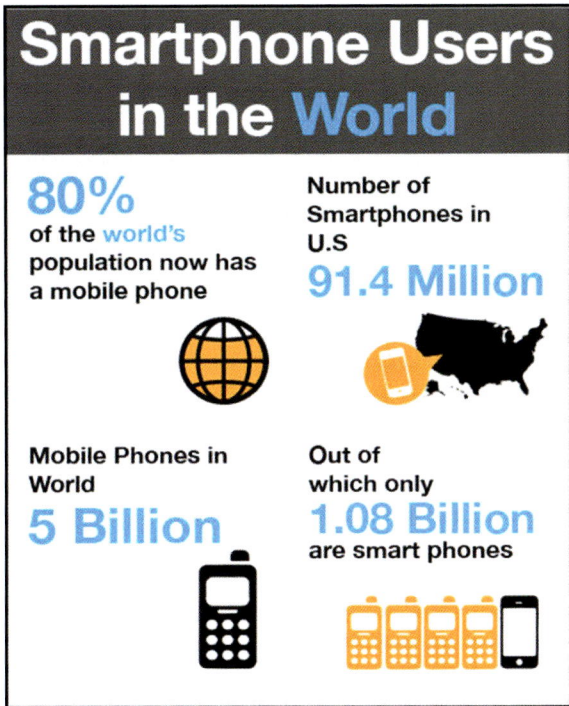

Fig. 11.3 Smartphone users in the world (Aboelwafa et al. 2013)

11.6.2 Qualitative Research

We used the following techniques in conducting qualitative research:

- In-depth interviews.
- Word association.
- Focus groups.
- Group dynamics.
- Thought leaders and their effect on others.

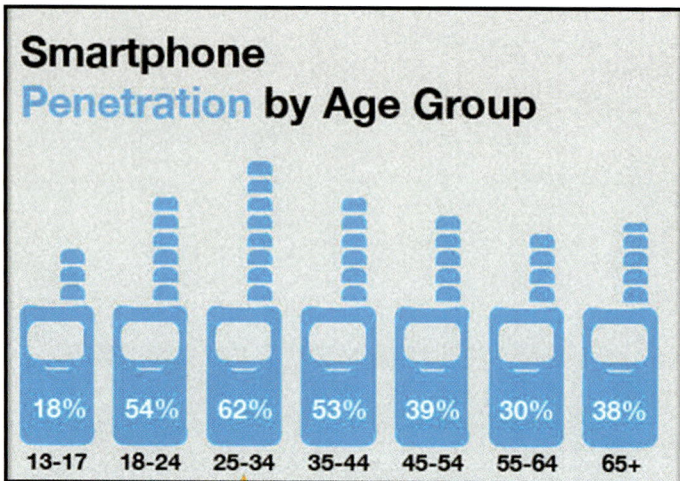

Fig. 11.4 Smartphone penetration by age group (Aboelwafa et al. 2013)

- Probing techniques.
- Positive reinforcement.
- Thought-Cloud (tag cloud or word cloud).

11.6.3 Quantitative Research

We used the following techniques in conducting quantitative research:

- Survey.
- Non-probability sampling methods (convenience, snowball).

  We asked participants some of the following sample questions:

- On average, how much time [quality time] on a weekday do you spend together with your partner?
- On average, how much time [quality time] on a weekend day do you spend together with your partner?
- How important are these shared activities in your relationship with your partner: playing sports, playing games, parenting, cooking, shared hobbies, watching movies/TV, going out with friends, messaging, time with family, traveling, date night, display of affection, and talking with your partner.
- What other activities do you think are effective to strengthen your relationship?
- Here are some elements that may be part of your relationship. Please rate these elements in order of importance for your relationship: affection, agreement/support, communication, fun, shared activities, and objectives/goals.
- Do you use a smartphone?

- Do you use mobile applications?
- Did you pay for any of these applications?
- How much time every day do you spend actively using mobile applications?
- Let's imagine we present to you a mobile application that aims to improve your relationship with your partner through your use of a set of functions. Which functions below do you think are most useful in this kind of application: gift store, games, sharing drawings, tips about relationship, rewards, financial tracker, partner's mood, random suggestions of activities, wish list, video call, shared calendar, sending love notes, and taking pictures or videos?
- How important are the following criteria in your decision to buy a relationship app: tips/advices from specialists, aesthetic design, price, efficiency in improving your relationship, fun interactions, functionality, and privacy?
- Would you be willing to pay (in USD) for the application? If yes, how much?

11.7 Results and Conclusions

Among the core principles of marriage, the most cited during these interviews for qualitative research were the following: activities, being together, sharing, community, friends, communication, honesty, trust, and support.

We also asked people about ideas of applications for the Marriage Machine. Among the answers, one can find birthday reminders, suggestions, coupons, calls, messaging, calendar, hugs, pictures, gifts, events, wishes, and attention.

The following figures represent the results of the quantitative research.

The first two figures represent the answers posed to the participants about the quality of time spent together with the partner on weekdays and weekend days (Figs. 11.5 and 11.6).

Most participants answered that they spend more than 3 h together on weekdays and more than 4 h on weekend days.

We also asked about the order of importance of shared activities in a person's relationship with his/her partner (Fig. 11.7):

According to the respondents, the most important shared activities in a relationship are:

- "Talking with your partner"
- "Display of affection"

Affection and communication appear to be the most important components of a relationship for the majority of respondents. Affection and communication expressed most of the responses. Other activities cited include sex, going shopping, going to church, and having fun.

We also gathered information about the ranking, in order of importance, regarding elements of relationships (Fig. 11.8):

Affection and communication are considered very important for the relationship, according to the respondents.

Time Together: Weekday

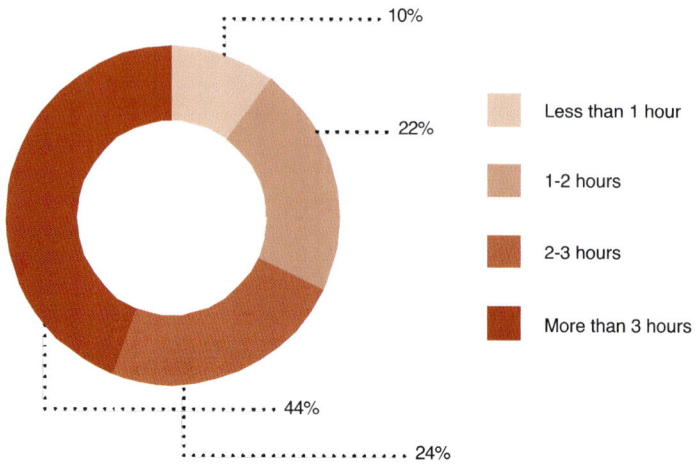

Fig. 11.5 Chart representing participant's opinions about the quality of time spent together with the partner on weekdays (Arora et al. 2014)

Time Together: Weekend

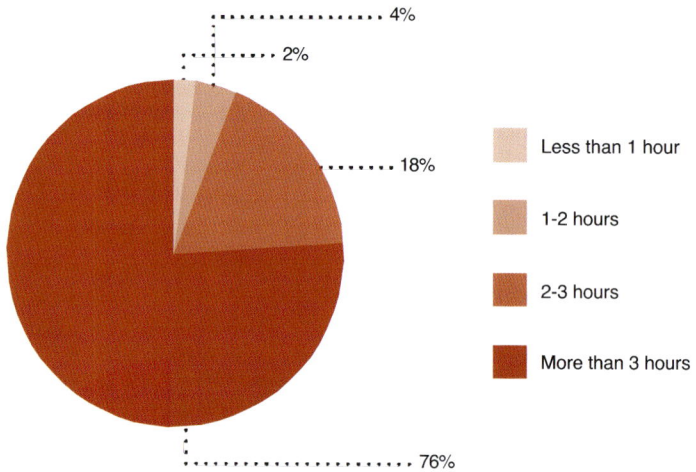

Fig. 11.6 Chart representing participants' opinions about the quality of time spent together with the partner on weekend days (Arora et al. 2014)

Important shared activities in your relationship

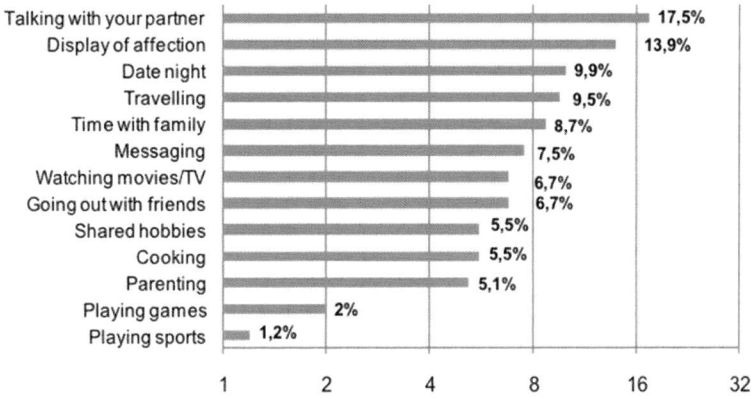

Fig. 11.7 Chart representing participants' ranking of the most important activities practiced together (Arora et al. 2014)

Elements in order of importance for your relationship

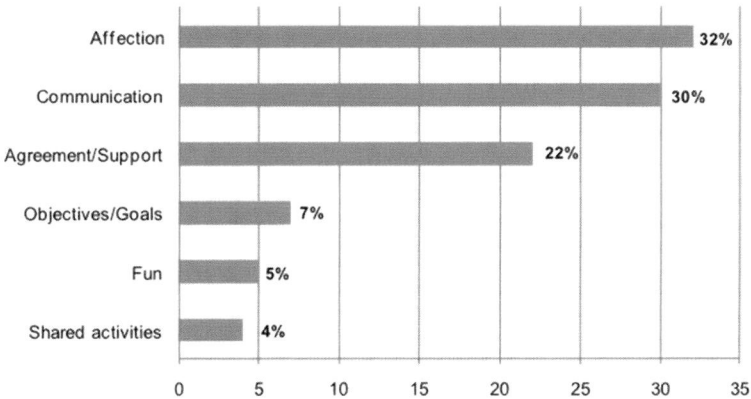

Fig. 11.8 Elements in order of importance for your relationship (Arora et al. 2014)

We asked other questions about the applications (Fig. 11.9).

Most of the respondents had already paid for a mobile application (Fig. 11.10).

More than 90 % of respondents spend more than half an hour per day on mobile applications, and most of the respondents said they spend 1–2 h with mobile applications (Fig. 11.11).

More specifically about the Marriage Machine, from all the features we offered to the respondents, the most appreciated features were, in order of importance, "shared calendars," "taking pictures or videos," "shared love notes," and "wish list" (Fig. 11.12).

Fig. 11.9 Question about
if the participants paid for
an application (Arora et al.
2014)

Did you already pay for a mobile app?

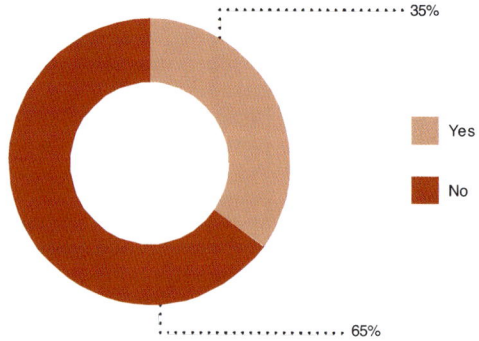

.... 35%

Yes

No

.... 65%

Time spent with mobile app

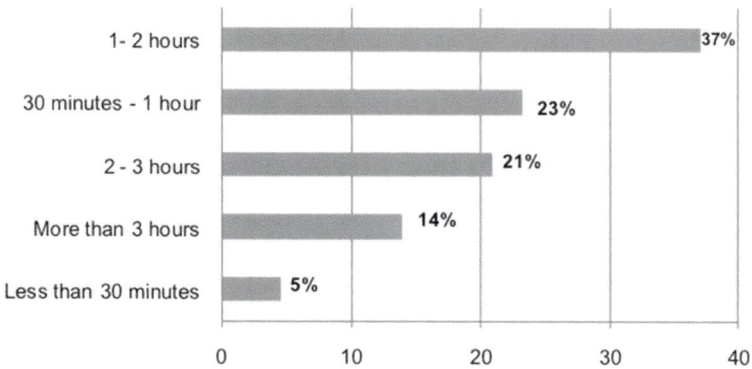

Time	%
1- 2 hours	37%
30 minutes - 1 hour	23%
2 - 3 hours	21%
More than 3 hours	14%
Less than 30 minutes	5%

0   10   20   30   40

Fig. 11.10   Time spent with mobile app (Arora et al. 2014)

Among the concerns about buying the Marriage Machine application, privacy seemed to be the main issue for the respondents. The functionality of the application and fun interactions are the second and third most important criteria to buy the application, respectively (Fig. 11.13).

More than 65 % of the interested respondents would be willing to pay more than one dollar for the application.

The above marketing research data influenced our thinking as we progressed through the product development.

Features that are most useful for Marriage Machine

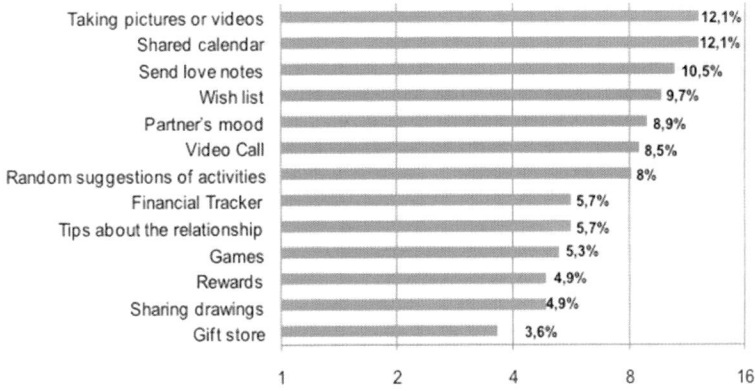

Fig. 11.11   Features that are most useful for Marriage Machine (Arora et al. 2014)

Criteria to buy a relationship app

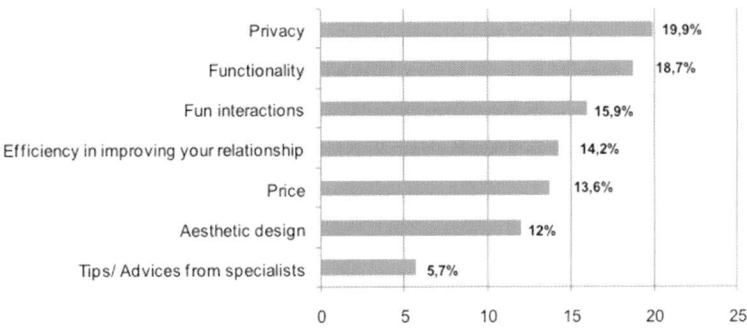

Fig. 11.12   Concerns about buying a relationship application (Arora et al. 2014)

Price Marriage Machine App

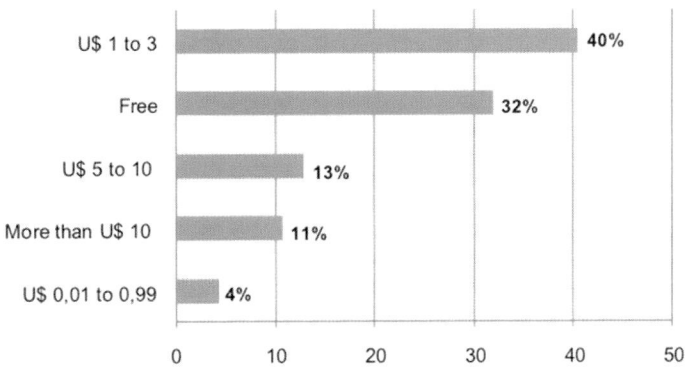

Fig. 11.13   Price that participants were willing to pay for such an application (Arora et al. 2014)

11.8   Competitive Analysis

Before undertaking conceptual and perceptual (visual) designs of the Marriage Machine prototype, we carried out comparison studies of competitive products. AM+A studied in detail ten smartphone applications. Through screen comparison and customer review analyses, AM+A derived these applications' major benefits and drawbacks. This in-depth analysis helped us to develop further initial ideas for the Marriage Machine's detailed functions, data, information architecture (metaphors, mental model, and navigation), as well as look-and-feel (appearance and interaction). The following summarize our findings about these products as of mid-2014.

Avocado

• Offers joint virtual activities for couples (Fig. 11.14).

Pros

• Shared calendars, photos, and to-do lists on a single platform.
• Send drawings with custom stickers.

Cons

• No home screen/dashboard.
• The core functions of Avocado are basic: product does not distinguish itself from others.

Better Marriages

• Audiovisual and written material on marriage (Fig. 11.15).

Pros

• Contains comprehensive articles, tips, and advice.
• Synergy with online forums and discussions.

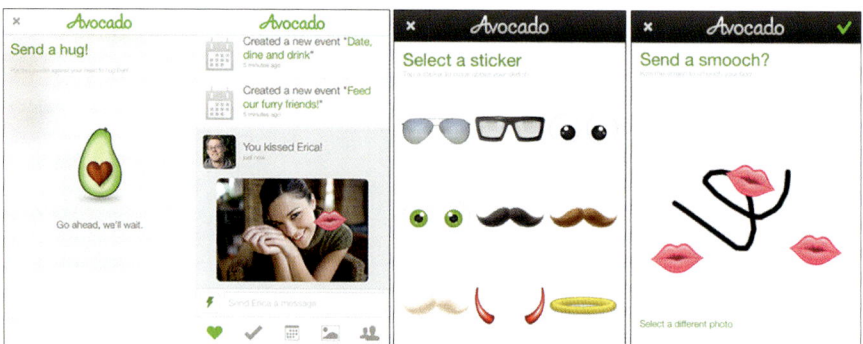

Fig. 11.14 Example screenshots of the Avocado screen as of 2014 (Images: Avocado Software, Inc.)

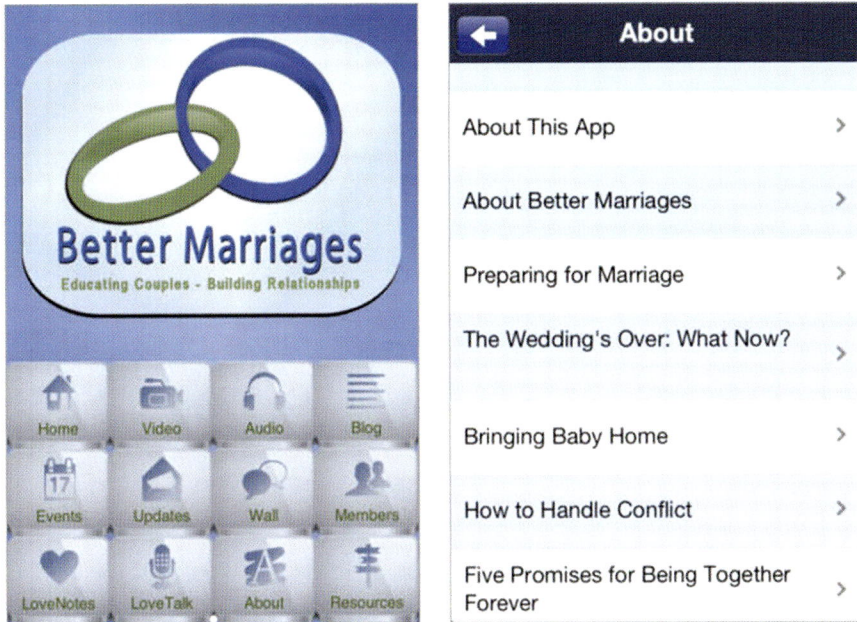

Fig. 11.15 Example screenshots of the Better Marriages app screens as of 2014 (Images: The Association for Couples in Marriage Enrichment)

Cons

- No functions for couples to interact.
- Large chunks of text not broken down: users not motivated to read.

Between

- South Korean app which provides private space for couples to share moments (Fig. 11.16).

Pros

- Well-thought-out navigation.
- Includes weather updates.
- Detailed shared calendar.

Cons

- Shared function is limited: no audio and video.
- Chat function is limited: no video call.
- No finance management and to-do list.

Casamento Feliz

- Brazilian app which provides tips and advice about relationships (Fig. 11.17).

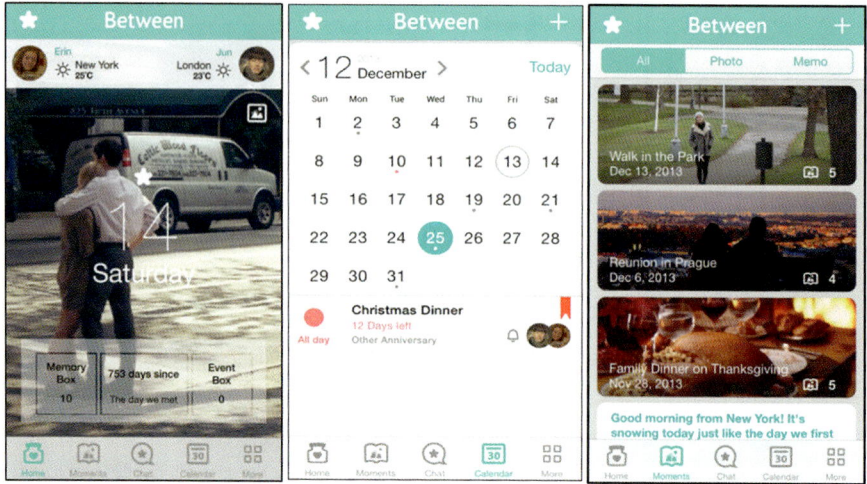

Fig. 11.16 Example screenshots of the Between app as of 2014 (Images: VCNC Corp.)

Fig. 11.17 Example screenshots of the Casamento Feliz app in 2014 (Images: Casamento Feliz)

Pros

- Contains useful and well-categorized information.

Cons

- Large chunks of text not broken down: users not motivated to read.
- Contains almost no functions.
- Text size badly proportioned.

Couple

- Enables one to stay in touch with partner privately (Fig. 11.18).

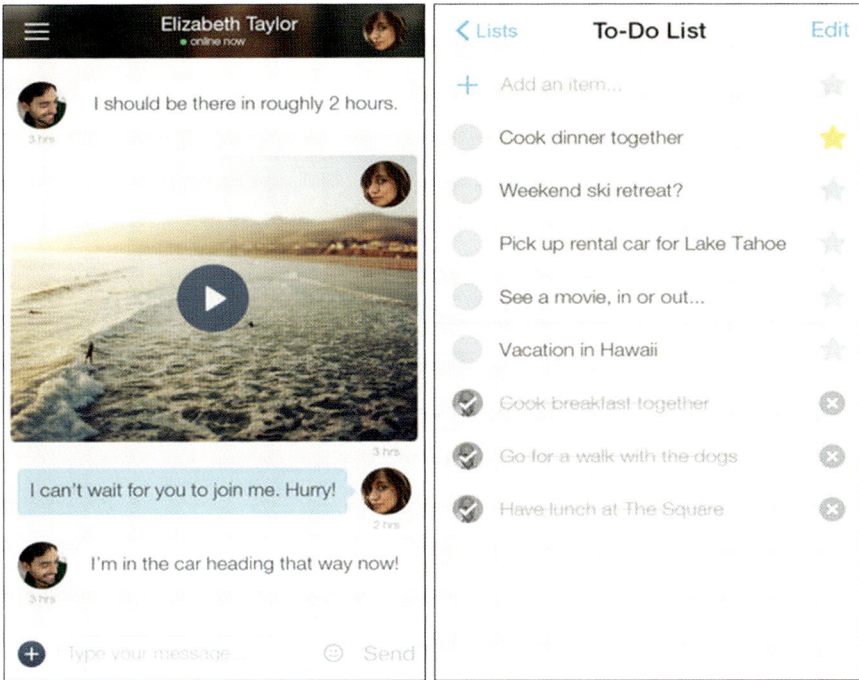

Fig. 11.18   Example screenshots of the Couple app in 2014 (Images: TenthBit, Inc.)

Pros

- Send audio, video, and photos.
- Shared to-do list.

Cons

- Shared calendar shows list instead of calendar layout.
- Unable to personalize/edit photos.
- Nostalgic and random photos are stored in same "Moments" folder.
- Limited store products: only stickers.

Hub: Helps organize and plan family activities together (Fig. 11.19).

Pros

- Family planning functions: lists, calendar, and activities.
- Check completed items on list.

Cons

- Shared function is limited: no audio, video, and photos.
- Not many functions for couples to interact.

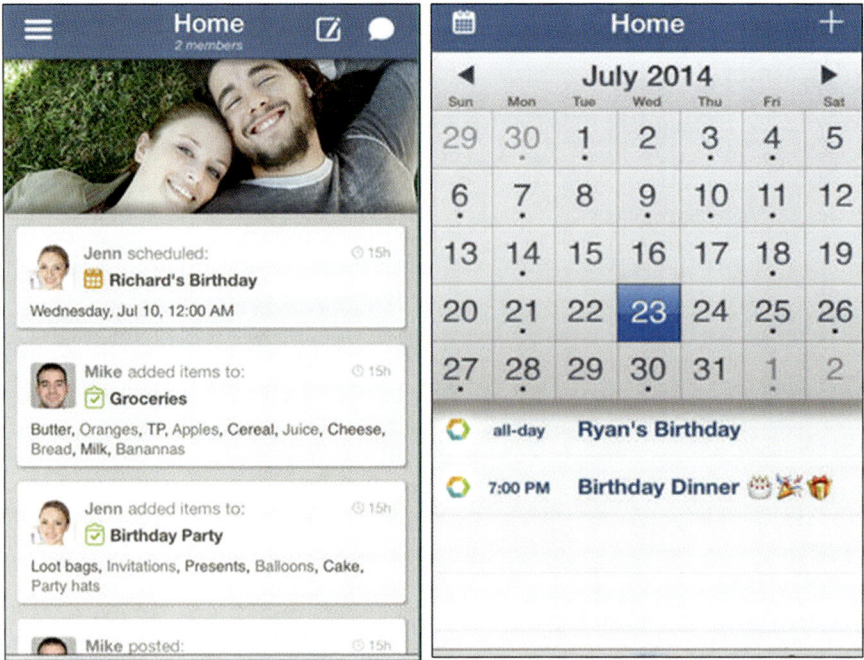

Fig. 11.19 Example screenshots of the Hub app in 2014 (Images: Hub)

LoveByte

• A Singaporean app which provides couples with a shared time line (Fig. 11.20).

Pros

• Tutorial introduction.
• Store contains merchandise: can buy gifts for partner.
• Ease of navigation.

Cons

• No shared calendar and financial planner.
• Shared function is limited: no audio and video.
• Chat function is limited: no video call.

Love Maps

• Offers detailed information about one's partner (Fig. 11.21).

Pros

• Contains insightful just-in-time knowledge to enhance relationships, sourced from 20 years of research at Gottman Institute.

Cons

• Rudimentary user interface.

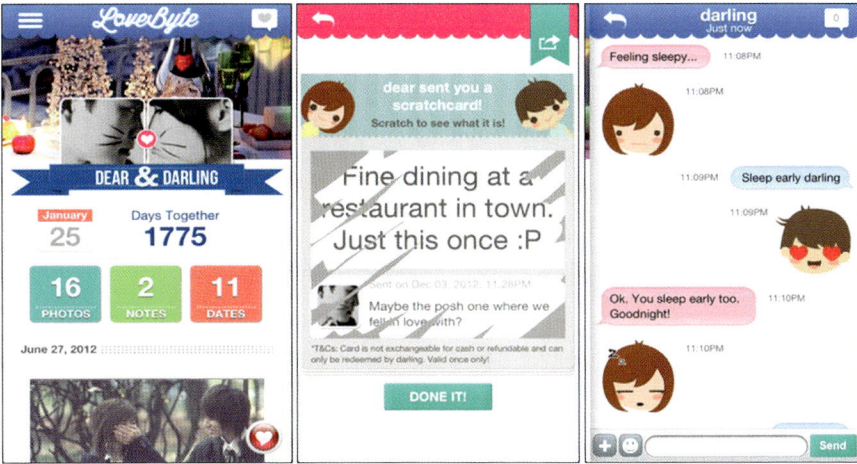

Fig. 11.20 Example screenshots of the LoveByte app in 2014 (Images: mig33 Pte Ltd.)

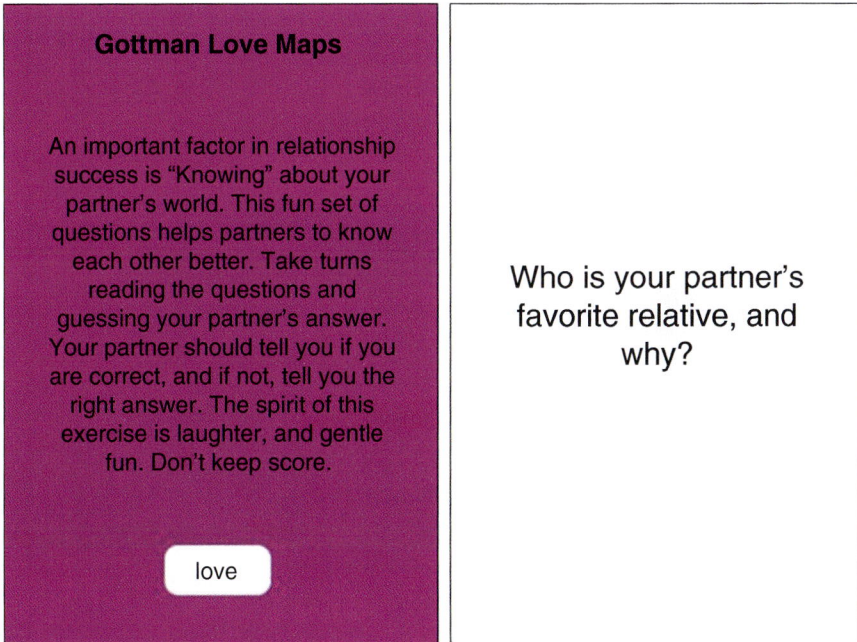

Fig. 11.21 Example screenshots of the Love Maps app as of 2014 (Images: The Gottman Institute)

- Lack of persuasion design to act on insights.

Swiitt

- A Taiwanese app which provides couples with a shared time line (Fig. 11.22).

Pros

- Tutorial introduction.
- Store contains merchandise: user can buy gifts for partner.
- Ease of navigation.

Cons

- No shared calendar and financial planner.
- Shared function is limited: no audio and video.
- Chat function is limited: no video call.

Twyxt

- Facilitates a couple's shared moments together with fun (Fig. 11.23).

Pros

- Fun interactions for chat.
- Well-designed shared calendar with photo options.
- Save favorite quotes and photos in "Keepsake" folder.
- Check completed items on list.

Fig. 11.22 Example screenshots of the Swiitt app as of 2014 (Images: Swiitt Computing, Inc.)

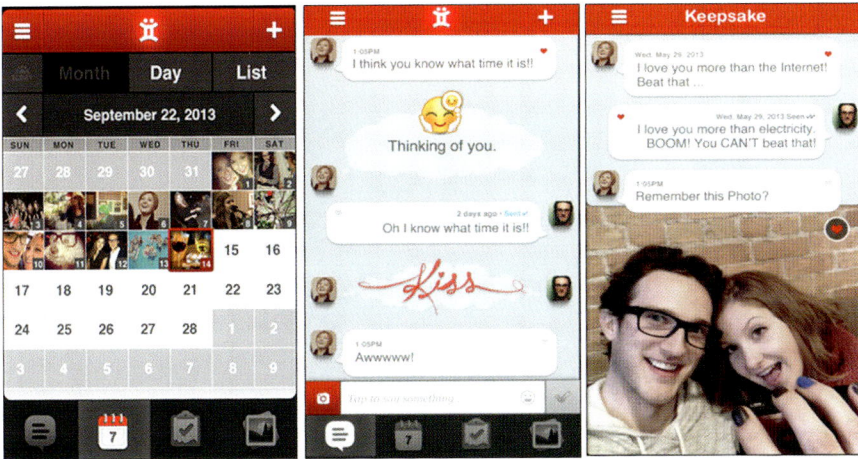

Fig. 11.23 Example screenshots of the Twyxt app as of 2014 (Images: Life of Two, Inc.)

Cons

- No home screen/dashboard.
- Shared function is limited: no audio and video.

11.9   Conclusions

Many mobile applications already exist in the field of marriage and relationship maintenance, most of which emphasize the private connection between the members of the couple. However, few if any of the applications focus on a holistic approach to help users realize and achieve success in their relationships. The Marriage Machine should find an opportunity in its personal, intercultural learning and will thereby distinguish itself from existing applications.

11.10   Personas and Use Scenarios

11.10.1   Introduction

Because people experience marriage in many different ways, we wished to demonstrate use scenarios for relatively different functions of the Marriage Machine with different personas or user profiles. Personas and use scenarios have been described earlier in Chap. 1. This Machine required us to consider two people within one Persona Couple, which is quite unusual.

11.10.2   Persona Couple 1: Cheryl and Brady

Slogan: "It was tough, but now that we're working towards a goal, we're glad we saved for our life together."

Cheryl Liu, 29, and Brady Johnson, 30, are high school sweethearts from Los Angeles, California. After high school, Cheryl attended UCLA and is now the director of large nonprofit organization. Brady decided to skip college to pursue his passion as an amateur surfer and now works as a part-time surfing instructor. They have been married for 8 years, but recently, they have been feeling that their marriage is heading towards a rough patch. Brady feels that Cheryl works too much and spends too little time with him. Cheryl feels like she's the only one doing the chores around the house and providing financially. Brady is a big spender who likes to go on extravagant trips chasing swells around the world, while Cheryl is constantly trying to save money so they can finally purchase their first home and have kids.

During the drive home from one of his morning classes, Brady's phone vibrates—it's a message from the Marriage Machine application reminding him that it's his turn this week to run the errands. He taps the notification, which opens the application to the "grocery list" shared-list tab. It looks like Cheryl updated "yogurt and cereal" to the list a few hours ago. Instead of heading home, he decides to make a quick detour at Trader Joe's. They are out of cereal, but they have yogurt and blueberries. As he's checking out, he checks off the items he was able to buy and Cheryl receives a notification in the grocery list and sends a love note to Brady.

The couple has also been using the application to track their finances. With it tied to their credit cards, updating the dashboard to reflect spending patterns, Brady admits it's made him more willing to make sacrifices without resenting that surf trip to Australia he had to give up. While on her lunch break, Cheryl opens up the app—to her surprise, they are now at 28 % progress in saving up for the home they've been dreaming of.

11.10.3 Persona Couple 2: John and Christina

Slogan: "We forgot what it felt like to date again."

John Stevens, 35, and Christina Green, 35, are what people might call the new crop of Manhattan power couples. Christina is a partner at a large law firm, and John is managing director of a midsized investment bank. They met as undergraduates at Dartmouth University and live in Greenwich, CT, with their two children.

Due to their high-profile positions, their work entails long hours and frequent travel. They often hire a nanny during the week if both are away, but on the weekends, they spend most of their time with their children. Between parenting and their careers, Christina and John have forgotten what it feels like to be a couple.

While at home with the kids after an intense litigation case, Christina's phone vibrates—the wallpaper on her phone changes to a picture of John at the Buckingham Palace. John is in London with a client for the week, and she knows that it's going to be a busy week for him—he has landed an IPO deal with a promising biotech startup. Christina takes a picture with their kids during dinner, and they choose a funny frame in the gallery function to send to John. The application reminds

Christina that they haven't gone on a date in a while now and suggests that she plan something special for the both of them when John returns from his trip. She uses the app's "Date Planner," which pulls information from their shared interests and their availability according to their shared calendar to suggest an activity—dinner at Jean-Georges followed by the "Book of Mormon," a play at Eugene O'Neill Theatre on Saturday night. She sends the date invite to John and uses the nanny list feature to find someone to take care of their children for Saturday night.

11.10.4   Persona Couple 3: Miguel and Amanda

Slogan: "We needed to better understand each other and build our marriage."

Miguel Jimenez, 27, and Amanda Hernandez, 26, are two newlyweds who live in Miami Beach, Florida. They had been dating for only 8 months until Miguel proposed during their vacation in Europe. They are a fun, hip couple and are heavily involved in the art scene in Miami. Miguel is a freelance architect and Amanda is a fashion jewelry design who runs her own boutique.

After their wedding last year, the couple purchased a loft and moved in together. The first year was full of passion, dinner parties with other couples, and road trips. As they are hitting the one-and-a-half-year mark, their honeymoon phase is wearing off. They've had to grapple with very real questions about how they envision spending the rest of their marriage. While they enjoy spending time together, they admit they did not share much about their hopes, habits, dreams, and goals before marriage.

Sometimes these things become a problem between them and bring up some disagreements.

During the commute to work, Amanda opens up the application to Games, which prompts her to answer a few questions about her husband. "Who is your partner's favorite relative, and why?" Amanda realizes she has no idea and begins to realize that she needs know more about Miguel. She navigates to tips and advices in the application and starts to search helpful information about how to improve the couple's communication. During dinner that night, Amanda brings up this question to Miguel, and the two start talking more about their extended families over wine, which then transitions into a discussion about whether they'd want to adopt or have children.

11.10.5 Persona Couple 4: Ryan and Jessica

Slogan: "We wanted to stay connected despite the distance."

Ryan Carter, 29, and Jessica Jones, 26, are a couple in a long-distance relationship. Ryan is a military officer in the US Army Special Operations and was deployed to Iraq last year. The couple have been married for 4 years and have a 2-year-old daughter.

As a military spouse, Jessica understands that Ryan's duty comes first. In addition to her job as a waitress in Berkeley, CA, Jessica takes care of their daughter, Silvia, full-time. They both have had to make many sacrifices to make their relationship work and experience the loneliness that comes with physical distance. And because their time zones and schedules are so different, they have been struggling to find ways to stay in touch.

On her lunch break at Jupiter's restaurant/bar, Jessica opens the application and sees that it is almost midnight in Baghdad. It's 37 °F and drizzling. She glances over to the news section and reads up on the recent surge in sectarian violence in Baghdad, with at least 33 people having been killed in a series of bomb attacks this week. She notices that Ryan has updated his status that he is out in the field this week and so she knows they won't be able to video call until the weekend. Missing him greatly, Jessica sends a video message of her and their daughter and a virtual kiss.

Ryan wakes up the next day with a notification from the app, informing him that Jessica has sent a video message and a virtual kiss. After watching the video, he checks their shared calendar and is reminded that Jessica's birthday is coming up soon. With about a half an hour to spare before 07:00 morning formation, Ryan uses the gift shop to browse for a present for her.

11.11   Information Architecture

The following text describes the information architecture of the first version of the Marriage Machine. In designing the information architecture (IA) for the Marriage Machine, AM+A began by examining the IA for past Machines, including the Green Machine, the Health Machine, the Money Machine, the Driving Machine, and the Innovation Machine (see earlier citations). We discovered an overarching model for the IA that permeated each of the past machines. In this model, there are five primary "modules" or branches of the IA, each of which is described below. While we altered slightly the details of each module to fit the needs and requirements for the Marriage Machine, we maintained the same general approach (see image below). The basic modules are described below. The accompanying figure shows the first draft of the information architecture.

11.11.1   Dashboard

The Dashboard module is similar to a landing page for a Machine. It provides an overview into the status of the user's behavior change. Here, the user gets a view of his/her objectives/goals and where he or she stands in achieving those objectives/ goals. (Objectives are broad, long-range intentions; goals are specific, measurable states with time frames.)

11.11.2   Road Map

The journey map, road map, or Process View module is where the user gets more high-level views or a "road map" of the journey and more details regarding each objective/goal. The user sees the progress being made, as well as the next steps in achieving a particular objective/goal.

11.11.3   Social Network

The Social Network module is an integral part of behavior change in modern software. Users engage in focused, subject-matter-based connections with friends, family, and/or like-minded people that either share similar goals or wish to support others in achieving behavior-change objectives.

11.11.4   Tips/Advice

The Tips/Advice module provides focused knowledge about a given topic to give users insight into the habits they wish either to get rid of or adopt.

11.11.5   Incentives

The Incentives module presents users with fun and engaging ways to change their behavior. Gamification has proven to be a powerful tool in encouraging users to try an application, even with virtual incentives, although in some cases, real incentives are provided. In addition, a leaderboard allows a user to compare his/her progress with others, tapping in the competitive nature of the human mind to create behavior change.

11.11.6   Information Architecture Draft and Final Versions

AM+A adapted the basic structure of the past Machines in creating the Marriage Machine information architecture. However, due to particular complexities that were unique to the Marriage Machine, we provided a different menu navigation. Nevertheless, the five high-level basic modules are retained, with some modules being partially distributed across the information architecture. AM+A constructed several draft versions of the information architecture, which are presented in the accompanying figures (Figs. 11.24, 11.25, 11.26, and 11.27).

11.11.7   Objectives

The Marriage Machine will contain several key components. The many detailed functions are grouped into three "tabs": a dashboard (1), side panel (2), and tips/advice (3). The sections below explain each tab.

11.11.8   Tab 1: Dashboard

The Marriage Machine will contain a dashboard, which can be considered as the "home page" of the mobile application. The dashboard will contain the most important information users may need at any time during their daily lives. Note: all of the following elements will be optional for Marriage Machine users, which means that users can personalize and customize the dashboard according to their needs and interests, by deciding which of the features to show and which ones to hide.

11.11.9   Current Time and Date

11.11.9.1   Daily Updates

An in-app "news" source for users to access the following information:

- *Marriage health:* This function depicts the relative "health" of a couple's marriage, using a scale from 0 % to 100 %. Percentages are based on a complex algorithm that takes into account the following components of the couple's marriage: finances, frequency of application use, performances on games and quizzes, number of items purchased in the gift store, number of love notes sent, average mood of each partner, and number of virtual kisses/hugs sent. The attribution of a percentage in the marriage health can stimulate users to discover ways to improve that number and to proceed in their learning process.
- *Partner's mood:* This function allows the user to view his/her partner's mood.
- *My mood:* This function allows the user to view/edit his/her own mood.
- *Weather/Time/Local News:* This function displays the weather, time, and local news information of each partner's current location, allowing the couple to stay in sync with one another. Each can customize which information is shown and how.

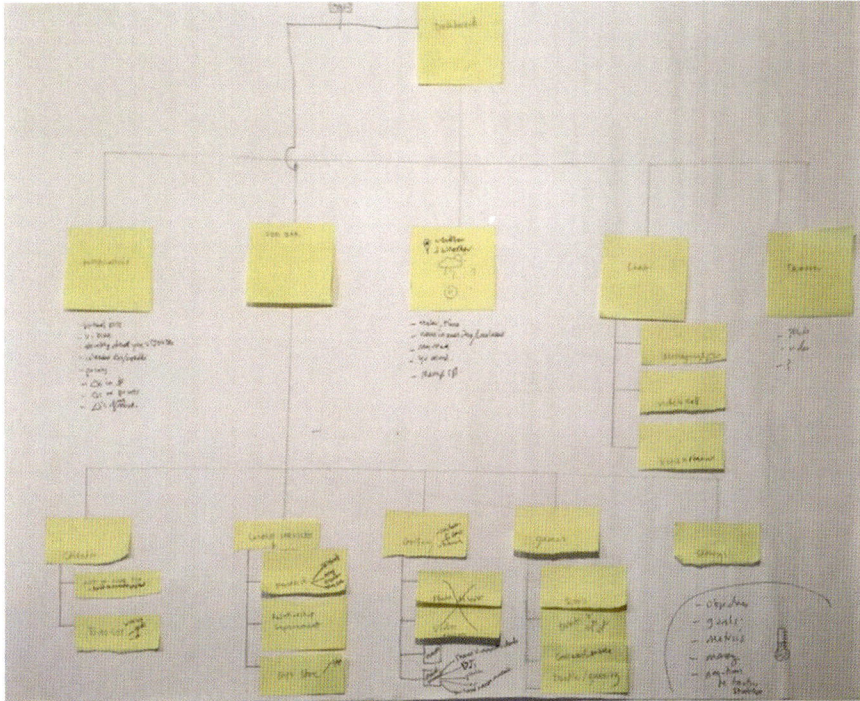

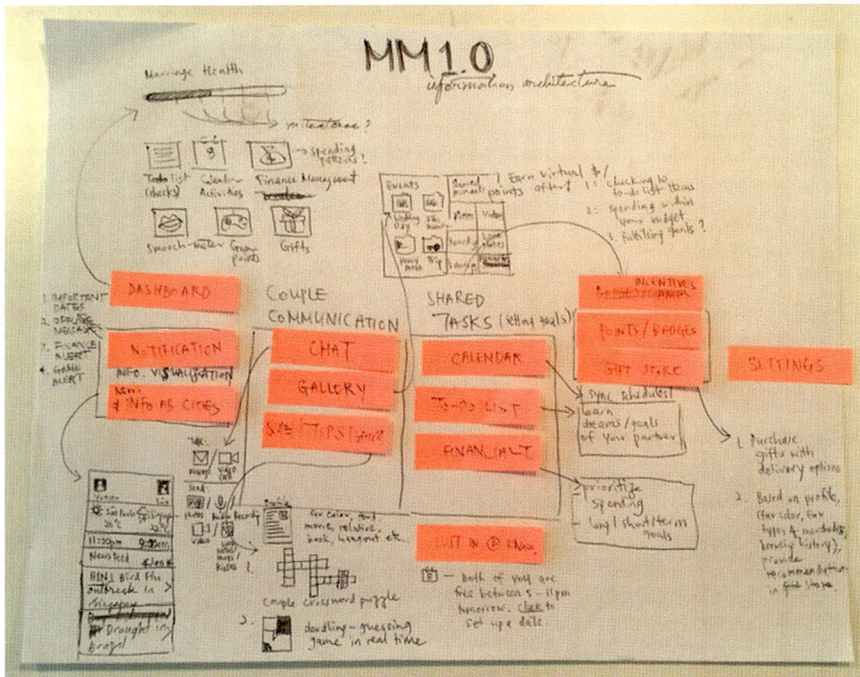

Fig. 11.24 Information architecture, draft version

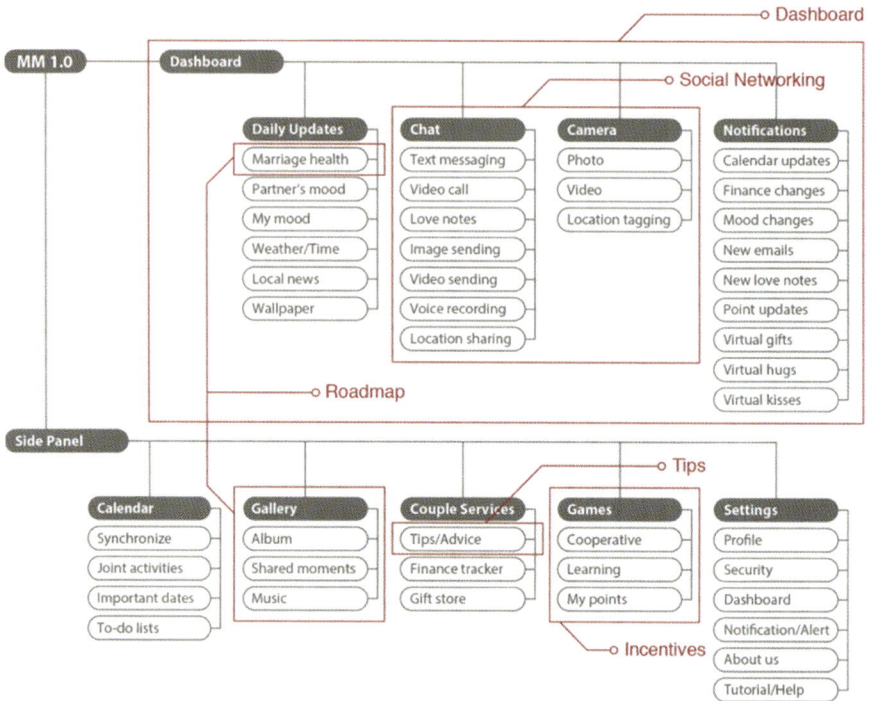

Fig. 11.25 Information architecture of the Marriage Machine, final version, showing equivalent primary modules, such as Dashboard, Roadmap, Social Networking, Tips, and Incentives

- *Shared Wallpaper:* This function allows the user to edit his/her wallpaper on the dashboard, which then causes the partner's wallpaper to sync with this change.

11.11.9.2   Chat

Couples can communicate with each other via the chat icon. Chat supports the following modes of communication:

Text Messaging

- Doodle.
- Virtual displays of affection (virtual hugs/kisses).
- Screen Freeze: temporarily "freezes" the partner's screen, such as during a heated argument, forcing the partner to tap on the screen repeatedly until "the ice is broken."
- Send offline messages.
- View availability status.

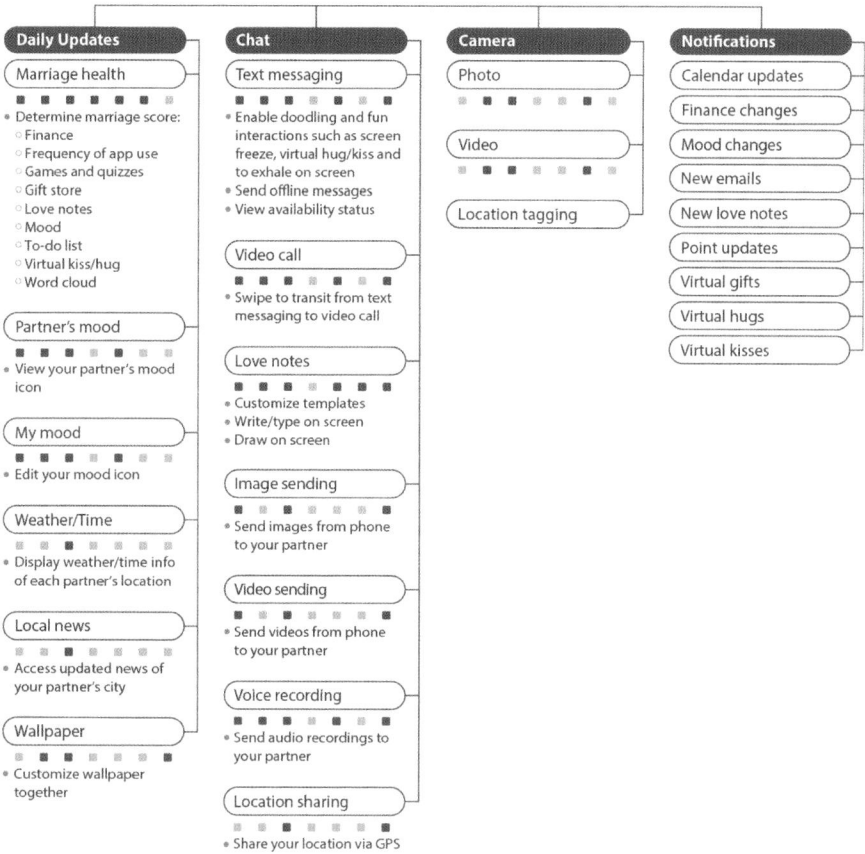

Fig. 11.26 Information architecture of the Marriage Machine, final version, showing detailed menus of the Dashboard

Video Call

- Users can quickly swipe to toggle between text messaging to video calls.

Love Notes

- Users can create custom templates by writing/typing/drawing on their own screen.

Image Sending

- Send images from phone to your partner.

Video Sending

- Send videos from phone to your partner.

Voice Recording

- Send videos from phone to your partner.

Calendar

Synchronize
- Sync your calendar activities with your partner's automatically

Joint activities
- Dining
- Entertainment
- Sports
- Travel
- Volunteering
- Option to update nannies contact info for quick reference
- Just-in-time tip: detect free time slots; prompt users to engage in joint activities

Important dates
- Just-in-time tip: remind users of partners birthdays, anniversaries, and any other special dates

To-do lists
- Create multiple lists for grocery, necessities, shopping, and tasks
- Divide chores between couple
- Check boxes when done
- Just-in-time tip: prompt users near deadlines

Gallery

Album
- Search and save photo/video in folders with option to title, date, tag, and describe events
- Enable photo upload or download
- Add filters/frames on images
- Create slide show with music

Shared moments
- Categorize according to media type:
  ○ All doodles sent via text messaging
  ○ All love notes sent on chat
  ○ All photos sent on chat
  ○ All videos sent on chat
  ○ All audio recordings sent on chat
- Just-in-time tip: prompt users to create album after completing joint activities

Music
- Send and save music files
- Create shared playlists

Couple Services

Tips/Advice
- Provide expert knowledge:
  ○ Enhance your love map
  ○ Nurture fondness
  ○ Turn toward each other
  ○ Let partner influence you
  ○ Resolve conflicts
  ○ Create shared meaning
  ○ Cultivate excitement
  ○ Improve communication

Finance tracker
- Link credit cards to app to track spending patterns
- Set long-term/short-term objectives and goals
- Provide metrics to measure these objectives and goals
- Just-in-time tip: notify when credit goals/limits are approaching

Gift store
- Purchase gifts from online merchandise store
- Link to existing online shops (e.g. retail, romantic gifts, music, books, etc...)
- Deliver products in romantic gift wrap
- Create wish lists
- Earn points for discounts
- Just-in-time tip: suggest gifts based on partner's browsing/purchase history, and partner's favorites from wish list and preferences

Games

Cooperative games
- Play together in real-time
- Drawing-guessing game: random word generated based on category
- Hangman: user inputs word; partner guesses

Learning games
- Play individually
- Crossword puzzle: decode puzzle made up of info about partner
- Quiz: answer personal questions about partner and his/her family

My points
- View scoreboard to check current points
- Earn points via completed joint activities, to-do lists, love notes, games, virtual hugs/kisses, and finance tracker
- Earn points to buy virtual gifts such as:
  ○ Flowers
  ○ Diamond rings
  ○ Romantic balloons
  ○ Cakes
  ○ Teddy bears
- Earn points to upgrade fun interactions and virtual gifts
- Just-in-time tip: prompt users to buy virtual gifts when points are sufficient

Settings

Profile
- Edit profile photo
- Input personal data and preferences

Security
- Change password
- Mobile security for credit card details
- Shared password

Dashboard
- Change wallpaper
- Location (GPS/Manual)
- Location-based news
- My mood
- Partner's mood
- Weather/Time

Notification/Alert
- Alert sound/vibrate
- Lock screen alert
- Message alert pop-up
- Message preview

About us
- Developer info
- Version info

Tutorial/Help
- Contact us/Feedback
- FAQ/Troubleshooting
- How to use
- Introductory tutorial
- Privacy/security policy
- Terms of service

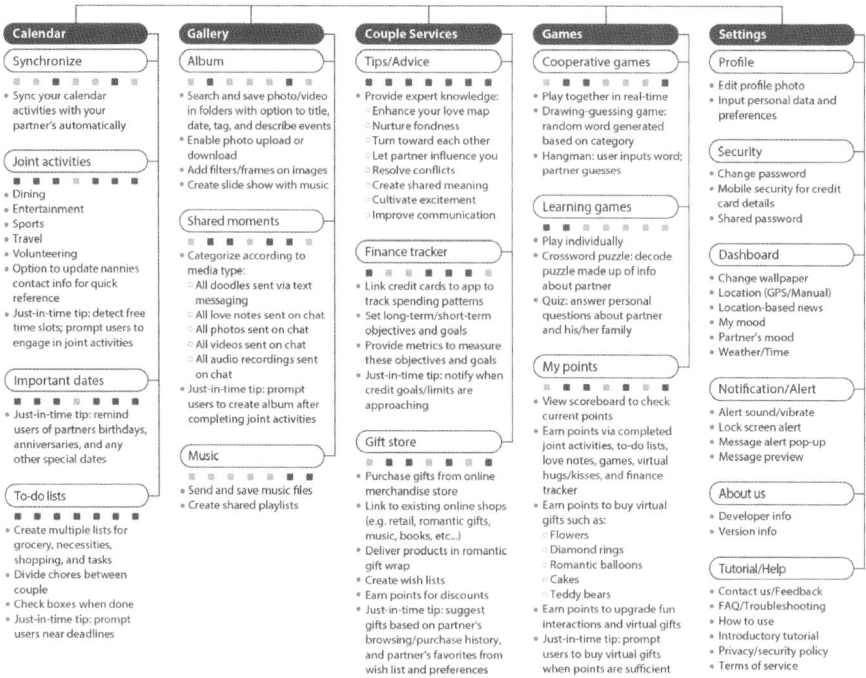

Fig. 11.27   Information architecture of the Marriage Machine, detailed menus of the Side Panel

Location Sharing

Share your location via GPS.

11.11.9.3   Camera

This function helps users capture and share their marriage moments via personal photos and video on the Marriage Machine.

11.11.9.4   Notifications

This function provides updates to the user regarding the activity of his/her partner. Marriage Machine supports updates for the following activities:

- Calendar updates.
- Finance changes.
- Mood changes.
- New emails.
- New love notes.

- Point updates.
- Virtual gifts.
- Virtual hugs.
- Virtual kisses.

11.11.9.5   Tab 2: Side Panel

The Marriage Machine also supports a side panel, which provides access to other functions of the Marriage Machine.

11.11.9.6   Calendar

Synchronize

- Automatically syncs the user's calendar activities with the user's partner's automatically.

Joint Activities

- The machine supports just-in-time functionality, detecting free time slots and prompting users to engage in joint activities such as dining, entertainment, sports, travel, volunteering, and option to update nannies' contact info for quick reference.

Important Dates

- The machine supports just-in-time functionality, reminding users of partners' birthdays, anniversaries, and any other special dates.

To-do lists

- Users can create multiple lists for groceries, necessities, shopping, and tasks. The lists allow the couple to divide chores and check boxes when they are finished. The machine supports just-in-time functionality, prompting users when task deadlines are near.

11.11.9.7   Gallery

Album

- Search and save photo/video in folders with option to title, date, tag, and describe events.
- Enable photo upload or download.
- Add filters/frames on images.
- Create slide show with music.

Shared Moments

This function allows users to "look back" on the memories they have shared together. The application supports just-in-time functionality to prompt users to create an album after completing joint activities. Users can categorize according to media type (all doodles sent via text messaging, all love notes sent on chat, all photos sent on chat, all videos sent on chat, all audio recordings sent on chat, etc.)

Music

- Send and save music files.
- Create shared playlists.

11.11.9.8   Couple Services

The Marriage Machine offers a suite of services designed to help couples in many aspects of their marriage.

Tips/Advice

Provides expert knowledge to couples on ways they can:

- Enhance their love map.
- Nurture fondness.
- Turn towards each other.
- Resolve conflicts.
- Create shared meaning.
- Cultivate excitement.
- Improve communication.

Finance Tracker

Users link their credit cards to application to track spending patterns. They can set long-term/short-term objectives and goals and view metrics to measure these objectives and goals. The application supports just-in-time functionality to notify users when credit goals/limits are approaching.

Gift Store

- Purchase gifts from online merchandise store.
- Link to existing online shops (e.g., retail, romantic gifts, music, books, etc.).

- Deliver products in romantic gift wrap.
- Create wish lists.
- Earn points for discounts.
- Just-in-time tip: suggest gifts based on partner's browsing/purchase history.
- Partner's favorites from wish list and preferences.

11.11.9.9   Games

The users' marriage and sharing experiences will be enhanced and enriched in the form of games. This form of interactive game with friends will serve as a major incentive: it will have a stimulating effect on users and will increase their motivation to create and share marriage moments. For example, the attribution of points to redeem at the gift shop represents a strong incentive and increases use frequency. The integration of games will, at the same time, add a component of fun and play to the mobile application and will thus make the experience not only informative and reflective but also entertaining.

Cooperative Games

These games are played together in real time:

- Drawing-guessing game: random word generated based on category.
- Hangman: user inputs word; partner guesses.

Learning Games

These games are played individually:

- Crossword puzzle: decode puzzle made up of info about partner.
- Quiz: answer personal questions about partner and his/her family.

My Points

Users can view a scoreboard to check their current points. The application supports just-in-time functionality to prompt users to buy virtual gifts when points are sufficient:

- Earn points via completed joint activities, to-do lists, love notes, games, virtual hugs/kisses, and finance tracker.
- Earn points to buy virtual gifts such as flowers, diamond rings, romantic balloons, cakes, teddy bears.
- Earn points to upgrade fun interactions and virtual gifts.

11.11.9.10   Settings

Profile

- Edit profile photo.
- Input personal data and preferences.

Security

- Change password.
- Mobile security for credit card details.
- Shared password.

Dashboard

- Change wallpaper.
- Location (GPS/manual).
- Location-based news.
- My mood.
- Partner's mood.
- Weather/Time.

Notification/Alert

- Alert sound/vibrate.
- Lock screen alert.
- Message alert pop-up.
- Message preview.

About us

- Developer info.
- Version info.

Tutorial/Help

- Contact us/Feedback.
- FAQ/Troubleshooting.
- How to use.
- Introductory tutorial.
- Privacy/security policy.
- Terms of service.

11.11.10 Tips

Marriage Machine users will be assisted and accompanied by a mobile marriage "advisor." Based on users' characteristics, needs, and preferences, on usage behaviors, and on the respective usage environment, the advisor provides personalized and tailor-made offers and recommendations.

11.12 Designs

Based on the information architecture, AM+A prepared initial design sketches of some key screens (Figs. 11.28, 11.29, 11.30, and 11.31).

Fig. 11.28 *Left:* Marriage Machine Dashboard. *Right:* Marriage Machine Side Panel

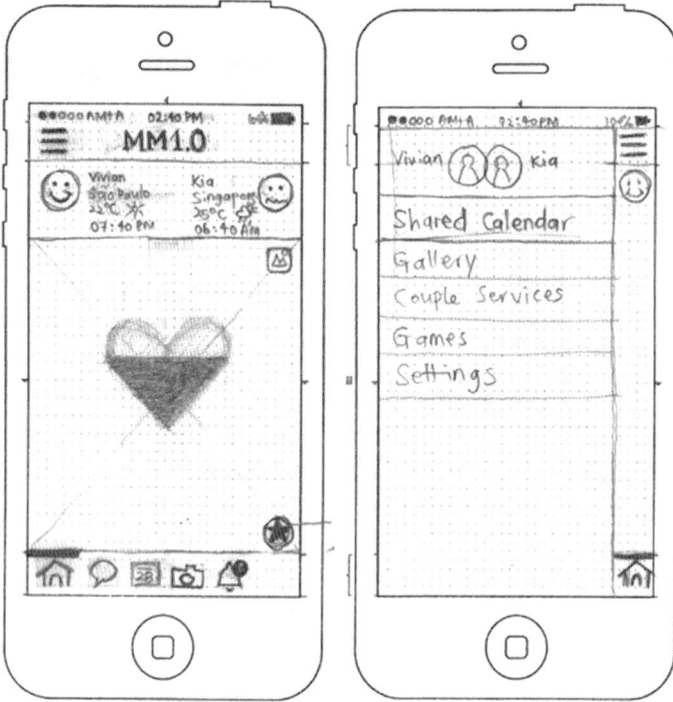

Fig. 11.29 *Left:* Shared Calendar. Star icon represents one just-in-time tip. *Right:* Shared Calendar Month View

Fig. 11.30 *Left:* Chat function. *Middle:* Additional functions within Chat. *Right:* Types of interactions available once "Affection" button is pressed

Fig. 11.31 *Left:* Love Note is a type of interaction within chat. *Right:* Example of a Love Note

11.12.1   Revised Screen Designs

Based on all previous efforts, AM+A designed revised versions of many significant screens. These are described below:

11.12.2   Landing Screen

The Landing Screen typifies the main menu and application instruction throughout the Marriage Machine (Fig. 11.32).

11.12.3   Side Panel

The Side Panel provides additional features for the user (Fig. 11.33).

Fig. 11.32 Landing screen

Fig. 11.33 Side Panel

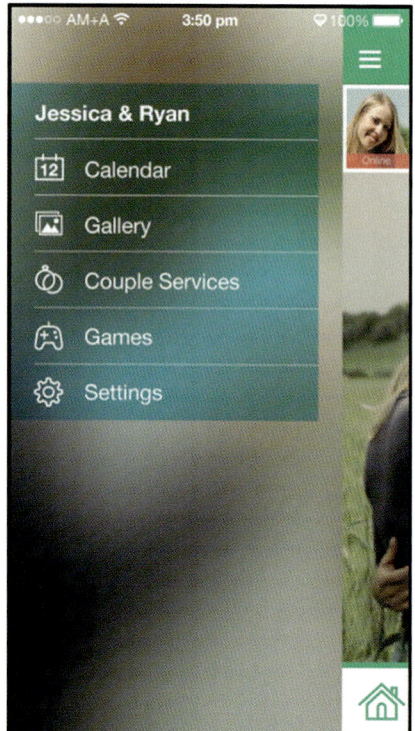

11.12.4   Shared Calendar

The Shared Calendar allows both parties of the relationship to view his or her partner's schedule. It also acts as a personal assistant by sending date suggestions if both parties are free (Fig. 11.34).

11.12.5   Chat

In addition to normal chat functionality, the Chat function in the Marriage Machine allows the user to send customized interactions, such as Love Notes and Virtual Hugs and Virtual Kisses (Fig. 11.35).

11.13   Usability Testing

11.13.1   Introduction

The Marriage Machine project (version 1.0) of 2014 researched, analyzed, designed, and evaluated effective ways to foster a shift to greater relationship success among committed couples (regardless of gender, legal, or religious circumstances) by changing people's daily behavior in the short and in the long term. The main objective of the Marriage Machine is to motivate and persuade couples to open themselves up to techniques of daily practice and interaction with their partners, to make their relationships with each other deeper, more personally enriching, and educational.

Usability tests are conducted to assess the user experience of a design, a system of metaphors, mental models, navigation, interaction, and appearance and identify navigation and usability issues. These tests help to determine the extent to which a user interface facilitates routine task completion. We tested the current Marriage Machine prototype with a small group of users. We asked them to complete a series of routine tasks, rate the app, and give any comments. We analyzed these test sessions to identify limitations and potential ways to improve the mobile application.

AM+A designer/analysts conducted the usability test by showing screenshots of the Marriage Machine from an iPad display. A laptop was used to record the answers, comments, and navigation choices. A camera was used to capture the participants' interaction. The results of each participant's navigational choices, task completion, comments, overall satisfaction ratings, questions, and feedback are summarized in this usability-test evaluation report.

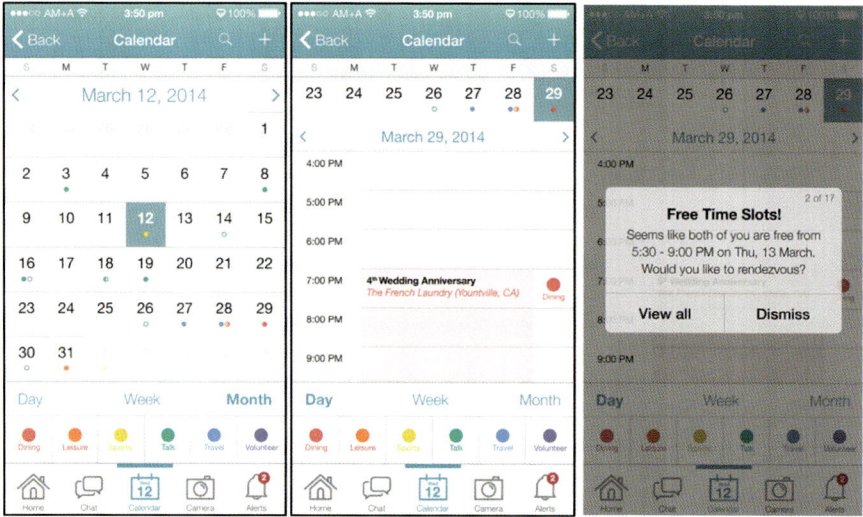

Fig. 11.34 Shared calendar screens

11.13.2   Executive Summary

Six adults, ages 22 to 30, participated in the usability test. To ensure reliable results, we chose the participants from the age group of the application's target market. Each session lasted for an average of 13 min. In general, all participants found the Marriage Machine clear, simple, and easy to use. 83 % of the respondents found the application beautiful and/or friendly.

Four out of six of the participants were uncertain about some aspects but interested to know what some functions on the dashboard meant. These functions are Ryan and Jessica, Local News, and Tips.

11.13.3   Methodology

The test administrator contacted the participants via emails and phone calls to inform them of the scope of the test and to confirm their availability and participation. Participants scheduled appropriate dates and times. The total time taken for each session was approximately 13 min each. During each session, the test administrator explained the scope of the test and described the tasks. The relevant screenshots were then showed to the participants, and the participants were asked to complete several task scenarios and to answer some questions.

Section 1: *First impressions—(a) quick look (15 seconds)*

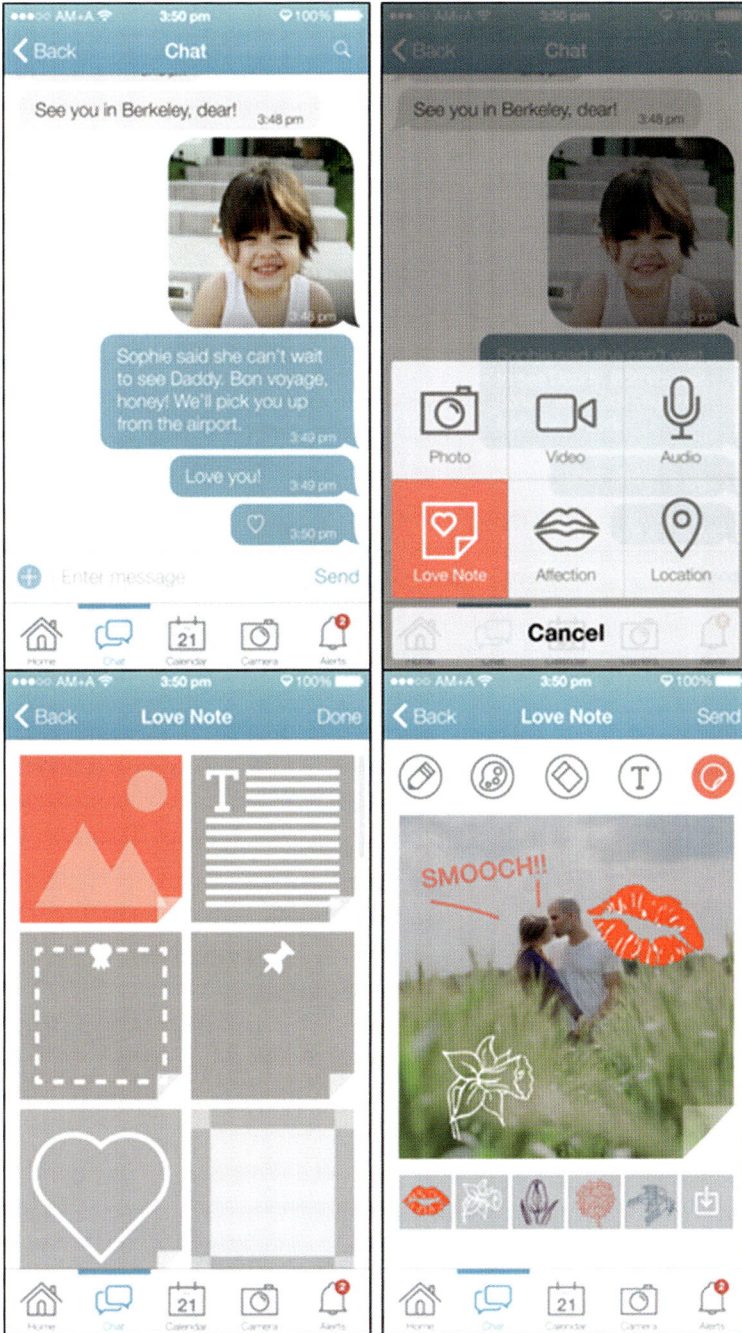

Fig. 11.35 Chat function screens

After the participants looked at the dashboard for the first time, the administrator asked several questions regarding their impressions of the Marriage Machine, such as the following:

- For what/whom do you think the application is designed?
- How do you feel about the look and feel of the screens?
- What do you think about the organization/structure of the application content?
- What do you think of the colors?

Section 2: *First impressions—(b) second look*

Next, the participants were asked to look at the dashboard in detail and we asked the following:

- Which function would you like to start with? Why? What would you expect?
- Do you think it is easy to navigate the application content?

Section 3: *Navigating from the dashboard*

After the first two tasks were completed, the administrator gave some navigation tasks to the participants.

Section 4: *Posttest questions*

After several navigation tasks, the test administrator asked the participants the following questions:

- What are your likes/dislikes and any other recommendations?
- Would you like to use this mobile application?
- Would you recommend this application to someone?

Finally, the participants were asked to rate the application using a 5-point scale (5 = excellent, 3 = okay, and 1 = very poor).

11.13.4 Participants

All participants were between 22 and 30 years old. Six participants were scheduled over the three testing dates. Out of the six participants, three were males and three females (Fig. 11.36).

11.13.5 Relationship Status

Participants were asked to select their relationship status as shown in the table below (Table 11.1).

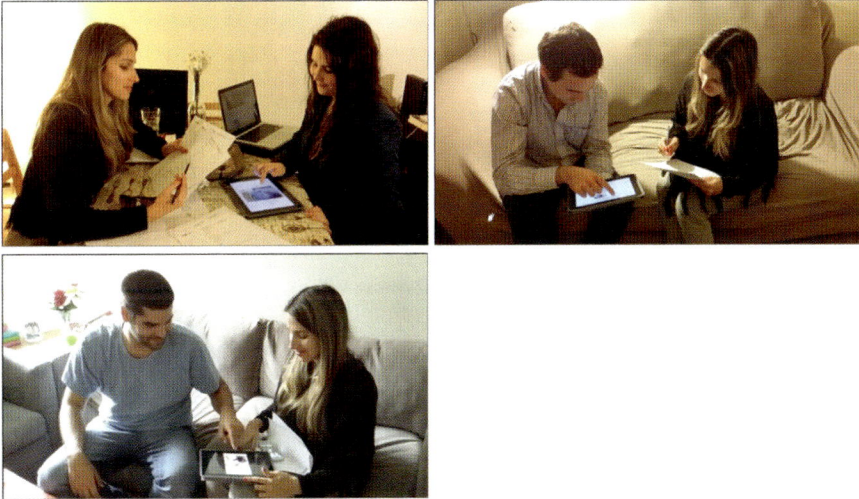

Fig. 11.36 User testers for the Marriage app (Images: AM+A, used with participants' permission)

Table 11.1 Participant relationship status

Single	Committed	Total
1	5	6

11.13.6 Task Scenarios

Participants were asked to perform the following tasks:

- Navigate from *Dashboard* to *Side Panel*.
- Find *Settings* and activate *Shared Password*.
- Return to *Dashboard* from *Settings*.
- Find the button that leads to *Chat* function.
- Add a *Love Note*.
- Choose the template with options to add background photo.
- Add a sticker and type a message on the photo.

11.13.7 Results

Section 1*: First impressions—(a) quick look (15 seconds)*

All the participants agreed that the interface of the Marriage Machine looked well designed and friendly.

- When the participants were asked about what they think the application is for, the participants thought it was designed either for:
- Users to find a partner and start a relationship.
- Couples currently in a committed relationship.

  33.3 % of the participants answered (a) while the other 66.7 % replied (b).
  How they feel about the look and feel of the screens (Fig. 11.37):

- What they think about the organization/structure of application content:

  100 % of the participants felt that the contents on the dashboard were well structured and could anticipate the corresponding functions that the five icons at the bottom would lead them to. The ease of navigating the application is confirmed in the task completion table below.

- What they think of the colors:

  100 % of the participants agreed that the color scheme of the application looks pleasant.

Section 2: *First impressions—(b) second look*

Which function would they like to start with? Why? What they would expect (Table 11.2).

Fig. 11.37 Participants' feelings about the application

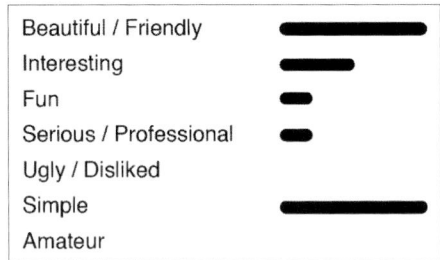

| Beautiful / Friendly |
| Interesting |
| Fun |
| Serious / Professional |
| Ugly / Disliked |
| Simple |
| Amateur |

Table 11.2 Table of participant expectations

Participant	Function	Explanation
1	Ryan and Jessica	Curious about what this function offers
2	Chat	Eager to communicate with partner
3	Chat	Interested to explore the chat functions
4	Ryan and Jessica	Curious about what this function offers
5	Chat	Curious to find out whether there are more functions compared to other messaging apps
6	Ryan and Jessica	To view Ryan and Jessica's relationship

- Whether they think it is easy to navigate the application content:

100 % of the participants felt that the application would be easy to navigate because the information on the dashboard seems clean and well structured.

Section 3: *Navigating from the dashboard*

- All participants successfully completed:
- Task 1 (navigate from *Dashboard* to *Side Panel*).

- Task 2 (find *Settings* and activate *Shared Password*).

- Task 3 (return to *Dashboard* from *Settings*).

- Task 4 (find the button that leads to *Chat* function).

- Task 6 (choose the template with options to add background photo).

- Task 7 (add a sticker and type a message on the photo).

One participant was not able to complete Task 5 (add a *Love Note*).
Similarly, only one participant did not complete Task 7 (add a sticker and type a message on the photo).

11.13.8   Task Completion

The following table (Table 11.3) shows the success/completion rates of each task.

11.13.9   Errors

AM+A captured the number of errors participants made while trying to complete the task scenarios:

- Task 5: Add a *Love Note*.

Table 11.3 Table of task completion

Participant	Task 1	Task 2	Task 3	Task 4	Task 5	Task 6	Task 7
1	✓	✓	✓	✓	✓	✓	✓
2	✓	✓	✓	✓	✓	✓	✓
3	✓	✓	✓	✓	✓	✓	✓
4	✓	✓	✓	✓	✓	✓	✓
5	✓	✓	✓	✓	✓	✓	✓
6	✓	✓	✓	✓	✓	✓	✓
Success	6	6	6	6	5	6	5
Completion rates	100 %	100 %	100 %	100 %	83 %	100 %	83 %

Participant 4 (P4) was able to perform all the tasks except to add a *Love Note:* Because P4 thought there was a specific button to create a *Love Note*.

- Task 7: Add a sticker and type a message on the photo.

Participant 2 (P2) was able to perform all the tasks except finding the sticker function since this task required the participants to deal with two different functions. P2 added that he has never used an application that allowed users to add stickers and add text on an image.

11.13.10   Average Time Taken

The chronometer recorded the time taken for the usability tests of each participant. The average time taken to complete the test was 12,6 min.

11.13.11   Total Time Taken

The following table (Table 11.4) shows the total time taken for each participant to complete the entire usability test.

Table 11.4 Table of time for each participant to complete the entire usability test

Participant	Time (min)
1	12.4
2	18.1
3	8.7
4	10.4
5	15.3
6	11
Average	12.6

Table 11.5 Tables of participants' likes/dislikes and recommendations

Like Most	Like Least	Recommendations for improvement
Clean design Simple interface Easy to use	Participants did not mention anything negative. All of them were satisfied with the app	Functions such as Ryan and Jessica, Local News, and Tips can be explained more

Table 11.6 Likes and dislikes

Question	Yes	No	Percentage agree
Would you like to use this mobile application?	5	1	83 %
Would you recommend this mobile application to someone?	6	0	100 %

Table 11.7 Table of overall recommendation of one participant

Recommendation	Justification
Add an introductory tutorial to describe all functions of the Marriage Machine	Some of the participants could not understand what some functions on the dashboard meant: *Ryan and Jessica*, *Local News*, and *Tips*

Section 4: *Posttest questions*

- What are your likes/dislikes and any other recommendations? (Tables 11.5 and 11.6).

11.13.12   Overall Ratings

After participants answered all questions and completed all tasks, they were asked to rate on the scale of 1 to 5 (5 = excellent, 3 = okay, and 1 = very poor).
67 % rated 4, and 33 % rated 5.

11.13.13   Recommendations

One participant was very impressed with the Marriage Machine and gave additional recommendations (Table 11.7).

11.13.14   Conclusion

Most of the participants found the Marriage Machine mobile application clean, simple, organized, and easy to use. To ensure that optimal user experience is achieved, AM+A will consider these recommendations and continue to listen to the Marriage Machine's users in its ensuing prototypes.

11.14   Conclusions

The Marriage Machine is conceived to motivate and persuade couples to open themselves up towards techniques of daily practice and interaction with their partners, with the goal of making their relationships with each other deeper, more enjoyable, more personally enriching, and educational. The application incorporates persuasive and motivational elements in order to stimulate users to reflect on and change their behavior. Based on user tests, we found that 67 % of our participants rated the application "very good" and 33 % rated the application "excellent." With this information in mind, we made no immediate changes to the screen designs.

AM+A plans to continue development of the Marriage Machine through future design workshops and as opportunities for commercial development arise.

Acknowledgements   For the Marriage Machine, the author acknowledges the primary assistance of Mr. Kia Hwee Chew, Ms. Vivian Lemes, and Mr. Kenneth So, AM+A Designer/Analyst.

In addition, the author thanks Prof. Bob Steiner, University of California/Berkeley Extension, International Diploma Program, and his graduate students (Aditi Arora, Fernando Bittar, and Caroline Verghote) for their assistance in carrying out market research, including surveys and interviews.

The author acknowledges the following publication (Marcus 2015) based on the AM+A white paper, on which this chapter is based:

Marcus, Aaron (2015). "The Marriage Machine" Proceedings, Design, User Experience, and Usability, 2–7 August 2015, Los Angeles, in Marcus, Aaron, Editor, *User Experience Design for Diverse Interaction Platforms and Environments, Lecture Notes in Computer Science*, Volume 8519, 2015, in press. Springer, London.

Further Reading

Achor S (2010) The happiness advantage: The seven principles of positive psychology that fuel success and performance at work. Crown Publishing Group, New York

Bernstein E (2013) How to be a better conversationalist. Wall Street J. Retrieved from http://online. wsj.com/. Accessed 21 Aug 2013

Buhalis D (2003) eTourism: Information technology for strategic tourism management. Prentice Hall, Harlow

Coelho A, Dias L (2011) A mobile advertising platform for etourism. In: Law R, Fuchs M, Ricci F (eds) Information and communication technology in tourism 2011. Springer, New York, pp 203–214

Córdova JV, Gee CG, Warren LZ (2005) Emotional skillfulness in marriage: Intimacy as a mediator of the relationship between emotional skillfulness and marital satisfaction. J Soc Clin Psychol 24:218–235

Eco U (1970) Semiologie des messages visuels. Communication 16:11–52

Eco U (1976) A theory of semiotics. Indiana University Press, Bloomington

Eco U (1985) The semantics of metaphor. In: Innis R (ed) Semiotics: An introductory anthology. Indiana University Press, Bloomington, p 262

Frey BS (2010a) Experience sampling Method ESM). http://www.bsfrey.ch/articles/C_516_ Marriage_and_Public_Choice_LV.pdf

Frey BS (2010b) Marriage: A revolution in economics (Munich lectures in economics). The MIT Press, Cambridge, MA

Gottman J, Gottman J (2011) How to keep love going strong: 7 principles on the road to happily ever after. YES! Mag 56:38–39

Innis RE (1985) Semiotics: An introductory anthology. Indiana University Press, Bloomington

Karney B (2010) Keeping marriages healthy, and why it's so difficult. Retrieved from http://www.apa.org/science/about/psa/2010/02/sci-brief.aspx

Lakoff G, Johnson M (1980) Metaphors we live by. The University of Chicago Press, Chicago

Law R, Fuchs M, Ricci F (2011) Information and communication technologies in tourism 2011. Springer, Vienna

Levi-Strauss C (2000) In: Jacobson C, Schoepf B (eds) Structural anthropology. Basic Books, New York

Löfgren O (2002) On holiday: A history of vacationing. University of California Press, Berkeley

Luo S, Klohnen EC (2005) "Assortative Mating and Marital Quality in Newlyweds: A Couple-Centered Approach," Shanhong Luo and Eva C. Klohnen, University of Iowa. J Pers Soc Psychol 88(2):304–326

MacCannell D (1999) The tourist. A new theory of the leisure class. University of California Press, Berkeley

Marcus A (1998) Metaphors in User-Interface Design. ACM SIGDOC 22(2):43–57

Marcus A (2002) Information visualization for advanced vehicle displays. Inf Vis 1(2):95–102

Maslow AH (2006) Motivazione e personalità. (Ed. 11). Armando, Roma

Peirce CS (1933) Existential graphs. In: Hartshorne C, Weiss P (eds) Collected papers of Charles Sanders Peirce (Vol. 4). Harvard University Press, Cambridge, MA, pp 293–470

Tristram C (2001) The next Computer interface. Technol Rev 104(10):53–59

Wagner M, Armstrong N (2003) Field guide to gestures: How to identify and interpret virtually every gesture known to man. Quirk Books, Philadelphia

References

Aboelwafa K, Panth M, Roggio A, Yolgecti B (2013) The happiness machine: marketing research. University of California/Berkeley Extension, International Diplomoa Program, Marketing Research Class Project Report, 26 June 2013

American Community Survey/American FactFinder (2012) http://factfinder.census.gov/faces/tableservices/jsf/pages/productview.xhtml?src=bkmk. Checked 11 Sept 2015

Arora A, Bittr F, Verghote C (2014) The marriage machine: marketing research. University of California/Berkeley Extension, International Diploma Program, Marketing Research Class Project Report, 06 March 2014

Cialdini RB (2001a) The science of persuasion. Sci Am 284:76–81

Cialdini RB (ed) (2001b) Influence: Science and practice, 4th edn. Allyn and Bacon, Boston

Cohn D (2013) Love and marriage. Pew Research. Social and Demographic Trends. http://www.pewsocialtrends.org/2013/02/13/love-and-marriage/. Accessed 13 Feb 2013

Fogg BJ (2003) Persuasive technology: Using computers to change what we think and do. Morgan Kaufmann, Amsterdam

Fogg BJ, Eckles D, Bogost I, Consolvo S, Holmen E, Spasojevic M, Ulm J, Tanguay S, Walker S, White S (2007) Mobile persuasion: 20 perspectives on the future of behavior change. Stanford University Press, Palo Alto

Hartson R, Pyla PS (2012) The UX book. Process and guidelines for ensuring a quality user experience. Elsevier, Waltham

Marcus A (2011) The health machine. Inf Des J 19(1):69–89

Marcus A (2012a) The money machine. Helping Baby Boomers Retire. User Exp 11(2):24–27

Marcus, A. (2012b). *The story machine: A mobile application to change family story-sharing behavior.* Workshop paper presented at CHI conference on human factors in computing systems. (5–10 May 2012). ACM, New York

Marcus A et al (2013a) The innovation machine. In: : Marcus A (Ed.) (2013) Proceedings, design, user experience, and usabilty conference, Las Vegas, NV, 20–25 July 2013, pp. 67–76. Springer, London

Marcus A et al (2013b) The driving machine. In: : Marcus A (Ed.) (2013) Proceedings, design, user experience, and usabilty conference, Las Vegas, NV, 20–25 July 2013, pp. 140–149. Springer, London

Marcus A et al (2013c) The learning machine. In: : Marcus A (Ed.) (2013) Proceedings, design, user experience, and usabilty conference, Las Vegas, NV, 20–25 July 2013, pp. 247–256. Springer, London

Marcus A (2014a) The happiness machine: mobile behavior change. Proceedings of the design, user experience, and usability conference, design, user experience, and usability 2014, 20–24 June 2014, Iraklion Crete Greece. In: Marcus A (ed) User experience design for diverse interaction platforms and environments, Lecture notes in computer science, vol 8518, pp 258–268. Springer, London

Marcus A (2014b) The happiness machine. In: Proceedings of the user-experience professionals association/China and user friendly conference 2014, Wuxi, China, 13–16 November 2014. Unnumbered pages available on memory stick

Marcus A (2015) The marriage machine. In: Proceedings of the design, user experience, and usability, 2–7 August 2015, Los Angeles. In: Marcus A (ed) User experience design for diverse interaction platforms and environments, Lecture notes in computer science, vol 8519 (in press). Published by London: Springer

Marcus A, Jean J (2009) The green machine: Going green at home. Inf Des J 17(3):233–243

Maslow AH (1943) A theory of human motivation. Psychol Rev 50:370–396

Tsapelas I, Aron A, Orbuch T (2009) Marital boredom now predicts less satisfaction 9 years later. Psychol Sci 20(5):543–545

Chapter 12
Conclusion

12.1 Concluding Remarks

The ten case studies presented here are the result of 5 years of effort by approximately 30 dedicated Designer/Analysts at Aaron Marcus and Associates, Inc., most of them interns gaining professional expertise and experience during 3-month visits. Their primary activity was to work on these unfunded, pro bono projects. Many of them became quite interested in the subject matter and the approach, and some even changed their professional directions accordingly to move towards mobile user-experience design and the particular perspective of mobile persuasion design.

We embarked on these projects because we believed in the value of the approach, the importance of the subject matter for each Machine, and the potential benefits that these Machines, taken to the next step of commercial development, could bring to industry, technology, and society. We have shared this information from the beginning for each one, and are gratified to see interest in taking some of the ideas further, independently of AM+A.

Each of these projects is something like the birth of a new child. We have watched them come into being, begin to crawl, then walk, and eventually run. We are eager to see what the future holds for further developments in this area of technology, communication, and society. Especially interesting will be the adaptation of the Machines to different cultures, age groups, and usage communities (personal vs. enterprise/organization).

There are possible dangers, of course, in the potential misuse of this approach, as mentioned earlier in the book. Nevertheless, we believe that with some warnings, some education in ethics, and more attention to cultures, sustainability, and social-political implications of such development and the "quantified self" of our lives (as news reports record daily), all stakeholders can move forward with awareness and caution.

Many exciting developments, we feel, lie ahead in this area. We look forward to the future of mobile persuasion design.

© Springer-Verlag London 2015   659
A. Marcus, *Mobile Persuasion Design*, Human–Computer
Interaction Series, DOI 10.1007/978-1-4471-4324-6_12